Advancing Precision Agriculture With AI and IoT

Rashid Abbas Khan
University of South Africa, South Africa

Vishnu D. Rajput
https://orcid.org/0000-0002-6802-4805
Southern Federal University, Russia

Asghari Bano
University of Wah, Pakistan

IGI Global
Scientific Publishing
Publishing Tomorrow's Research Today

Vice President of Editorial — Melissa Wagner
Director of Acquisitions — Mikaela Felty
Director of Book Development — Jocelynn Hessler
Production Manager — Mike Brehm
Cover Design — Jose Rosado

Published in the United States of America by
IGI Global Scientific Publishing
701 East Chocolate Avenue
Hershey, PA, 17033, USA
Tel: 717-533-8845 | Fax: 717-533-7115
Website: https://www.igi-global.com E-mail: cust@igi-global.com

Library of Congress Cataloging-in-Publication Data

Names: Rajput, Vishnu D. editor | Khan, Rashid Abbas, 1994- editor | Asghari Bano, 1954- editor
Title: Advancing precision agriculture with AI and IoT / edited by Vishnu Rajput, Rashid Abbas Khan, Asghari Bano.
Description: Hershey, PA : IGI Global Scientific Publishing, 2026. | Includes bibliographical references and index. | Summary: "This book examines how artificial intelligence and the Internet of Things are transforming agriculture through data-driven precision, automation, and sustainability. Bridging theory and practice, it explores smart farming technologies, ethical considerations, and policy implications for achieving climate-resilient, efficient, and resource-optimized food production systems"-- Provided by publisher.
Identifiers: LCCN 2026002230 (print) | LCCN 2026002231 (ebook) | ISBN 9798337372570 hardcover | ISBN 9798337372587 paperback | ISBN 9798337372594 ebook
Subjects: LCSH: Precision farming | Artificial intelligence--Agricultural applications | Internet of things
Classification: LCC S494.5.P73 A23 2026 (print) | LCC S494.5.P73 (ebook) | DDC 630.2/085--dc23/eng/20260204
LC record available at https://lccn.loc.gov/2026002230
LC ebook record available at https://lccn.loc.gov/2026002231

British Cataloguing in Publication Data
A Cataloguing in Publication record for this book is available from the British Library.

Table of Contents

Detailed Table of Contents

This chapter presents an overview of the evolution of global agriculture, from traditional knowledge-based farming to mechanized, industrial, and Green Revolution systems. It examines the benefits of these transitions for crop yields, food security, and economic growth, as well as their environmental and socio-economic challenges, including soil degradation, biodiversity loss, and resource depletion. The development of Agricultural 2.0 and 3.0, with mechanization, chemical inputs, and early digital tools, is analyzed, leading to Precision Agriculture 4.0, which integrates artificial intelligence, IoT, advanced analytics, and autonomous systems. Emerging trends such as Agriculture 5.0, cognitive farming, digital twins, and blockchain-based supply chains are also discussed. The chapter emphasizes the integration of traditional knowledge with advanced technologies to build sustainable, resilient, and inclusive agricultural systems for future food security.

Precision Agriculture is a transformation of the agricultural industry; agricultural data is being generated and utilized rather than depending on experience. The chapter emphasizes that smart farming systems involve the use of technologies such as GPS,

Internet of Things, artificial intelligence, drones, and robotics to be more productive, sustainable, and efficient in terms of using resources. Some of the major components are data management, predictive analytics driven by AI, and intelligent decision-support systems, which are essential in the optimization of inputs and reduction of environmental impact. It attempts to solve the issues of expenses, confidentiality of data, and digital illiteracy. Successful implementations are presented in case studies, where the importance of startups and SMEs is stressed in the area of innovation. Future arguments will discuss how AI can be used to generate recommendations and block chain to promote transparency, self-driving machines, and how Precision Agriculture can support climate-smart agriculture.

Chapter 3
Artificial Intelligence in Farming: Transforming Crop Management and Resource Allocation 57
Tauqeer Ahmed Qadri, College of Graduate Studies, University of South Africa, Pretoria, South Africa

Artificial Intelligence (AI) is transforming agriculture by improving precision, efficiency, and sustainability. Key technologies like machine learning, deep learning, computer vision, and remote sensing enable early pest and disease detection, real-time crop monitoring, and precision practices such as variable-rate inputs, soil mapping, and autonomous machinery. AI-driven decision support integrates climate, soil, and market data for yield prediction and harvest planning. Case studies highlight scalable applications, while challenges include technical limits, data privacy, ethics, and socioeconomic barriers. AI offers a path to resilient, sustainable, and globally secure food systems.

Chapter 4
Data-Driven Farming: Turning Data Into Smarter Decisions and Higher Yields 85
Nadia Tahseen, Lahore College of Women University, Pakistan
Muhammad Waqas, Chinese Academy of Agricultural Sciences, China
Muhammad Iftikhar, University of Education, Lahore, Pakistan
Andrew Tapiwa Kugedera, Morgenster Teachers College, Zimbabwe

Global agriculture is facing unprecedented challenges intensifying from population growth, climate change, soil degradation, new pests, lack of labour and pressure on natural resources. These constraints have contributed to unsustainable agriculture leading to decline in productivity. This chapter gives an overall view about data driven farming and the revolutionizing role of artificial intelligence (AI) in modern farming. The chapter explore various new trends in agriculture such as Internet of Things (IoTs), sensors, and drones to mention a few. Precision agriculture and AI tools that maximise smart irrigation, robotics, and resource optimization are also discussed. The integration of real-time data and automation has the capacity

The integration of Artificial Intelligence (AI) and the Internet of Things (IoT) is transforming crop health management by enabling predictive, precise, and sustainable agriculture. Traditional manual and reactive monitoring methods are increasingly inadequate for addressing climate variability, pest outbreaks, nutrient deficiencies, and abiotic stresses. AI–IoT systems combine high-resolution sensors, drones, edge computing, and advanced analytics to provide continuous real-time monitoring of soil, plant, and environmental conditions. Machine learning, deep learning, and computer vision support early detection of diseases, pests, and stress while enabling data-driven decision making through predictive models and digital twins. Decision-support systems translate complex data into actionable recommendations for irrigation, nutrient management, and targeted pest control, while integration with automation and robotics improves operational efficiency. Despite significant economic and environmental benefits, adoption remains limited by connectivity gaps, costs, etc.

In this chapter, the author discusses the paradigm shift in modern agriculture that is motivated by the need to have sustainable food production and the use of resources efficiently through smart farming. Smart farming combines IoT, AI, ML, big data analytics, drones, and robotics and is used to increase productivity and environmental sustainability. Making sense of data is turning real-time sensor data into actionable insights in order to monitor the conditions of soil, crop development, weather conditions, and the health of livestock. Precision farming allows for the process of irrigating, fertilizing, pest control, and harvesting, which minimizes the amount of inputs wasted. Experiment results demonstrate machine learning algorithms as the best classifiers using the precision, recall, accuracy and F1-score that play a key role in getting smarter farming to work. The model is stable, as the accuracy of disease detection is 98.26% and validation outcomes are more proficient. Agriculture 5.0 will encourage transparency, efficiency, and the use of less water, energy, and chemicals.

Plant diseases significantly impact global food security, causing yield losses and reduced crop quality. Traditional detection methods are labor-intensive, slow, and hindering timely interventions. This chapter explores how AI, deep learning, and IoT technologies can revolutionize plant disease detection in sustainable agriculture. A review of recent research and experiments shows that convolutional neural networks (CNN), transfer learning models like ResNet and EfficientNet, and multispectral imaging can all be used to classify crops like rice, maize, and wheat with 90% to 98% accuracy. The integration of IoT sensors and cloud-based platforms enables real-time monitoring, early diagnosis, and optimized pesticide use, fostering sustainable practices. AI-driven disease detection helps make better predictions about crop yields, has less of an effect on the environment, and makes it easier to make decisions based on data. The chapter emphasizes the need for understandable AI systems and accessible tools for farmers to ensure equitable benefits and digital transformation in agriculture.

Preface

Agriculture today faces unprecedented challenges. Climate change, soil degradation, water scarcity, and a growing global population are placing immense pressure on food production systems. In response, innovative technologies such as Artificial Intelligence (AI) and the Internet of Things (IoT) are reshaping the way we cultivate, manage, and sustain crops. This book, *Precision Agriculture 4.0: AI and IoT for Sustainable Farming*, explores the transformative role of these technologies in creating smarter, more resilient, and environmentally responsible agricultural systems.

The goal of this book is to bridge the gap between theory and practice. It combines research-driven insights with real-world applications, showcasing how AI algorithms and IoT-enabled systems can optimize resource use, improve crop management, and enhance farm productivity. Readers will find discussions on precision irrigation, predictive analytics for crop yields, smart farming systems, automation through robotics and drones, and ethical and legal considerations in technology adoption.

This book is intended for a wide audience, including researchers, students, farmers, agribusiness professionals, policymakers, and technologists interested in the future of sustainable agriculture. By providing an integrated perspective on AI and IoT applications, it highlights innovative strategies that can address pressing challenges such as food security, resource scarcity, and environmental sustainability.

We hope this work serves as a practical and scholarly resource for advancing knowledge, inspiring innovation, and promoting the adoption of sustainable agricultural technologies worldwide. Our sincere gratitude goes to mentors, colleagues, and institutions whose guidance and support have made this work possible.

ORGANIZATION OF THE BOOK

Chapter 1: The Evolution of Agriculture: From Traditional Methods to Precision Farming

In this opening chapter, the editors invite readers on a sweeping journey across the history of agriculture, tracing its transformation from traditional, knowledge-intensive practices to the mechanized systems of the Green Revolution and onward to today's digitally enabled paradigms. The authors carefully examine the productivity gains and food security advances achieved through Agricultural 2.0 and 3.0, while offering a balanced critique of the environmental and socio-economic costs, including soil degradation and biodiversity loss. The narrative culminates in a compelling exploration of Agriculture 4.0 and emerging Agriculture 5.0 frameworks—where artificial intelligence, IoT, autonomous systems, digital twins, and blockchain reshape the farming landscape. Importantly, the chapter underscores that the future of agriculture lies not in abandoning tradition, but in integrating indigenous knowledge with advanced technologies to build resilient, inclusive, and sustainable food systems.

Chapter 2: Decoding Precision Agriculture: A New Era in Farming Innovation

This chapter provides a foundational understanding of Precision Agriculture as a decisive shift from experience-based farming to data-driven decision-making. The contributors unpack the architecture of smart farming systems, examining the integration of GPS, IoT, AI, robotics, and advanced analytics into cohesive management platforms. Beyond technical innovation, the authors confront pressing concerns such as data ownership, digital literacy, and economic barriers. Through illustrative case studies, they highlight the catalytic role of startups and SMEs in accelerating innovation. The chapter also looks forward to AI generated recommendations, blockchain-enabled transparency, and autonomous machinery, positioning Precision Agriculture as a cornerstone of climate-smart and sustainable farming.

Chapter 3: Artificial Intelligence in Farming: Transforming Crop Management and Resource Allocation

Here, the transformative role of artificial intelligence in reshaping agricultural practice takes center stage. The chapter details how machine learning, deep learning, computer vision, and remote sensing technologies enable early disease detection, real-time monitoring, and optimized resource allocation. By integrating climate, soil, and market data into predictive models, AI-driven systems empower farmers with actionable insights for yield forecasting and harvest planning. While celebrating scalable successes, the author also engages critically with ethical, technical, and socio-economic limitations. The chapter positions AI not merely as a tool for efficiency, but as a strategic pathway toward resilient and globally secure food systems.

Chapter 4: Data-Driven Farming: Turning Data into Smarter Decisions and Higher Yields

Responding to mounting pressures from climate change, population growth, and resource scarcity, this chapter presents data-driven farming as an urgent and transformative response. The authors explore the convergence of IoT devices, drones, sensors, and AI analytics in optimizing irrigation, input management, and robotic operations. With particular attention to smallholder constraints, they outline barriers such as infrastructure gaps and high implementation costs. The chapter concludes with a thoughtful roadmap for policymakers and researchers, advocating inclusive strategies to ensure that the benefits of AI-enabled farming extend across diverse agricultural contexts.

Chapter 5: Revolutionizing Crop and Soil Monitoring through IoT-Enabled Farming

This chapter delves deeply into the technological backbone of IoT-enabled agriculture. The authors demonstrate how interconnected sensors, edge–cloud computing, and advanced analytics continuously monitor soil and plant parameters, enabling early detection of stress factors and precise agronomic interventions. By integrating AI and machine learning with automated control systems, IoT platforms translate data streams into real-time action. The chapter's strength lies in its demonstration of applicability across scales—from smallholder plots to controlled-environment agriculture—highlighting tangible gains in yield, efficiency, and environmental stewardship.

Chapter 6: Transforming Irrigation Systems with IoT for Water Conservation and Efficiency

Water management emerges as a defining challenge of modern agriculture in this focused exploration of IoT-enabled irrigation systems. The authors present a comprehensive review of sensor technologies, communication protocols, and data architectures that underpin precision irrigation. By introducing AI, machine learning, digital twins, and nano-sensors, the chapter showcases innovative pathways to reduce water waste and enhance climate resilience. Particularly compelling is the discussion of low-cost and community-based IoT solutions, offering scalable models for water-smart agriculture in diverse socio-economic settings.

Chapter 7: AI- and IoT-Driven Approaches for Crop Health Monitoring and Management

In this chapter, predictive and preventive crop health management is advanced through the synergy of AI and IoT technologies. The authors illustrate how high-resolution sensors, drones, and advanced analytics facilitate continuous environmental monitoring and early detection of pests, diseases, and nutrient deficiencies. Decision-support systems and digital twins convert complex datasets into precise recommendations, while automation enhances operational efficiency. Although adoption barriers persist—especially in connectivity and affordability—the chapter convincingly demonstrates how integrated AI–IoT systems can elevate sustainability and productivity simultaneously.

Chapter 8: Predicting and Managing Crop Health through Artificial Intelligence and IoT

This contribution highlights the measurable performance gains of AI and machine learning models in crop health diagnostics. Drawing on experimental findings demonstrating high classification accuracy, the authors present smart farming as a data-centric ecosystem integrating IoT, big data analytics, robotics, and drones. Precision interventions in irrigation, fertilization, and pest control are shown to significantly reduce input waste. The chapter situates these innovations within the broader trajectory toward Agriculture 5.0, emphasizing transparency, efficiency, and responsible resource use as defining features of next-generation farming systems.

Chapter 9: AI and Computer Vision for Plant Disease Detection in Sustainable Agriculture

Focusing specifically on plant pathology, this chapter explores the power of deep learning and computer vision in disease detection. The authors review the performance of convolutional neural networks, transfer learning models, and multispectral imaging in classifying staple crops with high accuracy. The integration of IoT sensors and cloud platforms enables real-time diagnosis and optimized pesticide application. Importantly, the chapter advocates for explainable AI and accessible tools, ensuring that digital transformation remains equitable and farmer-centered.

Chapter 10: Farm Automation Empowered by Artificial Intelligence for Enhanced Efficiency

This chapter charts the transition from mechanized systems to AI-driven autonomous farming architectures. By detailing sensors, cloud-edge infrastructures, robotics, and decision-support systems, the authors outline how multisource data streams are integrated into closed-loop control systems. Applications ranging from autonomous planting to precision harvesting illustrate both environmental and economic benefits. The chapter also addresses interoperability, cybersecurity, regulatory frameworks, and economic feasibility, presenting a balanced assessment of the promises and constraints of large-scale automation.

Chapter 11: Robotics and Drones: Shaping the Future of Agriculture with Automation

Expanding the automation discourse, this chapter synthesizes advancements in robotics, drones, AIoT ecosystems, and cloud analytics. The authors describe how autonomous platforms enable site-specific spraying, robotic harvesting, and intelligent monitoring, reducing agrochemical dependency and optimizing resource use. Economic analyses underscore long-term profitability, even as the chapter confronts social implications such as labor displacement and technological divides. The result is a forward-looking assessment of next-generation autonomous farming systems.

Chapter 12: Ethical Considerations of Artificial Intelligence and Automation in Agricultural Decision Making

In a timely and thought-provoking contribution, this chapter shifts attention from technological capability to ethical responsibility. As AI increasingly informs irrigation timing, outbreak prediction, and crop grading, questions of data ownership, transparency, accountability, and rural livelihoods take center stage. The author urges stakeholders to move beyond theoretical ethics toward actionable governance frameworks. By foregrounding the human dimensions of automation, this chapter anchors the book's technological discourse in social responsibility.

Chapter 13: Addressing Global Challenges and Overcoming Barriers to the Widespread Adoption of Precision Agriculture Technologies

This chapter situates Precision Agriculture within the broader context of global food insecurity, climate change, and resource depletion. The authors categorize adoption barriers—economic, technological, and knowledge-based—highlighting disparities between developed and developing systems. Emerging tools such as AI-driven predictive analytics, blockchain transparency mechanisms, robotics, and digital twin ecosystems are presented as transformative yet unevenly distributed innovations. The chapter ultimately frames Precision Agriculture not only as a technological shift, but as a strategic development pathway toward inclusivity and sustainability.

Chapter 14: Embracing AI and IoT to Ensure a Sustainable Agricultural Future for Global Food Security

The contributors of this chapter present AI and IoT as twin pillars of a sustainable agricultural future. Advanced sensors, satellite imagery, and real-time analytics enable site-specific management of soil health, water, and crop diseases. Applications in controlled environments such as greenhouses and vertical farms further illustrate efficiency gains and waste reduction. Yet the authors remain attentive to sociocultural, infrastructural, and economic barriers, particularly for smallholders. Policy reform, education, and technological accessibility are advanced as critical levers for unlocking global food security.

Chapter 15: A Vision for the Future: The Next Wave of Innovation in Precision Agriculture

The volume concludes with a visionary synthesis of emerging frontiers in precision agriculture. Integrating AI, robotics, hyperspectral imaging, nano-enabled inputs, biodegradable sensors, and digital twins, the chapter envisions interoperable and adaptive farm ecosystems. The authors explore research gaps in AI generalizability, robotic coordination, and equitable deployment, while articulating scalable pathways for farms of all sizes. This closing contribution reaffirms the book's central thesis: that the next wave of innovation must harmonize technological sophistication with environmental stewardship and social inclusion to secure the future of global agriculture.

Chapter 1
The Evolution of Agriculture: From Traditional Methods to Precision Farming

Rashid Abbas khan
http://orcid.org/0000-0002-5438-6274
University of Education, Lahore, Pakistan

Asghari Bano
http://orcid.org/0000-0003-3800-1909
University of Wah, Pakistan

ABSTRACT

This chapter presents an overview of the evolution of global agriculture, from traditional knowledge-based farming to mechanized, industrial, and Green Revolution systems. It examines the benefits of these transitions for crop yields, food security, and economic growth, as well as their environmental and socio-economic challenges, including soil degradation, biodiversity loss, and resource depletion. The development of Agricultural 2.0 and 3.0, with mechanization, chemical inputs, and early digital tools, is analyzed, leading to Precision Agriculture 4.0, which integrates artificial intelligence, IoT, advanced analytics, and autonomous systems. Emerging trends such as Agriculture 5.0, cognitive farming, digital twins, and blockchain-based supply chains are also discussed. The chapter emphasizes the integration of traditional knowledge with advanced technologies to build sustainable, resilient, and inclusive agricultural systems for future food security.

DOI: 10.4018/979-8-3373-7257-0.ch001

INTRODUCTION

Agriculture has been the cornerstone of human civilization; it has always evolved according to changing demands posed by demographic increase, environmental constraints and technological innovation. From the earliest domestication of flora and fauna to the latest data-driven farming systems, agriculture has undergone some major changes that have shaped societies, economies, and landscapes, both natural and otherwise (Mešić et al., 2024). Each historical transition reflects a change in relationships between humans and the land - first it was the manual labor and the indigenous way; then it was the mechanization and the use of chemicals; now it is artificial intelligence (AI), the Internet of Things (IoT), robotics and automation (Figure 1) (Islam et al., 2025). Comprehending this evolutionary path is fundamental in appreciating the problems and opportunities of the contemporary phase, where the global demand for food will increase as the supply of natural resources is rapidly dwindling.

Figure 1. Key components and benefits of precision farming, integrating AI, IoT, and modern agricultural technologies. These innovations enhance crop production, optimize resources, protect the environment, and support sustainable farming practices

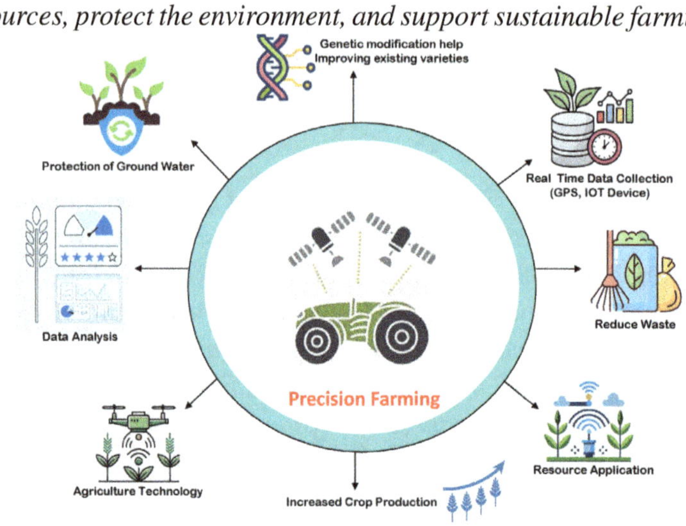

1.1.1. Background and Importance of Agricultural Evolution

The agricultural evolution is closely tied to man's survival and global evolution. Traditional agriculture was based on locally based knowledge, hand labor and natural ecological processes. While these systems supported early societies, they were often plagued by low productivity, high vulnerabilities to climatic uncertainty and with low scalability (Mastretta-Yanes et al., 2024). The twentieth century brought major revolutionary scientific developments, including high yielding varieties, the use of synthetic fertilizers and mechanized farming, which together formed the basis of the Green Revolution. These innovations were a huge boost to food production and hunger in many areas. However, the environmental ramifications, including soil degradation, over-water consumption, loss of biodiversity and greenhouse gas emissions, exposed the need for more sustainable ways (X. Chen, 2025). Agricultural evolution is not therefore something of a past story but is also a key basis for future strategies to balance productivity and environmental management. As climate change has increased and as natural resources have become scarcer, the transformation towards smarter, efficient and more precise agriculture methods has been an indispensable tool. As shown in Figure 2, agriculture has evolved from simple manual practices to advanced intelligent farming systems, with each technological phase contributing to improved precision, sustainability, and efficiency

Figure 2. A visual journey that reflects the remarkable evolution of agriculture from simple manual practices to advanced intelligent farming systems. Each technological era adds a new layer of innovation, guiding farming toward greater precision, sustainability, and efficiency in the modern world

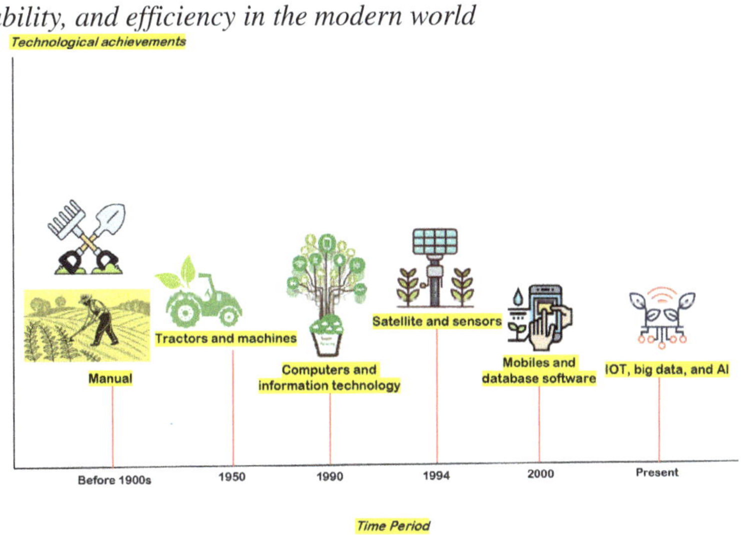

1.1.2. Key Transitions in Global Agriculture

The history of global agriculture can be split into several key phases. The first transition - from hunter-gatherer to settled cultivators - formed the foundation for the early civilizations. The second big change was the advent of mechanization which brought in tractors, harvesters and irrigation systems which made labor more efficient than ever before (Grinin et al., 2024). The Green Revolution was the third big transition with biochemical innovations to drastically improve yields. The advent of precision agriculture is the fourth and great change. Unlike past transitions that mainly boosted the efficiency of physical labor or increased efficiency of chemical inputs, precision agriculture combines data and powerful analytics in real time with automated decision-making (Hamdan et al., 2022). This change has made farming redefine itself from a uniform, input-heavy system to a site-specific, optimized and environmentally conscious system. Today, agriculture is at the edge of Agriculture 4.0 - a phase characterized by Artificial Intelligence driving insights, internet of things (IoT) enabling monitoring, autonomous farming machinery and predictive intelligence (Dayioğlu & Türker, 2021). As shown in Figure 3, continuous advancements in tools, mechanization, automation, and smart technologies have transformed agriculture into a more efficient and sustainable system.

Figure 3. A comprehensive depiction of the evolution of agriculture from ancient manual practices to the era of intelligent digital farming. Each stage reflects the steady advancement of tools, machinery, automation, and smart technologies that continue to shape modern sustainable agriculture

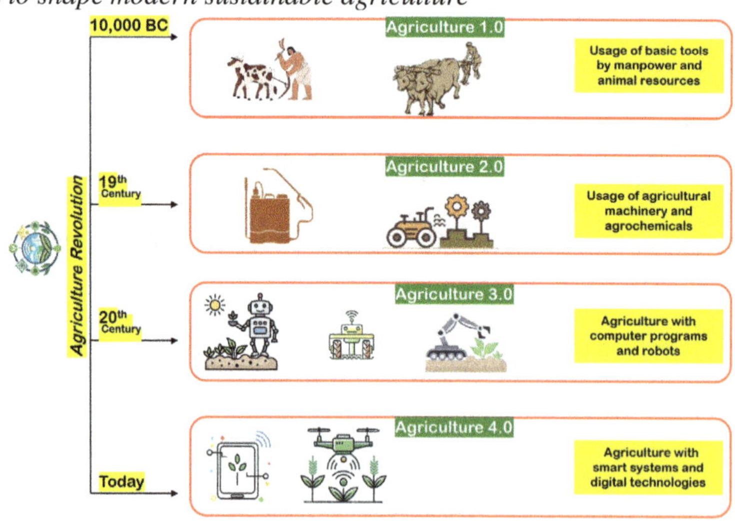

1.1.3. Role of Digital Technologies in Modern Food Systems

Digital technologies have become central to the complex challenges of modern food systems. IoT sensors give constant information about soil moisture, nutrients, plant health, climatic and livestock behavior. AI algorithms analyze these data streams to come up with precise recommendations and to detect diseases in early stages, yield predictions and allocation of resources (Rajak et al., 2023). Drones and remote sensing are being used to map the heterogeneity of fields in detail and robotics increase efficiency in planting, weeding, and harvesting. These technologies collectively increase productivity, minimize waste of inputs, reduce impact on the environment and increase resilience to climate induced stresses (Guebsi et al., 2024). Moreover, digital platforms have made the lock-in of markets, transparency of supply chains and real-time advisory systems to farmers easier, which in turn contribute to food security and sustainable development.

This chapter gives an overview of the development of agriculture from the traditional systems that have been based on the indigenous knowledge of people, to the highly advanced precision farming systems of today. It draws attention to some of the major shifts that have defined world agriculture, explores the role of the emerging technologies to transform food production, and lays the basis for understanding Precision Agriculture 4.0. By tracing this historical and technological course the chapter sets the stage for the rest of the chapters in the book, which explore many of them in more depth, covering AI, IoT, robotics, automation, and other cutting-edge tools that drive sustainable agriculture in the twenty-first century.

1.2 TRADITIONAL AGRICULTURE SYSTEM

1.2.1 Characteristics of Subsistence Farming

Subsistence farming comprises one of the oldest agricultural enterprises, and one of the most common agricultural activities worldwide, and is generally not performed to meet the needs of commercial markets but to meet the dietary needs of households or local communities. It is generally defined by small scale, limited resources and is one of the most influenced by the natural environment around them (Mishra et al., 2019). Farmers in the area grow a wide range of crops, most often cereals, legumes and vegetables, to try to ensure year-round domestic food production. Livestock is kept in small quantities and used simultaneously as sources of both nourishment and draught power and for farmyard manure (Ramana et al., 2025). Decision-making in subsistence systems is usually based on local experience, seasonal cycles and instantaneous family needs, rather than on formal scientific analysis or decisions based on market regime imperatives. These systems are usually characterized by low external inputs, low mechanization, and a high level of crop diversity, thus being the result of a risk management strategy as well as an expression of cultural dietary preferences (Habib et al., 2025).

1.2.2 Indigenous Knowledge Systems

Indigenous knowledge forms the basis of the traditional agricultural systems, which provides farmers with empirically corroborated strategies of land stewardship, meteorological forecasting, seed selection and pest management. This knowledge is passed down from one generation to the next through oral traditions, learning from observation, and experiences (Malapane et al., 2024). Illustrative cases include the use of phenological patterns such as the activity of birds or the flowering pattern of a plant to predict the rainfall regime, the use of traditional seed storage methods to maintain germination viability over time and the use of botanical extracts to manage plant pest infestations. Indigenous practices often consider ecology, the conservation of biological diversity and the rational use of natural resources as primary concerns (Varsamis et al., 2025). They are fundamentally contextual based on local edaphic conditions, microclimates and cultural heritage, and many are very much in line with modern agroecological principles. The importance of indigenous knowledge has received a new international attention as climate change intensifies and farmers search for resilient, low-cost, or low-input methods of doing agriculture (Ghorbani et al., 2021).

1.2.3 Manual Labor, Animal Traction, and Rudimentary Tools

Prior to the start of mechanization, classical farming practice was mainly based on manual labor, primitive tools and animal pulling. Common tools included hand hoes, sickles, wooden ploughs, mattocks and locally adapted tools which were engineered to work with local soil characteristics. Animal traction, mostly by oxen, water buffaloes, donkeys or camels, was necessary for land preparation and transport operations (Fao & Weber, n.d.). These systems required large amounts of labor, and the work was physically exhausting, as the sowing, weeding, irrigation and harvesting had to be done physically. The operational efficiency of such tools was limited and often limited the area that could be put to cultivation and the number of crops grown in a year's cycle. Nonetheless, the low technological requirements made such methods available and cost-effective for rural, agrarian societies and they continue to be used in modern settings in areas where mechanical methods are prohibitive financially or geographically (Kaur et al., 2023).

1.2.4 Soil Fertility and Crop Rotation Practices

Management of soil fertility under traditional agricultural systems is dependent on the operation of natural biogeochemical processes and the supply of organic inputs. Farmers use compost, animal manure, green manures, crop waste and fallowing to maintain the nutrient balance in the soil (Al-Shammary et al., 2024). Crop rotation is part of a common approach, when in alternating seasons cereals, legumes, and fodder crops are planted, to control the increase of pests, prevent nutrient deficiencies, and improve soil structure. Leguminous species like the cowpea (*Vigna unguiculata*) or the mung bean (*Vigna radiata*) or the clover (*Trifolium* spp) enable biological nitrogen fixation which contributes to the soil fertility (i.e. in the absence of synthetic inputs) (Al-Musawi et al., 2025). In some areas, shifting cultivation or "slash - and - burn" type agriculture were once used as a form of soil renewal; however, such farming techniques are now discouraged because of their ecological degradation. These practices collectively reflect a complex understanding of ecological cycles, although they tend to limit agricultural intensification.

1.3 GREEN REVOLUTION AND INDUSTRIAL AGRICULTURE

1.3.1 Emergence of High-Yielding Crop Varieties

The development of high-yielding varieties (HYVs) of important staple crops such as wheat, rice and maize formed a watershed during global agriculture during the middle of the XX century with the emergence of the Green Revolution. These HYVs were carefully bred for such traits as semi-dwarf height, increased response to fertilizer application, and resistance to lodging (von der Goltz et al., 2020). The application of these improved genotypes helped countries, mainly in Asia and Latin America, to significantly enhance the productivity of crops without having to expand existing agricultural land. For example, the semi-dwarf wheat cultivars revolutionized production in India and Pakistan, and IR8 and its descendants were responsible for yielding gains not seen before in Southeast Asia (Altaf et al., 2021). HYVs thus became the cornerstone of industrial agriculture which led to global food production following a paradigm which is more and more dominated by technological and input responsiveness.

1.3.2 Synthetic Fertilizers, Pesticides, and Irrigation Expansion

The success of HYVs was inextricably related to the generalization of synthetic fertilizer, pesticide and improved irrigation infrastructure. Nitrogen-, phosphorus-

and potassium-based fertilizers had significant effects on the growth rate and grain filling potential of crops. Chemical pesticides-including herbicides, insecticides, and fungicides one of the earliest developments in golf courses in the short-solution approach to pest and disease control, the achievement of uniformity and stability of crops (Chang et al., 2024). At the same time, the expansion of irrigation with the advent of tube wells, canals and much later modern pumping, made it possible to have multiple cropping cycles per annum. Regions such as the Indo-Gangetic Plains benefited immensely from such interventions and they were able to switch from food deficit to food surplus zones. This triad of inputs (HYVs, agrochemicals and irrigation) characterized the intensification model of industrial agriculture and transformed national food systems (Jain et al., 2021).

1.3.3 Mechanization and Monoculture Farming

Mechanization became a feature of industrial agriculture, which replaced manual labor and animal traction with tractors, harvester-machines, threshers and associated equipment. This shift in technology allowed farmers to farm bigger lands and to be less dependent on labor and more efficient. Mechanization also helped to create monoculture systems, where vast stretches of land were used to grow only one type of crop (e.g. wheat, rice, cotton or maize). Monoculture made the farm easier to manage and allowed the application of fertilizers, pesticides and irrigation to be uniform (Bhooshan et al., 2024). Nonetheless, it reduced crop diversity and increased dependency on external support. The speed at which machinery spread in this area - especially in North America, Europe and some areas of Asia - brought about the rapid modernization of agriculture, but at the same time caused aggravating socio-economic differences as smallholders found it difficult to compete with capital-intensive operations (Sharafi et al., 2025).

1.3.4 Major Achievements: Increased Yields, Food Security, Economic Growth

The Green Revolution produced some spectacular worldwide consequences. Crop yields in many nations more than doubled as a result, and thus famine was greatly reduced. Nations such as India and Pakistan moved from chronic food problems to self-sufficiency in grain production of wheat and rice. Improved agricultural productivity led to rural economic growth and job opportunities in input markets, seed industries and food processing, etc. (John & Babu, 2021). Caloric availability improved and contributed to reduced hunger and favorable nutritional outcomes in many regions of the world. Industrial agriculture also opened the way for agricultural trade on a global basis, allowing the surplus producing countries to enter export markets. In sum, the Green Revolution was a crucial element in sustaining a rapidly growing population worldwide and the foundation of the modern-day food supply chain (Said et al., 2025)

1.3.5 Criticisms: Soil Degradation, Biodiversity Loss, Environmental Pollution

Despite the achievements made by the Green Revolution, it was criticized primarily due to its environmental and ecological consequences. Excessive application of synthetic fertilizers led to the imbalances of soil nutrients, reduction of soil organic matter, and increase of salinity and water logging in the irrigated areas. Monoculture farming and the extensive use of pesticides hastened biodiversity decline, which at the time destroyed beneficial insects, pollinators, and microorganisms in the soil (Pozza & Field, 2020). Excess irrigation led to ground water depletion and contamination, especially in regions dependent on tube wells. Chemical runoff of fields led to eutrophication of rivers and lakes in a threat to aquatic ecosystems. Moreover, the dependence on HYVs of limited genetic diversity multiplied the susceptibility to the newly arising pests, diseases and climate stressors. These challenges highlighted the limitations of input intensive agriculture and concerned the world about the long-term viability of the Green Revolution model (Hakami et al., 2024).

1.4 AGRICULTURAL 2.0: MECHANIZATION AND EARLY MODERNIZATION

1.4.1 Adoption of Tractors and Mechanical Tillage

Agricultural 2.0 marks the period when mechanization started to change the conventional farming landscapes into more structured and commercially viable production systems. The use of tractors replaced manual labor and the use of animals to pull agricultural machinery, allowing farmers to use more land with greater speed and accuracy. Mechanical tillage helped with ploughs, disc harrows and rotavators increased soil aeration, suppressed weeds and prepared the seedbed (Razu et al., 2024a). These must have been technological developments that substantially alleviated the drudgery related to agricultural activities and increased timing of operations, especially in key periods such as sowing and harvesting. In many areas, the tractor became a symbol of agricultural modernization and allowed for several cropping seasons per year promoting an increase in farm productivity as a whole (Huang & Wang, 2024).

1.4.2 Introduction of Chemical Inputs and Improved Seeds

Concomitant with mechanization, was the early modernization characterized by the widespread use of chemical fertilizers, pesticides, superior seed varieties, etc. The availability of nitrogen, phosphorus and potassium fertilizers meant that producers were able to replace nutrients in the soil very quickly and to achieve much higher yields of production (Penuelas et al., 2023). Chemical pesticides provided for more reliable control of pests, weeds and diseases compared to traditional remedies. Improved seeds - including hybrids and varieties that had been designed through breeding to have higher levels of productivity, uniformity, and stress tolerance - increased the stability of production systems. Although these early innovations were not as transformative as the succeeding Green Revolution high yielding varieties (HYVs), they created the foundation for the development of the Input-Responsive Agriculture and encouraged a shift from subsistence to market-oriented farming (Vermelho et al., 2024).

1.4.3 Post-Harvest Mechanization: Threshers and Harvesters

Post-harvest mechanization was a major force of the Agricultural 2.0, especially during the advent of threshers, combining harvesters, mechanical dryers and grain cleaners. These implements have largely diminished the post-harvest losses, increased the quality of the grain, and reduced the labor requirement (Nath et al., 2024). Threshers helped producers to easily and quickly handle large tons of grain, replacing time-consuming manual threshing. Combine harvesters combine cutting, threshing and cleaning functions in one operation, thus greatly speeding up the harvesting process. Post-harvest mechanization also consolidated value chains by standardizing the quality of product and making better storage facilities possible and reducing spoilage as well as widening the commercial grain markets (Kaur et al., 2023).

1.4.4 Expansion of Commercial Farming

Mechanization and improved inputs were the catalyst for the change from traditional subsistence farming to commercial agriculture. Larger farms were able to reap the most benefits from mechanization because of economies of scale and the ability to mass produce staple crops, cash crops, and industrial crops such as cotton, sugarcane, and oilseeds (Mirzakhaninafchi et al., 2022). Commercial farming brought investment in infrastructure (farm machinery dealerships, rural roads, storage facilities and input supply chains) with it. Market oriented production systems developed where producers sold surpluses to urban centers and agro-processing industries. This commercial growth tied agriculture to national and global markets and, therefore, contributed to the economic growth of rural areas and strengthened agro based industries (GC et al., 2020).

1.4.5 Socio-Economic Impacts

Agricultural 2.0 had huge impacts on the rural socio-economic landscapes. Mechanization made less use of manual labor, diverting human resources to other-line employment, but also helped to displace people from the labor force in some areas. A rise in income levels occurred for the adopters of modern technologies, but the smallholders experienced challenges competing with the larger mechanized farms (Ma et al., 2024). Gender roles changed because technological devices such as mechanical tools often took over work that had been previously done by women and relate to their economic participation. The development of commercial agriculture favored rural-urban migration, a stronger integration into nation markets, and the emergence of agribusiness sectors (Masset et al., 2023). Nevertheless, differences in access to mechanization and inputs led to such socio-economic inequalities as persist to this day. These are important results that highlight the need for equitable and inclusive approaches as agriculture spends the next few years on its way to digital and precision-based agriculture.

1.5 AGRICULTURAL 3.0: DIGITALIZATION AND DATA-DRIVEN FARMING

1.5.1 ICTs in Agriculture

Agricultural 3.0 denotes the shift in a strictly mechanized and input-intensive paradigm to one that is progressively better informed by information and communication technologies (ICTs). In this stage, cell phones, personal computers, Internet access and digital instrumentation penetrated agricultural industry decision making (Rashid et al., 2025). ICTs transformed the way farmers receive weather forecasts, market prices, pests warnings, fertilizer application advice and government services. The extension systems started using the digital platform which offers interaction in training and provision of instructional videos and advisory messages to the end users (Byamukama et al., 2025). The development of mobile based advisory services and e-agriculture platforms made available in specific countries and especially in developing ones allowed making the farmer decision-making evidence-based, minimizing the risk of production, and maximizing the use of resources. ICTs therefore offered the backbone infrastructure through which the further sophisticated digital and precision technologies were developed (Thilakarathne et al., 2025).

1.5.2 Decision Support Systems (DSS)

Decision Support Systems (DSS) became a critical tool of analysis in Agricultural 3.0, combining data, models, and convenient interfaces to the decisions in farm-management. The first DSS platforms included crop scheduling, irrigation schedule, nutrient and pest prediction. These systems utilized past data, observational data in the field and climatological data to come up with customized advice (Dhanaraju et al., 2022). DSS tools helped to close this knowledge gap between scientific investigation and field practice by offering real time actionable knowledge. Although these applications were initially limited by a lack of data and the computational power to handle it, they demonstrated the importance of data-driven agriculture and provided the foundation of more advanced analytics of prediction, AI-based advisory systems, and real-time monitoring tools that typify Precision Agriculture 4.0 (Ataei et al., 2024).

1.5.3 Geographic Information Systems (GIS) and Remote Sensing

With the introduction of Geographic Information Systems (GIS) and remote-sensing, a breakthrough in spatial analysis and field-level surveillance was made. GIS was used to make spatial heterogeneity of soil properties, crop performance, land-use changes and water availability easier to visualize and analyze, which was beneficial to farmers, researchers, and policy makers (Louzada et al., 2025). Remote sensors, which included satellite imagery, unmanned aerial vehicles and multispectral sensors provided real time measures of crop health, vegetation indices (e.g., the Normalized Difference Vegetation Index) and stress indicators and estimates of yield. These abilities led to the substitution of homogeneous with field management, through the clarification of intra-field diversity (Q. Chen et al., 2025). As a result, GIS-based platforms were integrated into government agencies and agricultural institutions when preparing irrigation regimes, managing natural resources, and monitoring agricultural landscapes within the regional and national levels. Together, GIS and remote-sensing have formed the basis of precision mapping systems, variable-rate application systems, and automated monitoring systems (Rabie et al., 2025).

1.5.4 Early Automation and Robotics

The initial automation and robotics came up in this time as precursors to the advanced autonomous systems that are dominant today. The new wave of automation was embodied in systems of automated irrigation, mechanized weeding, greenhouse climate control and robotic milking. Their main goals entailed to reduce labor representations, increase operational accuracy, and raise the productivity (Bazargani & Deemyad, 2024). Precise seeders and planters with electronic control helped to place the seeds in a uniform manner, which strengthened the establishment of crops. Robotic feeders and automated health-monitoring machines were developed in the animal husbandry industry that enhanced the welfare and productivity of animals. Although limited by high prices and the complexity of the technology, early robotics demonstrated the viability of automated repetitive tasks, increased accuracy, and the ability to incorporate sensors and sensor arrays into mechanical subsystems into the agricultural system (Azmi et al., 2021).

1.5.5 Limitations of Conventional Digitalization

Even though Agricultural 3.0 had significant transformative potential, it did not receive adoption because it had several major limitations. The introduction of numerous digital tools required good access of internet channels, the consistency of power supply, and the degree of technical performance that was not evenly distributed, especially in the rural and developing regions (Cao et al., 2025). The available information and communications technologies (ICTs) often provided generalized recommendations as opposed to field-based guidance and as such, this limitation limited its applicability in different agroecological environments. The absence of interoperability between devices, platforms as well as data repositories resulted in incoherent systems that created significant integrational problems for farmers. Also, small-scale farmers could not access advanced digital infrastructures because of their high investment costs (Díaz et al., 2025). The traditional digitalization initiatives were also insufficient in addressing real-time sensing, automation, and predictive analytics, which is why they were not sufficient to address the new challenges of climatic variability, resource depletion, and supply-chain disrupters. All these limitations led to the need for an improved, connected, and smart agricultural paradigm, which eventually developed into Precision Agriculture 4.0 (Enescu et al., 2025).

1.6 TRANSITION TO PRECISION AGRICULTURE

1.6.1 Technological Foundations and Core Approaches

This phase marks the shift to the traditional, homogenized management of farms to a more technology-driven and data-driven decision-making paradigm. Precision agriculture is based on the axioms of the application of relevant inputs at the right time and place (Marinello et al., 2023). Its main components include Site-Specific Crop Management (SSCM), which supports agronomic processes and heterogeneity of the field; Variable Rate Technology (VRT), which presents automated and differentiated methods to lay down seeds, fertilizers, and pesticides; and smart sensors which collect real-time information about the soil, crop, and the environment. Together, these technologies lead to a combined solution enhancing monitoring, prediction, and precision of operation (Mirzakhaninafchi et al., 2022).

1.6.2 Benefits, Efficiency Gains, and Global Adoption Trends

The advantages of precision agriculture over traditional farming include but are not limited to a high level of resource-use efficiency, reduced environmental effects, and enhanced yield stability. Live monitoring and accurate input control will save money and lower wastage and encourage climate-wise production (Kumari et al., 2023). The adoption trends vary across the world, the developed regions surpass the developing ones because of their highly mechanized systems and well-developed infrastructure, but the developing countries are running faster with affordable sensors, mobile apps, and government-supported digital agriculture projects. Nevertheless, inequality notwithstanding, precision agriculture is becoming increasingly popular across the globe as one of the avenues towards sustainable, resilient, and productive food systems (Okonta & Vukovic, 2024).

Table 1. Evolution of agriculture from traditional systems to precision agriculture 4.0

Agricultural Era	Key Characteristics	Technologies Used	Benefits	Limitations / Challenges	References
Traditional / Subsistence Agriculture (Agriculture 1.0)	Manual labor, indigenous knowledge, mixed cropping, soil fertility maintained through natural means	Hand tools, animal traction	Low cost, environmentally friendly, culturally rooted	Low productivity, climate-dependence, labor intensive	(Hamadani et al., 2021)
Green Revolution & Industrial Agriculture	High-yielding varieties, monoculture, chemical-intensive farming	HYVs, synthetic fertilizers, pesticides, large-scale irrigation	Increased food production, reduced hunger, economic growth	Soil degradation, biodiversity loss, pollution	(John & Babu, 2021; Zimdahl, 2022)
Agriculture 2.0 (Mechanization Era)	Large-scale farming, motorized operations, post-harvest mechanization	Tractors, harvesters, threshers, improved seeds	Labour efficiency, higher yields, commercialization	High cost, fossil fuel dependence, rural job reductions	(Razu et al., 2024b)
Agriculture 3.0 (Digital Era)	Initial data use, remote sensing, early automation	ICT tools, GIS, DSS, satellite imaging, basic robotics	Better monitoring, informed decisions, improved planning	Limited real-time data, digital literacy barriers	(Abiri et al., 2023)
Precision Agriculture (Agriculture 4.0)	Site-specific management, IoT and AI-driven decisions, automation	IoT sensors, drones, VRT, AI, big data, cloud computing	Higher efficiency, lower inputs, climate-smart farming	Costly technologies, data privacy issues	(Pérez et al., 2024)
Agriculture 5.0 (Future Vision)	Fully autonomous systems, cognitive robotics, transparent supply chains	AI + robotics, digital twins, blockchain, quantum computing	Autonomous farms, higher precision, global resilience	Ethical concerns, cybersecurity, technology gaps	(Bissadu et al., 2025; Fraser & Campbell, 2019)

1.7 PRECISION AGRICULTURE 4.0: AI AND IOT INTEGRATION

Precision Agriculture 4.0 is the latest phase of agri-technological innovation, where all devices are interconnected, artificial intelligence, robotics, and big-data analytics are combined to become intelligent, autonomous, and highly efficient agricultural systems (Javaid et al., 2022). It is not just a digitalization epoch, but agro-ecosystems at such a level of hyper-connectivity, with artificial intelligence serving as their driving engine, will be able to sense, predict, automate, and make informed decisions in real-time. A combination of these technologies enables the agronomists to maximize the use of resources and enhance productivity as well as enhance resilience to climatic and market volatilities (Natsir et al., 2025).

1.7.1 Internet of Things (IoT) in Agriculture

Internet of things (IoT) is what has now become the foundation of modern precision agriculture. IoT networks connect various Agri-equipment, such as sensors, equipment, irrigation systems, unmanned aerial vehicles, and livestock-monitoring units, to cloud infrastructures, which provide the opportunity to carry out continuous data collection, remote command execution, and automated operation (Choudhary et al., 2025).

1.7.2 Wireless Sensor Networks

WSNs are an important part of agricultural IoT applications. These networks include nodes that are spatially distributed and continually observe parameters like soil moisture, temperature, humidity, nutrient concentrations, wetness of the leaves and microclimatic factors. WSNs use low-power communication protocols, such as LoRaWAN, ZigBee and NB-IoT to support long-range connectivity. Real-time sensing that is always present enhances accuracy in nutrient control, irrigation planning and prognostication of diseases, and transforms farms into big data places (Rajak et al., 2023).

1.7.3 Smart irrigation systems

This is the combination of soil-moisture sensors, actuated valves and weather-dependent algorithms to optimize water allocation. These types of systems turn on irrigation when needed and regulate the time and rates of flow based on the needs of crops. Embarking on drip irrigation plans coupled with AI will significantly reduce the amount of water wastage and prevent excessive irrigation (Chowdhury et al., 2022). In addition, the information about weather stations, evapotranspiration models, and AI-powered irrigation controllers can also increase the water-use efficiency, which makes smart irrigation an essential feature of drought-prone regions

1.7.4 Soil and Plant health monitoring

IoT -sensors can measure soil PH, electrical conductivity, organic matter content and nutrient turnover, thereby providing real-time data on soil vitality. Plant-health-monitoring devices measure chlorophyll concentration, canopy temperature, and stress biomarkers using optical, thermal, and hyperspectral sensing modalities (Kumari et al., 2025). These tools identify predisposing factors of nutrient deficiencies, water stress, and disease outbreak, hence simplifying corrective measures and reducing crop losses. The production of handheld sensors, in-field sensors and aerial sensor produces a multi-level monitoring network (Yan et al., 2025).

1.8 ARTIFICIAL INTELLIGENCE (AI) APPLICATIONS

Artificial Intelligence is the analytic engine which forms the basis of Precision Agriculture 4.0. AI models process large volumes of information collected by sensors, drones, satellites, and farm records therefore creating accurate forecasts and automatic decision-making guidelines (Rashid et al., 2025).

1.8.1 Machine Learning for Crop disease detection

Machine-learning (ML) models are questioning soil properties, weather patterns, and past yields and harvesting patterns to predict nutrient needs, irrigation patterns, pest development, and crop development patterns. Decision support systems (DSS) based on the application of ML help the agronomist to identify the best planting window, dosage of fertilizer and crop cultivars. These analytical products increase the consistency of yields and reduce the risk of production (Chaudhry & Rashid, 2011).

Deep-learning systems, in particular, convolutional neural networks (CNNs), enable extremely precise detection of ailments and pest infestations in plants by leaf and aerial imagery. Models that have been trained on large, annotated datasets are able to differentiate between nutrient deficiencies, fungal attacks as well as insect attacks. Mobile applications based on clouds have now allowed farmers to diagnose crop issues real-time by merely putting evidence in the form of photographs (Shoaib et al., 2025).

1.8.2 AI in yield forecasting

The AI systems absorb the multispectral satellite images, climatic conditions, land parameters, and farming activities to forecast the crop yields in an exceptionally accurate way. These predictions guide the supply-chain planning, insurance underwriting and risk management policies. Random forests, long short-term memory (LSTM) networks, and gradient-boosting machines are some of the popular machine-learning methods used in large-scale yield estimation (Aslan et al., 2024).

1.9 ROBOTICS AND AUTOMATION

Robotics is changing the labor-intensive agricultural operations into automated and precise and highly efficient processes.

1.9.1 Drones for mapping and spraying

Multispectral and thermal crop mapping, stress assessment, and yield estimation Multispectral and thermal remote sensing are conducted using Unmanned aerial vehicles. Pesticides and foliar nutrients are sprayed with high accuracy using drones even when there is a problematic topography such as orchards and hilly lands. Live drone analytics reduces the use of chemicals and aid in the quick supervision of the fields (Sharma et al., 2025).

1.9.2 Harvesting robots

The AI vision systems installed on harvest-automation robots detect ripe produce thus enabling selective picking. Such robots are used to speed up harvest throughput, reduce after harvest losses, and run all day, thus alleviating agricultural workforce shortages. Examples of representative systems are strawberry pickers, tomato harvesters and robotic grape cutters (Carbone et al., 2025).

1.9.3 Weed-control robots and AI guided spraying

The self-driving weed boats use their self-driving abilities with vision sensors and AI program to distinguish between crops and weeds, allowing them to be removed mechanically or a herbicide can be applied precisely where it is required. AI-controlled sprayers release chemicals only on identified weeds and this minimizes the use of herbicides significantly. Such technologies facilitate sustainability through exposure to a limited amount of chemicals, and reduced environmental pollution (Upadhyay et al., 2024).

1.10 CLIMATE-SMART PRECISION AGRICULTURE

1.10.1 Managing Climate Variability Through Digital Decision Support

Climate-wise precision agriculture incorporates the use of sophisticated digital technologies to help farmers deal with growing climate uncertainty, weather extremities and uncertain crop situations. Real-time detection of droughts, floods, heatwaves, pest infestation and disease threats are made possible by early-warning systems that are powered by satellite data, IoT networks and machine-learning algorithms (Dey et al., 2025). These systems strengthen preparedness and give useful and timely alerts and recommendations. Digital climate advisory services Digital climate advisory services are provided via mobile, cloud services, or automated messaging to translate complex climate forecasts into practical field-level advice (to aid in decisions about planting, irrigation, fertilizer, and crop protection). With the integration of accurate information and localized weather information, farmers achieve the ability to respond and mitigate the loss of yield in the event of a climate crisis (Charalampopoulos et al., 2024).

1.10.2 Carbon-Smart Practices and Greenhouse Gas (GHG) Monitoring

Climate mitigation is also reinforced by precision technologies with carbon smart farming. AI-based analytics, aerial imaging and IoT-enabled soil sensors track soil organic carbon changes, nitrogen fluxes as well as greenhouse gas emissions with a high level of spatial and temporal accuracy. These information streams reinforce the carbon farming strategies of specific carbon and emission reduction (e.g., optimized fertilizer application, reduced tilling, cover crops, and improved residue management) (N. Ahmed & Shakoor, 2025). High-accuracy nutrient management, fertilizer placement and irrigation systems, VRT and automated systems cut down nitrous oxide emissions and increase nutrient-use efficiency. Precision agriculture establishes avenues through which farmers are able to join in carbon markets, gain incentives, and have their GHG cuts count towards global emission reduction goals through constant monitoring, quantification and verification of GHG reductions (M. T. Ahmed et al., 2025).

1.10.3 Socio-Economic Impacts of Precision Agriculture

Precision agriculture is transforming the socioeconomic environment of the world farming by changing the dynamics of labor, transforming the livelihoods of smallholders and creating new models of digital inequality. Automation, robotics and AI-driven technologies will decrease the amount of labor necessary to be done manually, and at the same time increase the demand of a more technically skilled workforce, changing the distribution of labor and necessitating specific capacity building (Bayar et al., 2025). Precision agriculture has enormous potential to boost yield, decrease input wastage, and enhance profitability in the case of smallholder farmers (especially in developing countries), but the initial cost of technologies, inadequate digital literacy, and infrastructural limits often prevent its use, further separating the technologically progressive farms and their traditional counterparts. This digital divide is further enhanced by the inequalities in the access to internet, access to sensors and machinery and inequitable distribution of extension services (Padhiary et al., 2025). According to recent cost-benefit studies, the net effect of precision technologies is positive, despite some negative impacts on the economy, and it is expected to yield beneficial economic outcomes over time due to enhanced nutrient-use efficiency, lessened loss of inputs, and enhanced climate resiliency, but the issue with capital-intensive technologies remains a major barrier. Ethical issues are also raised when AI-controlled farming brings up the issue of data ownership, privacy of farmers, algorithm bias, and the ability of corporations to control online space (Sanyaolu & Sadowski, 2024). All this socio-economic data points to the necessity of encompassing policies, affordable technologies, and digital ecosystems that are just so that precision agriculture can be helpful to everyone in the farming community.

1.10.4 Challenges and Future Directions

Precision agriculture is still developing at a high rate, but it is limited to its broad adoption due to several challenges that have persistently been associated with it. The sensors, drones, automated machines and software have high initial investment costs, which restrict accessibility, particularly to smallholder farmers and resource-poor areas (Bayar et al., 2025). Technical know-how and insufficient training in the use of technology, even where such is accessible, acts as a barrier to success in the use and maintenance of technology in the long run. Poor infrastructure such as unreliable internet connection, power outages, and absence of cloud computing, also hinder real-time data gathering and digitalization (Gkrimpizi et al., 2023). Cybersecurity and data privacy concerns are also another issue because farms are becoming more interconnected, increasing the risk of unauthorized access of data, data manipulation, and misuse of algorithms. There are also policies and regulatory bottlenecks that limit innovation, investment and technology diffusion, including poor data governance regime, sluggish government approvals and the lack of monetary incentives. Those impediments highlight the necessity to use multi-level interventions that involve financial aid, digital literacy initiatives, and effective policy changes (Diro et al., 2025).

In the future, Agriculture 5.0 paints a picture of a fully autonomous, connected and smart farming environment in the future. The autonomous farms will combine robots, self-driving tractors, and autonomous drones that can work in the field continuously and coordinately without the involvement of few people. A combination of AI, robotics, and high-level sensing is called cognitive agriculture and is going to allow machines to learn, adapt, and make decisions in real-time, enhancing accuracy and sustainability (Saiz-Rubio & Rovira-Más, 2020). Virtual replicas of farms will be created in the form of digital twins that will simulate crop growth, soil dynamics, and interactions with the environment so that farmers could test their strategies prior to implementation. Quantum computing has transformative potential to process complex agricultural data, which can be used to predict climate, pest outbreaks and crop performance over ultra-fast predictive model (Subeesh & Chauhan, 2025). The transparency, traceability and trust in food supply chains will be enhanced using blockchain technologies that will aid in food safety and verification of authenticity. Collectively, these new innovations are on track to an even more sustainable, autonomous, data-driven, and resilient agricultural paradigm of the next generation.

1.11 CONCLUSION

The history of agriculture development is the shift towards industrialization, digitalization and precision technology, a persistent endeavor of humanity to enhance food production and resource efficiency and increase environmental resilience. Every age brought significant innovations but also presented some weaknesses on which further progress was based. The combination of the old knowledge and the latest technologies will provide a middle ground in the advancement, as in the future, agricultural systems will be ecologically oriented and technologically empowered. With the world experiencing even more complicated issues like climate change, population increase, and resource depletion, a hybrid model that does not ignore the native ways but adopts a scientific innovative one becomes more necessary. This hybrid model is not only more sustainable, but it also improves the flexibility of agricultural populations in a wide range of socio-economic settings.

In the future, the vision of global agriculture will be focused on innovation of sustainable technologically powered agriculture based on real-time data, automation, and intelligent decision-support technologies. Precision Agriculture 4.0: This is the future of this transformation, and it will be based on AI, IoT, robotics, and big data, allowing more accurate use of inputs, lessening environmental footprints, maximizing yields, and improving climate resilience. It has a pivotal contribution to global food security and provides scalable solutions to enable more food to be produced using fewer resources, reducing risks and uncertainties. With the potential to fulfill all the capabilities of digital technologies and provide farmers across the globe with equal opportunities, Precision Agriculture 4.0 can lead to a prompt shift to a resilient, productive, and future-oriented agricultural environment.

REFERENCES

Abiri, R., Rizan, N., Balasundram, S. K., Shahbazi, A. B., & Abdul-Hamid, H. (2023). Application of digital technologies for ensuring agricultural productivity. *Heliyon*, *9*(12), e22601. DOI: 10.1016/J.HELIYON.2023.E22601 PMID: 38125472

Ahmed, M. T., Ahmed, M. W., & Kamruzzaman, M. (2025). A systematic review of explainable artificial intelligence for spectroscopic agricultural quality assessment. *Computers and Electronics in Agriculture*, *235*. Advance online publication. DOI: 10.1016/J.COMPAG.2025.110354

Ahmed, N., & Shakoor, N. (2025). Advancing agriculture through IoT, Big Data, and AI: A review of smart technologies enabling sustainability. *Smart Agricultural Technology*, *10*, 100848. DOI: 10.1016/J.ATECH.2025.100848

Al-Musawi, Z. K., Vona, V., & Kulmány, I. M. (2025). Utilizing Different Crop Rotation Systems for Agricultural and Environmental Sustainability: A Review. *Agronomy 2025, Vol. 15, 15*(8). DOI: 10.3390/AGRONOMY15081966

Al-Shammary, A. A. G., Al-Shihmani, L. S. S., Fernández-Gálvez, J., & Caballero-Calvo, A. (2024). Optimizing sustainable agriculture: A comprehensive review of agronomic practices and their impacts on soil attributes. *Journal of Environmental Management*, *364*, 121487. DOI: 10.1016/J.JENVMAN.2024.121487 PMID: 38889650

Altaf, A., Gull, S., Shah, A. Z., Faheem, M., Saeed, A., Khan, I., Al Amin Miah, A., Zhu, X., & Zhu, M. (2021). ADVANCED GENETIC STRATEGIES FOR IMPROVING RICE YIELD. *Journal of Global Innovations in Agricultural Sciences*, *9*(4), 167–172. DOI: 10.22194/JGIAS/9.9520

Aslan, M. F., Sabanci, K., Aslan, B., Aslan, M. F., Sabanci, K., & Aslan, B. (2024). Artificial Intelligence Techniques in Crop Yield Estimation Based on Sentinel-2 Data: A Comprehensive Survey. *Sustainability 2024, Vol. 16, 16*(18). DOI: 10.3390/SU16188277

Ataei, P., Takhtravan, A., Gheibi, M., Chahkandi, B., Faramarz, M. G., Wacławek, S., Fathollahi-Fard, A. M., & Behzadian, K. (2024). An intelligent decision support system for groundwater supply management and electromechanical infrastructure controls. *Heliyon*, *10*(3), e25036. DOI: 10.1016/J.HELIYON.2024.E25036 PMID: 38317976

Azmi, H. N., Hajjaj, S. S. H., Gsangaya, K. R., Sultan, M. T. H., Mail, M. F., & Hua, L. S. (2021). Design and fabrication of an agricultural robot for crop seeding. *Materials Today: Proceedings*, *81*(2), 283–289. DOI: 10.1016/J.MATPR.2021.03.191

Bayar, J., Ali, N., Cao, Z., Ren, Y., & Dong, Y. (2025). Artificial Intelligence of Things (AIoT) for Precision Agriculture: Applications in Smart Irrigation, Nutrient and Pest Management. *Smart Agricultural Technology, 101629*. Advance online publication. DOI: 10.1016/J.ATECH.2025.101629

Bazargani, K., & Deemyad, T. (2024). Automation's Impact on Agriculture: Opportunities, Challenges, and Economic Effects. *Robotics (Basel, Switzerland), 13*(2). Advance online publication. DOI: 10.3390/ROBOTICS13020033

Bhooshan, N., Raman, M. S., Gupta, S., Suyal, G., Singh, A., & Sharma, A. (2024). Revolutionizing agriculture: Role of agricultural mechanization and global trends in farming technology. *Current Science, 126*(10), 1209–1216. DOI: 10.18520/CS/V126/I10/1209-1216

Bissadu, K. D., Sonko, S., & Hossain, G. (2025). Society 5.0 enabled agriculture: Drivers, enabling technologies, architectures, opportunities, and challenges. *Information Processing in Agriculture, 12*(1), 112–124. DOI: 10.1016/J.INPA.2024.04.003

Byamukama, W., Ssemakula, E., Ssekandi, J., Shaman, A., Businge, P., Kalibwani, R., Atukunda, C., & Turyasingura, B. (2025). A Review of ICT Integration in Agricultural Extension Services. A Global Perspective. *East African Journal of Agriculture and Biotechnology, 8*(1), 292–312. DOI: 10.37284/EAJAB.8.1.849

Cao, A., Guo, L., & Li, H. (2025). Understanding farmer cooperatives' intention to adopt digital technology: Mediating effect of perceived ease of use and moderating effects of internet usage and training. *International Journal of Agricultural Sustainability, 23*(1). Advance online publication. DOI: 10.1080/14735903.2025.2464523

Carbone, G., Gurtatta, A. S., & Malyshev, D. (2025). Development and implementation of a hybrid visual prediction algorithm for robotic smart tomato harvesting. *Engineering Applications of Artificial Intelligence, 161*, 112261. DOI: 10.1016/J.ENGAPPAI.2025.112261

Chang, X., He, H., Cheng, L., Yang, X., Li, S., Yu, M., Zhang, J., & Li, J. (2024). Combined Application of Chemical and Organic Fertilizers: Effects on Yield and Soil Nutrients in Spring Wheat under Drip Irrigation. *Agronomy 2024, Vol. 14, 14*(4). DOI: 10.3390/AGRONOMY14040655

Charalampopoulos, I., Droulia, F., Charalampopoulos, I., & Droulia, F. (2024). A Pathway towards Climate Services for the Agricultural Sector. *Climate 2024, Vol. 12, 12*(2). DOI: 10.3390/CLI12020018

Chaudhry, Z., & Rashid, H. (2011). ISOLATION AND CHARACTERIZATION OF RALSTONIA SOLANACEARUM FROM INFECTED TOMATO PLANTS OF SOAN SKESAR VALLEY OF PUNJAB. *Pakistan Journal of Botany*, *43*(6), 2979–2985.

Chen, Q., Vaudour, E., Richer-de-Forges, A. C., & Arrouays, D. (2025). Spectral indices in remote sensing of soil: Definition, popularity, and issues. A critical overview. *Remote Sensing of Environment*, *329*, 114918. DOI: 10.1016/J.RSE.2025.114918

Chen, X. (2025). The role of modern agricultural technologies in improving agricultural productivity and land use efficiency. *Frontiers in Plant Science*, *16*, 1675657. DOI: 10.3389/FPLS.2025.1675657/XML PMID: 41036393

Choudhary, V., Guha, P., Pau, G., & Mishra, S. (2025). An overview of smart agriculture using internet of things (IoT) and web services. *Environmental and Sustainability Indicators*, *26*, 100607. DOI: 10.1016/J.INDIC.2025.100607

Chowdhury, S., Sen, S., & Janardhanan, S. (2022). *Comparative Analysis and Calibration of Low Cost Resistive and Capacitive Soil Moisture Sensor.* https://arxiv.org/abs/2210.03019

Dayioğlu, M. A., & Türker, U. (2021). Digital transformation for sustainable future-agriculture 4.0: A review. *Tarim Bilimleri Dergisi*, *27*(4), 373–399. DOI: 10.15832/ANKUTBD.986431

Dey, S., Widanagamage, N., Achar, S., Debangshi, U., Palla, S., Kim, J., & Jha, G. (2025). *Precision Agriculture Tools, Techniques, and Future Directions for Climate Resilience*. 89–115. DOI: 10.1007/978-981-96-7699-6_5

Dhanaraju, M., Chenniappan, P., Ramalingam, K., Pazhanivelan, S., Kaliaperumal, R., Dhanaraju, M., Chenniappan, P., Ramalingam, K., Pazhanivelan, S., & Kaliaperumal, R. (2022). Smart Farming: Internet of Things (IoT)-Based Sustainable Agriculture. *Agriculture 2022, Vol. 12, 12*(10). DOI: 10.3390/AGRICULTURE12101745

Díaz, J., Quiñonez, Y., De-la-Hoz-Franco, E., Butt-Aziz, S., Mercado, T., Salcedo, D., Díaz, J., Quiñonez, Y., De-la-Hoz-Franco, E., Butt-Aziz, S., Mercado, T., & Salcedo, D. (2025). Information and Communication Technologies Used in Precision Agriculture: A Systematic Review. *AgriEngineering 2025, Vol. 7, 7*(6). DOI: 10.3390/AGRIENGINEERING7060167

Diro, A., Kaisar, S., Saini, A., Fatima, S., Hiep, P. C., & Erba, F. (2025). Workplace security and privacy implications in the GenAI age: A survey. *Journal of Information Security and Applications*, *89*, 103960. DOI: 10.1016/J.JISA.2024.103960

Enescu, F. M., Dr guşin, S. A., Boştinaru, R. N., Marica, M. C., Nicolăescu, C. G., Frîncu, N. C., & Bizon, N. (2025). Energy market: A review of current solutions, trends, issues, and future directions based on the blockchain, internet of things and artificial intelligence. *Results in Engineering*, *28*, 107736. DOI: 10.1016/J.RINENG.2025.107736

Fao, , CWeber, , R. (n.d.). *Policy Analysis*. Advance online publication. DOI: 10.4060/cc9152en

Fraser, E. D. G., & Campbell, M. (2019). Agriculture 5.0: Reconciling Production with Planetary Health. *One Earth*, *1*(3), 278–280. DOI: 10.1016/j.oneear.2019.10.022

GC, R. K., Hall, R. P., GC, R. K., & Hall, R. P. (2020). The Commercialization of Smallholder Farming—A Case Study from the Rural Western Middle Hills of Nepal. *Agriculture 2020, Vol. 10, 10*(5). DOI: 10.3390/AGRICULTURE10050143

Ghorbani, M., Eskandari-Damaneh, H., Cotton, M., Ghoochani, O. M., & Borji, M. (2021). Harnessing indigenous knowledge for climate change-resilient water management – lessons from an ethnographic case study in Iran. *Climate and Development*, *13*(9), 766–779. DOI: 10.1080/17565529.2020.1841601

Gkrimpizi, T., Peristeras, V., & Magnisalis, I. (2023). Classification of Barriers to Digital Transformation in Higher Education Institutions: Systematic Literature Review. *Education Sciences 2023, Vol. 13, 13*(7). DOI: 10.3390/EDUCSCI13070746

Grinin, L., Grinin, A., & Korotayev, A. (2024). *Global Technological Transformations Since the Stone Age: Theory and History*. 57–106. DOI: 10.1007/978-3-031-56764-3_2

Guebsi, R., Mami, S., & Chokmani, K. (2024). Drones in Precision Agriculture: A Comprehensive Review of Applications, Technologies, and Challenges. *Drones 2024, Vol. 8, 8*(11). DOI: 10.3390/DRONES8110686

Habib, M., Singh, S., Jan, S., Jan, K., & Bashir, K. (2025). The future of the future foods: Understandings from the past towards SDG-2. *NPJ Science of Food*, *9*(1), 138. DOI: 10.1038/S41538-025-00484-X PMID: 40651949

Hakami, R. A., Naser, R. S. M., El-Bakkali, M., Othman, M. D. M., Yahya, M. S., Raweh, S., Mohammed, A., & Belghyti, D. (2024). Groundwater quality deterioration evaluation for irrigation using several indices and geographic information systems: A case study. *Desalination and Water Treatment*, *320*. Advance online publication. DOI: 10.1016/J.DWT.2024.100645

Hamadani, H., Rashid, S. M., Parrah, J. D., Khan, A. A., Dar, K. A., Ganie, A. A., Gazal, A., Dar, R. A., & Ali, A. (2021). *Traditional farming practices and its consequences. Microbiota and Biofertilizers* (Vol. 2). Ecofriendly Tools for Reclamation of Degraded Soil Environs., DOI: 10.1007/978-3-030-61010-4_6

Hamdan, M. F., Noor, S. N. M., Abd-Aziz, N., Pua, T. L., & Tan, B. C. (2022). Green Revolution to Gene Revolution: Technological Advances in Agriculture to Feed the World. *Plants*, *11*(10), 1297. DOI: 10.3390/PLANTS11101297 PMID: 35631721

Huang, W., & Wang, X. (2024). The Impact of Technological Innovations on Agricultural Productivity and Environmental Sustainability in China. *Sustainability 2024, Vol. 16, 16*(19). DOI: 10.3390/SU16198480

Islam, M. T., Sepanloo, K., Woo, S., Woo, S. H., & Son, Y. J. (2025). A Review of the Industry 4.0 to 5.0 Transition: Exploring the Intersection, Challenges, and Opportunities of Technology and Human–Machine Collaboration. *Machines 2025, Vol. 13, 13*(4). DOI: 10.3390/MACHINES13040267

Jain, M., Fishman, R., Mondal, P., Galford, G. L., Bhattarai, N., Naeem, S., & Lall, U., Balwinder-Singh, & DeFries, R. S. (2021). Groundwater depletion will reduce cropping intensity in India. *Science Advances*, *7*(9). Advance online publication. DOI: 10.1126/SCIADV.ABD2849 PMID: 33627418

Javaid, M., Haleem, A., Singh, R. P., & Suman, R. (2022). Enhancing smart farming through the applications of Agriculture 4.0 technologies. *International Journal of Intelligent Networks*, *3*, 150–164. DOI: 10.1016/J.IJIN.2022.09.004

John, D. A., & Babu, G. R. (2021). Lessons From the Aftermaths of Green Revolution on Food System and Health. *Frontiers in Sustainable Food Systems*, *5*, 644559. DOI: 10.3389/FSUFS.2021.644559 PMID: 34212131

Kaur, B., Mansi, Dimri, S., Singh, J., Mishra, S., Chauhan, N., Kukreti, T., Sharma, B., Singh, S. P., Arora, S., Uniyal, D., Agrawal, Y., Akhtar, S., Rather, M. A., Naik, B., Kumar, V., Gupta, A. K., Rustagi, S., & Preet, M. S. (2023). Insights into the harvesting tools and equipment's for horticultural crops: From then to now. *Journal of Agriculture and Food Research*, *14*, 100814. DOI: 10.1016/J.JAFR.2023.100814

Kumari, S., Ali, N., Dagati, M., & Dong, Y. (2025). IoT-Enabled Soil Moisture and Conductivity Monitoring Under Controlled and Field Fertigation Systems. *AgriEngineering 2025, Vol. 7, 7*(7). DOI: 10.3390/AGRIENGINEERING7070207

Kumari, S., Raturi, S., Kulshrestha, S., Chauhan, K., Dhingra, S., András, K., Thu, K., Khargotra, R., & Singh, T. (2023). A comprehensive review on various techniques used for synthesizing nanoparticles. *Journal of Materials Research and Technology*, *27*, 1739–1763. DOI: 10.1016/J.JMRT.2023.09.291

Louzada, R. O., Bergier, I., Bolfe, É. L., & Barbedo, J. G. A. (2025). Integrating GIS and remote sensing for soil attributes mapping in degraded pastures of the Brazilian Cerrado. *Soil Advances*, *3*, 100044. DOI: 10.1016/J.SOILAD.2025.100044

Ma, W., Zhou, X., Boansi, D., Horlu, G. S. A., & Owusu, V. (2024). Adoption and intensity of agricultural mechanization and their impact on non-farm employment of rural women. *World Development*, *173*, 106434. DOI: 10.1016/J.WORLDDEV.2023.106434

Malapane, O. L., Musakwa, W., & Chanza, N. (2024). Indigenous agricultural practices employed by the Vhavenda community in the Musina local municipality to promote sustainable environmental management. *Heliyon*, *10*(13), e33713. DOI: 10.1016/J.HELIYON.2024.E33713 PMID: 39040358

Marinello, F., Zou, X., Liu, Z., Zhu, X., Zhang, W., Qian, Y., Li, Y., Karunathilake, E. M. B. M., Le, A. T., Heo, S., Chung, Y. S., & Mansoor, S. (2023). The Path to Smart Farming: Innovations and Opportunities in Precision Agriculture. *Agriculture 2023, Vol. 13*, *13*(8). DOI: 10.3390/AGRICULTURE13081593

Masset, E., Kapoor Malhotra, S., Gupta, N., Bhandari, R., White, H., MacDonald, H., Puskur, R., Singaraju, N., & Sharma Waddington, H. (2023). PROTOCOL: The impact of agricultural mechanisation on women's economic empowerment: A mixed-methods systematic review. *Campbell Systematic Reviews*, *19*(3), e1334. DOI: 10.1002/CL2.1334 PMID: 37361554

Mastretta-Yanes, A., Tobin, D., Bellon, M. R., von Wettberg, E., Cibrián-Jaramillo, A., Wegier, A., Monroy-Sais, A. S., Gálvez-Reyes, N., Ruiz-Arocho, J., & Chen, Y. H. (2024). Human management of ongoing evolutionary processes in agroecosystems. *Plants, People, Planet*, *6*(6), 1190–1206. DOI: 10.1002/PPP3.10521

Mešić, A., Jurić, M., Donsì, F., Maslov Bandić, L., & Jurić, S. (2024). Advancing climate resilience: technological innovations in plant-based, alternative and sustainable food production systems. *Discover Sustainability 2024 5:1*, *5*(1), 423-. DOI: 10.1007/S43621-024-00581-Z

Mirzakhaninafchi, H., Singh, M., Dixit, A. K., Prakash, A., Sharda, S., Kaur, J., Nafchi, A. M., Mirzakhaninafchi, H., Singh, M., Dixit, A. K., Prakash, A., Sharda, S., Kaur, J., & Nafchi, A. M. (2022). Performance Assessment of a Sensor-Based Variable-Rate Real-Time Fertilizer Applicator for Rice Crop. *Sustainability 2022, Vol. 14, 14*(18). DOI: 10.3390/SU141811209

Mishra, P., Singh, P. P., Singh, S. K., & Verma, H. (2019). Sustainable agriculture and benefits of organic farming to special emphasis on PGPR. *Role of Plant Growth Promoting Microorganisms in Sustainable Agriculture and Nanotechnology*, 75–87. DOI: 10.1016/B978-0-12-817004-5.00005-1

Nath, B., Chen, G., O'Sullivan, C. M., & Zare, D. (2024). Research and Technologies to Reduce Grain Postharvest Losses: A Review. *Foods*, *13*(12). Advance online publication. DOI: 10.3390/FOODS13121875 PMID: 38928816

Natsir, M. H., Mahmudy, W. F., Tono, M., & Nuningtyas, Y. F. (2025). Advancements in artificial intelligence and machine learning for poultry farming: Applications, challenges, and future prospects. *Smart Agricultural Technology*, *12*, 101307. DOI: 10.1016/J.ATECH.2025.101307

Okonta, D. E., & Vukovic, V. (2024). Smart cities software applications for sustainability and resilience. *Heliyon*, *10*(12), e32654. DOI: 10.1016/J.HELIYON.2024.E32654 PMID: 39183850

Padhiary, M., Kumar, A., & Sethi, L. N. (2025). Emerging technologies for smart and sustainable precision agriculture. *Discover Robotics 2025 1:1, 1*(1), 6-. DOI: 10.1007/S44430-025-00006-0

Penuelas, J., Coello, F., & Sardans, J. (2023). A better use of fertilizers is needed for global food security and environmental sustainability. *Agriculture & Food Security*, *12*(1). Advance online publication. DOI: 10.1186/S40066-023-00409-5

Pérez, M. Á. G., Guerrero-González, A., Rodríguez, F. J. C., Leon, I. M. M., & Abrisqueta, F. A. L. (2024). Precision Agriculture 4.0: Implementation of IoT, AI, and Sensor Networks for Tomato Crop Prediction. *Buletin Ilmiah Sarjana Teknik Elektro*, *6*(2), 172–181. DOI: 10.12928/BISTE.V6I2.10954

Pozza, L. E., & Field, D. J. (2020). The science of Soil Security and Food Security. *Soil Security*, *1*, 100002. DOI: 10.1016/J.SOISEC.2020.100002

Rabie, A. B., Elhag, M., Subyani, A., Rabie, A. B., Elhag, M., & Subyani, A. (2025). Remote Sensing, GIS, and Machine Learning in Water Resources Management for Arid Agricultural Regions: A Review. *Water 2025, Vol. 17, 17*(21). DOI: 10.3390/W17213125

Rajak, P., Ganguly, A., Adhikary, S., & Bhattacharya, S. (2023). Internet of Things and smart sensors in agriculture: Scopes and challenges. *Journal of Agriculture and Food Research, 14*, 100776. DOI: 10.1016/J.JAFR.2023.100776

Ramana, M. V., Kumari, C. P., Karthik, R., Alibaba, M., Reddy, G. K., Chiranjeevi, K., Kumar, M. S., Chandra, M. S., Ravisankar, N., Bhatt, R., Gaber, A., & Hossain, A. (2025). Integrated Farming Systems Improve the Income of Small Farm Holdings—An Overview of Earlier Findings in the Indian Context. *Food and Energy Security, 14*(2). Advance online publication. DOI: 10.1002/FES3.70064

Rashid, A. Bin, Kausik, A. K., Khandoker, A., & Siddque, S. N. (2025). Integration of Artificial Intelligence and IoT with UAVs for Precision Agriculture. *Hybrid Advances, 10*, 100458. DOI: 10.1016/J.HYBADV.2025.100458

Razu, Md. A. H., Md. Faruk Shah, & Muhammad Bayezid Husain. (2024a). A study on the impact of agricultural mechanization and practices of traditional knowledge in agriculture in jhenaidah district, Bangladesh. *Humanities Journal, 2*(1), 2025002. DOI: 10.31893/HUMANITJ.2025002

Razu, Md. A. H., Md. Faruk Shah, & Muhammad Bayezid Husain. (2024b). A study on the impact of agricultural mechanization and practices of traditional knowledge in agriculture in jhenaidah district, Bangladesh. *Humanities Journal, 2*(1), 2025002. DOI: 10.31893/HUMANITJ.2025002

Said, Z., Vigneshwaran, P., Shaik, S., Rauf, A., & Ahmad, Z. (2025). Climate and carbon policy pathways for sustainable food systems. *Environmental and Sustainability Indicators, 27*, 100730. DOI: 10.1016/J.INDIC.2025.100730

Saiz-Rubio, V., & Rovira-Más, F. (2020). From Smart Farming towards Agriculture 5.0: A Review on Crop Data Management. *Agronomy 2020, Vol. 10, 10*(2). DOI: 10.3390/AGRONOMY10020207

Sanyaolu, M., & Sadowski, A. (2024). The Role of Precision Agriculture Technologies in Enhancing Sustainable Agriculture. *Sustainability 2024, Vol. 16, 16*(15). DOI: 10.3390/SU16156668

Sharafi, S., Mottaghi, F., & Omidvari, F. (2025). Tracking agricultural transformation: Fertilizer efficiency and crop production trends worldwide (1961–2022). *Cleaner and Circular Bioeconomy, 12*, 100191. DOI: 10.1016/J.CLCB.2025.100191

Sharma, H., Sidhu, H., Bhowmik, A., Sharma, H., Sidhu, H., & Bhowmik, A. (2025). Remote Sensing Using Unmanned Aerial Vehicles for Water Stress Detection: A Review Focusing on Specialty Crops. *Drones 2025, Vol. 9, 9*(4). DOI: 10.3390/DRONES9040241

Shoaib, M., Sadeghi-Niaraki, A., Ali, F., Hussain, I., & Khalid, S. (2025). Leveraging deep learning for plant disease and pest detection: A comprehensive review and future directions. *Frontiers in Plant Science*, *16*, 1538163. DOI: 10.3389/FPLS.2025.1538163/BIBTEX PMID: 40061031

Subeesh, A., & Chauhan, N. (2025). Agricultural digital twin for smart farming: A review. *Green Technologies and Sustainability, 100299*. Advance online publication. DOI: 10.1016/J.GRETS.2025.100299

Thilakarathne, N. N., Abu Bakar, M. S., Abas, P. E., & Yassin, H. (2025). Internet of things enabled smart agriculture: Current status, latest advancements, challenges and countermeasures. *Heliyon*, *11*(3), e42136. DOI: 10.1016/J.HELIYON.2025.E42136 PMID: 39959477

Unasylva - Vol. 11, No. 1 - Shifting cultivation. (n.d.). Retrieved November 27, 2025, from https://www.fao.org/4/x5382e/x5382e03.htm

Upadhyay, A., G C, S., Zhang, Y., Koparan, C., & Sun, X. (2024). Development and evaluation of a machine vision and deep learning-based smart sprayer system for site-specific weed management in row crops: An edge computing approach. *Journal of Agriculture and Food Research*, *18*, 101331. DOI: 10.1016/J.JAFR.2024.101331

Varsamis, G., Merou, T., Alexandropoulou, I., Menti, C., Karapatzak, E., Varsamis, G., Merou, T., Alexandropoulou, I., Menti, C., & Karapatzak, E. (2025). Seed Germination Ecology and Dormancy Release in Some Native and Underutilized Plant Species with Agronomic Potential. *Agriculture 2025, Vol. 15, 15*(20), 2139. DOI: 10.3390/AGRICULTURE15202139

Vermelho, A. B., Moreira, J. V., Akamine, I. T., Cardoso, V. S., & Mansoldo, F. R. P. (2024). Agricultural Pest Management: The Role of Microorganisms in Biopesticides and Soil Bioremediation. *Plants*, *13*(19), 2762. DOI: 10.3390/PLANTS13192762 PMID: 39409632

von der Goltz, J., Dar, A., Fishman, R., Mueller, N. D., Barnwal, P., & McCord, G. C. (2020). Health Impacts of the Green Revolution: Evidence from 600,000 births across the Developing World. *Journal of Health Economics*, *74*. Advance online publication. DOI: 10.1016/J.JHEALECO.2020.102373 PMID: 33002797

Yan, J., Wu, H., Diao, Z., Miao, Y., Zhang, B., & Zhao, C. (2025). Recent Developments and Applications of Crop Disease Detection, Prediction, and Early Warning: A review. *Engineering*. Advance online publication. DOI: 10.1016/J.ENG.2025.10.032

Zimdahl, R. L. (2022). The conduct of agricultural science. *Agriculture's Ethical Horizon*, 17–33. DOI: 10.1016/B978-0-12-823667-3.00006-9

Chapter 2

Decoding Precision Agriculture: A New Era in Farming Innovation

Muhammad Waqas
http://orcid.org/0009-0009-8126-9709
State Key Laboratory of Cotton Bio-Breeding and Integrated Utilization, Institute of Cotton Research, Chinese Academy of Agricultural Sciences, Anyang, China

Muhammad Nadeem
State Key Laboratory of Cotton Bio-Breeding and Integrated Utilization, Institute of Cotton Research, Chinese Academy of Agricultural Sciences, Anyang, China

Burhan Khalid
College of Plant Science and Technology, Huazhong Agricultural University, Wuhan, China

Hafiz Ghulam Nabi
Frontiers Science Centre for Molecular Design Breeding, Beijing Key Laboratory of Crop Genetic Improvement, College of Agronomy and Biotechnology, China Agricultural University, Beijing, China

ABSTRACT

Precision Agriculture is a transformation of the agricultural industry; agricultural data is being generated and utilized rather than depending on experience. The chapter emphasizes that smart farming systems involve the use of technologies such as GPS, Internet of Things, artificial intelligence, drones, and robotics to be more productive, sustainable, and efficient in terms of using resources. Some of the major components are data management, predictive analytics driven by AI, and intelligent decision-support systems, which are essential in the optimization of inputs and reduction of environmental impact. It attempts to solve the issues of expenses, confidentiality of

DOI: 10.4018/979-8-3373-7257-0.ch002

data, and digital illiteracy. Successful implementations are presented in case studies, where the importance of startups and SMEs is stressed in the area of innovation. Future arguments will discuss how AI can be used to generate recommendations and block chain to promote transparency, self-driving machines, and how Precision Agriculture can support climate-smart agriculture.

Figure 1. Graphic abstract

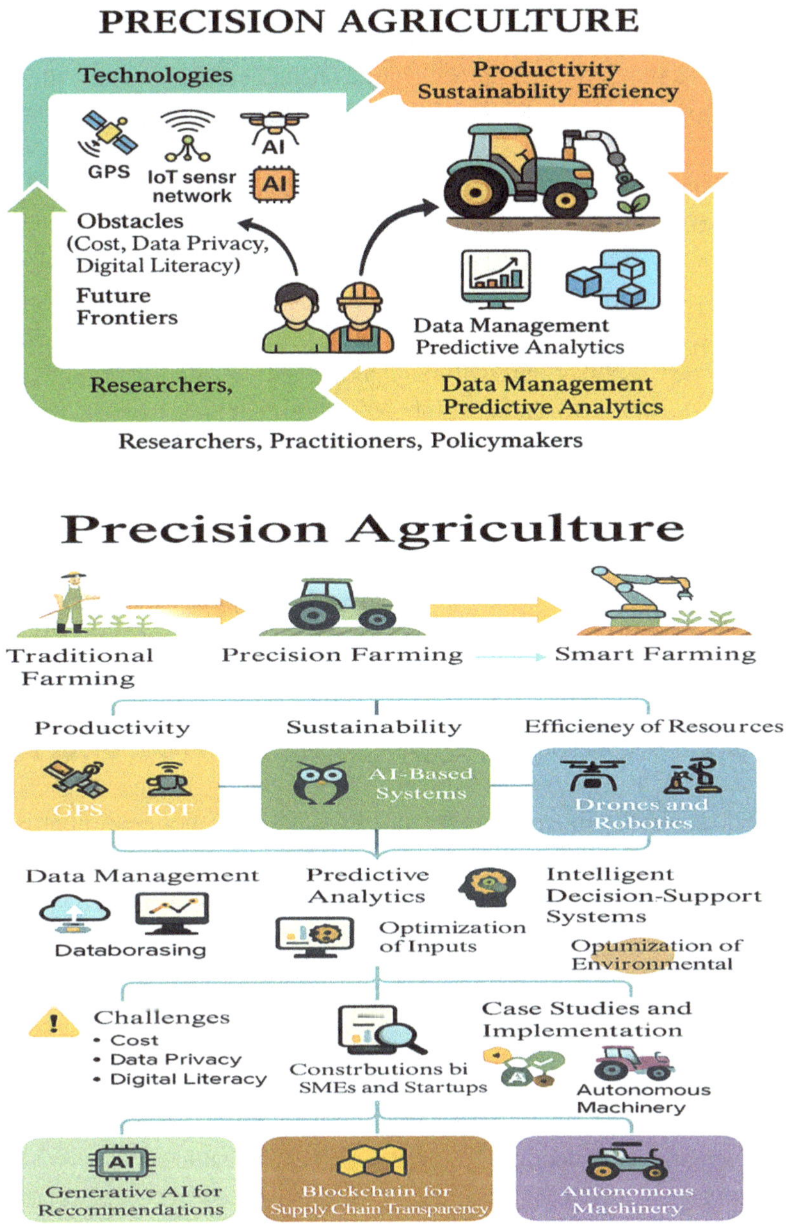

1. INTRODUCTION

Precision Agriculture (PA) is a paradigm shift in the management of farms that no longer follows a standardized approach based on the field, but rather, a site-specific, data-driven approach. International Society of Precision Agriculture (ISPA) defines it as a management technique which employs information technology to gather, process and analyze temporal, spatial and individual data. The essence of it is to increase the effectiveness of resource utilization, productivity, quality, profitability, and sustainability of agricultural production (Saha et al., 2025). This is done by the combined application of a set of technologies, such as Global Positioning Systems (GPS), Geographic Information Systems (GIS), Internet of Things (IoT), sensor networks, drones, robotics, and artificial intelligence (AI) to measure and respond to variability in an intra-field (Delfani et al., 2024; Ferrández-Pastor et al., 2018). When the civilian supply of GPS became possible in the 1980s, the earliest steps toward the possibility of field mapping accurately were made, owing to the genesis of PA.

Yield monitors and variable rate technology (VRT) came into existence in the 1990s, enabling the differentiated entry application (Lowenberg-DeBoer and Erickson, 2019). Precision Agriculture is currently a digital ecosystem on its own. The crossroads between remote sensing, AI, and IoT have transformed the field of thought to intelligent, AI-powered analytics. These systems can not only predict but also prescribe and allow fully automated and proactive responses to changing field conditions (Gupta and Kumar, 2025). Finally, the metamorphosis is the only necessary reaction to severe pressures of the global environment: a continuous population increase, climate change, and exhaustion of nonrenewable resources (Surur et al., 2025).

As a solution to the 21st-century environment, Precision Agriculture (PA) has grown out of the niche activity to become a core part of sustainable intensification, a process indicated by its estimated market value of more than 12 billion dollars worldwide by 2025. There is a triple depth of its meaning. PA creates potent economic yields and profitability through a reduction in the input costs. At the same time, it offers a needed environmental benefits as the foundation of climate intelligent agriculture with its reduced agricultural footprint and resource saving by ultra-efficient input application. Lastly, it generates valuable social worth as global food security is solidified, and the farmers are empowered with data information to stabilize production (Said et al., 2025).

This chapter will help to formulate a complete and modern model of Precision Agriculture (PA). It will start with the covering of main principles and technologies that are the backbone of PA starting from the basics such as GPS and sensors to the high-tech AI-based analytics and robots. The following analysis will then focus on key advantages of these technologies including increased sustainability features, economic returns; other issues such as high startup costs, data privacy concerns, digital literacy issues will also be provided as key barriers for adoption. In order to base this discussion on practice, the chapter will provide real-life case studies of different global settings of how well and how scalable PA-implementation is. Lastly, it will address the emerging trends and the future direction of the field and developments of integrating generative AI, Blockchain as a visibility tool in the supply chain, and next-generation autonomous machines as a carbon footprint monitoring tool and climate-resilient farming. In thus exploration, the chapter will also reveal how Precision Agriculture is redefining fundamentally the 21st century agriculture by leveraging on the power of data driven understanding in simultaneously promoting an independent trend of achieving the dual goals of global food security and environmental sustainability.

2. CORE TECHNOLOGIES AND INNOVATIONS IN PRECISION AGRICULTURE

Precision Agriculture (PA) has a transformative potential to integrate the most basic technologies to form a closed-loop system in which data is gathered, analyzed, and automated to make decisions. This technological framework is used to transform conventional agriculture into a data-driven, adaptive and intelligent enterprise (Soussi et al., 2024).

Precise positioning and mapping are the basis of PA, provided by Global Navigation Satellite Systems (GNSS), which offer the geospatial texture of all site-specific functions. This canvas is made alive with the help of multi-layered sense architecture. Satellite remote sensing is a high-frequency, macro-scale monitoring of crop health and moisture stress, with the application of a global satellite image in agriculture set to increase 45 percent in 2023-2025 (Huang et al., 2023). In line with this, Unmanned Aerial Vehicles (UAVs or drones) provide extremely high-resolution images, making it possible to identify both pests and nutrient deficiencies at a sub-5cm resolution, and nowadays, with sophisticated fleets, it is possible to cover more than 1,000 hectares in a single day (Alsadik et al., 2024; Aslan et al., 2022). At the same time, in-field Internet of Things (IoT) sensor networks can deliver real-time and ground-truth soil moisture, temperature, and nutrient levels data, which can generate a continuous flow of information that can be immediately used to start an automated irrigation or alert system (Ansari and Vidyarthi, 2025). This closed-loop system, where data collection, transmission, AI-assisted analytics, and automated actuation develop a continuous feedback cycle for sustainable farm management, as illustrated in *Figure 2*.

Figure 2. Closed-loop precision agriculture system where IoT sensors and Multispectral/Hyperspectral Cameras gather data from the field, LoRaWAN/5G sends the data to computer platforms for AI/ML analysis and autonomous tractors, drones, robotics and variable rate technology apply prescriptions. The feedback loop enhances sustainability, increases productivity (~+30%) and also decreases the use of pesticides (~80%)

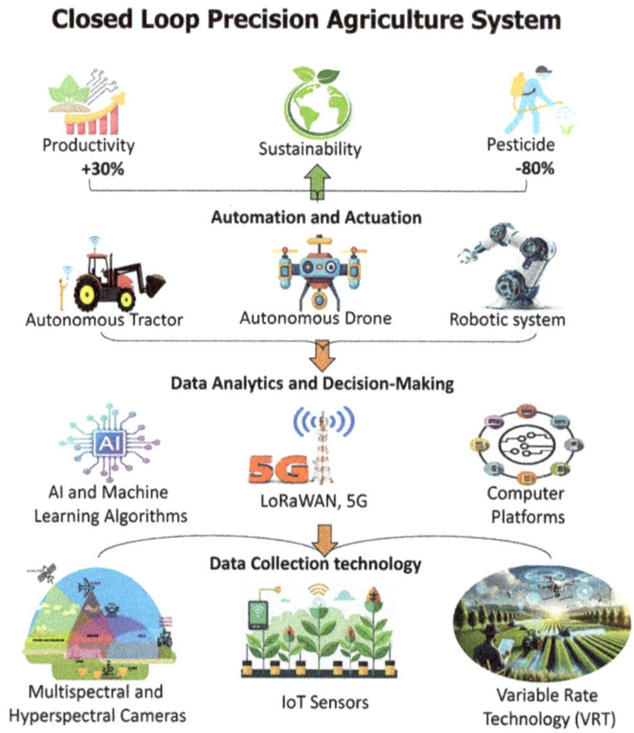

Artificial Intelligence (AI) and Machine Learning (ML) transform the huge and unstructured data streams of such sources into actionable insights. The drastic move towards descriptive analytics to predictive and prescriptive models is currently underway. It can be done with the help of computer vision algorithms, especially Convolutional Neural Networks (CNNs), which detect the stresses of crops before they are recognizable by the human eye (Javidan et al., 2024). Moreover, by combining historical data, current weather, and satellite images, the ML models will be able to produce accurate yield forecasts and forecast pest outbreaks to intervene proactively. This potential will boost worldwide crop productivity by as much as 30 percent by 2025 and take farm management beyond a responsive to a predictive science (Gupta and Kumar, 2025).

Robotics and automation represent the end, physical expression of this cycle of data. Self-driving tractors, self-driving implements, and other autonomous machinery have been developed to conduct accurate field operations such as plowing and seeding without employing a large number of human workers. In addition to big machines, specialized agricultural robots (Agribots) are also underway: machine-vision weeders are being used to distinguish crops and weeds so that machines can do weeding or spray them with particular herbicides, eliminating over 80 percent of herbicides, and robotic harvesters with sophisticated vision and soft grippers are being used to combat serious labor shortages in crops with high value (Upadhyay et al., 2024). This is an optimal integration of GNSS enabled sensing with AI based analytics and Robotic actuation, which makes an intelligent and adaptive agricultural system. It is a closed loop that supports an agricultural paradigm of unexampled efficiency, productivity and sustainability.

3. APPLICATION AREAS AND BENEFITS OF PRECISION AGRICULTURE

Precision Agriculture is worth its technological set due to the utility of the information it provides to the crops, which turns into tangible gains during the crop production process. All these applications translate into a self-sustaining cycle, integrating economic benefit, environmental sustainability and improved productivity.

Nevertheless, one of the major drivers is its fine-tuning of the critical inputs.

Variable Rate Technology (VRT): This technology uses Global Positioning System (GPS) and Artificial Intelligence (AI) to generate prescription maps for optimizing application of fertilizers, pesticides and seeds, considering the site-specific conditions instead of applying the same treatment across the entire fields. Studies show that this approach can help save as much as 30 percent in fertilizer and pesticide application, directly addressing the problem of low Nitrogen Use Efficiency (low nutrient use efficiency), which currently is estimated at around 55 percent around the world. For example, sensor-based irrigation system can efficiently irrigate only where and when needed, leading to 20-50% less water used compared with conventional irrigation system; a crucial adaptation for areas that are prone to droughts (Kumar et al., 2024). *Figure 1* shows complete cycle enabling benefits of using these technologies.

This effective regulation directly increases the crop yield and quality. By ensuring that plants receive optimal nutrition and are well protected without being under stress, farmers achieve optimal performance in terms of genetic potentials. Moreover, yield maps and predictive analysis can help in making decisions on planting density and harvest time, which will lead to more homogenous crops and better overall yield. This field-to-plant-level management transformation is what makes sustainable intensification, which permits cultivating more food on the same area of the land (Aslan et al., 2024).

Sustainability implications of the paradigm shift make PA one of the pillars of climate-smart and regenerative agriculture. The radical decrease in the overuse of chemicals reduces the amount of nutrient runoff into waterways, besides reducing the amount of greenhouse gases emitted through the use of fertilizers. Besides, such practices as the accurate management of nutrients help to improve the health of soil, promoting the presence of organic matter and the activities of microorganisms. The most prominent application is monitoring carbon footprint, where the PA technologies are applied to monitor and confirm carbon sequestration and carbon emissions in the soil to be able to enter into carbon markets and generate a new source of income through sustainable practices (Okoli and Birkenberg, 2024).

These efficiencies can be economically converted into irresistible cost-saving and enhanced profitability for farmers. The cost of initial technology may be expensive, but it will save on its operations in the long term. Less money is spent on fertilizers, pesticides, water, and fuel, which decreases the cost of operation directly. The economic and environmental cases are as solid as each other when supplemented by higher and steadier yields, and the profit margin is stronger and the business model more stable, making the case of PA economic as mighty as the environmental (Papadopoulos et al., 2024). A summary of comparative benefits of key PA technologies across different crop types and environments is presented in *Table 1.*

Table 1. Comparative benefits of PA technologies across crop types and environments

Technology	Key Application	Primary Benefit	Example Crop and Impact
Variable Rate Technology (VRT)	Site-specific application of fertilizer, lime, and seeds	Optimizes input use, boosts yield, and improves soil health	**Corn/Wheat:** Applying nitrogen only where needed, increasing yield by 5-10% while reducing fertilizer use by 15-20%
Sensor-based Irrigation	Using soil moisture sensors to automate water application	Conserves water and energy, prevents over/under-watering	**Cotton/Vegetables:** Reducing water consumption by 20-50% while maintaining or improving yield quality
Drone-based Scouting and Spraying	High-resolution crop health monitoring and targeted spraying	Enables early pest/disease detection and drastically reduces chemical drift	**Vineyards/Orchards:** Spot-spraying for fungal diseases, reducing pesticide volume by up to 90% compared to broadcast spraying
AI-Powered Weed Detection	Using computer vision to distinguish crops from weeds	Enables mechanical weeding or spot-spraying, reducing herbicide reliance	**Soybeans/Carrots:** Robotic weeders can reduce herbicide use by over 80%, lowering costs and environmental impact

continued on following page

Table 1. Continued

Technology	Key Application	Primary Benefit	Example Crop and Impact
Yield Monitoring and Mapping	GPS-linked sensors on harvesters that create yield maps	Provides data for understanding variability and planning for the next season	**All Arable Crops:** Identifying underperforming zones to investigate causes (e.g., soil compaction, drainage) and take corrective action
Carbon Monitoring	AI and satellite data to track soil carbon and emissions	Verifies sustainability for carbon markets and eco-certifications	**Broadacre Crops/Perennials:** Quantifying carbon sequestration for voluntary carbon credits, creating a new income stream

To conclude, Precision Agriculture takes farming to a new level of being an intelligent and responsive rather than a reactive practice. Their benefits are closely linked: optimising inputs increases sustainability which in itself increases soil health and profitability which then feeds into soil health and so on - resulting in positive feedback that supports the livelihoods of farmers and also contributes to planetary health.

Data Management, Modeling, and Decision Support Systems

The key to Precision Agriculture is a complex value chain of data that transforms the raw data from a variety of inputs into actionable intelligence. This ecosystem is the pipe-dream between technology capabilities and farm effectiveness to enable farm management decision making, which characterizes modern-day agriculture (Sanyaolu and Sadowski, 2024).

This system is based on a network of multi-layered sources of data that provides another perspective on the field conditions. Satellite remote sensing offers regular and macro-scale information on crop health and moisture stress through vegetation indices, including NDVI (Judith et al., 2025). Besides this, Unmanned Aerial Vehicles (UAVs) are able to capture ultra-high-resolution multispectral and hyperspectral images, which display early pest infestation or disease before the naked eye can detect it (Matese et al., 2024). Meanwhile, soil sensor networks of Internet of Things (IoT) can offer real-time ground-truth measurements of moisture, temperature and nutrient levels and automated weather stations can offer hyper-local weather information that would be needed by operational planning and predictive modelling (Ou et al., 2025).

The Big Data in the area of agriculture is the volume, velocity and nature of data generated, and thus storage, processing, and analysis can be considered a critical issue. These heterogeneous data now constitute an indispensable component of the infrastructure as they entail cloud computing platforms and data lakes to offer the scalability and computing capability to do computations on the said datasets. A transition to API-based integration platforms allowing the combination of data in real-time with satellites, drones, sensors, and machinery into one unified platform that can be analyzed holistically and do away with major interoperability challenges will be among the trends that are bound to become mandatory in 2025 (Ahmed and Shakoor, 2025).

The rich environment of data is being populated with advanced crop yield simulation and predictive models, and being run by machine learning (ML) and deep learning. The complex characteristics and latent relationships can be detected automatically by the advanced algorithms, such as the Convolutional Neural Networks (CNNs), which can detect the spatial patterns, and the Long Short-Term Memory (LSTM) networks, which can detect the temporal patterns of weather and sensor data. These models are also more often used to make predictions, or prediction analytics, including to make specific management recommendations, like the optimal irrigation regimes or even nutrient regimes (Zhu, 2024).

The output of this value chain of data is a series of Decision Support Systems (DSS) and farm management software, which convert the multi-dimensional data and model outputs into action plans for the farm. They are evolving into centralized nervous systems of the farm, bringing all the parts of the PA data ecosystem together into zoning the field, creating variable rate application plans, tracking resources, and keeping track of finances (da Silveira et al., 2025). They symbolize the working point where data-based insights are translated into the practical farming decision-making process between data gathering and data action in the field, so that the agricultural results can be the greatest. Leading data platforms and software solutions currently available in PA are summarized in *Table 2*.

Table 2. Leading data platforms and software solutions in precision agriculture

Platform Name	Core Technology and Key Features	Primary Applications	Unique Selling Points
Farmonaut	Satellite imagery, AI advisory, IoT integration, Blockchain traceability	Real-time crop/soil monitoring, yield forecasting, supply chain traceability, sustainability reporting (carbon footprint)	Affordable, scalable solution for diverse users, from smallholders to large agribusinesses; strong focus on sustainability metrics
Climate FieldView (Bayer)	Sensor fusion, AI, extensive compatibility with farm equipment	Field data mapping, yield prediction and analysis, precision seeding and fertilization	Deep integration with a wide range of machinery; powerful data-driven input optimization tools
John Deere Operations Center	Cloud-based platform, real-time data from connected equipment	Equipment monitoring and diagnostics, yield tracking, field operations management	Seamless connection with John Deere machinery; robust remote diagnostics and fleet management

continued on following page

Table 2. Continued

Platform Name	Core Technology and Key Features	Primary Applications	Unique Selling Points
Trimble Ag Software	GPS/GNSS focus, GIS, variable rate technology (VRT)	Precision guidance, field mapping, VRT application of inputs, yield monitoring	Long-standing expertise in positioning technology; strong solutions for guidance and variable-rate control
Granular (Corteva)	Farm business focus, data analytics, financial management tools	Farm financial planning and analysis, field-level profitability tracking, operational management	Bridges the gap between agronomic data and business/ financial decision-making for farmers
IBM's The Weather Company	AI, Big Data, proprietary global weather models	Hyper-local weather forecasting, irrigation planning, risk assessment for pest/disease	Unmatched, high-accuracy weather and climate risk analytics tailored for agricultural planning

Even with that transformational potential, these data management systems have implementation challenges, such as data silos, interoperability of systems and data privacy. The future is moving towards decentralized and smarter systems, and research is being put into architectures of IoRT (Internet of Robotic Things), where IoT sensor and drone data are used to feed the autonomous robot behaviors on the field. With the birth of hybrid AI models that bring out a lot of machine learning techniques, in the future, even more useful and innovative solutions to the complex problems of agriculture can be expected.

4. CHALLENGES AND LIMITATIONS OF PRECISION AGRICULTURE

The great potential of the use of Precision Agriculture (PA) may be restricted by the multifaceted range of issues that may even make the application of the new technologies not only technically feasible but also economically, socially and ethically feasible. These barriers are critical issues that need to be addressed in an effort to place PA in world agricultural systems in an equitable and sustainable way.

The main challenge however remains the high initial investment of PA technologies. GPS-based machines, sensor networks and data subscription services are expensive and a major obstacle for smallholder farmers and communities in developing countries. This unequal economic distribution is a danger in creating a digital divide where operations with higher resource endowments will accumulate more speed in their productivity while small-scale farms struggle to keep up. This is worsened by the digital divide that persists and a systematic review in 2025 confirms that the technical complexity of PA systems demands specialized skills to operate them, to interpret the data produced, and the steep learning curve to operate them in the absence of appropriate training and support infrastructure (Ahmad et al., 2025).

Together with economic and technical aspects, PA poses severe questions about the privacy and security of data and ethical control. Recent agriculture management systems are collecting vast amounts of sensitive operation data including yield maps, input application, and business information. This raises serious concerns with regard to ownership of data, control and the possibility of abuse by technology providers, agribusinesses, or other third parties. Such granular information consolidated creates ethical issues regarding who will ultimately receive the benefits of such digital assets and how they can be utilized to alter market forces or subvert the control of the farmer (Gardezi et al., 2024).

Moreover, even though PA is praised due to its environmental sustainability, trade-offs and unintended consequences have to be noted. Production, implementation, and disposal of electronic parts and sensors are the factors that lead to the use of resources and increase the amount of e-waste. Threats that efficiency gains might unwillingly stimulate cultivation in ecologically sensitive places or may foster the growth of new addictions to proprietary systems and non-renewable resources also exist (Gibson, 2013). These ecological issues are linked to more general socio-economic effects, such as possible unemployment through automation, concentration of agricultural knowledge in proprietary digital technologies, possibly undermining the wisdom of traditional farming, and the further increase in structural inequalities in the rural population. *Figure 1* visually represents this data value chain. Briefly, key challenges in PA adoption, along with potential solutions are summarized in *Table 3*.

Table 3. Summary of challenges with potential mitigation strategies

Challenge Category	Specific Challenges	Potential Mitigation Strategies
Economic Barriers	High initial technology costs; Limited access for smallholders; Uncertain return on investment	Development of scalable, affordable solutions (e.g., mobile-based platforms, drone services); Government subsidies and financing programs; Cooperative ownership models; Clear demonstration of long-term cost-benefit analysis
Technical and Capacity Barriers	System complexity; Interoperability issues between platforms; Digital literacy gaps; Lack of technical support	User-centered design principles; Standardization of data protocols; Comprehensive training programs; Digital extension services; Establishment of local support networks
Data Governance and Ethics	Data privacy concerns; Questions of data ownership; Security vulnerabilities; Ethical use of algorithms	Transparent data agreements; Farmer-centric data governance frameworks; Robust cybersecurity measures; Ethical AI guidelines and oversight; Legislative protection of farmer data rights
Socio-Environmental Impacts	E-waste from technology refresh cycles; Potential job displacement; Knowledge consolidation in proprietary systems; New resource dependencies	Sustainable technology design and recycling programs; Reskilling initiatives for agricultural workers; Open-source platforms and knowledge sharing; Lifecycle assessment of PA technologies; Balanced approach combining traditional and digital knowledge

The solution to these multi-layered problems must involve a multi-stakeholder approach. Policymakers should come up with regulatory mechanisms in which interests of farmers are safeguarded and enhancing innovation is achieved. The technology companies should focus on affordability, interoperability, and solutions that were designed with ethics. The agricultural extension services should be reformed in order to close the digital literacy gap, and the research institute should increase the study of the unintended socio-ecological consequences of technological adoption.

5. CASE STUDIES AND REAL-WORLD IMPLEMENTATIONS

The hypothetical possibility of Precision Agriculture is finally confirmed in the various world applications that reveal both practical effects in various agricultural backgrounds, scales and production systems. These practical applications indicate the way PA technologies are being modified to meet region-specific issues and provide quantifiable economic, environmental, and social returns.

5.1 North America: Large-Scale Efficiency and Sustainability

In vast grain production systems of North America, precision technologies have developed into a tool that was once optional but is now becoming integral to profitable and sustainable agriculture. Combining yield monitoring, soil mapping and Variable Rate Technology (VRT) has enabled farmers to record outstanding efficiencies (Getahun et al., 2024). Precision nutrient management systems to be installed in the US corn belt have shown 15-20% less application of fertilizers and no decline or growth in yields, directly overcoming the dire situation of low Nitrogen Use Efficiency at about 55% around the world (Wang et al., 2025). These systems use AI-assisted prescription maps, which combine the data of soil sensors, satellite images, and previous yield patterns to maximize the placement of inputs, and in the process minimize the cost of production and the environmental impact caused by the reduction of nutrient run-off.

5.1.1 Israel: Overcoming Extreme Resource Limitation

With an acute shortage of water, Israeli agriculture has led the way in coming up with more accurate water management technologies that have seen desert-like landscapes being converted into productive farming areas. The sensor-guided drip irrigation systems combined with AI-managed monitoring have proven to be the most efficient of all when it comes to water use, showing up to 90% of water saving as compared to the conventional flood irrigation techniques (Morchid et al., 2025). Firms such as Netafim have come up with advanced irrigation platforms that combine soil moisture measurements of the soil, evapotranspiration rates, and plant stress sensors to provide millimeter-accurate water. Not only has this method conserved the limited water resources but has also enhanced quality of crops and stability of yield in undesirable environment to grow crops, establishing an international standard of water-saving methods in agriculture (Jenkins and Block, 2024).

5.1.2 India: Digital Agriculture to the Smallholder

The example of the Indian agricultural landscape illustrates how practical innovations can be used to support smallholder farmers with the help of precision technologies and innovative business solutions of the government. The digital application of AI-based advice via a mobile phone has assisted limited-resource farmers to apply superior farming intelligence. Their combination of satellite images, all-weather information and soil sensor data has helped farmers to reduce their input costs by 50 percent (15-20 percent) and reduce chemical use by 25 percent (25 percent) and achieve higher results in changing climatic patterns. The success of these applications highlights the need to develop scalable cost effective solutions that have the ability to bridge the technology access gap of small scale producers who constitute the backbone of most food systems in the world (Balkrishna et al., 2023).

5.1.3 Netherlands: Precision in Controlled Environment

Dutch greenhouse agriculture is seen as the ultimate of precision farming based on controlled environments, where IoT sensor networks and AI systems can produce optimized microclimates to produce high-value products. Significant systems of sensors constantly measure soil moisture, air temperature, humidity, light intensity and indicators of plant health and send it to central artificial intelligence systems that turn the irrigation, air conditioning, and lighting on and off automatically. This closed loop model has proven to save 30 percent of water and increased yields more than 20 percent and practically cleared pesticide application due to integrated pest management systems made possible by constant monitoring. The Dutch model is a perfect representation of how accuracy technologies can optimize the production rate per a restricted amount of land and reduce environmental effects (Karanisa et al., 2022).

5.2 Emerging Innovations: The SME and Startup Ecosystem

An exciting international ecosystem of Small and Medium-sized Enterprises (SMEs) and startups is propelling the fast innovation in precision agriculture. Farmonaut and other companies are bringing satellite tracking and AI-powered pest prediction to small-scale operations, and other companies are developing blockchain applications to provide visibility to the supply chain. Developing startups are addressing certain sustainability issues; an example is CH_4 Global has created seaweed-based feed supplements that are able to lower methane emissions in livestock by as much as 90% (Watson, 2025). Such agile innovators can often be found to fill key gaps that larger companies have left, creating solutions tailored to the context and new business models such as "PA-as-a-Service" that bring advanced technologies to a larger audience.

5.3 Supporting Frameworks: Policy and Market Development

The policy frameworks and changing market structures are boosting the development of precision agriculture. The precision agriculture sector worldwide is set to obtain in excess of 12 billion in the coming 2025, with more than 60 percent of the big farms anticipated to have advanced PA solutions (Precision Agriculture Market Growth 2025, 2025). Governments around the world are now actively promoting this shift through specific programs, such as the large U.S. federal investments in the research and development of PA, or state-level precision horticulture programs in India (Mondal and Basu, 2009; Office, 2024). At the same time, strategic corporate-startup alliances and investments are transforming the market, e.g., by embedding IoT sensors into conventional irrigation machines to form more holistic and interoperable farm management systems. The detailed description of AI-assisted or PA adoption and their outcomes are illustrated in *Figure 3*.

Figure 3. Global adoption of precision agriculture appraising regional adoption along the continents such as IoT, GPS, Drones, Artificial Intelligence (AI), VRT, etc. The map is country-specific about innovators -- those that are providing smart irrigation, robotics, cloud analytics and mobile tools that are enhancing yield, efficiency and sustainability of inputs around the globe

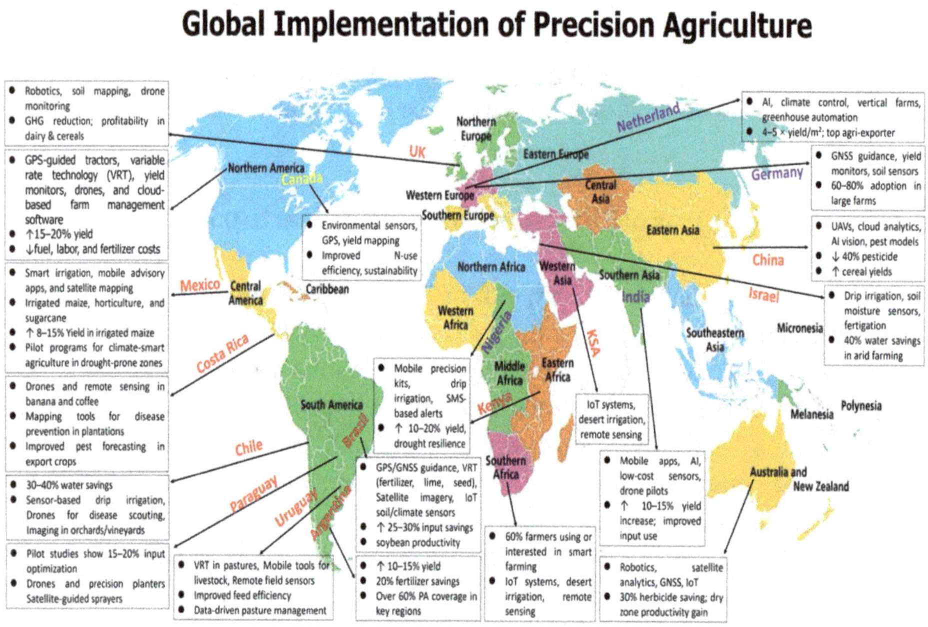

5.4 Lessons Learned and Best Practices

An overview of the world experiences of implementation of Precision Agriculture (PA) indicates several important lessons and best practices that can be used to make a successful adoption. Context-specific adaptation is one of the most important aspects, technological solutions should be adapted to the specifics of the local environment, the availability of the resources and cultural customs, and not be universal and one-size-fits-all. Another point, which is also crucial, is the development of participation and inexpensive organizational concepts. Precision agriculture innovations have been facilitated by approaches including PA-as-a-Service platforms, mobile-first applications, and cooperative ownership structures that have helped smallholder farmers, who make up the majority of food producers in the world, to take part in the innovations and benefit from them (Vijayakumar et al., 2025).

The other crucial thing learnt is that, to become successfully adopted in technology, digital literacy must be in place. Introduction of PA tools is a successful process that needs elaborate locally embedded training and support systems. The capacity building process should be aimed at improving the technical skills of farmers as well as making them confident in their ability to use digital tools to make decisions. Moreover, a solid economic value is the best motivator to embrace. When a farmer can observe some concrete returns, e.g. a lower input cost, higher yield, or better access to a high-value market, they are much more likely to adopt a practice of precision agriculture (Vijayakumar et al., 2025).

The combination of data gathered by several different sources is also central. The integrated data systems, which combine data collected by satellites, IoT sensors, weather stations, and agricultural devices, have more analytical capabilities and actionable insights than the single-point solutions. Lastly, Joint innovation ecosystems linking governments, research institutes, technology suppliers, and organizations of farmers have been identified to be the most effective in helping to create a sustainable innovation, as well as expand practices that have proven successful (Dhanaraju et al., 2022). Taken altogether, these teachings form some useful strategic advice to researchers, policymakers, and technology developers who aim to optimize the advantages of precision agriculture in various farming settings.

6. FUTURE PROSPECTS AND EMERGING TRENDS

Precision Agriculture is on the brink of a new technological revolution as there is convergence in disciplines of artificial intelligence, robotics, and biological sciences. The following decade will see the paradigm shift of responsive monitoring to fully independent, autonomous, and self-optimizing agriculture systems capable of simultaneously enabling the problem of productivity, sustainability, and climate resilience to be tackled.

6.1 Artificial Intelligence and Analytics Next Generation

The discovery of AI in agriculture is developing to be more predictive, then generative, and finally prescriptive. The systems of generative AI are becoming virtual agronomists that can combine the multifaceted environmental conditions, market dynamics, and operational limits to generate dynamic management prescriptions via natural language interfaces (Shahriar et al., 2025). Such systems are capable of modeling reaction of crops to several management situations so that farmers can have a view of what will happen before making decisions. At the same time, it is reinforcement learning algorithms that are developing dynamic decision-making schemes to constantly optimize strategies on a field based on feedback and no longer using fixed prescriptions but dynamic interventions that would adapt to the evolving conditions on the field.

6.2 Sophisticated Sensing and Digital Twins

Sensor technology is being developed in terms of affordability, autonomy and comprehensiveness. Hyperspectral imaging improvements that have been made by satellites and drones allow revealing biochemical plant data notwithstanding the classical vegetation indices, and early acquaintances of nutrient deficiencies and disease stress in pre-monitored levels have become feasible. Such developed sensing functions drive the creation of agricultural digital twins, simulation models of the farming process where the real-time data streams are combined with the process-based model to simulate crops, soil interactions, and flow of resources. These digital twins can be used as testbeds of management strategies and make possible scenario planning of climate adaptation, fundamentally transforming the way the decisions of farming are made and assessed (Ram et al., 2024; Singh et al., 2023).

6.3 Autonomous Systems and Robotics

The automation trend moves towards the single-task robots of the automation to the integrated autonomous system. Current fleet coordination algorithms can coordinate the work of heterogeneous robotic systems - swarms of drones to monitor coordination with autonomous ground vehicles to provide targeted interventions. The introduction of multi-purpose robotic platforms, which can carry out sequential tasks (weeding, fertilizing, monitoring) on the same field pass can be viewed as a major efficiency breakthrough. The regulatory frameworks are also undergoing changes to facilitate this change, with future EU machinery directives defining the requirements of unmonitored autonomous operations with remote supervision functions, which will enable commercial-scale deployment (Abbas et al., 2023; Almalki and Angelides, 2024).

6.4 Farming Climate Strategy and Carbon Agriculture

Precision Agriculture is paying more emphasis on measuring and improving sustainability. AI-based carbon monitoring systems which use remote sensing are facilitating the field scale measurement of soil organic carbon and greenhouse gas emissions, which form the basis of verification of carbon credit markets. The fact that ecosystem services can be integrated as metrics into farm management software enables the optimization of the management systems to lead to better environmental performance, such as biodiversity conservation, water quality protection, and climate mitigation. These functionalities make PA the enabling technology for the regenerative agriculture system and climate-smart food production systems (Beisekenov et al., 2025).

6.5 Bio-Digital Integration and Advanced Genetics

The digital technologies and biological sciences is a frontier with the transformative potential. Phyto-sensing systems based on plant electrophysiological response give direct access to the plant's response to stress, creating a chance of responsive management (Stolarz, 2025). At the same time, genetic mapping technologies are also connecting genomic data to spatial management data so that the varieties can be matched to spatial data not only to the region but also to the particular soil and microclimate environments in a field. This bio-digital convergence is expected to open new possibilities of accuracy in the management of agriculture (Ahmed, 2024). *Figure 1* provides a visual summary of the integrated approach observed in case studies.

6.6 Knowledge Systems and Interoperability

The future of agricultural technology is shifting to open and interconnected systems that would promote cooperation and data exchange at all stages of farming. Recommendations and decision-making are being constantly improved with the assistance of new collective intelligence systems, which are based on the graphs of agricultural knowledge, which relate agronomic research, observations of the farmers, and data on field performance. Simultaneously, the development of API protocols and data standardization is dismantling the interoperability barriers that used to divide the work of various types of equipment, platforms, and applications. This increased technological base is also facilitating the rise of the decentralized science platforms where farmers are able to experiment and share the results, as well as contribute to agricultural knowledge production directly (Gordon and Rudin, 2022). A combination of these innovations is creating a more connected, transparent and collaborative future of farming.

6.7 Implementation Roadmap and Research Frontiers

It is expected that the evolution of Precision Agriculture (PA) will occur in a series of technological maturity levels, each level will be a major leap toward a complete, autonomous, and sustainable farming system. Short-term (1-3 years) developments will probably focus on prevalent use of generative AI-based advisory platforms, the fast development of networks of IoT to include smallholders, and the first commercialization of collaborative robots to facilitate agricultural activities in the field. The further development of highly complex digital models of agricultural systems, commonly known as digital twins, in the medium term (3-7 years) will present real-time monitoring, predictive modeling, and overall system optimization in the future. Throughout this stage, autonomous farming ecosystems will become fully mature after pilot projects are implemented on a large-scale commercial scale, resulting in high productivity, profitability, and resource efficiency (Xu et al., 2024).

In the long term (710 years), the agricultural industry will be more and more characterized by highly self-regulating, autonomous systems that are able to combine biological mechanisms and digital intelligence. Such profound convergence will make it possible to have circular agricultural ecosystems, in which energy, nutrients, and materials are regularly recycled in order to promote environmental sustainability (Liu et al., 2025). The main areas of research will be the design of explainable and transparent agricultural AI systems, the design of energy-independent field robots, new technologies of carbon monitoring and sequestration in the soil, and the development of inclusive socio-technical systems to achieve equitable access to innovation. Finally, Precision Agriculture will no longer serve as an instrument to streamline conventional processes but will become the foundation of a new paradigm of sustainable, resilient, and intelligent food production systems in the age of well-developed agricultural technology.

7. CONCLUSION

The Precision Agriculture is a paradigm shift in the way man is interacting with the food production process and management of the fields is no longer homogenized as it is now more data-driven approach and it is streamlining agricultural systems to solutions that were unimaginable before. The technological foundation of this change as described in this review has been the use of GPS and IoT sensors to AI analytics and robotics as well as shown how all of these can be applied collectively to solve the three-pronged dilemma of productivity, sustainability, and economic viability in the agricultural sector. The statistics above show that PA is a good opportunity in realizing sustainable intensification which will assist farmers in producing higher amounts of food with minimum resources and minimizing environmental impacts significantly.

The Precision Agriculture is revolutionary in far beyond the added efficiency. PA is moving towards a less responsive monitoring and more predictive analytics and prescriptive management, where new generative AI, autonomous systems and bio-digital integration capabilities will alter agricultural decision-making, as we have discussed. Carbon surveillance coupled with the quantification of ecosystem services is what makes PA the enabling technology of climate resistant food system and regenerative agriculture interventions. These are not the technological novelties, but the keys to adapting to the connectedness of population growth, climate change and inadequacy of resources.

However the possibilities of this agricultural revolution can be realized only with the collaboration in different spheres. The research fraternity need to aim at developing cost effective and sustainable solutions that can be made available to small holder farmers that provide a substantial proportion of the world food. This also entails the evolution of explainable AI code, sensor network growth and integrated platforms capable of wrapping up complex information into attainable information. The policymakers have a very significant role to play, which is developing the conducive regulatory frameworks, encouraging the sustainable practices through use of carbon markets, ensuring data privacy and security and investing in rural digital infrastructure. It is vital that the role of schools and extension programs should be an immediate role in bridging the digital divide among farmers by participating in a broad-based educational curriculum to enable them learn how to fully utilize these game-changing technology.

The urgency and the need to do something about it is clear. The technology providers have to pick up on the interoperability standards and develop business models which would allow precision tools to be financially viable across the agricultural range. Value chains established by food companies and retailers should be able to reward sustainable production practices established though precision technologies. Above all, farmers will need to be put as a major player of this technological transformation and their knowledge and experience will determine the creation and use of digital solutions.

At the doorstep of Agriculture 5.0, which is marked by the human-machine partnership and ecological unity, Precision Agriculture is not only promising technical solutions but also a model of how to redefine our food systems as the symbiotic units of the planetary welfare. The decisions of today in research funding, policy-making and adopting new technology will help make or break this potential. With the ability to utilize the strength of the precision technologies plus a reasonable (as well as ethical) access and application, we can grow a future where the agricultural innovation can support both the population and the planet with sustenance over the coming generations.

REFERENCES

Abbas, A., Zhang, Z., Zheng, H., Alami, M. M., Alrefaei, A. F., Abbas, Q., Naqvi, S. A. H., Rao, M. J., Mosa, W. F. A., Abbas, Q., Hussain, A., Hassan, M. Z., & Zhou, L. (2023). Drones in Plant Disease Assessment, Efficient Monitoring, and Detection: A Way Forward to Smart Agriculture. *Agronomy (Basel)*, *13*(6), 1524.

Ahmed, M. (2024). Genetic Mapping and Marker-Assisted Selection: Tools for Accelerating Crop Improvement. *Adv. Crop Sci. Technol.*, *12*(5), 1–2.

Ahmed, N., & Shakoor, N. (2025). Advancing agriculture through IoT, Big Data, and AI: A review of smart technologies enabling sustainability. *Smart Agricultural Technology*, *10*, 100848.

Almalki, F. A., & Angelides, M. C. (2024). A synthesis of machine learning and internet of things in developing autonomous fleets of heterogeneous unmanned aerial vehicles for enhancing the regenerative farming cycle. *Computing*, *106*(12), 4167–4192.

Alsadik, B., Ellsäßer, F. J., Awawdeh, M., Al-Rawabdeh, A., Almahasneh, L., Oude Elberink, S., Abuhamoor, D., & Al Asmar, Y. (2024). Remote Sensing Technologies Using UAVs for Pest and Disease Monitoring: A Review Centered on Date Palm Trees. *Remote Sensing*, *16*(23), 4371.

Ansari, S. A., & Vidyarthi, V. K. (2025). Use of Internet of Things in water resources applications: Challenges and future directions: a critical review. *Discov. Internet Things*, *5*(1), 96.

Aslan, M. F., Durdu, A., Sabanci, K., Ropelewska, E., & Gültekin, S. S. (2022). A Comprehensive Survey of the Recent Studies with UAV for Precision Agriculture in Open Fields and Greenhouses. *Applied Sciences (Basel, Switzerland)*, *12*(3), 1047.

Balkrishna, A., Pathak, R., Kumar, S., Arya, V., & Singh, S. K. (2023). A comprehensive analysis of the advances in Indian Digital Agricultural architecture. *Smart Agricultural Technology*, *5*, 100318.

Beisekenov, N., Banakinaou, W., Ajayi, A. D., Hasegawa, H., & Tadao, A. (2025). Remote sensing-based soil organic carbon monitoring using advanced machine learning techniques under conservation agriculture systems. *Smart Agricultural Technology*, *11*, 101036.

da Silveira, F., Barbedo, J. G. A., Baysoy, M. K., & Caba, Z. T. (2025). Digital Technologies Transforming the Agri-Food Value Chain in the Era of Agriculture 4.0. In Hassoun, A. (Ed.), *Food Industry 4.0 and Food Sustainability* (pp. 125–145). Springer Nature Switzerland.

Delfani, P., Thuraga, V., Banerjee, B., & Chawade, A. (2024). Integrative approaches in modern agriculture: IoT, ML and AI for disease forecasting amidst climate change. *Precision Agriculture*, *25*(5), 2589–2613.

Dhanaraju, M., Chenniappan, P., Ramalingam, K., Pazhanivelan, S., & Kaliaperumal, R. (2022). Smart Farming: Internet of Things (IoT)-Based Sustainable Agriculture. *Agriculture*, *12*(10), 1745.

Ferrández-Pastor, F. J., García-Chamizo, J. M., Nieto-Hidalgo, M., & Mora-Martínez, J. (2018). Precision Agriculture Design Method Using a Distributed Computing Architecture on Internet of Things Context. *Sensors (Basel)*, *18*(6), 1731. PMID: 29843386

Gardezi, M., Joshi, B., Rizzo, D. M., Ryan, M., Prutzer, E., Brugler, S., & Dadkhah, A. (2024). Artificial intelligence in farming: Challenges and opportunities for building trust. *Agronomy Journal*, *116*(3), 1217–1228.

Getahun, S., Kefale, H., & Gelaye, Y. (2024). Application of Precision Agriculture Technologies for Sustainable Crop Production and Environmental Sustainability: A Systematic Review. *TheScientificWorldJournal*, (1), 2126734. PMID: 39421732

Gibson, R. B. (2013). Avoiding sustainability trade-offs in environmental assessment. *Impact Assessment and Project Appraisal*, *31*(1), 2–12.

Gordon, W. J., & Rudin, R. S. (2022). Why APIs? Anticipated value, barriers, and opportunities for standards-based application programming interfaces in healthcare: Perspectives of US thought leaders. *JAMIA Open*, *5*(2), ooac023. PMID: 35474716

Gupta, G., & Kumar Pal, S. (2025). Applications of AI in precision agriculture. *Discov. Agric.*, *3*(1), 61.

Huang, G., Du, S., & Wang, D. (2023). GNSS techniques for real-time monitoring of landslides: A review. *Satellite Navigation*, *4*(1), 5.

Javidan, S. M., Banakar, A., Rahnama, K., Vakilian, K. A., & Ampatzidis, Y. (2024). Feature engineering to identify plant diseases using image processing and artificial intelligence: A comprehensive review. *Smart Agricultural Technology*, *8*, 100480.

Jenkins, M., & Block, D. E. (2024). A Review of Methods for Data-Driven Irrigation in Modern Agricultural Systems. *Agronomy (Basel)*, *14*(7), 1355.

Judith, J., Tamilselvi, R., Beham, M. P., Lakshmi, S. S. P., Panthakkan, A., Mansoori, S. A., & Ahmad, H. A. 2025. Remote Sensing Based Crop Health Classification Using NDVI and Fully Connected Neural Networks. arXiv preprint arXiv:2504.10522.

Karanisa, T., Achour, Y., Ouammi, A., & Sayadi, S. (2022). Smart greenhouses as the path towards precision agriculture in the food-energy and water nexus: Case study of Qatar. *Environment Systems & Decisions*, *42*(4), 521–546.

Liu, Q., Yu, R., Suo, H., Cai, Y., Chen, L., & Jiang, H. (2025). Autonomous Driving in Agricultural Machinery: Advancing the Frontier of Precision Agriculture. *Actuators*, *14*(9), 464.

Lowenberg-DeBoer, J., & Erickson, B. (2019). Setting the Record Straight on Precision Agriculture Adoption. *Agronomy Journal*, *111*(4), 1552–1569.

Matese, A., Czarnecki, J. M. P., Samiappan, S., & Moorhead, R. (2024). Are unmanned aerial vehicle-based hyperspectral imaging and machine learning advancing crop science? *Trends in Plant Science*, *29*(2), 196–209. PMID: 37802693

Mondal, P., & Basu, M. (2009). Adoption of precision agriculture technologies in India and in some developing countries: Scope, present status and strategies. *Progress in Natural Science*, *19*, 659–666.

Morchid, A., Et-taibi, B., Oughannou, Z., Alami, R. E., Qjidaa, H., Jamil, M. O., Boufounas, E.-M., & Abid, M. R. (2025). IoT-enabled smart agriculture for improving water management: A smart irrigation control using embedded systems and Server-Sent Events. *Scientific African*, *27*, e02527.

Office, U. S. G. A. 2024. Precision Agriculture: Benefits and Challenges for Technology Adoption and Use. U.S. Government Accountability Office, Washington, DC. Report No. GAO-24-105962.

Okoli, A. O., & Birkenberg, A. (2024). Monitoring soil carbon in smallholder carbon projects: Insights from Kenya. *Climatic Change*, *177*(9), 143.

Ou, G., Chen, Y., Han, Y., Sun, Y., Zheng, S., & Ma, R. (2025). Design and Experiment of an Internet of Things-Based Wireless System for Farmland Soil Information Monitoring. *Agriculture*, *15*(5), 467.

Papadopoulos, G., Arduini, S., Uyar, H., Psiroukis, V., Kasimati, A., & Fountas, S. (2024). Economic and environmental benefits of digital agricultural technologies in crop production: A review. *Smart Agricultural Technology*, *8*, 100441.

Ram, B. G., Oduor, P., Igathinathane, C., Howatt, K., & Sun, X. (2024). A systematic review of hyperspectral imaging in precision agriculture: Analysis of its current state and future prospects. *Computers and Electronics in Agriculture*, *222*, 109037.

Saha, S., Kucher, O. D., Utkina, A. O., & Rebouh, N. Y. (2025). Precision agriculture for improving crop yield predictions: A literature review. *Frontiers in Agronomy*, *7*, 1566201.

Said, Z., Vigneshwaran, P., Shaik, S., Rauf, A., & Ahmad, Z. (2025). Climate and carbon policy pathways for sustainable food systems. *Environmental and Sustainability Indicators*, *27*, 100730.

Sanyaolu, M., & Sadowski, A. (2024). The Role of Precision Agriculture Technologies in Enhancing Sustainable Agriculture. *Sustainability*, *16*(15), 6668.

Shahriar, S., Corradini, M. G., Sharif, S., Moussa, M., & Dara, R. (2025). The role of generative artificial intelligence in digital agri-food. *Journal of Agriculture and Food Research*, *20*, 101787.

Singh, A., Rajput, V. D., Varshney, A., Ghazaryan, K., & Minkina, T. (2023). Small Tech, Big Impact: Agri-nanotechnology Journey to Optimize Crop Protection and Production for Sustainable Agriculture. *Plant Stress*, *10*, 100253.

Soussi, A., Zero, E., Sacile, R., Trinchero, D., & Fossa, M. (2024). Smart Sensors and Smart Data for Precision Agriculture: A Review. *Sensors (Basel)*, *24*(8), 2647. PMID: 38676264

Stolarz, M. (2025). Integration of Plant Electrophysiology and Time-Lapse Video Analysis via Artificial Intelligence for the Advancement of Precision Agriculture. *Sustainability*, *17*(12), 5614.

Surur, F. M., Mamo, A. A., Gebresilassie, B. G., Mekonen, K. A., Golda, A., Behera, R. K., & Kumar, K. (2025). (in press). Unlocking the power of machine learning in big data: A scoping survey. *Data Sci. Manag.*

Upadhyay, A., Zhang, Y., Koparan, C., Rai, N., Howatt, K., Bajwa, S., & Sun, X. (2024). Advances in ground robotic technologies for site-specific weed management in precision agriculture: A review. *Computers and Electronics in Agriculture*, *225*, 109363.

Vijayakumar, S., Murugaiyan, V., Ilakkiya, S., Kumar, V., Sundaram, R. M., & Kumar, R. M. (2025). Opportunities, challenges, and interventions for agriculture 4.0 adoption. *Discover Food*, *5*(1), 1–35.

Wang, L., Gao, J., & Qureshi, W. A. (2025). Evolution and Application of Precision Fertilizer: A Review. *Agronomy (Basel), 15*(8), 1939.

Watson, E. 2025. CH4 Global teams up with Mitsubishi on livestock methane reduction. AgFunderNews. Available online: https://agfundernews.com/ch4-global-teams-up-with-mitsubishi-on-livestock-methane-reduction (accessed on 17 February 2026).

Xu, J., Cui, Y., Zhang, S., & Zhang, M. (2024). The evolution of precision agriculture and food safety: A bibliometric study. *Frontiers in Sustainable Food Systems*, *8*, 1475602.

Zhu, Y. (2024). Application of a QPSO-optimized CNN-LSTM model in water quality prediction. *Discover Water*, *4*(1), 100.

Chapter 3
Artificial Intelligence in Farming: Transforming Crop Management and Resource Allocation

Tauqeer Ahmed Qadri
http://orcid.org/0009-0000-9617-4512
College of Graduate Studies, University of South Africa, Pretoria, South Africa

ABSTRACT

Artificial Intelligence (AI) is transforming agriculture by improving precision, efficiency, and sustainability. Key technologies like machine learning, deep learning, computer vision, and remote sensing enable early pest and disease detection, real-time crop monitoring, and precision practices such as variable-rate inputs, soil mapping, and autonomous machinery. AI-driven decision support integrates climate, soil, and market data for yield prediction and harvest planning. Case studies highlight scalable applications, while challenges include technical limits, data privacy, ethics, and socioeconomic barriers. AI offers a path to resilient, sustainable, and globally secure food systems.

INTRODUCTION

Agriculture is facing a deep change in response to the rising global demand for food, the increasing variability of climate, the reduction of resources and the need for sustainable production systems. Conventional agrarian practices, mainly based on the generic management practices coupled with manual decision making, are often found to be limited in dealing with spatial and temporal complexity of the modern agricultural environment. Diversity in soil characteristics, weather conditions, pest

DOI: 10.4018/979-8-3373-7257-0.ch003

dynamics and response of crops require more accurate, adaptive and data-driven management approaches (Mba et al., 2025).

Artificial Intelligence (AI) has become a key facilitator of this change as it enables enormous and often complex agricultural data analysis and processing it into usable insights. From the growth, production, processing, and all the way to distribution, warehousing and food, AI assists in predictive analytics and real-time decision-making throughout the crop production cycle through technologies such as machine learning, deep learning, computer vision and robotics. As a result, these capabilities can help us identify crop stress in an early process, better forecast the yields, optimal technologies for irrigation scheduling, and efficient nutrient and pest management (Ali et al., 2025).

The concept of AI plus precision agriculture tools like remote sensing, internet of things (IoT) devices, geographic information systems, and automated farm machinery has boosted the capacity of implementing site specific management practices to a great extent (K. Sharma & Shivandu, 2024). Such integration encourages efficient use of resources by avoiding input wastage, increase in productivity, and environmental impacts due to excessive use of fertilizers, water and pesticides. Consequently, there is a gradual move in agricultural systems towards intelligent, automated and climate-resilient agricultural production models (Omotayo et al., 2025).

Despite its potential to transform the industry, adoption of AI in agriculture is governed by a number of technological, economic and institutional factors. Data accessibility, lack of infrastructure, high cost of implementation and digital literacy gaps continue to pose major challenges especially in smallholder farming contexts. Into these constraints a multidisciplinary collaboration, policy support, and building capacities is needed to ensure equitable and sustainable deployment of the AI technologies (Younas et al., 2025).

This chapter examines the role of Artificial Intelligence in enhancing crop management and optimizing agricultural resource use through applications such as yield prediction, irrigation, nutrient management, and pest detection within precision and climate-smart farming systems. It further evaluates emerging AI-driven technologies, their contribution to productivity and sustainability, and the key technical and socio-economic challenges influencing their adoption in modern agriculture.

1. CONCEPTUAL FRAMEWORK OF AI IN AGRICULTURE

Artificial intelligence in agriculture refers to computational methods that learn patterns from multi-source farm data (e.g., sensors, imagery, weather, and machinery logs) to support prediction and decision-making. Machine learning, deep learning, computer vision, and remote-sensing analytics underpin smart farming systems by translating raw data into site-specific recommendations that improve resource-use efficiency and crop productivity (Mustaza et al., 2025).

1.1. Machine Learning (ML) Fundamentals

Machine learning is the basis of analytical engine of AI enabled agriculture. Unlike the traditional rule-based programming paradigms, machine learning algorithms are based on extracting knowledge from past and real time data sets on the agriculture area, and their prediction ability is increasingly improved with the addition of more data sets. Within the context of crop management, the field of machine learning (ML) algorithms have been used for many applications such as yield prediction, soil classification, disease diagnosis, irrigation optimization, as well as stress early detection (Waqas et al., 2025a).

In agricultural applications, machine learning is commonly organized into three paradigms. Supervised learning uses labelled datasets for classification and regression (e.g., disease recognition, nutrient-status prediction, and yield estimation). Unsupervised learning identifies patterns in unlabeled data to support management-zone delineation, soil heterogeneity assessment, and anomaly detection (Waqas et al., 2025). Reinforcement learning uses feedback-driven optimization and is increasingly applied to autonomous navigation, variable-rate irrigation, and greenhouse climate control (Cheema et al., 2025).

1.2. Deep Learning (DL) and Neural Networks

Deep learning is a subset of machine learning characterized by multi-layer neural-network architectures that can automatically extract complex features from raw data. In agriculture, it is particularly effective for high-dimensional inputs such as images, multi-spectral remote-sensing products, and time-series sensor data (Waqas et al., 2025).

Convolutional neural networks (CNNs) are widely used in agriculture because they capture spatial patterns in images and have been applied to disease detection, weed identification, fruit counting, plant phenotyping, and canopy-cover estimation. Recurrent neural networks (RNNs) and long short-term memory (LSTM) models are commonly used for sequential data such as weather and evapotranspiration forecasting and growth-stage prediction (Salka et al., 2025). Autoencoders and related architectures enable anomaly detection in sensor streams, helping identify irrigation-system faults, nutrient imbalances, or equipment failures. With increasing availability of farm-scale data, deep-learning approaches provide high accuracy and automation, reducing reliance on manual and subjective field scouting (Adkisson et al., 2021).

1.3. Computer Vision and Remote Sensing Analytics

Computer vision, and remote sensing analytics, is an integral tool to help convert the value of a visual representation and spectral data captured into actionable insights in agriculture. Computer vision is the process of automated interpretation of images and videos taken by drones, fixed cameras, mobile phones and robotic platforms. It makes it possible to carry out real time detection of disease, evaluation of leaf discoloration, fruit maturity, weed-crop discrimination, and monitoring of plant growth. By facilitating non-pendulous and high throughput evaluation, computer vision recommendations significantly lower labor requirements and improvement precision (Ghazal et al., 2024).

Remote sensing is a useful complement to computer vision and supplies large-scale space- and time-resolved data in crop- and soil condition using satellite imagery system and UAV-mounted sensors and multispectral or hyperspectral camera. Vegetation indices - such as: Normalized Difference Vegetation Index (NDVI), Enhanced Vegetation Index (EVI) and Soil Adjusted Vegetation Index (SAVI) - are extensively used to estimate crop vigor, biomass, chlorophyll content and stress conditions (Sharma et al., 2025). When combined with AI algorithms, remote sensing is crucial to aid in early detection of drought stress, nutrient deficiency, pest outbreaks or variability of crop yields, whose prediction can then be acted upon at the field or regional scale, with site-specific intervention. The collective outcome of these technologies has been to increase the precision, scalability, and sustainability of the modern changing crop monitoring system (Rashid et al., 2025).

2. AI FOR CROP MONITORING AND HEALTH DIAGNOSIS

Crop health monitoring and timely diagnosis of biotic and abiotic stresses play an important role in ensuring high productivity and optimal use of resources and promoting sustainable agricultural practices. Historically, crop surveillance was based on manual inspection in the field and on expert judgement, both of which are labor-intensive and time consuming, and are generally limited in spatial coverage (Yan et al., 2025). The advent of Artificial Intelligence (AI) has revolutionized this paradigm by enabling early detection of plant health problems and allowing numerous users to find early, accurate and scalable solutions for plant health, thus enabling farmers to take proactive steps. By combining forces of machine learning, deep learning, computer vision, and remote sensing, AI offers various tools to continuously and accurately monitor crops at various levels: from the plant level to tracking the general landscape (Rodríguez et al., 2024).

2.1. Early Detection of Pests and Diseases through AI

Immediate detection of pests and diseases is the key to averting yield losses and the need for frequent chemical interventions and application. AI systems use image-based datasets, spectrum information and environmental variables to identify infections before visible infection symptoms prevail. Convolutional neural networks (CNNs) and other forms of deep learning architectures can be used as they can identify subtle image information that may become visible to people in the form of discolored or dead leaves, changes in leaf morphology, or lesion formation (Monika T et al., 2025). Moreover, machine learning models can be trained in order to predict pest outbreaks based on historical data of pest infestations combined with weather variables and thus take precautionary measures against pests. Automation of the detection process makes it less reliant on subjective judgement of human inventors and enables timely, site-specific interception of the need for intervention using minimal application of chemical inputs and the least environmental impact (Kariyanna & Sowjanya, 2024). The general workflow for image-based plant disease detection is summarized in Figure 1.

Figure 1. This flowchart illustrates the process of plant disease detection using image acquisition, processing, feature extraction, and AI-based classification. It highlights the integration of machine learning and deep learning techniques for accurate disease classification

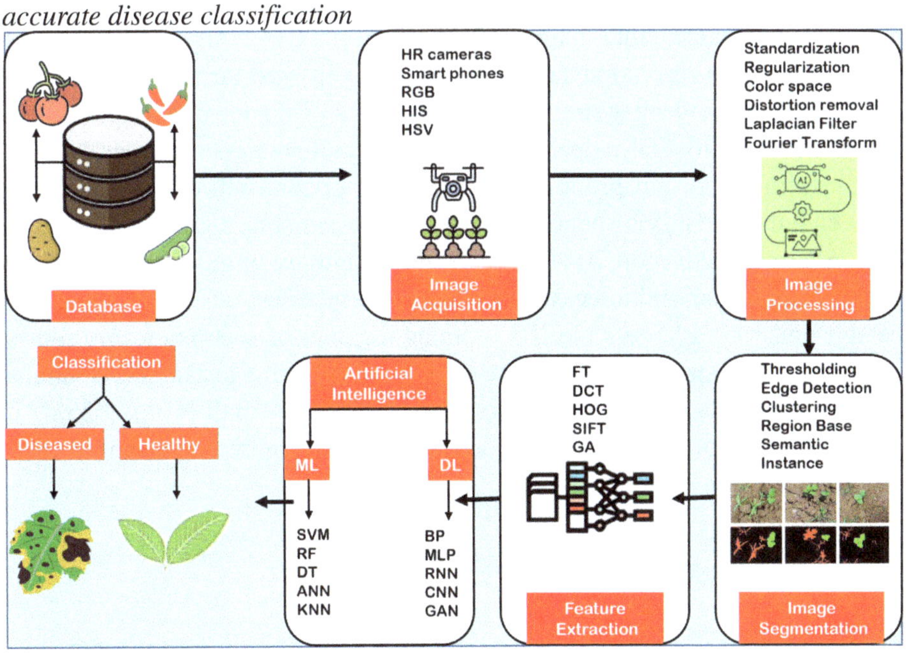

2.2. Stress Identification

Abiotic stresses, which include drought, nutrient deficiency, heat stress and salinity, are major constraints affecting crop growth, productivity and stability of yield (Khan, 2020) . Advanced artificial intelligence (AI) models use a combination of multidimensional data sets generated from a wide variety of sources such as remotely sensed vegetation indices, soil moisture measurements and climatic variables in order to detect and describe stress signatures on scales varying from the individual plant to entire fields (Paul et al., 2025). For example, the fusion of thermal imaging and time series analysis of the Normalized Difference Vegetation Index (NDVI) can be used for the early detection of incipient water stress, whereas hyperspectral reflectance analysis can help diagnose specific nutrient deficiency. Early identification of stress supports timely management interventions (cross-site) of best management practices, optimized irrigation planning, nutrients, and salinity finance strategies and enhance the use of resource efficiency and crop growth (Mushtaq et al., 2025).

2.3. Real-Time Monitoring through Drones, IoT, and Satellite Imagery

Real react crop monitoring is made possible this way through the integration of Artificial Intelligence, Unmanned Aerial Vehicles (UAVs), Internet of Things (IoT) sensor networks, and satellite imaging. UAVs with multispectral or hyperspectral cameras collect high-resolution canopy data which is passed through an artificial-intelligence model to identify stress, plot disease hazard and tree biomass. All from in-the-field execution of farm operations utilizing input prediction models, such as alarm data from oil moisture probes, temperature sensors, and nutrient meters, which generate cumulative data via a continuous data stream, assisting decision recognition systems (Sharma & Shivandu, 2024). Satellite products make significant contribution to regional monitoring and forecasting of yield to large disastrous events which are repeatable and with a large-scale. The coming together of such platforms allows for continuous, site-specific surveillance as well as facilitating instant interventions that maximize inputs and reduce environmental impacts (Tufail et al., 2025). An overview of a typical data-to-model workflow for crop-improvement analytics is shown in Figure 2.

Figure 2. This diagram demonstrates the use of machine learning for crop improvement through data collection, including phenotypic and genotypic data, and weather prediction. It highlights key applications like crop disease management, water management, species recognition, and crop yield prediction

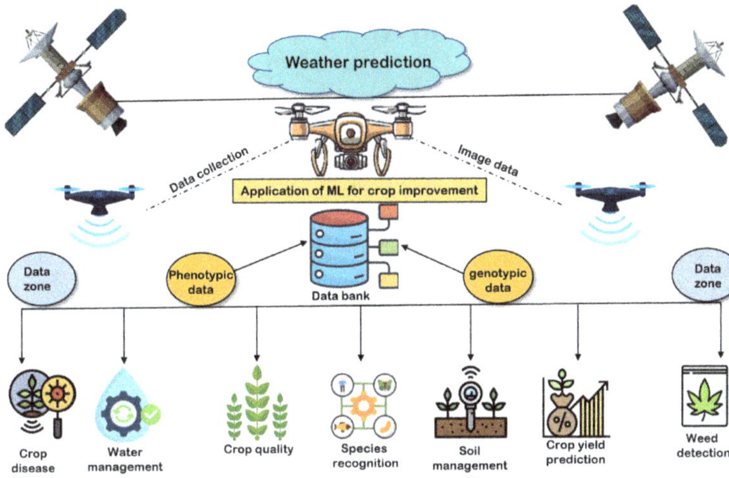

3. PRECISION AGRICULTURE POWERED BY AI

Precision agriculture (PA) is a move toward site-specific management with the goals of optimizing productivity and resource efficiency and sustainability for the environment. Wastage, less than optimal crop performance, and environmental degradation can be the result of applying inputs uniformly (e.g., water, fertilizers, and pesticides) (Gavhane et al., 2026). AI strengthens PA by integrating multi-source data (sensors, drones, satellites, and machinery) to generate actionable recommendations for targeted interventions and more sustainable input use (Wu & Zhong, 2025).

3.1. Variable Rate Technology (VRT) for Inputs

Variable Rate Technology (VRT) constitutes one of the corner stones of precision farming and enables dynamic modulation of the inputs like fertilizers, pesticides and irrigation in accordance with the exigencies for a particular field. AI algorithms analyze data about the property of the soil, the health of the crop and other environmental factors to generate prescription maps that determine the rate of application of inputs (Ahmed et al., 2024). Machine-learning models can identify areas in a field that require input and effectively utilize resources and prevent over-application. By combining VRT with real-time monitoring using drones and IoT sensors, agriculturists can use automated or semi-automated input application systems that can be used to improve yield potential while reducing input costs and environmental contamination (Sharma & Shivandu, 2024).

3.2. AI-Driven Soil Mapping and Nutrient Profiling

Understanding soil variability is indispensable for precision management of crop inputs. AI supports high-resolution soil mapping and nutrient profiling by integrating soil-sample observations, remote-sensing imagery, and environmental covariates. Machine-learning models can estimate nutrient availability, pH, organic-matter content, and other key physicochemical properties across heterogeneous fields, enabling targeted amendments and improved nutrient-use efficiency (Schweng et al., 2026). AI-driven soil monitoring also supports longer-term soil-health management and sustainable practices that help prevent soil degradation (Talaviya et al., 2020).

3.3. Precision Seeding, Irrigation, and Fertilization

Artificial intelligence goes beyond the rate of input in the precision agriculture, and includes operational tasks such as seeding, irrigation and fertilization. Through comprehensive meteorological analyses - including weather prediction, soil moisture measurements and crop growth analysis - predictive models allow AI systems to calculate the optimal planting density and sowing period based on the climatic conditions current on the ground (Padhiary et al., 2024). Fertilization schedules can be dynamically adjusted according to nutrient requirements at different stages of crop development, and thus ensure that each crop receives exactly the necessary quantities of nutrients at the right times (Adamo et al., 2025). The folding in of autonomous machinery and smart irrigation systems enables such prescriptions to be taken forward in real time, driving higher water and nutrients use efficiencies, and at the same time lowering the amount of labor. Collectively these AI-driven practices change conventional farming to a very efficient, responsive and sustainable enterprise (Wu & Zhong, 2025). Table 1 provides a consolidated overview of major AI application areas, enabling technologies, and associated benefits.

Table 1. Applications of AI in agriculture: Technologies, functions, and benefits

Application Area	AI Technologies Used	Key Functions	Benefits	References
Crop Monitoring & Health Diagnosis	Machine Learning, Deep Learning, Computer Vision, Remote Sensing	Early detection of pests/diseases, stress identification, real-time crop monitoring	Reduced yield losses, timely interventions, lower chemical use	(Yan et al., 2025)
Precision Agriculture	Variable Rate Technology (VRT), IoT Sensors, Predictive Analytics, Autonomous Machinery	Site-specific input application, precision irrigation and fertilization	Optimized resource use, increased productivity, reduced environmental impact	(Sharma et al., 2025)
Yield Prediction & Forecasting	Machine Learning, Deep Learning, Predictive Modeling	Yield estimation, harvest scheduling, resource planning	Improved planning, risk reduction, enhanced food security	(Shawon et al., 2025)
Supply Chain & Market Forecasting	Predictive Analytics, AI Decision Engines, IoT, Big Data	Market price prediction, storage optimization, logistics planning	Reduced post-harvest loss, cost efficiency, timely distribution	(Omotayo et al., 2025)
Resource Allocation	AI Models, Optimization Algorithms, Reinforcement Learning	Efficient water, fertilizer, and energy use	Reduced waste, cost savings, environmental sustainability	(Alsabt et al., 2024)
Robotics & Autonomous Systems	Autonomous Tractors, Harvesting Robots, Spraying Robots, Computer Vision	Field operations, harvesting, weed control, crop protection	Labor efficiency, precise operations, sustainable input use	(Priya et al., 2025)
Sustainable Agriculture & Environmental Management	AI Analytics, Remote Sensing, Predictive Modeling	Soil health monitoring, biodiversity management, carbon sequestration	Reduced environmental footprint, climate resilience, support for SDGs	(Mana et al., 2024)

4. AI IN YIELD PREDICTION AND FORECASTING

Accurate yield prediction is an important part of modern agriculture, and it enables interested parties, such as farmers, agribusinesses, and policymakers, to make fact-based decisions that relate to crop management, resource allocation, and market planning. Traditional methods of estimating yield information that rely a priori on historical records or on manual sampling are labor intensive, time consuming and prone to accuracy errors (Shawon et al., 2023). Artificial Intelligence (AI) provides sophisticated analytical tools that can question complex, multi-dimensional datasets that include meteorological variables, soil variables, agronomic activities, and remote sensing data in a way that provides very precise and time-relevant yield forecasts. Such type of prognostications makes it possible to make proactive interventions to optimize the productivity, reduce post-harvest losses and strengthen the economic viability of the agrarian systems (Demissie et al., 2026).

4.1. Machine Learning Models for Yield Estimation

Machine learning algorithms are the basis of many yield prediction models by helping to understand the relationships between historical yields and explanatory variables such as meteorological data, soil properties, crop management practices and remote sensed vegetation indices. Both supervised learning techniques (e.g., random forests, gradient - boosting) and deep learning techniques (e.g., convolutional neural networks and long short-term memory networks) have been used to learn about the spatial and temporal drivers of yield . (Benti et al., 2024). Accurate yield forecasting enables good planning at the farm level, coordination of the supply chain and risk management to provide good response to drought, heatwaves and other adverse conditions (Demissie et al., 2026).

4.2. Factors influencing: Climate, Soil, and Management

Yield is regulated by complex interaction of climatic, edaphic and management variables. Artificial-intelligence models assimilate meteorological data (temperature, rainfall, solar radiation, etc.) and, in relation to soil parameters (texture, moisture, content of nutrients, etc.) and agronomic practices (plant density, frequency of irrigation, fertilization regime), etc. By simultaneously taking these interacting factors into account, AI can simulate possible yields in a set of scenarios, assess danger of stress and give recommendations for adaptive management. The ability to ask powerful questions of high dimensional data sets allows precision agriculture interventions to optimize the potential yield while minimizing the consumption of both resources and the environmental footprint (Fan et al., 2025).

4.3. AI for Harvest Scheduling and Resource Planning

Beyond yield estimation, artificial intelligence (AI) promotes harvest (timing) and logistic planning by couple indices of maturity, weather forecast, labor availability and market structure constrictions. Predictive models can be used to know within which time periods there is a higher risk of losses e.g. due to rains, heat or frosts and producers will know which fields to focus on and where to optimize their labor and machinery use (Demissie et al., 2026). In horticultural systems, computer vision models help to perform an assessment inspection of readiness of the harvesting including aspects like fruit size and color, in order to ensure an even quality and mitigate come back losses (Zhang et al., 2022).

5. AI-ENABLED RESOURCE ALLOCATION

Efficient use of agricultural resources is a cornerstone of sustainable production. AI supports resource allocation by optimizing water, fertilizer, energy, and labor use based on crop demand, weather forecasts, soil status, and operational constraints. Optimization and reinforcement-learning approaches can recommend timing and quantities of inputs to reduce waste while maintaining yield and quality, thereby minimizing environmental impacts associated with overuse and runoff (Lan & Ban, 2025).

5.1. Optimizing Water, Fertilizer, and Energy through AI

AI-based irrigation management uses data from soil moisture sensors, weather forecasts, indexes for water lost by evapotranspiration, and crop development levels to determine accurate needs for water. Machine learning models and reinforcement learning algorithms could dynamically schedule the irrigation events and thereby limit water-waste, and prevent stress due to over or under irrigation (OĞUZTÜRK, 2025). Analogously, AI optimizes the application of fertilizers by examining the soil nutrient profile, nutrient uptake kinetics in the crop as well as past yield records to apply variable rate prescriptions, so that crops will receive the appropriate amount of nutrients at the optimal temporal juncture to take up nutrients through the plant. This methodology not only allows to improve nutrient-use efficiency, but it also reduces leaching and pollutant's pollution of the environment (Malik et al., 2025)

Energy consumption in agricultural enterprises is dominated by irrigation pumping, machinery operation, and (in protected cultivation) heating, ventilation, and cooling. AI can improve energy efficiency by forecasting demand, scheduling high-load operations, and optimizing equipment use under variable weather and tariff conditions. Predictive-maintenance models detect inefficiencies and impending failures in pumps, motors, and climate-control systems, reducing downtime and operational costs (Raj & Prahadeeswaran, 2025).

6. ROBOTICS AND AUTONOMOUS SYSTEM IN CROP MANAGEMENT

The use of robotics and autonomous systems in agriculture is expanding as costs decline and sensing and navigation accuracy improves. Robotic platforms can support repetitive and labor-intensive operations such as mechanical weeding, targeted spraying, and harvesting, while autonomous tractors and implements improve field-operation precision and safety (Anastasiou et al., 2025). Computer-vision and sensor fusion enable robots to perceive the crop environment, localize in the field, and execute precise manipulation or application tasks (Shah et al., 2025). Key components of an agricultural robotic system are summarized in Figure 3

Figure 3. This diagram showcases the components of an agriculture robot, integrating technologies like AI, bionics, sensors, and precision manipulation. It emphasizes the robot's autonomous navigation, image system, and internet connectivity, enabling efficient crop management through advanced automation

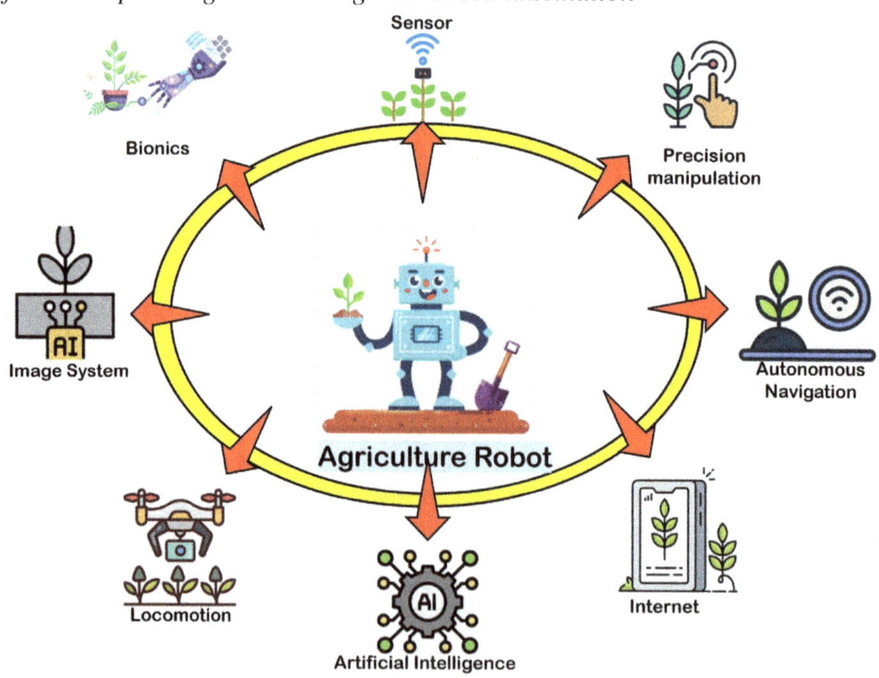

6.1. Autonomous Tractors and Field Robots

Autonomous tractors and field robots are designed to carry out critical agronomic functions (such as plowing, seeding, and fertilizing), without needing any direct human intervention. These machines use GPS positions, LiDAR, and AI algorithms to control exact positions, avoid bumps and ensure optimal coverage of the fields (Chen et al., 2026). Machine-learning models give these robots the ability to adapt to varying field conditions, soil type and crop variability. Consequently, through the automation of the basic tasks, autonomous tractors help reduce labor dependency while increasing the level of consistency in operations and complementing the efficiency of the large-scale farming process (D et al., 2026)

6.2. AI-Based Harvesting and Crop Management Robots

Beyond field preparation work, however, these robots with artificial intelligence are changing harvesting, weed management, and crop protection fields. Harvesting units make use of computer vision to identify ripe fruits and vegetables and thus allow for selective harvest that reduces mechanical damage and post-harvest losses (Bessaad et al., 2026). Concurrently, the use of robotic systems that have artificial intelligence (AI)-powered spraying and mechanical operations to weed can be used to precisely target weeds or agrochemical applications, while reducing the usage of chemicals and causing less adverse environmental impact. These integrated robotic solutions not only multiply the productivity and reduce the operating costs, but at the same time ensure the sustainable development of agriculture by optimizing the use of inputs and reducing soil compaction or crop injury (Prasanjana et al., 2025).

7. DECISION-SUPPORT SYSTEM IN SMART FARMING

Decision- Support Systems (DSS) are a central component of contemporary precision agriculture technologies that provide agronomists with the ability to make evidence-based data-driven determinations that will improve crop management, resource allocation and efficiency. These systems combine the assimilation of real time field metrics, predictive analytics, and artificial intelligence with the capacity to provide actionable guidance for the day-to-day farming operations (Mabhaudhi et al., 2023). Contrary to traditional management paradigms, which rely mainly on experiential judgement and intuition, DSS platforms aggregate complex datasets from a variety of diverse sources such as soil sensors, meteorological forecasts, market analytics, and satellite imagery all of which can then help farmers to solve environmental and economic contingencies in a proactive way (Manono et al., 2026).

7.1. AI-Driven Farm Management and Decision Engines

Contemporary DSS combine Farm Management Information Systems (FMIS) with Artificial Intelligence (AI) based decision engines to make agricultural operations streamlined. FMIS platforms collect, store, and display information from farm equipment, soil probes, irrigation systems and crop monitoring sensors, allowing a complete digital image of the farm conditions to be developed. AI algorithms use this information to make recommendations on how to conduct your daily operations such as irrigation schedules, fertilization rates or protocols for pest management and harvest or planting dates (Rajnish Jain, 2025).By taking into account weather predictions, market price movements and soil nutrient status, DSS help to ensure that decisions are both timely and location specific and reduce risk while maximizing productivity and profitability. This merge of AI with decision support structures enables the farmers to integrate with the precision agriculture practices effectively, thereby, enhancing the sustainability and efficiency of their operations in different farming systems (Mishra, 2025)

8. AI IN SUSTAINABLE AGRICULTURE

Sustainability has become a major goal of modern agriculture thanks to the urgent need to balance the production of food and environmental control and resilience to climatic variability. Artificial Intelligence (AI) has a revolutionary impact to promote sustainable agrarian techniques in teaching precise resource management, sustaining adverse ecological impacts and preserving ecosystem vitality (Omotayo et al., 2025). Through the combination of machine learning methods, computer vision, remote sensing and decision support mechanisms, AI helps the agronomists to optimize input deployment and to monitor environmental variables and to implement agronomic strategies in line with the constraints of sustainability and productivity (Demissie et al., 2026).

8.1. Enhancing Environmental Sustainability through AI

AI can contribute to climate-smart and environmentally sustainable agriculture by improving monitoring and management of soil, water, and biodiversity (Javed et al., 2025). By analyzing field-level and satellite-derived data, AI models can support erosion-risk assessment, nutrient-loss prediction, carbon accounting, and targeted interventions that reduce greenhouse-gas emissions and input leakage. These capabilities align with sustainable intensification goals by maintaining productivity while lowering the environmental footprint of agricultural systems (Thilakarathne et al., 2025).

8.2. Supporting Soil Health, Biodiversity, and Carbon Sequestration

AI is another way of augmenting long-term soil and ecosystem health. Machine learning models can track the dynamics of soil in terms of nutrients, moisture content and microbial activity and inform precision interventions for maintaining soil fertility and structure. By optimizing crop rotation, cover cropping and organic applications of inputs, artificial intelligence-based (AI) systems help to nurture soil biodiversity and reduce soil degradation (Mana et al., 2024). In addition, AI can be used to model carbon sequestration potential in agroecosystems to provide actionable recommendations for practices that can improve soil organic carbon sequestration, mitigate climate change and strengthen resilience of ecosystem functions. These capabilities make AI a useful tool to match agricultural practices and ecological sustainability goals (Schweng et al., 2026).

8.3. Contribution of AI to the UN Sustainable Development Goals (SDGs)

rtificial intelligence (AI) is being incorporated into the agriculture sector in a more and more of the working world, becoming directly connected to several United Nations Sustainable Development Goals (SDGs). Using the technology to boost the productivity of crops and reduce losses in food, AI drives SDG 2 - Zero Hunger. The precision application of the resources (fertilizers and water) is part of the SDG 12 - Responsible Consumption and Production and of SDG 13 - Climate Action, reducing environmental degradation and greenhouse gas emission (Petcu et al., 2024) Moreover, the monitoring of soil health, biodiversity and ecosystem services with the help of AI relates to SDG15 i.e. life on land that targets the sustainable management of land and biodiversity. Collectively, practices enabled by AI serve as a good example of how technological innovation can amplify global sustainability targets while at the same time increasing agricultural productivity and sustainability (Das et al., 2025).

9. CASE STUDIES OF AI APPLICATIONS IN AGRICULTURE

The practical implementation of artificial intelligence (AI) in agriculture can best be demonstrated by empirical case studies of success, challenges and lessons learned, both with respect to policy and practice. Across the world, AI has been used to enhance efficacy, lessen environmental impact, and boost decision-making in the farming system. Examining applications in developed as well as developing context leads to nuanced understanding as to how AI technologies can be adapted to different agro-ecological and socio-economic context (Singh et al., 2025; Talaviya et al., 2020).

9.1. AI Success Stories in Developed Countries

Developed countries have led the use of artificial intelligence (AI) in agriculture and this is mostly driven by their strong digital infrastructure, high-quality and standardized data repositories, and long-term investments in research, innovation and extension services (Kumari et al., 2025). Conversely, in many developing regions there are major limits to overcome, such as a lack of internet connectivity, poor levels of sensors, weak data management frameworks as well as a lack of technical capacity to operationalize data-driven decision-making tools. Such bottlenecks in their structure and institutions hinder effective implementation and scaling of AI based solutions in the agriculture sector. Bridging this digital and technological gap requires creating low-cost, scalable and inter-operable data acquisition platforms and context specific AI-models based on local agro-ecological and societal Self-costal conditions. Moreover, participatory approaches, including end-user input, are key requirements to guarantee usability, relevance and lasting adoption of decision support systems based on artificial intelligence (Al khatib & Alshaib, 2025).

9.2. Innovations from Developing Countries

Despite financial and infrastructural constraints, an increasing number of developing countries are increasingly taking up the use of artificial intelligence (AI)-enabled innovations in an effort to supplement agricultural productivity and climate resilience. Specifically, an advisory platform based on smart phones where image recognition technologies are combined with localized meteorological information has been implemented to provide farmers with recommendations about pest and disease control, optimal nutrient use and irrigation timing in real time at the site where it is installed (Zhou, 2025). Furthermore, field robotics and "see-and-spray" technologies are evaluated at the pilot scale that highlights the possibilities for an automated weed detection and targeting application of agrochemicals. Such precision systems enable plant-level discrimination between weeds and crops, and thereby reduce the inputs of chemicals, environmental contamination and decrease labor requirements. Collectively these new emerging applications highlight the importance of having scalable and cost-effectively developed AI tools to suit resource limited farming systems. The successful deployment of these, however, requires strengthened research-extension linkages, capacity-building activities and relevant institutional frameworks in supporting adoption in smallholder agricultural situations (Huda et al., 2026).

9.3. Comparative Insights and Lessons Learned

Comparative evidence points out that it is the developed economy where AI adoption is more because they have high availability of capital, reliable digital infrastructure, and better access to high-quality data and technical services (Mubarok et al., 2024). In developing economies, adoption is often based on low cost and mobile first and is supported by strong extension. Drafting of well covalentized policies and capacity building initiatives can help reduce these gaps and fast-track the inclusive uptake (Pienwisetkaew et al., 2025).

10. AI-INTEGRATED SUPPLY CHAIN AND MARKET FORECASTING

Artificial intelligence (AI) becomes increasingly relevant throughout the agricultural supply chain at the level of post-harvesting and storage, transportation, distribution and market forecasting. Predictive analytics are used to estimate and plan demand, inventory, and computer tomography (CT) and computer vision systems monitor produce quality and detect spoilage while in storage. In product management when coupled with traceability system, AI can provide more effective logistics efficiency, minimize post harvest losses by facilitating timely intervention and more responsive distribution network (Oyeyemi, 2022). By integrating AI approaches in the supply chain, stakeholders can create better transparency, minimize losses, as well as optimally control pricing patterns (Ghag et al., 2024). Machine learning is equipped with its own models to analyze historical pricing, weather anomalies and supply demand signals and make predictions of market movements and forecast procurement and distribution strategies. These tools are especially useful for perishable commodities, for which delays and mis-matching logistics quickly turn into waste (Javaid et al., 2023).

Beyond the forecasting of the market, AI helps in improving the handling and storage practices of the product after harvest by analyzing quality, predicting the spoilage risk and the optimal conditions for storage (Tadala & Fanny, 2024). For example, sensor networks and computer vision systems can monitor temperature, humidity, ethylene concentration and signs of good or poor quality and initiate a proper intervention at the right time. Such monitoring facilitates the advantageous storage prevention measures, lowering post-crop surpluses as well as enhancing the confidence in supply (Saki & Soori, 2026).

11. CHALLENGES AND BARRIERS TO AI ADOPTION IN AGRICULTURE

Although artificial intelligence (AI) has a considerable transformative potential, the diffusion of AI is held back by numerous technological, ethical and socio-economic challenges. These hindrances affect the accessibility, scalability and efficacy of AI based solutions especially in smallholder and resource poor contexts. Due to the biases in this perspective on AI, it is essential to understand these barriers carefully to determine an effective strategy to encourage equitable and sustainable incorporation of AI technologies into farming systems (Omotayo et al., 2025).

Technological and infrastructural constraints are major obstacles to AI application in agriculture, particularly in areas where connectivity is low, power supplies are unstable and sensor and computing capabilities do not exist. Moreover, locality bias: Model performance might decay when local environmental conditions deviate from those used in training media which support to require treasured, context-wise version of training and validation protocols (Majdalawieh et al., 2025).

Socio - economic barriers are especially vicious common to smallholders which affects their ability to adopt AI solutions due to initial costs, low digital literacy, and low access to credit. In an unplanned and untargeted form of inclusion and support, there is the risk that AI will be used to further perpetuate pre-existing inequalities in agricultural systems (Omotayo et al., 2025). Consequently, neat data governance standards, clarity of use of models and ethics of implementation of these technologies are imperative (Qamar, 2024).

Socio-economical barriers may continue to be strong for smallholder farmers, in which high initial pricing, inscribing digital literacy and access to credit hinder adoption. In the absence of examples of inclusive designing and focused support, AI may enhance existing disparity amongst farming systems (Dibbern et al., 2025; Wu & Zhong, 2025).

Mitigation of these barriers involves coordinated investment into rural connectivity and sensor infrastructure and the development of farmer-centered training and extension services and sense of cost-efficient and interoperable AI tools. Public-private collaborations, following the open data standards, and locally developed models including original farming communities ensure faster deployment as well as ensuring that the benefits will be equitably distributed amid diverse farming communities (Lidder et al., 2025).

CONCLUSION

Artificial intelligence is transforming agricultural systems with capability for precisely managing crops and optimizing use of resources and decision-making around the production to become predictive. Its association with technology of sensing, robotics, and data analytics have contributed towards a better monitoring, yield predictability, planning of irrigation, and management of pest attacks while reducing the inefficiency of inputs and environmental pressures. However, there is still a limitation to large scale adoption due to data gaps, high costs of implementation, digital inequities and low technical capacity in farming communities.

Future progress will hinge upon creating the arrangement of evolved AI structures and achievable costs, a interoperable data platform and a greatly effective expansion system that can change digital understandings into choices made in the field. Integration with biotechnology, digital twins, and climate related modelling will further contribute to part of the recipe for greater resilience and sustainability. Ultimately, AI will bring a significant contribution to global food security indirectly when all the necessary technological innovation, accompanied by inclusive governance, investment in infrastructure and farmer-centred capacity building are put in place for the success of the whole process.

REFERENCES

Adamo, T., Caivano, D., Colizzi, L., Dimauro, G., & Guerriero, E. (2025). Optimization of irrigation and fertigation in smart agriculture: An IoT-based microservices framework. *Smart Agricultural Technology, 11*, 100885. DOI: 10.1016/J.ATECH.2025.100885

Adkisson, M., Kimmel, J. C., Gupta, M., & Abdelsalam, M. (2021). *Autoencoder-based Anomaly Detection in Smart Farming Ecosystem*. https://arxiv.org/abs/2111.00099

Ahmed, H. A. ; G., Zeng, M.-D. ;, Applications, Y., Mushtaq, M. A., Ghulam, H., Ahmed, M.-D., & Zeng, Y. (2024). Applications of Artificial Intelligence in Wheat Breeding for Sustainable Food Security. *Sustainability 2024, Vol. 16, 16*(13). /DOI: 10.3390/SU16135688

Al khatib, A. M. G., & Alshaib, B. M. (2025). Smart Irrigation Systems: Optimizing Water Use with AI. *Transforming Agriculture through Artificial Intelligence for Sustainable Food Systems*, 73–93. /DOI: 10.1007/978-981-96-4795-8_5

Ali, Z., Muhammad, A., Lee, N., Waqar, M., & Lee, S. W. (2025). Artificial Intelligence for Sustainable Agriculture: A Comprehensive Review of AI-Driven Technologies in Crop Production. *Sustainability (Switzerland), 17*(5). Advance online publication. DOI: 10.3390/SU17052281/S1

Alsabt, R., Alkhaldi, W., Adenle, Y. A., & Alshuwaikhat, H. M. (2024). Optimizing waste management strategies through artificial intelligence and machine learning - An economic and environmental impact study. *Cleaner Waste Systems, 8*, 100158. DOI: 10.1016/J.CLWAS.2024.100158

Anastasiou, E., Ntakos, G., Kanakari, E., Bitsika, S., Gemtou, M., Katsaragakis, M., Soudris, D., Volioti, C., Arvanitou, E. M., Folina, M. T., Maikantis, T., Kanidou, E. P., Fountouli, M., Ampatzoglou, A., Tsiogkas, N., Villa-Henriksen, A., Pedersen, S. M., Tamirat, T. W., Milella, A., & Fountas, S. (2025). Applications of robotics and extended reality in agriculture: A review. *Smart Agricultural Technology, 12*, 101521. DOI: 10.1016/J.ATECH.2025.101521

Bessaad, N., Atsyo, S., Hanjie, D., Walkine, I., Dhillon, R., & He, L. (2026). Autonomous mowing in agriculture: Current status, needs, and future opportunities. *Smart Agricultural Technology, 13*, 101796. DOI: 10.1016/J.ATECH.2026.101796

Chen, Z., Yin, J., Farhan, S. M., Liu, L., Zhang, D., Zhou, M., & Cheng, J. (2026). A comprehensive review of obstacle avoidance for autonomous agricultural machinery in multi-operational environment. *Artificial Intelligence in Agriculture*, *16*(1), 139–163. DOI: 10.1016/J.AIIA.2025.10.001

D, D. C., AC, P. chandra, M, R., TS, B., MR, H., & B, S. (2026). *Artificial Intelligence enabled robotics and automation in modern agriculture*. /DOI: 10.21203/RS.3.RS-8601053/V1

Das, S., Dutta, S., Roy Choudhury, M., Garai, S., Mukherjee, S., Sengupta, S., Jana, S., Dey, S., Dhar, A., Dutta, S., & Awasthi, A. (2025). Regenerating rural soil and ecosystems: A 15-year systematic review of emerging methods and technologies. *The Science of the Total Environment*, *990*, 179926. DOI: 10.1016/J.SCITOTENV.2025.179926

Demissie, W. A., Sebastiani, L., & Rossetto, R. (2026). Integration of artificial intelligence and remote sensing for crop yield prediction and crop growth parameter estimation in Mediterranean agroecosystems: Methodologies, emerging technologies, research gaps, and future directions. *European Journal of Agronomy*, *173*, 127894. DOI: 10.1016/J.EJA.2025.127894

Dibbern, T., Romani, L., Massruhá, S., Dibbern, T., Romani, L., & Massruhá, S. (2025). Drivers and Barriers to Digital Agriculture Adoption: A Mixed-Methods Analysis of Challenges and Opportunities in Latin American. *Sustainability 2025, Vol. 17, 17*(8). /DOI: 10.3390/SU17083676

Ermias Benti, N., Mesfin, ·, Chaka, D., Addisu, ·, Semie, G., Warkineh, B., & Soromessa, · Teshome. (2024). Transforming agriculture with Machine Learning, Deep Learning, and IoT: perspectives from Ethiopia—challenges and opportunities. *Discover Agriculture 2024 2:1, 2*(1), 63-. /DOI: 10.1007/S44279-024-00066-7

Fan, X., Che, H., Wang, Y., Chen, D., & Hu, X. (2025). Effects of deep storage irrigation on summer maize yield formation and safety irrigation threshold for maximizing flood resources utilization. *Agricultural Water Management*, *319*, 109805. DOI: 10.1016/J.AGWAT.2025.109805

Gavhane, K. P., Kiran, P. R., Pradhan, N. C., Mandal, S., Kumari, K., & Parray, R. A. (2026). Precision agriculture and unconventional food sources. *Health, Nutrition and Sustainability*, 627–648. /DOI: 10.1016/B978-0-443-32920-3.00024-0

Ghag, N., Sonar, H., Jagtap, S., & Trollman, H. (2024). Unlocking AI's potential in the food supply chain: A novel approach to overcoming barriers. *Journal of Agriculture and Food Research*, *18*, 101349. DOI: 10.1016/J.JAFR.2024.101349

Ghazal, S., Munir, A., & Qureshi, W. S. (2024). Computer vision in smart agriculture and precision farming: Techniques and applications. *Artificial Intelligence in Agriculture, 13*, 64–83. DOI: 10.1016/J.AIIA.2024.06.004

González-Rodríguez, V. E., Izquierdo-Bueno, I., Cantoral, J. M., Carbú, M., & Garrido, C. (2024). Artificial Intelligence: A Promising Tool for Application in Phytopathology. *Horticulturae 2024, Vol. 10, 10*(3). /DOI: 10.3390/HORTICULTURAE10030197

Huda, S. S., Akhtar, A., Ahmed, E., Samiul Hoq, K. M., & Islam, M. N. (2026). Artificial intelligence in agriculture across south Asia: Technology adoption, improvements, and sustainability outcomes. *Sustainable Futures : An Applied Journal of Technology, Environment and Society, 11*, 101620. DOI: 10.1016/J.SFTR.2025.101620

Jain, R. (2025). Artificial Intelligence in Precision Agriculture: Advanced Systems for Crop Management and Farm Optimization. *International Journal of Scientific Research in Computer Science. Engineering and Information Technology, 11*(1), 1849–1857. DOI: 10.32628/CSEIT251112183

Javaid, M., Haleem, A., Khan, I. H., & Suman, R. (2023). Understanding the potential applications of Artificial Intelligence in Agriculture Sector. *Advanced Agrochem, 2*(1), 15–30. DOI: 10.1016/J.AAC.2022.10.001

Javed, K., Smagghe, G., Wang, Q., Javed, H., & Wang, Y. (2025). Artificial intelligence in crop protection: Revolutionizing agriculture for a sustainable future. *Information Processing in Agriculture*. Advance online publication. DOI: 10.1016/J.INPA.2025.12.003

Kanaga Priya, P., Reethika, A., Vijaykumar, G., & James Deva Koresh, H. (2025). Robotics-assisted precision and sustainable irrigation, harvesting, and fertilizing processes. *Hyperautomation in Precision Agriculture: Advancements and Opportunities for Sustainable Farming*, 217–226. /DOI: 10.1016/B978-0-443-24139-0.00018-7

Kariyanna, B., & Sowjanya, M. (2024). Unravelling the use of artificial intelligence in management of insect pests. *Smart Agricultural Technology, 8*, 100517. DOI: 10.1016/J.ATECH.2024.100517

Khan, R. A. (2020). Response of Wheat (Triticum aestivum L.) to Zinc Sulphate and Copper Sulphate under salt stress. *Pure and Applied Biology, 9*(4). Advance online publication. DOI: 10.19045/bspab.2020.90281

Kumari, K., Mirzakhani Nafchi, A., Mirzaee, S., & Abdalla, A. (2025). AI-Driven Future Farming: Achieving Climate-Smart and Sustainable Agriculture. *AgriEngineering 2025, Vol. 7, 7*(3). /DOI: 10.3390/AGRIENGINEERING7030089

Lan, J., & Ban, Q. (2025). The Farm-Level Economic and Environmental Benefits of Precision Agriculture Technology Adoption: A Meta-Analysis of Global Evidence. *Sustainability 2025, Vol. 17, 17*(24). /DOI: 10.3390/SU172411223

Lidder, P., Cattaneo, A., & Chaya, M. (2025). Innovation and technology for achieving resilient and inclusive rural transformation. *Global Food Security*, *44*, 100827. DOI: 10.1016/J.GFS.2025.100827

Mabhaudhi, T., Dirwai, T. L., Taguta, C., Sikka, A., & Lautze, J. (2023). Mapping Decision Support Tools (DSTs) on agricultural water productivity: A global systematic scoping review. *Agricultural Water Management*, *290*, 108590. DOI: 10.1016/J.AGWAT.2023.108590

Majdalawieh, M., Martins, C., Radi, M., Alaraj, M., & Khan, S. (2025). Precision agriculture in the age of AI: A systematic review of machine learning methods for crop disease detection. *Smart Agricultural Technology*, *12*, 101491. DOI: 10.1016/J.ATECH.2025.101491

Malik, P., Singh Tomar, A., Trivedi, A., Paliwal, A., & Goyal, A. (2025)... *Optimizing Fertilizer Application Using Machine Learning for Precision Agriculture.*, *13*, 3.

Mana, A. A., Allouhi, A., Hamrani, A., Rahman, S., el Jamaoui, I., & Jayachandran, K. (2024). Sustainable AI-based production agriculture: Exploring AI applications and implications in agricultural practices. *Smart Agricultural Technology*, *7*, 100416. DOI: 10.1016/J.ATECH.2024.100416

Manono, B. O., Mwami, B., Mutavi, S., & Nzilu, F. (2026). Precision Farming with Smart Sensors: Current State, Challenges and Future Outlook. *Sensors (Basel)*, *26*(3), 882. DOI: 10.3390/S26030882

Masud Cheema, J., Aqib, M., Elbeltagi, A., Rashid Saleem, S., Hussain, S., Wang, Z., Xiao, S., Wang, J., Parab, A., & Patel, S. (2025). Reinforcement Learning-Based Agricultural Fertilization and Irrigation Considering N2O Emissions and Uncertain Climate Variability. *AgriEngineering 2025, Vol. 7, 7*(8). /DOI: 10.3390/AGRIENGINEERING7080252

Mba, P. C., Njoku, J. N., & Uyeh, D. D. (2025). Enhancing resilience in specialty crop production in a changing climate through smart systems adoption. *Smart Agricultural Technology*, *11*, 100897. DOI: 10.1016/J.ATECH.2025.100897

Mishra, H. (2025). Artificial Intelligence Technologies in Agriculture. *AgriTech Revolution: Next-Gen Solutions and Challenges in Modern Agriculture*, 371–396. /DOI: 10.1007/978-981-95-1268-3_19

Monika, T, Thangadurai, K, Sivaguru, R, & Abdulkalamazad, G, & Lakshmi kanth R. (2025). AI-Driven Pest and Disease Detection in Smart Farming Systems. [IR-JAEM]. *International Research Journal on Advanced Engineering and Management*, *3*(06), 2315–2319. DOI: 10.47392/IRJAEM.2025.0365

Mubarok, M. U., Sari, M. I., Wibowo, Y. G., & Mathew, R. (2024). Comparative Study of Artificial Intelligence (AI) Utilization in Digital Marketing Strategies Between Developed and Developing Countries: A Systematic Literature Review. *Ilomata International Journal of Management*, *6*(1), 156–173. DOI: 10.61194/IJJM.V6I1.1534

Mushtaq, M. A., Ateeq, M., Ikram, M., Alam, S. M., Kaleem, M. M., Ashraf, M. A., Asim, M., Almutairi, K. F., Seleiman, M. F., & Shireen, F. (2025). Securing fruit trees future: AI-driven early warning and predictive systems for abiotic stress in changing climate. *Plant Stress*, *17*, 100953. DOI: 10.1016/J.STRESS.2025.100953

Mustaza, S. M., Mat Pauzi, N. A., Zainal, N., Mohd Zaman, M. H., & Mohamed Moubark, A. (2025). Artificial Intelligence in Precision Agriculture: A Review. *Jurnal Kejuruteraan*, *37*(2), 1025–1047. DOI: 10.17576/JKUKM-2025-37(2)-38

OĞUZTÜRK, G. E. (2025). AI-driven irrigation systems for sustainable water management: A systematic review and meta-analytical insights. *Smart Agricultural Technology*, *11*, 100982. DOI: 10.1016/J.ATECH.2025.100982

Omotayo, A. O., Adediran, S. A., Omotoso, A. B., Olagunju, K. O., & Omotayo, O. P. (2025). Artificial intelligence in agriculture: Ethics, impact possibilities, and pathways for policy. *Computers and Electronics in Agriculture*, *239*, 110927. DOI: 10.1016/J.COMPAG.2025.110927

Oyeyemi, B. B. (2022). Artificial Intelligence in Agricultural Supply Chains: Lessons from the US for Nigeria. *International Journal of Multidisciplinary Research and Growth Evaluation*, *3*(1), 1064–1074. DOI: 10.54660/.IJMRGE.2022.3.1.1064-1074

Padhiary, M., Saha, D., Kumar, R., Sethi, L. N., & Kumar, A. (2024). Enhancing precision agriculture: A comprehensive review of machine learning and AI vision applications in all-terrain vehicle for farm automation. *Smart Agricultural Technology*, *8*, 100483. DOI: 10.1016/J.ATECH.2024.100483

Paul, N., Sunil, G. C., Horvath, D., & Sun, X. (2025). Deep learning for plant stress detection: A comprehensive review of technologies, challenges, and future directions. *Computers and Electronics in Agriculture*, *229*, 109734. DOI: 10.1016/J.COMPAG.2024.109734

Petcu, M. A., Sobolevschi-David, M. I., Curea, S. C., & Moise, D. F. (2024). Integrating Artificial Intelligence in the Sustainable Development of Agriculture: Applications and Challenges in the Resource-Based Theory Approach. *Electronics 2024, Vol. 13, 13*(23). /DOI: 10.3390/ELECTRONICS13234580

Pienwisetkaew, T., Wongsaichia, S., & Ketkaew, C. (2025). Technology adoption in smart agricultural waste management among farmers: The role of simplicity and gamification. *Cleaner and Responsible Consumption*, *19*, 100344. DOI: 10.1016/J.CLRC.2025.100344

Prasanjana, V., Herath, I. S., Senadheera, U. E., Darshani, N. P., Suwan, B. V. S., Madusanka, C., Takimoto, Y., Amila, B., Hewawsam, C., Jayasanka, D. J., & Nakayama, T. (2025). Review of current robotic approaches for weed management in paddy cultivation. *Journal of Agriculture and Food Research*, *24*, 102516. DOI: 10.1016/J.JAFR.2025.102516

Qamar, A. (2024). Ethical Considerations in AI-Powered Decision-Making Systems. *Research Corridor Journal of Engineering Science, 1*(1), 90–102. https://researchcorridor.org/index.php/RCJES/article/view/236

Raj, M., & Prahadeeswaran, M. (2025). Revolutionizing agriculture: a review of smart farming technologies for a sustainable future. *Discover Applied Sciences, 7*(9), 937-. /DOI: 10.1007/S42452-025-07561-6/TABLES/9

Rashid, A. Bin, Kausik, A. K., Khandoker, A., & Siddque, S. N. (2025). Integration of Artificial Intelligence and IoT with UAVs for Precision Agriculture. *Hybrid Advances*, *10*, 100458. DOI: 10.1016/J.HYBADV.2025.100458

Saki, S., & Soori, M. (2026). Artificial intelligence, machine learning and deep learning in advanced transportation systems, a review. *Multimodal Transportation*, *5*(1), 100242. DOI: 10.1016/J.MULTRA.2025.100242

Salka, T. D., Hanafi, M. B., Rahman, S. M. S. A. A., Zulperi, D. B. M., & Omar, Z. (2025). Plant leaf disease detection and classification using convolution neural networks model: a review. *Artificial Intelligence Review 2025 58:10, 58*(10), 322-. /DOI: 10.1007/S10462-025-11234-6

Schweng, S., Bernardini, L., Keiblinger, K., & Kaul, H.-P.Jr, I. F., Lukač, N., Del Ser, J., & Holzinger, A. (2026a). What can artificial intelligence do for soil health in agriculture? *Computer Science Review*, *59*, 100832. DOI: 10.1016/J.COSREV.2025.100832

Shah, R., Doss, A. S. A., & Lakshmaiya, N. (2025). Advancements in AI-enhanced collaborative robotics: Towards safer, smarter, and human-centric industrial automation. *Results in Engineering*, *27*, 105704. DOI: 10.1016/J.RINENG.2025.105704

Sharma, H., Sidhu, H., Bhowmik, A., Sharma, H., Sidhu, H., & Bhowmik, A. (2025). Remote Sensing Using Unmanned Aerial Vehicles for Water Stress Detection: A Review Focusing on Specialty Crops. *Drones 2025, Vol. 9, 9*(4). /DOI: 10.3390/DRONES9040241

Sharma, K., & Shivandu, S. K. (2024). Integrating artificial intelligence and Internet of Things (IoT) for enhanced crop monitoring and management in precision agriculture. *Sensors International*, *5*, 100292. DOI: 10.1016/J.SINTL.2024.100292

Shawon, S. M., Barua Ema, F., Mahi, A. K., & Mohsin Sarker Raihan, M. (2023). Crop Yield Prediction: Robust Machine Learning Approaches for Precision Agriculture. *2023 26th International Conference on Computer and Information Technology, ICCIT 2023*. /DOI: 10.1109/ICCIT60459.2023.10441634

Shawon, S. M., Ema, F. B., Mahi, A. K., Niha, F. L., & Zubair, H. T. (2025). Crop yield prediction using machine learning: An extensive and systematic literature review. *Smart Agricultural Technology*, *10*. Advance online publication. DOI: 10.1016/J.ATECH.2024.100718

Singh, A., Singh, K. M., Kumari, T., Kumari, S., & Biswas, P. (2025). Artificial Intelligence in Agriculture: Case Studies and Success Stories. *Transforming Agriculture through Artificial Intelligence for Sustainable Food Systems*, 191–199. / DOI: 10.1007/978-981-96-4795-8_11

Tadala, N., & Fanny, C. (2024). IoT-Enabled crop storage monitoring system. *I-Manager's. Journal of Computational Science*, *11*(4), 43. DOI: 10.26634/JCOM.11.4.20654

Talaviya, T., Shah, D., Patel, N., Yagnik, H., & Shah, M. (2020). Implementation of artificial intelligence in agriculture for optimisation of irrigation and application of pesticides and herbicides. *Artificial Intelligence in Agriculture*, *4*, 58–73. DOI: 10.1016/J.AIIA.2020.04.002

Thilakarathne, N. N., Abu Bakar, M. S., Abas, P. E., & Yassin, H. (2025). Internet of things enabled smart agriculture: Current status, latest advancements, challenges and countermeasures. *Heliyon, 11*(3), e42136. DOI: 10.1016/J.HELIYON.2025.E42136

Tufail, R., Tassinari, P., & Torreggiani, D. (2025). Deep Learning Applications for Crop Mapping Using Multi-Temporal Sentinel-2 Data and Red-Edge Vegetation Indices: Integrating Convolutional and Recurrent Neural Networks. *Remote Sensing 2025, Vol. 17, 17*(18). /DOI: 10.3390/RS17183207

Waqas, M., Naseem, A., Humphries, U. W., Hlaing, P. T., Dechpichai, P., & Wangwongchai, A. (2025a). Applications of machine learning and deep learning in agriculture: A comprehensive review. *Green Technologies and Sustainability, 3*(3), 100199. DOI: 10.1016/J.GRETS.2025.100199

Wu, P., & Zhong, Y. (2025). Artificial intelligence in sustainable agriculture: Towards a socio-technical roadmap. *Smart Agricultural Technology, 12*, 101578. DOI: 10.1016/J.ATECH.2025.101578

Yan, J., Wu, H., Diao, Z., Miao, Y., Zhang, B., & Zhao, C. (2025). Recent Developments and Applications of Crop Disease Detection, Prediction, and Early Warning: A review. *Engineering*. Advance online publication. DOI: 10.1016/J.ENG.2025.10.032

Younas, M., Nadeem Akhtar, M., Batool, S., Owais, M., Sahar, S., & Anum, W. (2025). The Integration of Artificial Intelligence in Agriculture: Emerging Trends, Benefits and Challenges. *Journal of Asian Development Studies, 14*(1), 1316–1333. DOI: 10.62345/JADS.2025.14.1.105

Zhang, F., Zhang, W., Luo, X., Zhang, Z., Lu, Y., Wang, B., Zhang, F., Zhang, W., Luo, X., Zhang, Z., Lu, Y., & Wang, B. (2022). Developing an IoT-Enabled Cloud Management Platform for Agricultural Machinery Equipped with Automatic Navigation Systems. *Agriculture 2022, Vol. 12, 12*(2). /DOI: 10.3390/AGRICULTURE12020310

Zhou, Y. (2025). Artificial Intelligence-Based Crop Disease Identification Technology: Applications, Challenges, and Future Prospects. *Journal of Electronic Research and Application, 9*(6), 287–293. DOI: 10.26689/JERA.V9I6.13182

Chapter 4
Data-Driven Farming: Turning Data Into Smarter Decisions and Higher Yields

Nadia Tahseen
Lahore College of Women University, Pakistan

Muhammad Waqas
http://orcid.org/0009-0009-8126-9709
Chinese Academy of Agricultural Sciences, China

Muhammad Iftikhar
University of Education, Lahore, Pakistan

Andrew Tapiwa Kugedera
http://orcid.org/0000-0002-1700-6922
Morgenster Teachers College, Zimbabwe

ABSTRACT

Global agriculture is facing unprecedented challenges intensifying from population growth, climate change, soil degradation, new pests, lack of labour and pressure on natural resources. These constraints have contributed to unsustainable agriculture leading to decline in productivity. This chapter gives an overall view about data driven farming and the revolutionizing role of artificial intelligence (AI) in modern farming. The chapter explore various new trends in agriculture such as Internet of Things (IoTs), sensors, and drones to mention a few. Precision agriculture and AI tools that maximise smart irrigation, robotics, and resource optimization are also discussed. The integration of real-time data and automation has the capacity of improving climate-smart agriculture and reduce environmental degradation. Smallholder farmers have low adoption of these new trends due to high costs,

DOI: 10.4018/979-8-3373-7257-0.ch004

limited technical knowledge, and gaps in infrastructure. A sustainable roadmap is recommended so that researchers and policy makers can enforce AI into farming systems to improve food security.

6.1 INTRODUCTION

Agriculture faces a complex of interrelated challenges that pose a risk to global food security and the sustainability of agricultural production systems. Rapid population growth requires a huge expansion in agricultural production at a time when the amount of arable land is decreasing, and natural resources are coming under increased pressure. Climate change has led to an increase in the occurrence of extreme weather events - uncontrollable prolonged droughts, heat waves, floods, and unpredictable rainfall patterns - which affect crop growth and productivity directly (Viana et al., 2022). Emerging pests and diseases in new areas caused by changing climatic conditions account for billions of dollars in losses each year. Degradation of soils and lack of nutrients, along with decreasing biodiversity, further interrupt the resistance of agroecosystems. In many developing regions, constrained by a lack of access to advanced technologies, poor extension services, and rising costs of production, an additional dimension of complexity is added to this. These challenges are a reminder of the need to develop a paradigm shift from conventional farming practices to more intelligent, efficient, and adaptive farming systems (Shah et al., 2024).

It is in this shifting world that global agriculture is moving away from time-tested, experience-based decision-making processes to digital and artificial intelligence (AI) based systems that handle large amounts of data with precision. The digital revolution in agriculture is being defined by an integrated network of technologies such as Internet of Things (IoT), drones, satellite imagery, remote sensing, mobile applications, intelligent machinery, and cloud-based farm management platforms (Omotayo et al., 2025a). These systems produce continuous and high-resolution data sets on soil moisture, nutrient levels, pest activity, plant health, weather, and machinery performance. The analytical engine, artificial intelligence, takes these data streams, finds the patterns and anomalies, and converts them into something that you can do. Rather than having to rely only upon manual field scouting or generalized recommendations, farmers have, for the first time, access to precise and real-time information that is reflective of the true needs of their crops and soils (Kumar et al., 2024a).

The usefulness of real-time information, predictive analytics, and automation in agricultural decision-making is becoming more popular. Real-time information will assist farmers in discovering the issues at an early stage and responding in time to the threat and preventing any unjustifiable losses. Predictive analytics that enable one to foresee the stress associated with weather, the likelihood of the occurrence of diseases, or the change in yield, and, therefore, make informed choices by farmers even before the risk is increased (Bayar et al., 2025). Automation- through AI-powered tractors, drones, automated sprayers, robot weeders, and even automated irrigation systems to enhance the precision of the work and eliminate the necessity of human workers. These technologies not only improve efficiency but also reduce wastage and enhance environmental sustainability while supporting climate-smart agricultural practices too. By combining insights from AI and the management on the ground, farmers are able to meet the optimal allocation of resources, stability of yield, and lower production risks (Javaid et al., 2023).

As its journey of digital agriculture continues, the role of artificial intelligence is playing a major role in helping to define data-driven farming systems that are resilient, sustainable, and profitable. AI - driven platforms help to generate site-specific crop recommendations, enhance efficiencies of water and fertilizers, monitor stress, and ensure traceability across the production cycle. By cutting-edge analytics in conjunction with many decades of agronomic expertise, AI can provide a strong foundation to revolutionize agriculture into a more adaptive, sustainable, and knowledge-intensive sector (Aijaz et al., 2025).

The aim and scope of this chapter is to give an overview of how data-driven farming and artificial intelligence are changing the decision-making process and productivity of farming. Specifically, the chapter aims to explain important data sources, technologies for analysis, and artificial intelligence applications, while addressing some of the benefits, limitations, and opportunities for the future. It also presents a realistic action plan for the integration of AI in global farming systems, providing a basis for guidance to researchers, policy makers, and practitioners who want to catalyze the integration of intelligent agriculture (Paul-Chima Ugwu et al., 2025).

6.2 FOUNDATIONS OF DATA-DRIVEN FARMING

The primary form of agricultural transformation in the contemporary era is data-driven agriculture, as it assists the stakeholders to make informed decisions, which are timely and accurate to enable them to grade with real world condition. In contrast to the traditional farming, where the majority of the processes rely on the manual observations, intuition, and generalized advice, the information grounded on the digital technology became a part of the farming today, where the kind of information is generated in an unceasing and high-resolution manner, relying on a variety of sources (Marinello et al., 2023). This is a baseline that artificial intelligence (AI) systems can read the complex relationships among soil, crops, climate, and management practices that the productivity, sustainability, and resilience is enhanced. An adequate understanding of the different elements that constitute the framework of data-driven farming (the data sources, the analytical technology, and integration mechanisms) is essential to the potential realization of AI in the agrarian space.

6.3 KEY DATA RESOURCES IN MODERN AGRICULTURE

The effectiveness of data-driven farming is dependent on the availability, accuracy, and diversity of agricultural data. Contemporary farms produce huge amounts and types of data from physical and digital tools as well as automated farming equipment (Bettucci et al., 2024). One of the most important sources of agricultural data is field-deployed sensors. Soil sensors are used to determine the level of moisture, nutrient concentration, pH, salinity, and microbial activity, and thus enable accurate nutrient and irrigation management. Crop sensors measure plant health parameters of a plant, such as chlorophyll content, canopy temperature, stress response, etc. Crop sensor has given early warning about pests, diseases, or nutrient deficiency (Faqir et al., 2024). Weather sensors and automated stations provide real-time data for temperature, humidity, wind speed, and rainfall, which are necessary for planning planting periods, irrigation cycles, and pest control operations (Kumar et al., 2024b).

Aerial and satellite technologies have further increased the dimensions of agricultural data in space and time. Drones equipped with multispectral, hyperspectral cameras, and thermal cameras capture a detailed image of the state of the crop, variations in soil, and patterns of water distribution. These high-resolution images can help farmers spot variability within their fields and help diagnose problems that may not be obvious to the farmer at ground level (Ram et al., 2024). Satellite imagery, which is available at wider scales, is supporting monitoring of vegetation health and climate trends as well as large-scale crop performance over the long term. Together, drones and satellites enable one to ensure that there is eternal observation and accuracy in tracking the different (Karmakar et al., 2024).

Agricultural streams of data also receive a significant contribution of farm machinery telemetry and IoT networks. Contemporary tractors, harvesters, sprayers, and irrigation systems have in-built sensors and GPS to monitor the movement pattern, fuel consumption, use effectiveness, and application levels. These machines are connected to IoT networks that link them with ground-based sensors that aid in the exchange of data across the entire farm without any complications (Thilakarathne et al., 2025). Biological and environmental data, e.g., soil microbiome profile, pest dynamics, greenhouse gas emissions, etc. introduce further complexity to help in managing farms holistically. The various kinds of data together with the historical data and external data sources will create a robust foundation of AI -driven decision making.

6.4 AI-BASED TECHNOLOGIES THAT WILL FACILITATE SMART FARMING.

Advanced data-driven agriculture is built entirely on the basis of artificial intelligence technologies. Deep learning (DL) and machine learning (ML) algorithms are applied to analyze large and difficult data sets to determine the hidden patterns and relationships that would be difficult to determine manually. The ML and its methods assist in the yield of crops, identification of plant diseases, estimation of the irrigation requirements, and determination of the nutrient dynamics of the soil. Tasks such as disease detection, recognizing weeds using drone and satellite cameras, and canopies are most often performed with the help of DL models, which process information in accordance with images and sensors (Xu et al., 2021). Confusion by AI is a sub-branch of AI that permits visual data interpretation of camera, drones, and sensor data. It enables the automated identification of the pests, deficiency of nutrients, crop stage of growth, and anomalies in the fields with great precision. Computer vision is also making routine monitoring a highly automated and efficient process by extending it with real-time imaging, and an efficient and scalable process (Javaid et al., 2023).

Predictive modeling plays a key role in AI when used for farm management. Advanced tools combine weather forecasts, soil, and plant physiology to estimate the yield trends and identify stress and best management activities. Decision support systems (DSS) use these predictions along with user-friendly interfaces to give farmers recommendations on irrigation, fertilizer application, pest control, and machinery deployment. These platforms help bring complex analytics to simple actions that can be put into action (Nautiyal et al., 2025).

Cloud computing and edge computing provide the computer infrastructure necessary to manage large amounts of data in agriculture. Cloud platforms allow datasets to be stored and processed at scale and used to run intricate analyses without limitation. Edge computing, on the other hand, transfers data to the devices, enabling real-time decision making even in remote locations where connectivity is limited (Figure 1). Digital platforms and mobile applications bring these capabilities to the farmers front and center, including accessible, scalable, and interactive data-driven management tools (Yu et al., 2025).

Figure 1. Use of technology and devices in agriculture

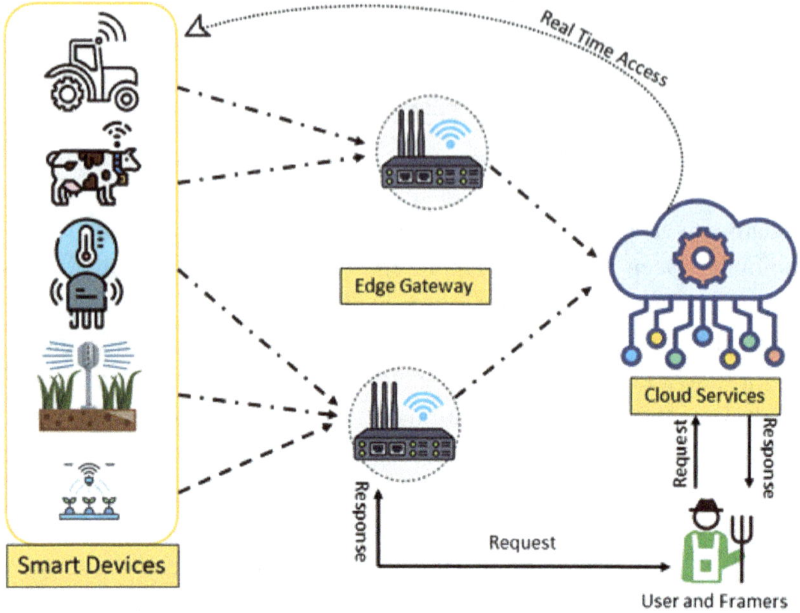

6.5 INTEGRATING DATA FOR REAL-TIME DECISION MAKING

The real potential of data-driven farming is the seamless integration of data over multiple transactions into an integrated analytics pipeline. This process usually consists of three major steps, which are data collection, data processing, and data analysis. Data collection begins with the use of sensors, drones, satellites, and machinery collecting ongoing information from fields. The data collected are then subject to processing (cleaning, filtering, structuring) to eliminate data errors, standardize formats, and prepare data sets for analysis. Finally, AI algorithms are used to analyze the data that has been processed to produce actionable insights, predictions, and recommendations (Eyitayo Raji et al., 2024; Subeesh & Chauhan, 2025).

One of the most innovative uses of the modern farm is the application of digital twins - virtual replicas of farms to simulate interactions of soil, crop, climate, and management. Digital twins enable farmers and researchers to experiment with scenarios, fine-tune strategies without the risks and costs of on-the-ground experimentation, and predict the results. These virtual representations combine real-time data with sophisticated models and can provide a dynamic decision-making environment (Kalyani et al., 2024).

Although this has improved, data standardization, interoperability, and interpretation are relevant problems. Various sensors, platforms, and equipment are usually incompatible formats that are not compatible in their integration. Inconsistency in data quality - i.e. the data might either contain no values or it might contain incorrect readings - could impact on the model reliability. Besides, the interpretation of insights made by AI is problematic to many farmers because of the absence of digital literacy or even training. The removal of these obstacles would involve the creation of standard data protocols, user-friendly and training to facilitate large-scale use (Cabrera et al., 2025).

6.6 ARTIFICIAL INTELLIGENCE CROP AND SOIL MANAGEMENT.

One of the most transformational spheres in digital agriculture is artificial intelligence - empowered management of crops and soil. Machine learning, computer vision, advanced sensor arrays, and predictive modeling have enabled practitioners to measure the health of crops and soil as well as the environment more precisely than ever before (Padhiary et al., 2025). These technologies are favorable to proactive management, minimise uncertainty, maximise productivity and sustainability. By providing site-specific recommendations and real-time insights, AI tools are transforming traditional approaches to agriculture and helping cultivators make evidence-based decisions to optimize resource allocation and strengthen system resilience. The following discussion indicates important uses of AI in crop monitoring, yield prediction, soil fertility management, irrigation optimization, and resource planning (climate-smart) (Lidder et al., 2025).

6.7 PRECISION CROP MONITORING AND YIELD PREDICTION

Precision crop monitoring based on AI involves using various forms of data, combined with satellite data, drone surveys, ground-based sensors, and meteorological data, to deliver constant and high-resolution updates of plant growth dynamics and health status. Central to this strategy are AI-dominated models of crop development from vegetation, which simulate the evolution of the crop under a range of environmental conditions and management situations (Padhiary et al., 2025). These models include photosynthetic activity, thermal patterns, soil moisture content, nutrient concentrations, and crop physiological parameters. By comparing what is being measured with what has been predicted by computer models, artificial intelligence (AI) systems can flag outliers at an early stage, pinpoint potential crop growth constraints, and adjust the decision-making process throughout the cropping season (Demissie et al., 2026).

Disease, pest, and stress detection is one of the advanced ways of AI in crop management. Computer vision (CV) and machine learning (ML) algorithms process imagery from drones and smartphones as well as field-mounted cameras to diagnose disease manifestation, insect infestation, nutrient deficiency, and abiotic stress, e.g., drought or heat. Unlike traditional scouting that is labor-intensive and often inconsistent, Artificial Intelligence (AI) tools are able to analyze thousands of images in a short amount of time, identify very small changes in diseases like leaf chromaticity or leaf texture, and identify disease with high fidelity (Figure 2). Early detection helps in timely intervention and hence yields losses, and the usage of pesticides is minimal (Anam et al., 2024).

AI is also playing an important role in predicting yield and optimizing the planting window. Yield prediction models combine remote detection data, soil characteristics, historical predicted yields, and weather pattern projections for prospective production outcomes. These forecasts guide the logistical planning of farmers, estimation of the market supplies, and efficient allocation of resources (Aslan et al., 2024). Moreover, machine learning algorithms analyze temperature trends, probabilities of rainfall, soil temperatures, and crop phenology to determine the best planting dates. Optimizing planting windows is a way to ensure that the crops have optimum climatic conditions during the critical stages of crop growth, and hence boosts not only the yield quantity but also the yield quality. By combining real-time monitoring and predictive analytics, AI enables crop management to be changed from a reactive to a predictive and preventative paradigm (Gao et al., 2025).

Figure 2. Precision agriculture using various tools

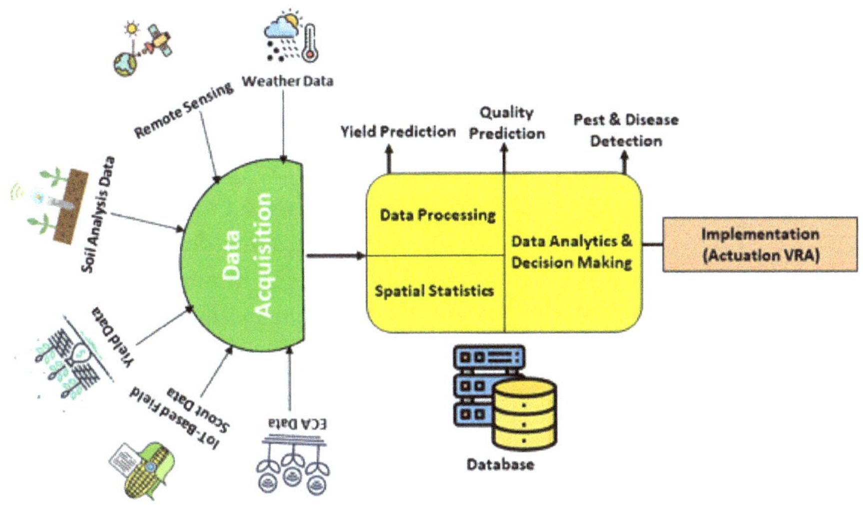

6.8 SMART SOIL HEALTH AND NUTRIENT MANAGEMENT

The fertility of soil is the cornerstone of agricultural productivity, and AI technologies are playing a role in the changing nature of soil management from broad-scale management to highly targeted and data-driven. AI-guided soil fertility maps combine the datasets of sensor arrays, lab results, drone imagery, and past land use records to become highly detailed spatial intelligence maps that reveal the variability of nutrients, organic content, and physical attributes of soil throughout the field (Mana et al., 2024). These maps facilitate practitioners to notice the micro-variations in soil fertility, which are often ignored by conventional sampling methodologies.

Data-driven fertilizer planning uses machine learning models to calculate which fertilizers to use, how much to use, and when to use them. Rather than applying uniform fertilizer allocations over wide areas of fields, AI systems make site-specific nutrient recommendations that are based on the needs of crops, nutrient status of soils, and projected weather patterns. Such recommendations limit fertilizer wastage, increase the nutrient-use efficiency, and reduce environmental problems such as nutrient leaching and greenhouse gas emissions (Liben et al., 2024). AI platforms also analyse the long-term soil trends, helping cultivators to adapt the nutrient management scheme to improve soil fertility over a period of time.

Water management is another important area where AI enhances sustainability and high productivity. AI-driven irrigation scheduling combines information on soil moisture, evapotranspiration, crop growth stages, and weather forecasts to provide optimal irrigation scheduling. Smart irrigation systems have real-time, automated adjustment of water applications to avoid under- or over-irrigation (OĞUZTÜRK, 2025). For example, machine learning algorithms can help predict a few days in advance the requirements of water-based irrigation cycles in order to manage them with greater efficiency. In this way, water consumption is minimized, crop water productivity is increased, and crops are more drought resilient. By combining information about the soil, weather conditions, and forecasting predictive models, AI enables cultivators to be precise and reliable in the management of irrigation (Parra-López et al., 2025).

6.9 RESOURCE OPTIMIZATION AND CLIMATE-SMART PRACTICES

The science of AI has become a powerful tool to allow the cultivator to use the least number of resources while maintaining or even increasing productivity. Through AI, water, fertilizer, and pesticide usage are cut down by recognizing the exact needs of crops and providing exact recommendations (Talaviya et al., 2020). For example, computer vision models are used to identify weeds and make targeted applications of herbicides, which helps to reduce the amount of chemical used compared to a blanket application. Similarly, AI-driven pest prediction systems help farmers to apply pesticides only when needed and in the right dosages to reduce environmental impact and production cost (Ehrampoosh et al., 2025).

Beyond input reduction, AI cut has improved sustainability through the manner in which it underpins sustainable practices (climate-smart) that build resilience against the vagaries of climate. AI models predict extreme weather events, such as droughts, heatwaves, and floods, which allow cultivators to implement proactive measures such as changing the time of planting, changing irrigation schedules, or deploying protective infrastructure. Machine learning algorithms process many years of climate and crop data to develop adaptation management practices to enhance long-term productivity in changing environmental conditions (Ayadi et al., 2025).

AI also enables the inclusion of environment-friendly practices such as conservation agriculture, precision irrigation, and carbon-smart management of soil. For instance, no-till systems using AI to manage their operation will maintain the soil structure and improve carbon sequestration. AI-enabled crop rotation and intercropping models will recommend combinations that make the best use of nutrient cycling, minimize pest pressure, and encourage biodiversity. These innovations help make certain that food production intensification goes hand in hand with environmental care (Mishra & Kayusi, 2025).

Real-life case studies show the significant impact of the optimization of resources through AI. In India, artificial intelligence (AI) - assisted irrigation systems have reduced water usage by as much as 40 percent in rice and sugarcane fields, and with little reduction in yields. In the United States, use of herbicides has been reduced by almost 90 per cent on maize and soybean farms using computer vision-enabled weed detection systems (M. B. Akanbi et al., 2025). In sub-Saharan Africa, AI-based weather advisory platforms help smallholder farmers make sound decisions on a variety of farming practices, such as planting and fertilizer application, and are leading to yield increases of 20-30 per cent. These examples from all over the world show that AI technologies are not limited to technologically advanced countries, but are more available and adaptable to various agricultural contexts (Paul-Chima Ugwu et al., 2025).

6.10 AI IN FARM MACHINERY, AUTOMATION, AND ROBOTICS

The automation paradigm of agriculture has been transformed by artificial intelligence in terms of taking the manual activities that were practiced in the traditional agricultural practice to very automated intelligence-based practice. Modern AI-powered equipment opposes these to the extent that it offers accuracy, effectiveness, and maneuverability, so that farms can work with minimal human interactions and at that stage to add more productivity to the same as well as the reliability of the same. Robots, automation and smart machinery saves labor, assists in the prompt field activity and provides consistency throughout the farming tasks (Ali et al., 2025). This technological development is gaining more significance with the world agriculture experiencing the problem of labor shortage, escalating production costs and pressure to produce more efficiently. The current part emphasizes how AI can enhance the speed of machinery automation by examining the use of autonomous machinery, intelligent farm management system, and the economic and operational merits of agriculture that is operated by robots (Aijaz et al., 2025).

6.10.1 Self-driven Machinery and Intelligent Equipment.

One of the advances in the AI in the agricultural domain is autonomous machinery. The AI-powered tractors, sprayers, and harvesters have GPS positioning, machine vision, and intelligent sensing, which allow the machines to accomplish complex tasks with accuracy. Self-driving tractors are able to navigate the fields, adjust their speed, remove obstacles, and operate independently without having to be controlled by a human operator (Subeesh & Mehta, 2021). These systems rely on the Machine Learning algorithms to interpret the sensor data so that the real-time adjustment of the soil condition and changes in the terrain could occur, including weather changes. AI-controlled nozzle sprayers target particular weeds or pest-infested zones rather than entire zones, which implies that they use less chemicals and produce fewer environmental effects (Zaigham Abbas Naqvi et al., 2025). Robotics is gaining a highly significant role in planting, weeding, and harvesting, which is generally a monumental task at hand. Planting robots can be controlled by AI, which can be used to lay the seeds correctly. at the most favorable depths and distances, and increases the germination and early vigor of vegetation. Robots that carry out weeding make use of computer vision to distinguish between the crops and the weeds. targeted or mechanical chemical treatment at the point of necessity (Palva et al., 2024). These systems significantly reduce the use of herbicides, and more sustainable herbicide weeds are possible. Harvesting robots: especially in horticulture and special crops. AI and robotic arms are used to recognize and examine ripe produce and estimate their size and maturity, and harvest fruits without damaging them. Strawberry, tomato, and apple-picking robots are already being deployed around the world to show high precision along with consistent performance (Sachithra & Subhashini, 2023).

AI-enabled machinery also promotes variable-rates operations-that is, the capability to change the rate of fertilizer, irrigation, and pesticide applications in various zones of the field. By combining data from sensors and machine learning, equipment will be able to apply inputs in areas where they are needed more, decreasing waste and making equipment more efficient. These constant, automated changes to improve resource use efficiency are part of making the farm more sustainable. As autonomous technologies develop, they are becoming increasingly compact and energy-conservative as well as adaptable to large and small-sized farm enterprises (Talaviya et al., 2020).

6.10.2 Intelligent Farm Management Systems

Intelligent farm management systems make up the network of digital systems connecting farm machinery, sensors, and analytics platforms into a single framework of operations. Automated decision-support platforms give recommendations on actionable crop management based on data from sensors in the field, satellite imagery, drones, and machine data. These platforms generally have AI algorithms that identify patterns to predict future scenarios and optimize field operations based on predictive patterns (Rajak et al., 2023). For example, farmers can turn their irrigation systems on, change their tractor routes, or get prompts about diseases to their crops right through their mobile devices (Kumar et al., 2024b).

A farm management system also allows automating the workflow. For example, an integrated system can offer automatic irrigation scheduled on the basis of soil moisture thresholds or automatic release of autonomous sprayers from alert alarms on pest detection. Machine learning enables such systems to become more sophisticated over time - to improve in terms of accuracy as more data is gathered and analyzed. Cloud computing enables both scalable processing and storage, which means that both smallholder farmers and large agribusinesses can equally utilize green technology advancements (Subeesh & Mehta, 2021).

In addition to this, digital tools enforce compliance and traceability through monitoring all. action in the farm - planting and spraying, as well as harvesting and transportation. This data is imperative to quality standards, certification policies, and regulatory limits in the market. By the way, which is the full visibility and control that the intelligent farm management systems provide strengthen the. openness, streamline business processes, and develop a digital ecosystem across the industry (Hasan et al., 2024).

6.10.3 Enhancing Efficiency and Reducing Labor Burden

One of the issues of agricultural systems facing labor shortage is automation. worldwide. Most nations are experiencing a diminished supply of agricultural workforce, due to urban migration, aging of the farming population, and the shifting preferences of the labor force. The solution to this is the AI-based machinery and robotics that reduce the need for manual labor in the tasks that are repetitive, physically demanding tasks, etc. Robotic tractors handle plowing, tilling, weeding, and spray robots do the same work, helping to eliminate the necessity to work on the field, and harvesting robots cover the gaps in the intensive sectors of the fruit and vegetable industry. This is not only for ensuring timely field work but also the reduction in the prices of labor and, therefore, the increase in the profitability of the farms in general (Bazargani and Deemyad, 2024; Jiang et al., 2025).

Machinery facilitated by AI has more economic returns than just saved working hours. By optimizing the use of inputs, helping to improve yield consistency and minimize operational errors, automated systems increase overall productivity and cost efficiency. Precision spraying, for example, reduces the cost of chemicals; automated irrigation reduces the water bill; and yield prediction tools enable companies to plan better at the market and reduce post-harvest losses. These resulting savings often pay for the upfront investment in AI-driven technologies, especially as equipment becomes more affordable and less expensive, and as parts of the equipment become more easily swapped out (Husein et al., 2024; Rashid & Kausik, 2024).

Operationally, automation has led to improvements in accuracy, repeatability, and timeliness -- factors which are important in getting high yields. Machines are not subject to fatigue, so the applications are uniform over large areas. Precision robotics results in less crop damage, waste, and ensures quality consistency (Table 6.1). The ability to work around the clock, day or night, is important to speed up operations in the field and avoid delays in crucial growth stages. This is consistent and creates more reliable production cycles and added resiliency in the face of workforce disruptions (Bazargani & Deemyad, 2024).

AI-powered automation is also a factor in reducing environmental sustainability. By using inputs more precisely and avoiding overuse of inputs, these systems can help reduce runoff, prevent soil degradation, and reduce greenhouse gas emissions. The transition to electric or hybrid autonomous machinery further reduces the carbon footprint of agricultural businesses, too. As the climate-savvy strategies of agriculture take hold, automation will be central to achieving efficient and low-impact production systems (Gamage et al., 2024).

Table 1. Key AI applications in data-driven farming: Technologies, data sources, and global impact

AI Application	Technologies Used	Primary Data Sources	Key Benefits	Global Examples	References
Precision Crop Monitoring	Machine Learning, Computer Vision, Drones	Satellite imagery, drone images, crop sensors	Early disease/pest detection, growth monitoring, yield optimization	Strawberry harvesting robots (Japan), UAV-based crop monitoring (USA)	Jia et al. (2025)
Yield Prediction & Forecasting	Predictive Modeling, AI Algorithms	Historical yields, soil data, and weather forecasts	Improved harvest planning, market alignment	Wheat yield prediction using ML (India)	Morchid et al. (2024)
Soil Health & Nutrient Management	ML, AI-guided Fertilizer Systems	Soil sensors, lab data, IoT devices	Optimal fertilization, improved soil fertility, reduced input costs	Variable-rate fertilizer application (USA)	Islam et al. (2023)
Smart Irrigation	IoT, ML, Edge Computing	Soil moisture sensors, weather data, evapotranspiration models	Water savings, crop stress reduction	AI-driven irrigation in rice fields (China)	Iqbal et al. (2024)
Autonomous Machinery & Robotics	AI-enabled Tractors, Robotic Arms, and GPS	Machinery telemetry, field sensors, and GPS data	Labor reduction, operational efficiency, precision planting/harvesting	Tomato-picking robots (Netherlands), Autonomous tractors (USA)	Zualkernan et al. (2023)

continued on following page

Table 1. Continued

AI Application	Technologies Used	Primary Data Sources	Key Benefits	Global Examples	References
Decision-Support Platforms	Cloud Computing, AI Analytics, Mobile Apps	Multi-source farm data, market trends, and climate data	Real-time recommendations, scenario planning	Climate-smart advisory apps (Kenya, India)	Omotayo et al. (2025b)
Resource Optimization & Sustainability	ML, AI Analytics, Digital Twins	Integrated soil, crop, climate, and operational data	Reduced inputs, environmental protection, and higher resilience	Smart farms using IoT and AI (Europe)	Subeesh & Chauhan (2025)

6.11 BIG DATA ANALYTICS AND DECISION-SUPPORT SYSTEM

The accelerated digitalization of the agricultural sector has produced large amounts of data covering soil and crop data, climate data, market trend data, and operational data. Big data analytics is the base structure of converting these streams of data into comprehensible information for decision making and, in turn, helps farmers make the right decision at the right time with the right amount (Abbasi et al., 2022). When combined with artificial intelligence, predictive modeling, and decision-support systems (DSS), big data makes planning of scenarios and risk management, as well as optimization of farm operations (Figure 3). Through the integration of these tools, modern agriculture's capacity to achieve greater efficiency, resilience, and sustainability can meet both of the twin imperatives of achieving increased productivity and environmental sustainability (Ahmed & Shakoor, 2025).

Figure 3. Big data use in agriculture

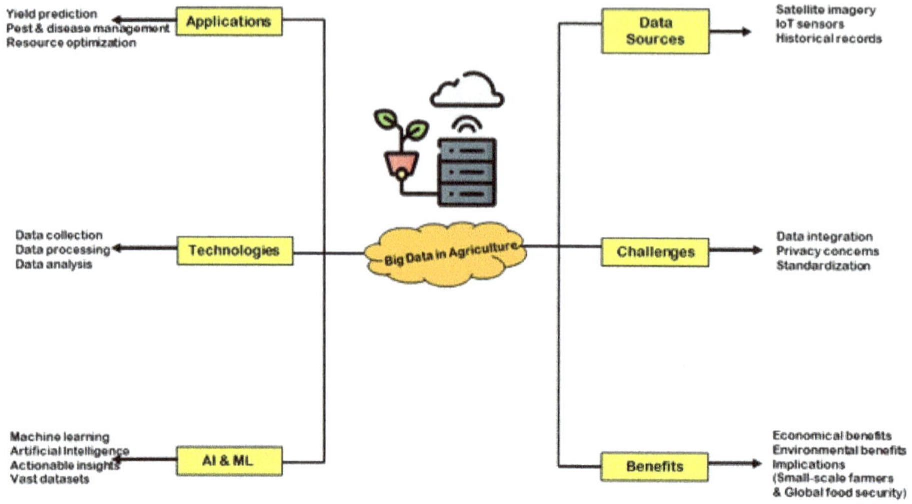

6.11.1 Predictive Analytics for Farm Planning

Predictive analytics uses past and real-time information to forecast key agriculture-related metrics such as weather, production, pest outbreaks, and markets. Weather Forecasting Modeling incorporates meteorological information and climate simulations for the predictive purposes of forecasting rain patterns, temperature changes, and extreme events. These forecasts assist in planting schedules and irrigation, and also in determining crop protection measures that minimize exposure to environmental risks (Rane et al., 2024). Similarly, decisions on farms to plan harvests, optimize logistics, and match production to market demand can be made using yield-prediction models that use historical yield data along with soil-fertility and crop-growth models to determine the possible production. The other significant application of predictive analytics is market forecasting. The analysis of the tendencies in commodity prices, demand, supply and economic indicators enables the analysis of the economy. AI systems assist farmers to make sound decisions when it comes to the decision of crops, their storage, and sale-time (Jabed and Azmi Murad, 2024). Risk-assessment models are also a useful one, support to the scenario-based planning as a tool of simulating various farming strategies in various economic, environmental and technological conditions. For example, a model can be used to predict the effect of drought on the maize yield under various irrigation practices and therefore to tell the different farmers which practice is most resilient. Scenario-based planning guarantees that decision-making is proactive rather than reactive, which may lessen the losses and improve the farm's profitability (Baluch et al., 2025).

6.11.2 Farm-Level Decision-Support Platforms

Decision support platforms make it easy for farmers to produce complex datasets and AI predictions into actionable recommendations that they can easily implement. These platforms combine operations from many sources of information, such as soil sensors, drones, satellite data, weather stations, and data telemetry from machinery, in a centralized interface. Cloud-based ecosystems possess the computing power required to store, process, and analyze huge datasets in near real-time (Aijaz et al., 2025). Application Programming Interfaces (APIs) are useful in allowing data exchange between different devices, sensors, and platforms with ease that promotes the interoperability of the agricultural technology landscape. AI-driven advisory systems integrated into such platforms provide in real time advisory for irrigation, fertilization, pest management, and crop management. For example, a farmer can be alerted to the emergence of pest infestation with recommendations on types and amounts of pesticides to be applied and methods of application based on the particular field conditions (Muhammed et al., 2024). Likewise, water use optimization based on soil moisture, rate of evapotranspiration and crop growth. stage AI powered irrigation recommendation. Integrated and mobile applications. dashboards help farmers to remotely control the operations and make dynamic alterations to the same. planning and make decisions, very few delays. By means of their synthesis of decision support platforms supplant operational predictive analytics and user-friendly interfaces. efficiencies, use of resources optimization, and yield stability to new heights (Senoo et al., 2024).

6.11.3 Data Privacy, Ownership, and Ethical Considerations

The opportunities of big data and AI are enormous, but they are immense concerns, as well. of government, confidentiality, and moral employment. Yield data, soil data, and other data generated in the farm level data, financial transactions, and farm operations, is in its very nature secret, and is thought to be proprietary. The issues of data ownership, consent and sharing agreements will require to be calculated to make sure that farmers manage their information (Omotayo et al., 2025b). Weak data governance may possess the propensity of causing abuse, unfair trade usage, or competitive disadvantages against the smallholder farmers. IT also is accountable in the implementation of AI is critical one. The algorithms that are made to make predictions, recommendations, and automations should be clear, articulate, and non-partisan that could be harmful to other farmers or areas. Ethical other considerations are the access to AI technologies in a fair manner, such that the digital divide an agribusiness-smallholder divide is reduced. Technology and policymakers. providers need to develop data security frameworks, interoperability standards and accountability systems that assist in establishing trust in the use of AI in agriculture (Atapattu et al., 2024). Moreover, Ethical implementation in AI also can result in a better environment of sustainability and social responsibility. For example, AI systems which are designed to restrict the usage of chemicals are also optimizing their use less and lessen their effect on the environment, a factor that can be used to make it climate-smart. farming practices. Education of the farmers and the extension workers on the use of the right and responsible use of AI platforms make these technological advantages reach high numbers, and reduce the negative influence of the disadvantaged. unintended consequences (Omotayo et al, 2025b).

6.11.4 Barriers to Adoption

AI-enhanced agriculture has a bright future, there is a range of concerns about it as a promise. In its adoption, it is faced with technical, financial, and infrastructural limitations. Lots of agricultural jurisdictions are victimized. due to being connected poorly, lacking reliable electricity, and sensor network on-demand. maintain AI systems. The capital cost involved in using smart machinery, unmanned. smallholder and underprivileged users are not allowed to fly, sensor, and digital platforms are prohibitive. resource-limited producers. Furthermore, there are interoperability problems which are common during integration. equipment of other manufacturers, and hence making farm management processes. difficult (Younas et al., 2025). The salient barriers are also digital illiteracy and lack of skill. The effective application of AI tools presupposes. an impressive degree of expertise in data analytics, interface systems and the principle of precision agriculture. The percentage of a good number of farmers especially in the developing countries have a low level of enjoyment. exposure to digital technologies thus restricting their capacity to give meaning to recommendations created. by artificial intelligence or control self-driving machines. Without the explicit training and extension assistance systems, these knowledge gaps have the potential to reduce the rate of adoption and the potential impact of AI solutions (Chowdhury et al., 2025).

6.11.5 Opportunities for Sustainable and High-Tech Agriculture

Emergent technological developments, nonetheless, under the influence of such constraints, offer. unworldly prospects of climate-sensitive and productive agriculture.

Climate-adaptive AI models already are becoming more predictive of extremes in weather, and maximize planting. dates, and to suggest techniques of adaptive management to be used to maintain yield in the face of the changing. conditions. These models can be employed to warn farmers of the risks and implement a preventive measure to enhance resilience to by integrating climate projections and crop growth simulations. climate shocks (Aijaz et al., 2025). The integration of AI with other related technologies such as nanotechnology, the Internet of Objects and advanced sensors also increase the possibility of high-precision agriculture. Nanotechnology assists in controlled discharge of fertilizers and agrochemicals, and consequently increases the nutrient use efficiency and reduces the environmental impact (Agrawal & Kumar, 2026). IoT networks can be used to monitor the conditions in the field at all times, and edge computing can facilitate making. choices, even in distant places, in real time! All of these innovations form an. ecosystem of inter synergy whereby AI is continually generating and analyzing data based on broadscale. data sets and optimized by the transformation of these data into actionable insights. allocation and utilization of resources at ecosystem scales, and both field, farm, and regional. scales (Ficili et al., 2025).

6.11.6 Strategic Roadmap for Global Data-Driven Agriculture

To achieve maximum opportunities of AI and data-driven agriculture holistic strategic. roadmap is of utmost significance. Digital agriculture should be facilitated and assisted by policy frameworks. by offering incentives to the use of this form of technology, the supply of is available. access to infrastructure to gather this information, as well as putting in place strong data governance standards to ensure that it remains private and security. The interdisciplinary work between governments and the private sector should be encouraged. The integrated solutions that can be created by the agronomists, data scientists, engineers, and policymakers are needed to address the issue. are responsive to the local conditions (Thilakarathne et al., 2025). Capacity building is also important. Farmer, extension, and other training programs. agribusiness managers can bring the digital literacy to a new level and assist them in the efficient. use of tools empowered by artificial intelligence. Programs such as knowledge hubs, demonstration farms, and others. online learning platforms have the potential to accelerate the acquisition of skills and trigger the uptake of. new agricultural practices in other agricultural environments. Standardization of data protocols and enabling the devices and platforms interoperability is another source of developing. smooth connection and scalability (Omotayo et al., 2025b). The future vision of an AI driven farming is a strong, productive, and sustainable farming. productivity and environmental watchdog systems of production. The farms will become smarter ecosystems, where AI will integrate sensor data, satellite data, etc.

and climate forecasts and predictive analytics to support particular interventions. Resource use will be enhanced, and labor loads will be decreased, and outputs will be capable of responding to world food. security requirements. Through offering a blend of technology, specific policy, and capacity building solutions, a data-grounded type of agriculture can be the keystone of. high-yielding, climate-resilient agriculture that is sustainable to the generations to come (Colizzi et al., 2025).

6.12 CONCLUSION

The artificial intelligence-based, data-driven farming works in a different way today in contemporary farming, and it has changed the traditional approaches to systematic, productive, and sustainable systems. The pressures of the constantly growing world food demand, climatic changes, soil erosion, and labour shortage have highlighted the inefficiencies of traditional agriculture and thus the necessity of technology-oriented solutions. AI empowers farmers to make better and proactive choices to make better use of many resources and generate a more efficient farming system, besides helping to sustain the environment. Intelligent agriculture, which relies on precise monitoring of crops, artificial intelligence (AI) fueled yield prediction, and early identification of diseases and pests allows addressing the losses and enhancing the performance of the crop in time. In the same manner, the decrease of the waste and environmental impact can also be guaranteed by soil fertility mapping with the assistance of AI, data-driven nutrient management, and intelligent irrigation systems that will guarantee optimal use of water, fertilizers, and agrochemicals. Robotics and autonomous equipment assists in streamlining the operations that are labor intensive and also enhancing the precision of the operation and has the possibility of farming with greater levels of precision and scalability. Integrated farm management: Coalescence of diverse data to be acted on to plan and make adaptive decisions according to the needs of the agronomy and the needs of the market: However, a number of challenges, such as high cost of technology, infrastructural barriers, and skills remain to be a major drawback to the general use of the same. Such hurdles have to be cleared this way through specific capacity-building efforts, facilitatory policy provisions, standard data protocols, and interdisciplinary efforts to cope with such hurdles. The potential of sustainable and high-tech agriculture is further increased by emerging opportunities, which entail climate-adaptive AI models, integration of IoT- and nanotechnology-based solutions. To sum up, AI-driven data-based agriculture is one solution in terms of resilient, prolific, and sustainable agriculture. With a combination of innovation and responsible implementation, farms will be converted into smart ecosystems with evidence-based decisions, which will allow optimization of resource use, achieve better yields, and be sustainable in the long run, and therefore contribute to the effective resolution of the global problem of food security and environmental issues.

REFERENCES

Abbasi, R., Martinez, P., & Ahmad, R. (2022). The digitization of the agricultural industry – a systematic literature review on agriculture 4.0. *Smart Agricultural Technology*, *2*, 100042. DOI: 10.1016/J.ATECH.2022.100042

Agrawal, K., & Kumar, N. (2026). Novel agricultural techniques: an artificial intelligence and machine learning approach. *Dietary, Sensory and Gastronomic Applications*, 39–61. https://doi.org/DOI: 10.1016/B978-0-443-33345-3.00006-7

Ahmed, N., & Shakoor, N. (2025). Advancing agriculture through IoT, Big Data, and AI: A review of smart technologies enabling sustainability. *Smart Agricultural Technology*, *10*, 100848. DOI: 10.1016/J.ATECH.2025.100848

Aijaz, N., Lan, H., Raza, T., Yaqub, M., Iqbal, R., & Pathan, M. S. (2025). Artificial intelligence in agriculture: Advancing crop productivity and sustainability. *Journal of Agriculture and Food Research*, *20*, 101762. DOI: 10.1016/J.JAFR.2025.101762

M. B. Akanbi, K. J. Adedotun, A. K. Raji, & I. K. Banjoko. (2025). Optimizing Water Resource Management in Agriculture Using AI-Powered Solar Irrigation Systems. *Journal of Science Innovation and Technology Research*. https://doi.org/DOI: 10.70382/AJSITR.V7I9.011

Ali, Z., Muhammad, A., Lee, N., Waqar, M., & Lee, S. W. (2025). Artificial Intelligence for Sustainable Agriculture: A Comprehensive Review of AI-Driven Technologies in Crop Production. *Sustainability (Switzerland)*, *17*(5). Advance online publication. DOI: 10.3390/SU17052281/S1

Anam, I., Arafat, N., Hafiz, M. S., Jim, J. R., Kabir, M. M., & Mridha, M. F. (2024). A systematic review of UAV and AI integration for targeted disease detection, weed management, and pest control in precision agriculture. *Smart Agricultural Technology*, *9*, 100647. DOI: 10.1016/J.ATECH.2024.100647

Aslan, M. F., Sabanci, K., Aslan, B., Aslan, M. F., Sabanci, K., & Aslan, B. (2024). Artificial Intelligence Techniques in Crop Yield Estimation Based on Sentinel-2 Data: A Comprehensive Survey. *Sustainability 2024, Vol. 16*, *16*(18). https://doi.org/DOI: 10.3390/SU16188277

Atapattu, A. J., Perera, L. K., Nuwarapaksha, T. D., Udumann, S. S., & Dissanayaka, N. S. (2024). Challenges in Achieving Artificial Intelligence in Agriculture. *Artificial Intelligence Techniques in Smart Agriculture*, 7–34. https://doi.org/DOI: 10.1007/978-981-97-5878-4_2

Ayadi, R., Forouheshfar, Y., & Moghadas, O. (2025). Enhancing system resilience to climate change through artificial intelligence: A systematic literature review. *Frontiers in Climate*, *7*, 1585331. DOI: 10.3389/FCLIM.2025.1585331/XML

Baluch, S. M., Wang, L., Faiz, M. A., Li, H., Chen, Y., Wang, L., & Li, M. (2025). Adaptation simulation and planning for crop yield under climate change: Integrating AquaCrop and DSSAT to project drought-induced yield risks in the Sanjiang Plain. *Agricultural Water Management*, *319*, 109818. DOI: 10.1016/J.AGWAT.2025.109818

Bayar, J., Ali, N., Cao, Z., Ren, Y., & Dong, Y. (2025). Artificial Intelligence of Things (AIoT) for Precision Agriculture: Applications in Smart Irrigation, Nutrient and Pest Management. *Smart Agricultural Technology*, *101629*. Advance online publication. DOI: 10.1016/J.ATECH.2025.101629

Bazargani, K., & Deemyad, T. (2024). Automation's Impact on Agriculture: Opportunities, Challenges, and Economic Effects. *Robotics (Basel, Switzerland)*, *13*(2). Advance online publication. DOI: 10.3390/ROBOTICS13020033

Bettucci, F., Sozzi, M., Benetti, M., & Sartori, L. (2024). A data-driven approach to agricultural machinery working states analysis during ploughing operations. *Smart Agricultural Technology*, *8*, 100511. DOI: 10.1016/J.ATECH.2024.100511

Cabrera, V. E., Bewley, J., Breunig, M., Breunig, T., Cooley, W., De Vries, A., Fourdraine, R., Giordano, J. O., Gong, Y., Greenfield, R., Hu, H., Lenkaitis, A., Niu, M., Noronha, E. A. F., & Sullivan, M. (2025). Data Integration and Analytics in the Dairy Industry: Challenges and Pathways Forward. *Animals 2025, Vol. 15, 15*(3). https://doi.org/DOI: 10.3390/ANI15030329

Chowdhury, K., Rahman, M. A., Mallik, S. K., Chowdhury, L. N., & Nova, N. A. (2025). Bridging the Global Digital Divide in Agriculture: The Role of AI in Equitable Technology Access. *International Journal of Sustainable Development and Planning*, *20*(6), 2469–2481. DOI: 10.18280/IJSDP.200616

Colizzi, L., Dimauro, G., Guerriero, E., & Lomonte, N. (2025). Artificial intelligence and IoT for water saving in agriculture: A systematic review. *Smart Agricultural Technology*, *11*, 101008. DOI: 10.1016/J.ATECH.2025.101008

Demissie, W. A., Sebastiani, L., & Rossetto, R. (2026). Integration of artificial intelligence and remote sensing for crop yield prediction and crop growth parameter estimation in Mediterranean agroecosystems: Methodologies, emerging technologies, research gaps, and future directions. *European Journal of Agronomy*, *173*, 127894. DOI: 10.1016/J.EJA.2025.127894

Ehrampoosh, A., Hettiarachchi, P., Koirala, A., Hassan, J., Islam, N., Ray, B., Nabi, M. N., Tolba, M., Mazid, A. M., Xu, C. Y., Ashwath, N., Dzitac, P., & Moore, S. (2025). Intelligent weed management using aerial image processing and precision herbicide spraying: An overview. *Crop Protection (Guildford, Surrey)*, *194*, 107206. DOI: 10.1016/J.CROPRO.2025.107206

Faqir, Y., Qayoom, A., Erasmus, E., Schutte-Smith, M., & Visser, H. G. (2024). A review of the application of advanced soil and plant sensors in the agriculture sector. *Computers and Electronics in Agriculture*, *226*, 109385. DOI: 10.1016/J.COMPAG.2024.109385

Ficili, I., Giacobbe, M., Tricomi, G., & Puliafito, A. (2025). From Sensors to Data Intelligence: Leveraging IoT, Cloud, and Edge Computing with AI. *Sensors (Basel)*, *25*(6). Advance online publication. DOI: 10.3390/S25061763

Gamage, A., Gangahagedara, R., Subasinghe, S., Gamage, J., Guruge, C., Senaratne, S., Randika, T., Rathnayake, C., Hameed, Z., Madhujith, T., & Merah, O. (2024). Advancing sustainability: The impact of emerging technologies in agriculture. *Current Plant Biology*, *40*, 100420. DOI: 10.1016/J.CPB.2024.100420

Gao, X., Liu, J., Lin, H., Javed, T., Yin, F., Chen, R., Wen, Y., Zhang, J., Yi, K., & Wang, Z. (2025). Using machine learning techniques to evaluate the impact of future climate change on wheat yields in Xinjiang, China. *Agricultural Water Management*, *317*, 109646. DOI: 10.1016/J.AGWAT.2025.109646

Hasan, H. R., Musamih, A., Salah, K., Jayaraman, R., Omar, M., Arshad, J., & Boscovic, D. (2024). Smart agriculture assurance: IoT and blockchain for trusted, sustainable produce. *Computers and Electronics in Agriculture*, *224*, 109184. DOI: 10.1016/J.COMPAG.2024.109184

Husein, M., Putro, H. P., Rajagukguk, J. R., & Putranto, K. E. (2024). The Role of Artificial Intelligence in Improving the Efficiency of the Company's Supply Chain. *International Journal of Engineering. Science and Information Technology*, *4*(4), 156–172. DOI: 10.52088/IJESTY.V4I4.596

Iqbal, N., Shahzad, M. U., Sherif, E. S. M., Tariq, M. U., Rashid, J., Le, T. V., & Ghani, A. (2024). Analysis of Wheat-Yield Prediction Using Machine Learning Models under Climate Change Scenarios. *Sustainability 2024, Vol. 16, 16*(16). https://doi.org/DOI: 10.3390/SU16166976

Islam, M. R., Oliullah, K., Kabir, M. M., Alom, M., & Mridha, M. F. (2023). Machine learning enabled IoT system for soil nutrients monitoring and crop recommendation. *Journal of Agriculture and Food Research*, *14*, 100880. DOI: 10.1016/J.JAFR.2023.100880

Jabed, M. A., & Azmi Murad, M. A. (2024). Crop yield prediction in agriculture: A comprehensive review of machine learning and deep learning approaches, with insights for future research and sustainability. *Heliyon, 10*(24), e40836. DOI: 10.1016/J.HELIYON.2024.E40836

Javaid, M., Haleem, A., Khan, I. H., & Suman, R. (2023). Understanding the potential applications of Artificial Intelligence in the Agriculture Sector. *Advanced Agrochem, 2*(1), 15–30. DOI: 10.1016/J.AAC.2022.10.001

Jia, H., Chen, W., Su, Z., Sun, Y., Qian, Z., & Huang, L. (2025). AI-Driven Cooperative Control for Autonomous Tractors and Implements: A Comprehensive Review. *AgriEngineering, 7*(11). Advance online publication. DOI: 10.3390/AGRIENGINEERING7110394

Jiang, L., Xu, B., Husnain, N., Wang, Q., Jiang, L., Xu, B., Husnain, N., & Wang, Q. (2025). Overview of Agricultural Machinery Automation Technology for Sustainable Agriculture. *Agronomy 2025, Vol. 15, 15*(6). https://doi.org/DOI: 10.3390/AGRONOMY15061471

Kalyani, Y., Vorster, L., Whetton, R., & Collier, R. (2024). Application Scenarios of Digital Twins for Smart Crop Farming through Cloud–Fog–Edge Infrastructure. *Future Internet, 16*(3). Advance online publication. DOI: 10.3390/FI16030100

Karmakar, P., Teng, S. W., Murshed, M., Pang, S., Li, Y., & Lin, H. (2024). Crop monitoring by multimodal remote sensing: A review. *Remote Sensing Applications: Society and Environment, 33*, 101093. DOI: 10.1016/J.RSASE.2023.101093

Kumar, V., Sharma, K. V., Kedam, N., Patel, A., Kate, T. R., & Rathnayake, U. (2024a). A comprehensive review of smart and sustainable agriculture using IoT technologies. *Smart Agricultural Technology, 8*, 100487. DOI: 10.1016/J.ATECH.2024.100487

Kumar, V., Sharma, K. V., Kedam, N., Patel, A., Kate, T. R., & Rathnayake, U. (2024b). A comprehensive review of smart and sustainable agriculture using IoT technologies. *Smart Agricultural Technology, 8*, 100487. DOI: 10.1016/J.ATECH.2024.100487

Liben, F., Abera, W., Chernet, M. T., Ebrahim, M., Tilaye, A., Erkossa, T., Degefie, D. T., Mponela, P., Kihara, J., & Tamene, L. (2024). Site-specific fertilizer recommendation using data-driven machine learning enhanced wheat productivity and resource use efficiency. *Field Crops Research, 313*, 109413. DOI: 10.1016/J.FCR.2024.109413

Lidder, P., Cattaneo, A., & Chaya, M. (2025). Innovation and technology for achieving resilient and inclusive rural transformation. *Global Food Security, 44*, 100827. DOI: 10.1016/J.GFS.2025.100827

Mana, A. A., Allouhi, A., Hamrani, A., Rahman, S., el Jamaoui, I., & Jayachandran, K. (2024). Sustainable AI-based production agriculture: Exploring AI applications and implications in agricultural practices. *Smart Agricultural Technology*, *7*, 100416. DOI: 10.1016/J.ATECH.2024.100416

Marinello, F., Zou, X., Liu, Z., Zhu, X., Zhang, W., Qian, Y., Li, Y., Karunathilake, E. M. B. M., Le, A. T., Heo, S., Chung, Y. S., & Mansoor, S. (2023). The Path to Smart Farming: Innovations and Opportunities in Precision Agriculture. *Agriculture 2023, Vol. 13, 13*(8). https://doi.org/DOI: 10.3390/AGRICULTURE13081593

Mishra, H., & Kayusi, F. (2025). Artificial intelligence and soil conservation: An overview. *Journal of Scientific Agriculture*, 229–249. https://doi.org/DOI: 10.25081/JSA.2025.V9.9661

Morchid, A., Jebabra, R., Khalid, H. M., El Alami, R., Qjidaa, H., & Ouazzani Jamil, M. (2024). An IoT-based smart irrigation management system to enhance agricultural water security using embedded systems, telemetry data, and cloud computing. *Results in Engineering*, *23*, 102829. DOI: 10.1016/J.RINENG.2024.102829

Muhammed, D., Ahvar, E., Ahvar, S., Trocan, M., Montpetit, M. J., & Ehsani, R. (2024). Artificial Intelligence of Things (AIoT) for smart agriculture: A review of architectures, technologies, and solutions. *Journal of Network and Computer Applications*, *228*, 103905. DOI: 10.1016/J.JNCA.2024.103905

Nautiyal, M., Joshi, S., Hussain, I., Rawat, H., Joshi, A., Saini, A., Kapoor, R., Verma, H., Nautiyal, A., Chikara, A., Ahmad, W., & Kumar, S. (2025). Revolutionizing agriculture: A comprehensive review of artificial intelligence applications in enhancing the properties of agricultural produce. *Food Chemistry: X*, *29*, 102748. DOI: 10.1016/J.FOCHX.2025.102748

OĞUZTÜRK, G. E. (2025). AI-driven irrigation systems for sustainable water management: A systematic review and meta-analytical insights. *Smart Agricultural Technology*, *11*, 100982. DOI: 10.1016/J.ATECH.2025.100982

Omotayo, A. O., Adediran, S. A., Omotoso, A. B., Olagunju, K. O., & Omotayo, O. P. (2025a). Artificial intelligence in agriculture: Ethics, impact possibilities, and pathways for policy. *Computers and Electronics in Agriculture*, *239*, 110927. DOI: 10.1016/J.COMPAG.2025.110927

Omotayo, A. O., Adediran, S. A., Omotoso, A. B., Olagunju, K. O., & Omotayo, O. P. (2025b). Artificial intelligence in agriculture: Ethics, impact possibilities, and pathways for policy. *Computers and Electronics in Agriculture*, *239*, 110927. DOI: 10.1016/J.COMPAG.2025.110927

Padhiary, M., Kumar, A., & Sethi, L. N. (2025). Emerging technologies for smart and sustainable precision agriculture. *Discover Robotics 2025 1:1, 1*(1), 6-. https://doi.org/DOI: 10.1007/S44430-025-00006-0

Palva, R., Kaila, E., García-Pascual, B., Bloch, V., Palva, R., Kaila, E., García-Pascual, B., & Bloch, V. (2024). Assessment of the Performance of a Field Weeding Location-Based Robot Using YOLOv8. *Agronomy 2024, Vol. 14, 14*(10). https://doi.org/DOI: 10.3390/AGRONOMY14102215

Parra-López, C., Ben Abdallah, S., Garcia-Garcia, G., Hassoun, A., Trollman, H., Jagtap, S., Gupta, S., Aït-Kaddour, A., Makmuang, S., & Carmona-Torres, C. (2025). Digital technologies for water use and management in agriculture: Recent applications and future outlook. *Agricultural Water Management, 309*, 109347. DOI: 10.1016/J.AGWAT.2025.109347

Paul-Chima Ugwu, O., Ogenyi, F. C., Ugo Alum, E., Hyginus Udoka Eze, V., Basajja, M., Nnenna Ugwu, J., Ugwu, C. N., Idu Ejemot-Nwadiaro, R., Ben Okon, M., Ikechukwu Egba, S., Daniel Ejim, U., & Daniel, U., paul-Chima Ugwu, okechukwu, ogenyi, F. C., Ugo alum, esther, hyginus Udoka eze, V., nnenna Ugwu, J., Ugwu, C., idu ejemot-nwadiaro, regina, … Daniel ejim, U. (2025). Implementing artificial intelligence and machine learning algorithms for optimized crop management: A systematic review on data-driven approaches to enhancing resource use and agricultural sustainability. *Cogent Food & Agriculture, 11*(1), 2569982. DOI: 10.1080/23311932.2025.2569982

Rajak, P., Ganguly, A., Adhikary, S., & Bhattacharya, S. (2023). Internet of Things and smart sensors in agriculture: Scopes and challenges. *Journal of Agriculture and Food Research, 14*, 100776. DOI: 10.1016/J.JAFR.2023.100776

Raji, E., Ijomah, T. I., & Eyieyien, O. G. (2024). Data-Driven decision making in agriculture and business: The role of advanced analytics. *Computer Science & IT Research Journal, 5*(7), 1565–1575. DOI: 10.51594/CSITRJ.V5I7.1275

Ram, B. G., Oduor, P., Igathinathane, C., Howatt, K., & Sun, X. (2024). A systematic review of hyperspectral imaging in precision agriculture: Analysis of its current state and prospects. *Computers and Electronics in Agriculture, 222*, 109037. DOI: 10.1016/J.COMPAG.2024.109037

Rane, J., Kaya, Ö., Mallick, S. K., & Rane, N. L. (2024). *Smart farming using artificial intelligence, machine learning, deep learning, and ChatGPT: Applications, opportunities, challenges, and future directions*. Generative Artificial Intelligence in Agriculture, Education, and Business., DOI: 10.70593/978-81-981271-7-4_6

Rashid, A. Bin, & Kausik, M. A. K. (2024). AI revolutionizing industries worldwide: A comprehensive overview of its diverse applications. *Hybrid Advances*, *7*, 100277. DOI: 10.1016/J.HYBADV.2024.100277

Sachithra, V., & Subhashini, L. D. C. S. (2023). How artificial intelligence is used to achieve the agriculture sustainability: Systematic review. *Artificial Intelligence in Agriculture*, *8*, 46–59. DOI: 10.1016/J.AIIA.2023.04.002

Senoo, E. E. K., Anggraini, L., Kumi, J. A., Karolina, L. B., Akansah, E., Sulyman, H. A., Mendonça, I., & Aritsugi, M. (2024). IoT Solutions with Artificial Intelligence Technologies for Precision Agriculture: Definitions, Applications, Challenges, and Opportunities. *Electronics 2024, Vol. 13, 13*(10). https://doi.org/DOI: 10.3390/ELECTRONICS13101894

Shah, M. H., Aktar, S. N., Pramanik, K., Pande, C. B., & Ali, G. T. (2024). The impact of climate change on plant diseases and food security. *Biocontrol Agents for Improved Agriculture*, 353–384. https://doi.org/DOI: 10.1016/B978-0-443-15199-6.00002-6

Subeesh, A., & Chauhan, N. (2025). Agricultural digital twin for smart farming: A review. *Green Technologies and Sustainability, 100299*. Advance online publication. DOI: 10.1016/J.GRETS.2025.100299

Subeesh, A., & Mehta, C. R. (2021). Automation and digitization of agriculture using artificial intelligence and internet of things. *Artificial Intelligence in Agriculture*, *5*, 278–291. DOI: 10.1016/J.AIIA.2021.11.004

Talaviya, T., Shah, D., Patel, N., Yagnik, H., & Shah, M. (2020). Implementation of artificial intelligence in agriculture for optimisation of irrigation and application of pesticides and herbicides. *Artificial Intelligence in Agriculture*, *4*, 58–73. DOI: 10.1016/J.AIIA.2020.04.002

Thilakarathne, N. N., Abu Bakar, M. S., Abas, P. E., & Yassin, H. (2025). Internet of things enabled smart agriculture: Current status, latest advancements, challenges and countermeasures. *Heliyon*, *11*(3), e42136. DOI: 10.1016/J.HELIYON.2025.E42136

Viana, C. M., Freire, D., Abrantes, P., Rocha, J., & Pereira, P. (2022). Agricultural land systems importance for supporting food security and sustainable development goals: A systematic review. *The Science of the Total Environment*, *806*, 150718. DOI: 10.1016/J.SCITOTENV.2021.150718

Xu, Y., Liu, X., Cao, X., Huang, C., Liu, E., Qian, S., Liu, X., Wu, Y., Dong, F., Qiu, C. W., Qiu, J., Hua, K., Su, W., Wu, J., Xu, H., Han, Y., Fu, C., Yin, Z., Liu, M., & Zhang, J. (2021). Artificial intelligence: A powerful paradigm for scientific research. *The Innovation*, *2*(4), 100179. DOI: 10.1016/J.XINN.2021.100179

Younas, M., Nadeem Akhtar, M., Batool, S., Owais, M., Sahar, S., & Anum, W. (2025). The Integration of Artificial Intelligence in Agriculture: Emerging Trends, Benefits and Challenges. *Journal of Asian Development Studies*, *14*(1), 1316–1333. DOI: 10.62345/JADS.2025.14.1.105

Yu, P., Teng, F., Zhu, W., Shen, C., Chen, Z., & Song, J. (2025). Cloud–edge–device collaborative computing in smart agriculture: Architectures, applications, and future perspectives. *Frontiers in Plant Science*, *16*, 1668545. DOI: 10.3389/FPLS.2025.1668545/XML

Zaigham Abbas Naqvi, S. M., Hussain, S., Awais, M., Tahir, M. N., Saleem, S. R., Al-Yarimi, F. A. M., Ashurov, M., Saidani, O., Khan, M. I., Wu, J., Wei, Z., & Hu, J. (2025). Climate-resilient water management: Leveraging IoT and AI for sustainable agriculture. *Egyptian Informatics Journal*, *30*, 100691. DOI: 10.1016/J.EIJ.2025.100691

Zualkernan, I., Abuhani, D. A., Hussain, M. H., Khan, J., ElMohandes, M., Zualkernan, I., Abuhani, D. A., Hussain, M. H., Khan, J., & ElMohandes, M. (2023). Machine Learning for Precision Agriculture Using Imagery from Unmanned Aerial Vehicles (UAVs): A Survey. *Drones 2023, Vol. 7*, *7*(6). https://doi.org/DOI: 10.3390/DRONES7060382

Chapter 5
Revolutionizing Crop and Soil Monitoring Through IoT-Enabled Farming

Dilawar Hassan
http://orcid.org/0000-0001-9422-8580
University of South Africa, South Africa

Nadia Tehseen
Lahore College for Women University, Pakistan

Urooj Farid
COMSATS University Islamabad, Pakistan

Muhammad Ehsan
Centro de Bachillerato Tecnológico Agropecuario, Mexico

ABSTRACT

The integration of Internet of Things (IoT) technologies is transforming agriculture from traditional input-based practices to data-driven, real-time, and automated production systems. IoT platforms integrate advanced sensors, connectivity infrastructures, and edge–cloud computing to continuously monitor soil moisture, nutrient status, pH, temperature, and plant physiological parameters, enabling early detection of biotic and abiotic stresses. When combined with artificial intelligence and machine learning models, heterogeneous data streams are analyzed to optimize irrigation, fertigation, pest management, and overall farm productivity. Automated control systems translate analytical insights into timely agronomic interventions,

DOI: 10.4018/979-8-3373-7257-0.ch005

reducing resource inputs, labor demands, and environmental impacts. Applications span large-scale farms, smallholder systems, and controlled environments, demonstrating improved yield, resource efficiency, and crop quality.

1. INTRODUCTION

Agriculture is undergoing a paradigmatic change where traditional methodologies are becoming less adequate in the face of the growing rate of adversities imposed on agriculture by the action of global environmental and socioeconomic pressure. Climate change always interferes with crop phenology, which is reflected in extreme temperature regimes, erratic precipitation regimes, changing pest regimes and prolonged drought and flood regimes. These perturbations not only impermeant yields but also affect the integrity of soils, nutrient cycling as well as water availability - all of which are crucial pillars of stable agricultural production (Mishra et al., 2025). Concurrently, the worldwide population is expected to grow to more than 9.7 bn by 2050, thus significantly increasing the need for food, feed and fiber. With limited arable land and natural resource depletion, achieving better productivity with sustainability has become a main global imperative (Gargaro et al.,2023).

Conventional methods for crop and soil monitoring mainly use manual scouting, visual assessments and random measurements. Although these methods have intrinsic value, they are plagued by human error, labor intensity and an inability to capture any continuous variations that are inherent in dynamic field environments. Soil moisture depletion, nutrient deficiencies, pest infestations and early signs of stress often go undetected until irreparable damage occurs. The growing conflict between manual observations and the requirements of modern agriculture has brought even greater need to more sophisticated monitoring tools (Chaudhari et al., 2022).

A prospect that the Internet of Things (IoT) allows to address this deficiency in a transformative way by providing continuous, high-resolution and data-driven information on the crop and soil status. Through inter-connected sensors with wireless networks and cloud-based analytics, intelligent systems for agriculture (IOTs) monitor parameters such as soil moisture, pH, electrical conductivity, nutrient concentrations, canopy temperature, plant health indices, micro climatical fluctuations and disease risk parameters (Choudhary et al., 2025a). The ability to record and send real-time information helps in making informed decisions about irrigation schedules, fertilization schedules, pest control, and resource optimization. For instance, autonomous operation of irrigation systems triggered when soil moisture levels drop below a certain threshold and nutrient sensors that help address over-application of fertilizer to curb incurred costs and environmental consequences (Lakhiar et al., 2024).

Beyond the data acquisition process, IoT platforms enable predictive modelling, warning and notifications of early warnings, automated decision making and integration with AI-driven advisory systems. This change of management from reactive to proactive strengthens the resilience of crops, improves the efficiency of using inputs and increases the overall farm productivity. Implemented either in open field, controlled environment, green house, orchard and other agriculture, IoT is part of the basic infrastructure for precision agriculture, climate-smart farming and sustainable intensification (Alawode et al., 2025a; Mushtaq et al., 2025a).

This chapter aims to critically analyze transformative impact of IoT technologies in crop & soil monitoring using real-time sensing, big-data analytics and automation. It presents an extensive overview of the IoT architecture, applications, benefits and challenges in modern agriculture. The discussion includes technological basics, empirical case studies and implementation challenges as well as potential research avenues going forward to provide the fundamentals of sustainable, data-driven agronomic systems.

2. IOT ARCHITECTURE AND CORE TECHNOLOGICAL FOUNDATIONS

The effectiveness of the Internet -of-Things (IoT) - enhanced agronomy is dependent on a solid technological architecture that enables continuous sensing, reliable communication, and smart data processing. In the context of agriculture - with highly variable environmental conditions, fit for limited resources - IoT systems must provide energy efficiency, durability, scalability, and operational resiliency of operations in remote or infrastructure-deficient locales. This section defines the basic elements and technological layers that form the backbone of the crop and soil monitoring systems based on IoT (Ali et al., 2023; Kumar et al., 2024a). As shown in Figure 1. IoT-enabled smart agriculture relies on real-time data acquisition, wireless data transfer, and intelligent data visualization to support precision farming.

Figure 1. Overview of the IoT-enabled smart agriculture architecture showing sensor-based data acquisition, wireless data transmission, analytical processing, and visualization to support precision farming

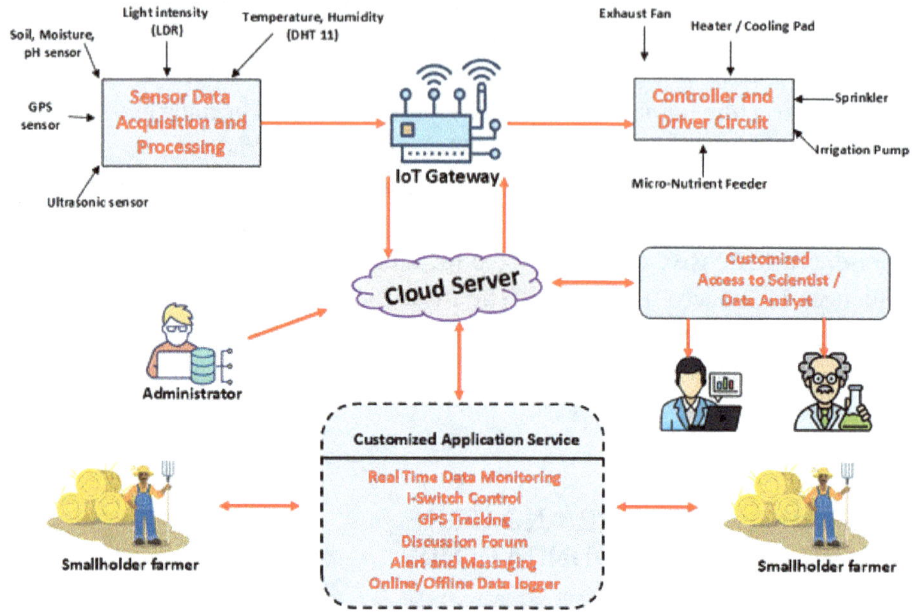

2.1. Core Components of Agricultural IoT Systems

The architecture model for cultivating IoT usually includes four main layers, including sensing devices, communication networks, data processing units, and application interfaces. At the foundational level lies sensors and a form of embedded hardware that measure soil, crop and environmental indicators of characteristics like soil moisture content, pH, nutrient concentrations, leaf temperature, humidity and atmospheric conditions (Choudhary, 2024). These devices are designed with low power consumption and long operational life and are resistant to adversities in the field. Microcontrollers and edge nodes - this refers to microcontrollers such as Arduino or ESP32 or small computers suh as Raspberry Pi - are local processing units, which can be used to perform some preliminary data filtering, compression, and calibration before the data is sent (Binu Mol T. V., 2024). As shown in Figure X, sensor nodes, communication gateways, and cloud-based processing units collectively enable real-time data acquisition and intelligent decision-making in agricultural IoT systems.

Superimposing the sensing layer is the connectivity layer to ensure smooth data exchange between the field nodes and the cloud platforms. The choice of communication protocol depends on the terrain, distance, bandwidth requirement and power limitation. Low power wide area network (LPWAN) like LoRaWAN, NB-IoT and Sigfox are preferred for large agricultural sites due to its long-range coverage and low energy requirements (Choudhary et al., 2025a). In turn, Wi-Fi, Bluetooth Low Energy (BLE) and Zigbee are used for short-range sensing networks generally used in greenhouses or small plots. Satellite networks and cellular networks provide connectivity in remote areas without the accessibility of terrestrial infrastructure, thus guaranteeing continuous wireless data flow even in difficult situations (Hosny et al., 2025).

Figure 2. Overview of an IoT-driven smart agriculture system integrating sensor networks, cloud computing, application services, and automated control mechanisms to support precision farming

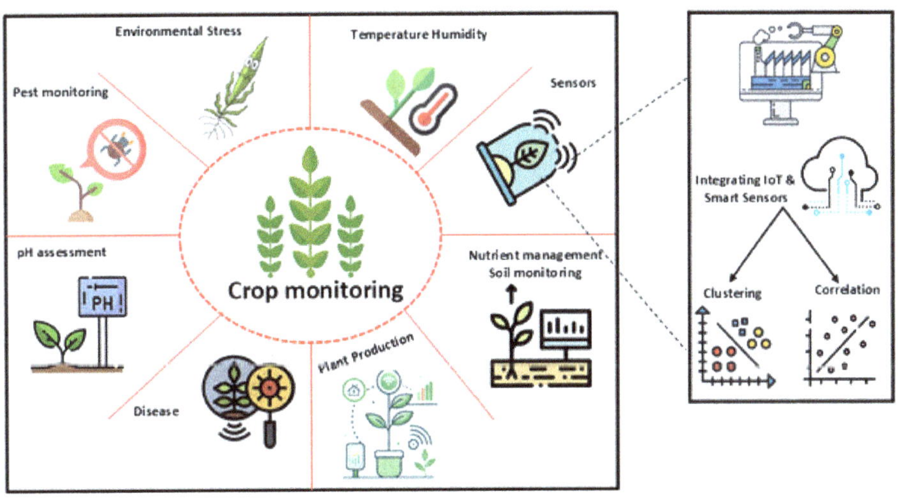

2.2. Sensor Technologies for Soil, Crop, and Environmental Monitoring

Sensors are the "eyes and ears" of the IoT enabled agriculture, which translate real-world conditions into quantifiable digital data. Soil sensors include capacitance and TDR based moisture probes, electrochemical type pH electrodes, nutrient ion selective sensors (nitrogen, phosphorus and potassium), conductivity meters and salinity detectors. These instruments provide important data for irrigation and fertilizing decisions. Advanced soil probes also measure soil organic matter, the carbon content, and the microbial activity by using optical and biosensing methods (Meshram et al., 2024).

For crop monitoring, there is use of multispectral and hyperspectral sensors which capture reflectance patterns indicating plant stress, plant chlorophyll content and disease progression. Infrared thermometers and thermal cameras detect changes in the temperature of plant canopies, and this reflects a first sign of water stress. Leaf wetness sensors help forecast disease models indicating conditions that are favorable for fungal infection. Moreover, stem diameter, sap flow sensors and dendrometers are used to monitor water dynamics and growth responses of plants in real time (Karmakar et al., 2024).

Environmental sensors (such as temperature, humidity, CO_2 concentration, rainfall, solar and wind speed) build a microclimate picture much needed for predictive modelling. Collectively, these sensors form a whole monitoring ecosystem in which the soil, plant and atmospheric data interact and give holistic information of the field condition (Chen et al., 2024).

2.3. Data Pipeline: From Acquisition to Intelligent Application

The IoT data lifecycle in agriculture is comprised of a well-organized pipeline that involves Acquisition Transmission Processing Application. The data must be acquired and then transmitted to the next stage - processing - which cleans up and extracts meaningful information data from the raw stream and used for actionable insights (Rathor et al., 2024). At the acquisition stage, the field sensors will collect high frequency data, and the data can be processed initially at the edge level for noise correction, outlier and volute removal. This step makes energy use more efficient and its response time shorter (Jafarbiglu & Pourreza, 2022).

The processed or raw data then transmitted through wireless connecting networks, cloud servers, gateways or the hybrid platforms. Cloud-based tools become available for the organization, storage and further analysis of the data, based on machine learning algorithms, in digital dashboards, and decision support systems. In many systems of today, edge computing is used to complement the cloud, as it makes it possible performing analytics in real time locally so possible responses can be made immediately, like, for example, actuating irrigation valves or issuing pest alerts (Majigi et al., 2025).

The last stage - that of application - converts the insights into action. The farmers view the data through mobile apps, dashboards or automated control systems. Visualizations are used to outline soil moisture trends, nutrient deficiencies, crop health variance and climate hazard arising from weather. Predictive models come up with irrigation schedules, fertilizer recommendations and warnings on disease risk (Sharafat et al., 2025). The smooth operations of this pipeline guarantee that IoT is not only part of the surveillance system but of the decision engine that can help increase productivity and reduce waste and support sustainable agricultural uses (Duguma & Bai, 2025).

3. SOIL MONITORING THROUGH IOT SYSTEMS

Soil is the main resource that controls curved productivity, nutrient cycling, and long-term agricultural production. Spatial and temporal heterogeneity of soil properties is the result of lateral variability of moisture dynamics, nutrient mobility, microbial processes and environmental stress. Conventional monitoring techniques - manual sampling and laboratory analysis - provide accurate information but are hampered by their manual and time-consuming characteristic and their inability to detect rapid changes on a field-level (Babla et al., 2022). Internet of Things (IoT) - enabled soil monitoring systems to overcome these limitations by providing real-time continuous high-resolution data to improve precision agriculture and optimized input application and decisions. By combining sensors, communication technologies and analytics, producers can take soil from being an opaque "black box" and turn it into a dynamic and measurable system (Rajak et al., 2023).

3.1. Digital Monitoring of Soil Moisture, Nutrients, Salinity, and pH

Soil moisture monitoring is one of the most popular IoT applications, as availability of water directly affects growth of plants, watering schedules and the plant's drought tolerance. IoT based moisture measuring devices such as capacitive, Time Domain Reflectometry (TDR) and Frequency Domain Reflectometry (FDR) provide high precision volumetric water content measurement. Real-time moisture map: Boom! Facilitation in optimizing drip and sprinkler irrigation system, wastage of water, water logging in rain feed-i.e., farming system (Laha et al., 2023). Monitoring nutrients, especially nitrogen (N), phosphorus (P) and potassium (K) is of vital importance for evidence-based fertilization. Contemporary IoT systems use ion -selective electrodes (ISEs), optical spectroscopy, and nanotechnology - enhanced sensors to measure nutrient concentrations in situ. These systems allow for the precise application of nutrients so as to meet the demands of the crop in question and ensure sustainable fertilizer applications with minimal loss to the environment (Reza et al., 2025).

Soil salinity monitoring is important in arid and semi-arid areas of agriculture. IoT sensors measure electrical conductivity (EC) which indicates salt concentration helping farmers in controlling irrigation water quality, avoiding sodality and planning crop rotation. Soil pH monitoring affects the availability of nutrients, microbial activity and chemical reactions in the soil. IoT enabled pH probes provide continuous profiles of the pH values providing information for lime or sulfur applications and long-term soil conditioning (Eltarabily et al., 2024).

3.2. Advanced Sensing for Biochemical, Biological, Redox, and Microbial Activity Indicators

Recent developments in soil sensing are moving away from physicochemical properties towards biological and biochemical indicators, which play a major role in becoming a driving force behind soil health. Biochemical sensors measure soil enzymes, like dehydrogenase, urease and phosphatase - some good indicators of nutrient cycling and decomposition of organic matter (Lee et al., 2020). These sensors are used to give assessments of soil fertility and microbial metabolic activity early warnings of soil being degraded. Biological sensors are used to detect certain biological markers, such as exudates from plant roots, metabolites produced by microorganisms, or plant stress hormones. Wearable biosensors that are implanted in the roots or near the root (probes) can directly read plant-soil interactions (Bharti et al., 2024).

Redox potential sensors help assess soil aeration condition, and waterlogged effect as well as anaerobic microbial activities like denitrification or methane production that play a major role in rice fields and in wetland agriculture. Microbial activity indicators such as CO2 flux, oxygen consumption or bio electrochemical indicators from microbial fuel cells are examples of emerging indicators of soil ecological functioning (Zhang & Furman, 2021). IoT-enabled bio electrochemical sensors convert microbial activity into electrical outputs which can be measured thereby providing an interesting solution to the real-time assessment of soil health. These advanced sensors help to better understand biological processes occurring in the soils and support nature based agricultural practices by monitoring life in soils (Bharti et al., 2024).

3.3. Spatial Variability Mapping Using Geo-statistics and Proximal Sensing Tools

Soil properties have inherent variability across the landscape due to variation in texture, topography, organic matter content and past land use. IoT systems used in combination with geostatistical tools such as kriging, interpolation models and digital soil mapping algorithms allow high resolution maps of soil variability (Álvarez-Herrera et al., 2025). Proximal sensing technologies, such as electromagnetic induction (EMI), ground-penetrating radar (GPR), multi-spectral and hyper-spectral imaging and portable x-ray fluorescence (XRF) increase spatial mapping by scanning fields in a non-invasive method. When combined with GPS tagged IoT sensors and remote sensing data, these tools produce accurate soil attribute maps for variable rate irrigation, fertilization and tillage. The integration of spatial data allows for site specific management, allowing each zone of the field to receive the appropriate management intervention that will maximize productivity and efficient use of resources (Pathirana et al., 2023).

3.4. Sensor Calibration, Placement Strategies, Redundancy, and Long-Term Performance

Accurate monitoring of the soil is based on correct calibration and installation of sensors. Calibration ensures that for measurement of soil properties (-reading from the sensor) relate to laboratory-measured soil properties under varying environment conditions. Soil texture, temperature, salinity and bulk density affect the sensor performance and require some routine adjustments. Sensor placement strategies are also very important (Satoh & Kakiuchi, 2021). Soil sensors should be placed at representative depths and locations - normally within root depths where soil moisture and nutrient changes are most profound. Vertical sensor arrays allow acquisition of multi depth measurements providing capture of infiltration pattern and leaching of nutrients. Sensor redundancy - it installs multiple sensors per site to detect failures and address anomalies and data integrity decreases, especially in difficult field conditions (Nimmo et al., 2025). Redundant networks increase reliability and allow for constant monitoring even in the face of disturbance from the environment or interference by wildlife or hardware degradation. Long term performance is required in terms of strong, weather resistant casings, antifouling coatings and low power operating modes. Solar power and energy harvesting technologies significantly increase operational life while self-diagnostic capabilities help to detect faults and perform predictive maintenance (Kull et al., 2025).

4. IOT-DRIVEN CROP MONITORING AND STRESS DETECTION

Effective crop monitoring has become a key part in maintaining agricultural productivity, reducing stress from climate change, as well as precision interventions. Conventional scouting techniques such as manual field visits and visual inspection as well as periodic sampling are often subjective, slow and limited in temporal resolution (Chattopadhyay et al., 2024). IoT empowered crop monitoring systems overcome these limitations by obtaining continuous, objective, high frequency information related to plant health, growth dynamics and signals that plants are under stress. By combining smart optical sensors, biosensors, microclimate probes, and automated analytics, the technologies of IoT transform crops into dynamic datasets which can tell us about a physiological change way before it is visible to the human eye (Hosny et al., 2025). Figure 3 explains an IoT-driven crop monitoring framework in which smart sensors enable real-time detection of environmental stress, pest and disease incidence, soil and nutrient status, and plant production dynamics through integrated data analytics.

Figure 3. Overview of an IoT-based crop monitoring framework integrating smart sensors for environmental, soil, and plant health monitoring with data analytics to support precision agriculture

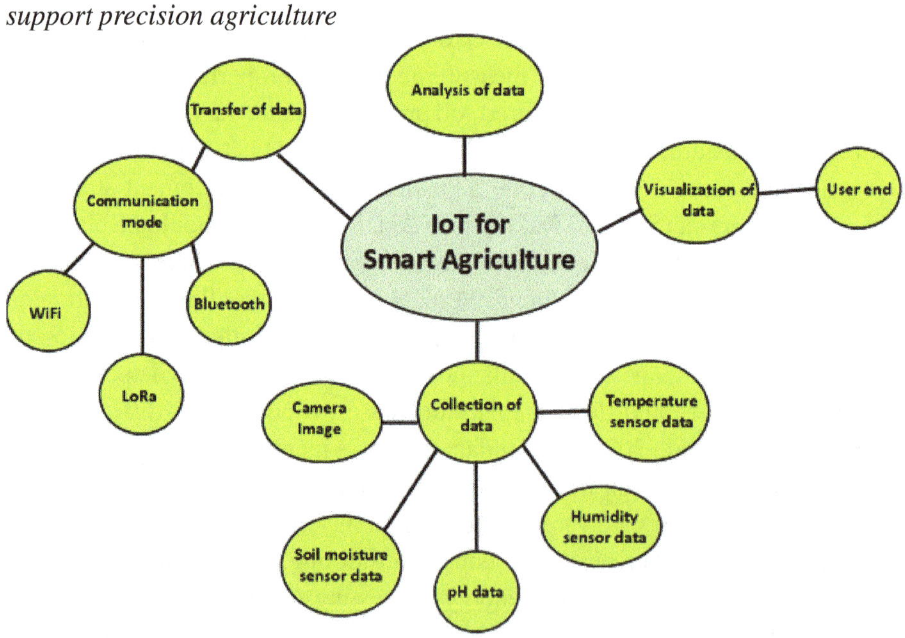

4.1. Multispectral, Hyperspectral, and Thermal Sensing for Plant Health Assessment

The use of optical remote sensing is one of the building blocks in the use of IoT for crop monitoring. Multispectral sensors, installed on field nodes, drones or satellites, acquire reflectance at specific wavelengths on sensors, i.e., red, near infrared (NIR) and red edge. These measurements are used to calculate vegetation indices like NDVI, SAVI, EVI and NDRE, which measures the vigor of plants, their chlorophyll content, accumulation of biomass and canopy coverage. Such indices provide the way to identify early the nutrient deficiencies, moisture stress and thinning of the canopy (Omia et al., 2023).

Hyperspectral sensors provide higher spectral resolution of hundreds of narrow bands. This capability enables the detection of certain subtle changes in biochemical and structural properties of crops such as pigment changes, water content changes, early disease onset, and stress-induced metabolic changes. When used with machine learning models, hyperspectral IoT devices can classify stress types and severities at a high level of precision (Shu et al., 2025).

Thermal imaging sensors measure the variation in the canopy temperature and give a direct indication for the transpiration rates and stomatal conductance. Turning off the pagination. Closing stomata, the plants under drought stress end up higher temperature for leaves. IoT connected thermal cameras produce real-time heat maps used to plan an irrigation schedule and predict drought and measure stress at the canopy level (Vieira & Ferrarezi, 2021). These sensors are also useful for detecting heat stress in conditions of extreme weather, especially in heat sensitive crops such as wheat, maize and cotton. The combination of both optical and thermal sensing provides a firm base for multi-dimensional plant health analysis that will allow growers to make evidence-based decisions during critical stages of growth (Pineda et al., 2021).

4.2. Canopy Temperature, Chlorophyll Fluorescence, Leaf Wetness, and Sap-Flow Systems

IoT based physiological sensing offers more direct information about the functioning of the plant. Thermal imaging systems used in conjunction with canopy temperature sensors measure micro - level temperature fluctuations in the crop canopy. Such information helps in identifying drought stress, the rate of evapotranspiration and irrigation needs (Dong et al., 2024).

Chlorophyll fluorescence sensors are quite sensitive instruments that measure photosystem II efficiency for advanced signs of stress in plants well before recognizable symptoms appear. IoT enabled fluorescence monitoring - photoinhibition, nutrient imbalances and environmental stressors like excessive radiation, or heat, can be detected. These sensors are finding more use in high value crops, such as vegetables and fruits and greenhouse systems (Arief et al., 2023).

Leaf wetness sensors track the accumulation of water on the surface of leaves which is a key parameter in predicting the outbreak of fungal disease. Many pathogens rust, mildews and blights for example - need certain leaf wetness periods to infect. Continuous leaf wetness data gathered from IoT networks aid fungicide usage to be optimized and avoid unnecessary chemical usage (Peddicord et al., 2025).

Sap-flow sensors measure water flow inside the plant stem and reflect water uptake rate by the plant, the transpiration rate, and water hydraulic functioning inside the plant. These sensors provide real time information about plant-soil-atmosphere interactions to aid in the fine-tuning irrigation scheduling and early warning of millions of people about drought stress. By acquiring physiological, biochemical and environmental signals, the physiological and environmental parameters measured by these IoT-enable sensing systems give a holistic picture of the performance and adaptive response of the crop under stress (Capurro et al., 2024).

4.3. Early Detection of Pests, Diseases, and Abiotic Stress Using Integrated Sensors

Early detection of stress is vital in keeping losses in yield at bay. IoT systems improve early warning capabilities by combining different types of sensors into decision supportive processes. Optical sensors are used to detect color changes such as lesions or anomalies in the canopy by diseases like bacterial spots, viral chlorosis or fungal wilt (Al-Atawi et al., 2023). Coupled with tools of machine vision and AI classifiers, IoT systems are now able to spot disease symptoms at incipient stages, hence providing timely treatment. Environmental sensors - including humidity, temperature and leaf wetness sensors - are used to calculate disease-risk indices of pathogens such as downy mildew, late blight or powdery mildew. Predictive models embedded on IoT platforms raise alerts for infection threshold not being met in time (Gawande et al., 2024).

In case of pest detection, IoT-enabled smart pheromone traps equipped with a camera is able to count the number of insects automatically, identify species, and track the dynamic of the population. Thereby, these systems decrease the need for manual trap inspecting and facilitate the strategies of integrated pest management. Abiotic stress (thermal waves, frost, salinity or nutrient imbalances) is identified using combinations of thermal data, spectral data, soil sensors readings and microclimate data (Santaera et al., 2025). IoT-enabled frost alert systems forecast the critical temperature drop and help growers in triggering protective mechanisms like irrigation/wind machines. Through early detection, IoT systems significantly decrease the crop losses and help to save the inputs and improve the time of the intervention (Choudhary et al., 2025a).

4.4. Data Fusion: Integrating Soil, Crop, and Weather Signals for Accurate Decisions

The real value of IoT enabled agriculture lies in its data fusion - in the integration of various data streams into useable and context-aware insights. Combining parameters of the soil (moisture, nutrients, temperature), crop physiology (NDVI, canopy temperature, fluorescence), and weather data (humidity, solar radiation and winds) creates a full picture of the plant performance (Dhanaraju et al., 2022).

Machine-like learning models based on fused datasets increase the precision of predictions relating to irrigation demand, nutrient deficiency, probability of a disease outbreak and growth stage changes. Spatial data fusion, utilizing the concept of proximal sensing, use of drones and in-the-field IoT nodes send precision to micro- and macro-scales (He et al., 2025).

Advanced platforms combine the capabilities of real-time analytics with those of automated autonomous controls (i.e., smart irrigation valves, fertigation pumps or climate control systems) forming the foundation of closed loop agricultural systems that independently respond to the needs of the plants. Data fusion helps in converting the isolated signals to coherent perceptions which result in holistic farm management optimizing productivity, minimizing the risk and also for sustainable use of resources (Adamo et al., 2025).

5. DATA ANALYTICS, AI, AND SMART DECISION SYSTEMS

The rise in real-time data coming from hundreds of thousands of soils, crops, environmental and machinery sensors in agriculture has sparked unprecedented levels of data and the Internet of Things (IoT). Nevertheless, the intrinsic value of IoT goes far beyond the acquisition of data because it lies within the complex ecosystem of data analytics that turns raw sensor data into meaningful actionable intelligence (Miller et al., 2025a). Data analytics, machine learning (ML), deep learning (DL) and intelligent decision systems make up the interpretive layer which takes the continuous streams of the sensors and turns them into predictive models, risk assessments and optimized guidance. This chapter helps to understand the analytical backbone of IoT enabled agriculture which emphasizes the application of insights that powered by artificial intelligence helps in ensuring precision agriculture, efficient use of resources, and sustainable farming of crops (Taye, 2023).

5.1. Machine Learning and Deep Learning for Agricultural Intelligence

Machine learning and deep learning have taken over as the analytical core of IoT enabled agriculture at the transmutation level of continuous sensor streams into actionable insights. In the field of irrigation forecasting, machine learning algorithms like random forests, neural networks, and the long short-term memory (LSTM) architecture use data on soil moisture patterns, temperature fluctuations of the plant's canopy, estimates of evapotranspiration and short-term weather forecasts to determine the best time and volume of water application (Miller et al., 2025a). These models significantly reduce waste, preventing water stress and yield loss. Similarly, nutrient scheduling is enhanced with the aid of AI - powered models that help combine soil nutrient profiles, plant spectral signatures and crop growth stage prediction to provide precise fertilizer recommendations. Deep learning systems that are trained on images of leaves can be used to identify early signs of nutrient deficiencies, while regression-based ML systems can be used to predict amounts of nutrients needed at good levels of accuracy. In the case of anomaly detection, abnormal patterns resulting from irrigation failure, soil degradation, pest emergence or sensor malfunction are detected by AI models (Mushtaq et al., 2025b). Autoencoders and clustering algorithms help to discover deviations from normative behavior and problems in the field could be identified early on. AI-driven disease forecasting combines imaging-based recognition, microclimate data and canopy reflectance data to observe the early indicators of disease and predict an outbreak condition. These intelligent systems move management from being reactive to that of being predictive and highly targeted decision-making (Alzakari et al., 2024).

5.2. Edge Computing and TinyML for Low-Latency Intelligence

The inception of edge computing and TinyML is changing the data processing locales in an ecosystem of digital farming. Traditional cloud-based systems are often burdened with latency issues, connectivity challenges, and huge amounts of data transmission requirements, and especially in remote agricultural areas. Edge computing caters for bringing processing capability closer to the field by allowing for the local analysis of data by IoT gateways, field controllers, and embedded devices instead of depending on cloud resources alone (Gookyi et al., 2024). This guarantees real-time responsiveness to critical events such as sudden temperature changes, rapid changes in soil moisture or disease triggering weather conditions. TinyML is an extension of this idea that involves running ultra-efficient machine learning models directly onto highly efficient, microcontrollers that run on low power. With TinyML, independent processing on the field deployed devices is possible, where leaf imagery can independently be analyzed, the presence of pests and anomalies can be identified, or control commands sent for irrigation, without continuous access to a network (Nsoh et al., 2024). These technologies have given farms the power to act autonomously to make instantaneous and reliable decisions, regardless of network coverage or bandwidth limitations. Consequently, edge and TinyML help to make agriculture more resilient, save money in communication costs and creates a more robust and decentralized intelligence architecture (Gong et al., 2025).

5.3. Data Quality, Preprocessing, Interoperability, and Visualization

Secure decision making requires quality decision data; therefore, data preprocessing and standardization are key parts of IoT driven agriculture. Sensor outputs are often noisy, incomplete or inconsistent because of interference in the environment, sensor calibration drift and variability in sensor hardware. Preprocessing techniques such as smoothing, removing outliers, recalibrate, normalization and aligning timestamps ensure only accurate and meaningful data gets into analytical models (Desikan et al., 2025). However, it is not sufficient to have data quality alone, it must be interoperable. The heterogeneity of the IoT devices and communication protocols used in agriculture usually leads to scattered data streams. Standards like MQTT, LoRaWAN, ISO 11783 and the OGC Sensor Things API serve to fix those gaps allowing devices from different manufacturers to talk to each other without any problem. After data is cleansed and harmonized, data visualization platforms turn complex data it analyzes into digestible information through dashboards, maps, charts and digital twins. These interfaces allow farmers to see real-time soil moisture or to know the pattern of stress in crops, visualize crop nutrient variability as well as see decisively about the risk of disease. Thus, visualization tools are an important link between advanced analytics and field-level decision making (Roccatello et al., 2025).

5.4. Explainable AI and Trustworthy Decision Frameworks

As AI becomes more ingrained in agricultural decision making, farmers want transparency on how they are made, and what is used to generate the prediction and recommendations. Explainable AI (XAI) helps in this regard by providing legible, explainable explanations on the model behavior. Techniques like SHAP values, LIME, feature importance rankings and attention mapping reveal which variables were involved - fluctuation of humidity, spikes of temperature or change of spectra - in a particular irrigation or disease alert (Akkem et al., 2025). Such insights help to build confidence in the user and help farmers to validate the outputs of the AI against their experiential knowledge. Beyond transparency, decision frameworks built to be trustworthy include uncertainty quantification, risk scoring and human-in-the-loop oversight to ensure decisions by the automated systems are consistent with the real-world conditions. Farmers keep control and get the benefit of AI-driven guidance through which they can make an informed judgment and are not blind to recommendations. By creating more interpretable and trustworthy AI, responsible AI makes a broader adoption of digital tools possible and creates the basis for responsible, ethical and user-centric digital agriculture (Abi, 2025).

Table 1. IoT applications, technologies, and benefits in modern agriculture

Application Area	IoT Technologies & Devices	Monitored Parameters	Primary Benefits	References
Soil Monitoring	Soil moisture sensors, pH probes, nutrient sensors, redox/ microbial sensors	Soil moisture, pH, nitrogen, phosphorus, potassium, microbial activity	Optimized irrigation & fertilization, enhanced soil health, early stress detection	(Rajak et al., 2023)
Crop Health & Stress Detection	Multispectral/hyperspectral cameras, thermal sensors, chlorophyll fluorescence devices, sap-flow sensors	Leaf temperature, chlorophyll content, leaf wetness, canopy stress indicators	Early detection of drought, nutrient deficiency, pests, and diseases; improved yield & quality	(Berger et al., 2022)
Irrigation & Fertigation Automation	Smart valves, pumps, IoT controllers, variable-rate fertigation systems	Soil moisture, crop water demand, nutrient status	Reduced water & fertilizer usage, precise input application, energy efficiency	(Adamo et al., 2025)
Precision Machinery & Robotics	Autonomous tractors, drones, robotic sprayers, GPS-guided machinery	GPS location, crop growth stage, environmental conditions	Labor reduction, accurate planting/ spraying, real-time field management	(Padhiary et al., 2024)
Controlled Environment Agriculture	Greenhouse sensors, climate controllers, hydroponic monitoring systems	Temperature, humidity, CO_2, light intensity, nutrient solution levels	Consistent crop quality, resource efficiency, year-round production	(Kumar et al., 2024a)
Emerging IoT Innovations	Biosensors, pathogen surveillance, digital twins, federated sensing networks	Plant metabolites, pathogen presence, multi-parameter environmental data	Proactive disease/pest management, predictive analytics, resilient farm operations	(Gund et al., 2025)
Data Analytics & Decision Support	Cloud/edge computing, AI/ ML algorithms, visualization dashboards	Integrated soil, crop, and weather data	Predictive irrigation, nutrient scheduling, anomaly detection, explainable AI recommendations	(Sharafat et al., 2025)

6. FROM SENSING TO ACTION: AUTOMATION AND SMART CONTROL SYSTEMS

6.1. Automated Irrigation, Fertigation, and Precision Machinery

The intrinsic value of Internet of Things (IoT) technologies in precision agriculture can be seen when data from sensors are used in real time and used to trigger an automated, intelligent intervention. Automated irrigation and fertigation systems are the ultimate example of the most important uses of IoT - driven control. Soil moisture and nutrient concentrations and other ambient environmental parameters are monitored continuously using sensor networks (which are distributed across the area) while dedicated controllers determine the optimum timing and volume of water or fertilizer application (Shahab et al., 2025). Variable-rate irrigation and fertigation technologies allow for spatially precise management that will ensure that the inputs used in each sub-zone of a field are the exact requirements or the capacity to minimize over-application, runoff and resource wastage. Beyond water and nutrient management, IoT guided precision machinery and robotics are used for anything from autonomous planting, seeding, targeted pesticide application, pruning and crop monitoring. By combining sensory inputs with a mix of GPS (global positioning systems), machine vision, and artificial intelligence decision making, autonomous tractors, drones, and robotic sprayers perform functions with high levels of accuracy, resulting in fewer manual labor and errors and an overall more efficient operation of the field (Kumar et al., 2025).

6.2. Reliability, Fail-Safes, and Energy-Optimized Actuation

For automation to be effective in a real-world agricultural setting then reliability and energy optimization are of prime consideration. Redundant sensors, backup control units and fault-tolerant software means that operations are not halted in the event of failures or malfunctions. Energy efficient actuators, low power motor controllers, and solar powered pumps help to increase operational life especially in remote or off-grid situations (Altammar, 2023). Real-time monitoring of the performance of actuators does enable predictive maintenance and optimization of the system and hence helps in preserving the robustness under dynamic field conditions. Fail-parts and emergency control mechanisms stop things either being damaged too much or equipment causing damage, both in terms of safety and productivity. By enabling real time data by translating, precise and automated interventions, IoT enabled control systems can close the loop between sensing, decision making, and action to help enable sustainable, resource efficient and high precision farming practices (Djidrov, 2024).

7. REAL-WORLD APPLICATIONS AND CASE STUDIES

7.1. Large-Scale Precision Farms

Large-scale precision farms are an excellent example of the transformational capability of the Internet of Things (IoT) in modern-day agricultural practice. In commercial solutions covering hundreds or even thousands of hectares, IoT platforms are used in which the parameters of scenes are monitored with soil moisture sensors, nutrient probes, weather stations, and imaging arrays from the air, in a continuous way. Data-driven decision-making helps us to schedule irrigation with precision, use variable rate of fertilizers, and exit pest control in time, which are resources for increased productivity (Choudhary et al., 2025b). Performance measures from these farms show considerable savings in water use, fertilizer use and fertilizer costs and at the same time yield consistency and crop quality. For example, precision farms in North America and Europe for maize and wheat boast of 20 - 30 per cent reduction in irrigation water usage and 10 - 15 per cent higher yield using optimized management via IoT. These results help to highlight the potential of large-scale deployment of IoT to make commercial agriculture a high efficiency, environmentally responsible business (Gamage et al., 2024).

7.2. Low-Cost IoT Solutions for Smallholder Farming Systems

Smallholder farmers, especially in developing regions, are often faced with challenges such as lack of capital, poor connectivity and technical knowledge etc. Low-cost IoT solutions have appeared to be overcoming these problems by providing scalable and cheap solutions. Some examples are for solar-powered soil moisture sensors, mobile-based irrigation alerts and low-cost weather stations that can be set up at very low cost (Choudhary et al., 2025a). These systems enable smallholders to implement precision irrigation and nutrient management strategies using equipment which need not be costly or involve fancy systems. Field studies in Asia, Africa, and Latin America have shown that the use of more efficient water by 15-25%, better survival of crops under drought conditions, and greater yields in even resource constrained situations are possible through developing low-cost IoT deployments. Through bridging the digital divide, these solutions make previously reserved practices for large commercial farms available to smallholders through data-driven practices (Monamodi et al., 2025).

7.3. Greenhouse and Controlled-Environment Deployments

IoT adoption in the greenhouses industry and controlled environment agriculture (CEA) is another area with great potential. Sensors for temperature, humidity, CO_2, light intensity, and nutrients solutions are combined with automatic actuators, to provide the right conditions for growth. Real-time monitoring and precise control allow for productivity, quality and energy efficiency improvement and decrease the need for manual intervention (Vargas et al., 2024). Hydroponic and vertical farming systems have relied on IoT tools to control circulation of water, nutrients and the environment to gain all year-round production cycles. Case studies in Europe and The Middle East demonstrate that IoT-enabled greenhouses can help increase the yield of crops by 30-40%, reduction of fertilizer by up to 20%, and ensure constant quality of produce, thus proving IoT to be scalable and effective in controlled settings (Thilakarathne et al., 2025).

8. FUTURE DIRECTIONS AND INNOVATION ROADMAP

The future of IoT enabled agriculture is promising and will be a paradigm of transformative innovation thanks to the technology behind them, such as sensor technology, artificial intelligence, and decision-making integrated systems. Next generation sensors have a lead position in this evolution (Miller et al., 2025a). Biodegradable and low-cost multi-parameter probes, which can simultaneously measure soil moisture, nutrient content, pH, temperature and plant metabolites, are being developed which reduces environmental impact and will produce high resolution data. Intelligent sensor networks with self-calibration, adaptive sampling and wireless communication offer the potential for continuous and autonomous sensing across a variety of cropping systems with improved reliability and scalability across large-scale commercial cropping systems or resource restricted smallholder cropping systems (Pitoro et al., 2025).

Artificial intelligence has become a major component of the operational intelligence of farms, compared to digital twins, predictive modeling and real-time simulations. Digital twins-which are virtual representations of entire farm ecosystems-incorporate soil, crop, climate, and management data to simulate growth scenarios, predict stress events, and optimize resource usage (Subeesh & Chauhan, 2025). AI-powered predictive monitoring helps detect water shortage, nutritional imbalance, pest infestations and disease outbreaks early on to intervene and take preventive actions, as opposed to reactive management. Coupled with cloud and edge computing, these tools can be used to provide low latency and actionable insights to improve resilience in the face of changing climatic conditions and to reduce input wastage (Kariyanna & Sowjanya, 2024).

Large scale deployment strategies will be critical to help spread the benefits of IoT to smallholder dominated areas, where financial constraints and lack of technical expertise and connectivity often create barriers to adoption. Scalable solutions such as low-cost modular sensors, mobile-enabled monitoring platforms, shared services of the IoT and localized training programs can help in the uptake without compromising precision nor reliability. Public-private partnerships and policy support are crucial as the foundation for setting up the required infrastructure, incentives and extension services for smallholders implementing IoT driven practices in a sustainable way (Massoud et al., 2026).

Strategic priorities of the researchers, industry, and policy makers include a focus on interoperability and standardization and ethical data governance to serve as a basis for widespread usability and trust. Researchers need to create sensors and AI models that are tailored to local agroecological conditions and industry stakeholders need to focus on scalable and affordable deployment and service models (Hussein et al., 2025). Policymakers have a key role in regulating the use of data privacy, encouraging responsible management of devices over a lifetime, and subsidizing or incentivizing to reduce the barriers of entry. Cross-sector collaboration will help scale innovation, improve access disparities, and increase the impact on the societal, economic and environmental impact of IoT enabled agriculture (Benchis et al., 2025).

9. CONCLUSION

The integration of Internet or IoT, of technologies into agriculture use is a paradigm shift in the care of crops and soil and provides unprecedented potential of precision, efficiency and sustainability. By combining real-time sensing, data analytics, automation, and AI-released decision-making, IoT brings together continuous amounts of soil, crop and environmental data for actionable insights, including enabling early detection of abiotic and biotic stresses, and enabling optimized irrigation, nutrient management and pest control. Advanced sensing modalities, such as multispectral, hyperspectral, thermal and biochemical probes, predictive models and data fusion combine to improve the accuracy of field level decision making while reducing waste of resources. Automation and intelligent control systems translate these insights into precise interventions in terms of improving efficiency of inputs, reducing labor demands and mitigating environmental impact. Virtual deployments from commercial to smallholder, and from large to smallholder solutions, and controlled environment agriculture, as well as emerging technology from biosensors, pathogen surveillance and federated sensing networks further expand the potential from IoT-enabled agriculture. Successful adoption, however, requires more than mere technological sophistication and should include consideration of the socio-economic, environmental and ethical dimensions, such as affordability, inclusivity, gender equity, e-waste management, data privacy and transparent decision-makers using artificial intelligence. Looking to the future, next generation innovations that are particularly biodegradable, multi-parameter sensors, AI-enabled digital twins, predictive farm simulation and creation of scalable deployment have set the course for resilient data-driven agriculture. By balancing the development of technology with sustainability, ethics, and unequal access, IoT enabled farming can change agricultural systems around the world, raise food security, optimizing resources, and foster ethical and ecologically responsible high efficiency food production that can address the challenges we face in our rapidly changing world.

REFERENCES

Abi, R. (2025). Ethical and Explainable AI in Data Science for Transparent Decision-Making Across Critical Business Operations. *International Journal of Research Publication and Reviews*, *6*(6), 50–72. DOI: 10.55248/GENGPI.6.0625.2126

Adamo, T., Caivano, D., Colizzi, L., Dimauro, G., & Guerriero, E. (2025). Optimization of irrigation and fertigation in smart agriculture: An IoT-based microservices framework. *Smart Agricultural Technology*, *11*, 100885. DOI: 10.1016/J.ATECH.2025.100885

Akkem, Y., Biswas, S. K., & Varanasi, A. (2025). Role of Explainable AI in Crop Recommendation Technique of Smart Farming. *International Journal of Intelligent Systems and Applications*, *17*(1), 31–52. DOI: 10.5815/IJISA.2025.01.03

Al-Atawi, A. A., Alyahyan, S., Alatawi, M. N., Sadad, T., Manzoor, T., Farooq-I-Azam, M., & Khan, Z. H. (2023). Stress Monitoring Using Machine Learning, IoT and Wearable Sensors. *Sensors (Basel)*, *23*(21), 8875. DOI: 10.3390/S23218875

Alawode, A., Blessing, A. O., & Chiamaka, O. T. (2025a). Integrating IoT and AI in Sustainable Agriculture to Mitigate Environmental Risk and Financial Misuse. *International Journal of Research Publication and Reviews*, *6*(6), 2810–2828. DOI: 10.55248/GENGPI.6.0625.2187

Ali, A., Hussain, T., Tantashutikun, N., Hussain, N., & Cocetta, G. (2023). Application of Smart Techniques, Internet of Things and Data Mining for Resource Use Efficient and Sustainable Crop Production. *Agriculture 2023, Vol. 13, 13*(2). DOI: 10.3390/AGRICULTURE13020397

Altammar, K. A. (2023). A review on nanoparticles: Characteristics, synthesis, applications, and challenges. *Frontiers in Microbiology*, *14*, 1155622. DOI: 10.3389/FMICB.2023.1155622

Álvarez-Herrera, J. G., Jaime-Guerrero, M., & Fernández-Pérez, C. J. (2025). Spatial Variability and Geostatistical Modeling of Soil Physical Properties Under Eucalyptus globulus Plantations. *Geomatics 2025, Vol. 5, 5*(3). DOI: 10.3390/GEOMATICS5030041

Alzakari, S. A., Aldrees, A., Umer, M., Cascone, L., Innab, N., & Ashraf, I. (2024). Artificial intelligence-driven predictive framework for early detection of still birth. *SLAS Technology*, *29*(6). Advance online publication. DOI: 10.1016/J.SLAST.2024.100203

Arief, M. A. A., Kim, H., Kurniawan, H., Nugroho, A. P., Kim, T., & Cho, B. K. (2023). Chlorophyll Fluorescence Imaging for Early Detection of Drought and Heat Stress in Strawberry Plants. *Plants 2023, Vol. 12, 12*(6). DOI: 10.3390/PLANTS12061387

Babla, M., Katwal, U., Yong, M. T., Jahandari, S., Rahme, M., Chen, Z. H., & Tao, Z. (2022). Value-added products as soil conditioners for sustainable agriculture. *Resources, Conservation and Recycling, 178*. Advance online publication. DOI: 10.1016/j.resconrec.2021.106079

Benchis, M. P., Shahzad, K., & Dan, S. (2025). Comparative analysis of blockchain adoption in the public and private sectors. A technology-organization-environment (TOE) framework approach. *Journal of Innovation & Knowledge, 10*(4), 100746. DOI: 10.1016/J.JIK.2025.100746

Berger, K., Machwitz, M., Kycko, M., Kefauver, S. C., Van Wittenberghe, S., Gerhards, M., Verrelst, J., Atzberger, C., van der Tol, C., Damm, A., Rascher, U., Herrmann, I., Paz, V. S., Fahrner, S., Pieruschka, R., Prikaziuk, E., Buchaillot, M. L., Halabuk, A., Celesti, M., & Schlerf, M. (2022). Multi-sensor spectral synergies for crop stress detection and monitoring in the optical domain: A review. *Remote Sensing of Environment, 280*, 113198. DOI: 10.1016/J.RSE.2022.113198

Bharti, A., Jain, U., & Chauhan, N. (2024). From lab to field: Nano-biosensors for real-time plant nutrient tracking. *Plant Nano Biology, 9*, 100079. DOI: 10.1016/J.PLANA.2024.100079

Binu Mol, T. V. (2024). A Study on Raspberry PI, Arduino and IoT with Its Application in E-Controlling UPS. *Journal of Electrical Systems, 20*(2), 2785–2791. DOI: 10.52783/JES.6266

Capurro, M. C., Ham, J. M., Kluitenberg, G. J., Comas, L., & Andales, A. A. (2024). A novel sap flow system to measure maize transpiration using a heat pulse method. *Agricultural Water Management, 301*, 108963. DOI: 10.1016/J.AGWAT.2024.108963

Chattopadhyay, D., & Balachandra, Y., P, A., Pagire, K. S., Sahu, T. K., Panotra, N., Manisankar, G., Charitha, K., & Singh, B. veer. (2024). Precision Agriculture Technologies for Early Detection of Crop Pests and Diseases. *Uttar Pradesh Journal of Zoology, 45*(20), 328–342. DOI: 10.56557/UPJOZ/2024/V45I204588

Chaudhari, S. K., Patra, A., Dey, P., Bal, S. K., Gorantiwar, S., & Parsad, R. (2022). Sensor based Monitoring for Improving Agricultural Productivity and Sustainability- A Review. *Journal of the Indian Society of Soil Science, 70*(2), 121–141. DOI: 10.5958/0974-0228.2022.00013.5

Chen, J., Paw, U. K. T., North, M., & Franklin, J. F. (2024). The contributions of microclimatic information in advancing ecosystem science. *Agricultural and Forest Meteorology*, *355*, 110105. DOI: 10.1016/J.AGRFORMET.2024.110105

Choudhary, A. (2024). Internet of Things: a comprehensive overview, architectures, applications, simulation tools, challenges and future directions. *Discover Internet of Things 2024 4:1*, *4*(1), 31-. DOI: 10.1007/S43926-024-00084-3

Choudhary, V., Guha, P., Pau, G., & Mishra, S. (2025a). An overview of smart agriculture using internet of things (IoT) and web services. *Environmental and Sustainability Indicators*, *26*, 100607. DOI: 10.1016/J.INDIC.2025.100607

Choudhary, V., Guha, P., Pau, G., & Mishra, S. (2025b). An overview of smart agriculture using internet of things (IoT) and web services. *Environmental and Sustainability Indicators*, *26*. Advance online publication. DOI: 10.1016/J.INDIC.2025.100607

Desikan, J., Singh, S. K., Jayanthiladevi, A., Bhushan, S., Rishiwal, V., & Kumar, M. (2025). Hybrid Machine Learning-Based Fault-Tolerant Sensor Data Fusion and Anomaly Detection for Fire Risk Mitigation in IIoT Environment. *Sensors 2025, Vol. 25*, *25*(7). DOI: 10.3390/S25072146

Dhanaraju, M., Chenniappan, P., Ramalingam, K., Pazhanivelan, S., Kaliaperumal, R., Dhanaraju, M., Chenniappan, P., Ramalingam, K., Pazhanivelan, S., & Kaliaperumal, R. (2022). Smart Farming: Internet of Things (IoT)-Based Sustainable Agriculture. *Agriculture 2022, Vol. 12*, *12*(10). DOI: 10.3390/AGRICULTURE12101745

Djidrov, M. Z. (2024). Challenges in Damage Detection for Real-time Monitoring and Predictive Maintenance Methodology. *SAR Journal - Science and Research*, 310–316. DOI: 10.18421/SAR74-04

Dong, Y., Sloan, G., & Chappuies, J. (2024). Open-source time-lapse thermal imaging camera for canopy temperature monitoring. *Smart Agricultural Technology*, *7*, 100430. DOI: 10.1016/J.ATECH.2024.100430

Duguma, A. L., & Bai, X. (2025). How the internet of things technology improves agricultural efficiency. *Artificial Intelligence Review*, *58*(2). Advance online publication. DOI: 10.1007/S10462-024-11046-0

Eltarabily, M. G., Amer, A., Farzamian, M., Bouksila, F., Elkiki, M., & Selim, T. (2024). Time-Lapse Electromagnetic Conductivity Imaging for Soil Salinity Monitoring in Salt-Affected Agricultural Regions. *Land 2024, Vol. 13*, *13*(2). DOI: 10.3390/LAND13020225

Gamage, A., Gangahagedara, R., Subasinghe, S., Gamage, J., Guruge, C., Senaratne, S., Randika, T., Rathnayake, C., Hameed, Z., Madhujith, T., & Merah, O. (2024). Advancing sustainability: The impact of emerging technologies in agriculture. *Current Plant Biology*, *40*, 100420. DOI: 10.1016/J.CPB.2024.100420

Gargaro, M., Murphy, R. J., & Harris, Z. M. (2023). Let-us investigate; A meta-analysis of influencing factors on lettuce crop yields within controlled-environment agriculture systems. *Plants*, *12*(14), 2623.

Gawande, A., Sherekar, S., & Gawande, R. (2024). Early prediction of grape disease attack using a hybrid classifier in association with IoT sensors. *Heliyon*, *10*(19), e38093. DOI: 10.1016/J.HELIYON.2024.E38093

Gong, R., Zhang, H., Li, G., & He, J. (2025). Edge Computing-Enabled Smart Agriculture: Technical Architectures, Practical Evolution, and Bottleneck Breakthroughs. *Sensors 2025, Vol. 25, 25*(17). DOI: 10.3390/S25175302

Gookyi, D. A. N., Wulnye, F. A., Arthur, E. A. E., Ahiadormey, R. K., Agyemang, J. O., Agyekum, K. O. B. O., & Gyaang, R. (2024). TinyML for smart agriculture: Comparative analysis of TinyML platforms and practical deployment for maize leaf disease identification. *Smart Agricultural Technology*, *8*, 100490. DOI: 10.1016/J.ATECH.2024.100490

Gund, R., Badgujar, C. M., Samiappan, S., & Jagadamma, S. (2025). Application of Digital Twin Technology in Smart Agriculture: A Bibliometric Review. *Agriculture 2025, Vol. 15, 15*(17). DOI: 10.3390/AGRICULTURE15171799

He, T., Li, M., & Jin, D. (2025). Deep learning-based time series prediction for precision field crop protection. *Frontiers in Plant Science*, *16*, 1575796. DOI: 10.3389/FPLS.2025.1575796/BIBTEX

Hosny, K. M., El-Hady, W. M., & Samy, F. M. (2025). Technologies, Protocols, and applications of Internet of Things in greenhouse Farming: A survey of recent advances. *Information Processing in Agriculture*, *12*(1), 91–111. DOI: 10.1016/J.INPA.2024.04.002

Hussein, J. B., Workneh, T. S., Kassim, A., Ntsowe, K., Melesse, S. F., & El-Mesery, H. S. (2025). A review on the impact of big data analytics in transforming agricultural practices, food processing, and preservation strategies. *Applied Food Research*, *5*(2), 101234. DOI: 10.1016/J.AFRES.2025.101234

Jafarbiglu, H., & Pourreza, A. (2022). A comprehensive review of remote sensing platforms, sensors, and applications in nut crops. *Computers and Electronics in Agriculture*, *197*, 106844. DOI: 10.1016/J.COMPAG.2022.106844

Kariyanna, B., & Sowjanya, M. (2024). Unravelling the use of artificial intelligence in management of insect pests. *Smart Agricultural Technology, 8*, 100517. DOI: 10.1016/J.ATECH.2024.100517

Karmakar, P., Teng, S. W., Murshed, M., Pang, S., Li, Y., & Lin, H. (2024). Crop monitoring by multimodal remote sensing: A review. *Remote Sensing Applications: Society and Environment, 33*, 101093. DOI: 10.1016/J.RSASE.2023.101093

Kull, K., Asad, B., Khan, M. A., Naseer, M. U., Kallaste, A., & Vaimann, T. (2025). Faults, Failures, Reliability, and Predictive Maintenance of Grid-Connected Solar Systems: A Comprehensive Review. *Applied Sciences 2025, Vol. 15, 15*(21). DOI: 10.3390/APP152111461

Kumar, A., Singh, S. R., Yadav, M. C., Yadav, V. K., Channi, A. S., Chouria, A., Kumar, N., Dall, H., & Nazir, T. (2025). The Role of Variable Rate Technology (VRT) in Modern Agriculture: A Review. *Journal of Advances in Biology & Biotechnology, 28*(6), 20–39. DOI: 10.9734/JABB/2025/V28I62370

Kumar, V., Sharma, K. V., Kedam, N., Patel, A., Kate, T. R., & Rathnayake, U. (2024a). A comprehensive review on smart and sustainable agriculture using IoT technologies. *Smart Agricultural Technology, 8*, 100487. DOI: 10.1016/J.ATECH.2024.100487

Laha, S. R., Pattanayak, B. K., Pattnaik, S., Mishra, D., Kumar Nayak, D. S., & Dash, B. B. (2023). An IOT-Based Soil Moisture Management System for Precision Agriculture: Real-Time Monitoring and Automated Irrigation Control. *Proceedings of the 4th International Conference on Smart Electronics and Communication, ICOSEC 2023*, 451–455. DOI: 10.1109/ICOSEC58147.2023.10276266

Lakhiar, I. A., Yan, H., Zhang, C., Wang, G., He, B., Hao, B., Han, Y., Wang, B., Bao, R., Syed, T. N., Chauhdary, J. N., Rakibuzzaman, Md., Lakhiar, I. A., Yan, H., Zhang, C., Wang, G., He, B., Hao, B., Han, Y., … Rakibuzzaman, Md. (2024). A Review of Precision Irrigation Water-Saving Technology under Changing Climate for Enhancing Water Use Efficiency, Crop Yield, and Environmental Footprints. *Agriculture 2024, Vol. 14, 14*(7). DOI: 10.3390/AGRICULTURE14071141

Lee, S. H., Kim, M. S., Kim, J. G., & Kim, S. O. (2020). Use of Soil Enzymes as Indicators for Contaminated Soil Monitoring and Sustainable Management. *Sustainability 2020, Vol. 12, 12*(19). DOI: 10.3390/SU12198209

Majigi, M. U., Idris, I., Abdulhamid, S. M., & Ikuesan, R. A. (2025). Big data transfer service architecture for cloud data centers: problems, methods, applications, and future trends. *Discover Computing, 28*(1), 163-. DOI: 10.1007/S10791-025-09682-3/TABLES/9

Massoud, M., Kineber, A., Elshaboury, N., Abunada, Z., Arashpour, M., Alatroush, M., Mostafa, S., & Alhusban, M. (2026). Identifying and assessing the internet of things implementation barriers for sustainable building projects: SEM-ANN approach. *KSCE Journal of Civil Engineering, 30*(1), 100310. DOI: 10.1016/J.KSCEJ.2025.100310

Meshram, A., Kadam, R., Shinde, G., & Khalge, M. (2024). *Sensor Based Agriculture and Horticulture Crop Production(Application of Sensors).* DOI: 10.20944/PREPRINTS202401.0544.V1

Miller, T., Mikiciuk, G., Durlik, I., Mikiciuk, M., Łobodzińska, A., Śnieg, M., Miller, T., Mikiciuk, G., Durlik, I., Mikiciuk, M., Łobodzińska, A., & Śnieg, M. (2025a). The IoT and AI in Agriculture: The Time Is Now—A Systematic Review of Smart Sensing Technologies. *Sensors 2025, Vol. 25, 25*(12). DOI: 10.3390/S25123583

Mishra, H., Gautam, S., Dwivedi, S., Raza, M. Y., Ali, H., & Ikram, S. (2025). *Evaluation of Climate Resilient Agriculture Practices in Subcontinent.* 225–278. DOI: 10.1007/978-3-032-00190-0_11

Monamodi, P., Ndoro, J. T., Matiwane, M. Ben, Monamodi, P., Ndoro, J. T., & Matiwane, M. Ben. (2025). The Impact of Innovative Irrigation System Use on Crop Yield Among Smallholder Farmers in Mbombela Local Municipality, South Africa. *Agriculture 2025, Vol. 15, 15*(16). DOI: 10.3390/AGRICULTURE15161755

Mushtaq, M. A., Ateeq, M., Ikram, M., Alam, S. M., Kaleem, M. M., Ashraf, M. A., Asim, M., Almutairi, K. F., Seleiman, M. F., & Shireen, F. (2025a). Securing fruit trees future: AI-driven early warning and predictive systems for abiotic stress in changing climate. *Plant Stress, 17*, 100953. DOI: 10.1016/J.STRESS.2025.100953

Mushtaq, M. A., Ateeq, M., Ikram, M., Alam, S. M., Kaleem, M. M., Ashraf, M. A., Asim, M., Almutairi, K. F., Seleiman, M. F., & Shireen, F. (2025b). Securing fruit trees future: AI-driven early warning and predictive systems for abiotic stress in changing climate. *Plant Stress, 17*, 100953. DOI: 10.1016/J.STRESS.2025.100953

Nimmo, J. R., Wiekenkamp, I., Araki, R., Groh, J., Singh, N. K., Crompton, O., Wyatt, B. M., Ajami, H., Giménez, D., Hirmas, D. R., Sullivan, P. L., & Sprenger, M. (2025). Identifying preferential flow from soil moisture time series: Review of methodologies. *Vadose Zone Journal, 24*(2), e70017. DOI: 10.1002/VZJ2.70017

Nsoh, B., Katimbo, A., Guo, H., Heeren, D. M., Nakabuye, H. N., Qiao, X., Ge, Y., Rudnick, D. R., Wanyama, J., Bwambale, E., & Kiraga, S. (2024). Internet of Things-Based Automated Solutions Utilizing Machine Learning for Smart and Real-Time Irrigation Management: A Review. *Sensors 2024, Vol. 24, 24*(23). DOI: 10.3390/S24237480

Omia, E., Bae, H., Park, E., Kim, M. S., Baek, I., Kabenge, I., & Cho, B. K. (2023). Remote Sensing in Field Crop Monitoring: A Comprehensive Review of Sensor Systems, Data Analyses and Recent Advances. *Remote Sensing 2023, Vol. 15, 15*(2). DOI: 10.3390/RS15020354

Padhiary, M., Saha, D., Kumar, R., Sethi, L. N., & Kumar, A. (2024). Enhancing precision agriculture: A comprehensive review of machine learning and AI vision applications in all-terrain vehicle for farm automation. *Smart Agricultural Technology, 8*, 100483. DOI: 10.1016/J.ATECH.2024.100483

Pathirana, S., Lambot, S., Krishnapillai, M., Cheema, M., Smeaton, C., & Galagedara, L. (2023). Ground-Penetrating Radar and Electromagnetic Induction: Challenges and Opportunities in Agriculture. *Remote Sensing, 15*(11). Advance online publication. DOI: 10.3390/RS15112932

Peddicord, L., Xavier, A., Cryer, S., Barr, J., Heijden, G. van der, Peddicord, L., Xavier, A., Cryer, S., Barr, J., & Heijden, G. van der. (2025). Scalable Prediction of Northern Corn Leaf Blight and Gray Leaf Spot Diseases to Predict Fungicide Spray Timing in Corn. *Agronomy 2025, Vol. 15, 15*(2). DOI: 10.3390/AGRONOMY15020328

Pineda, M., Barón, M., & Pérez-Bueno, M. L. (2021). Thermal imaging for plant stress detection and phenotyping. *Remote Sensing, 13*(1), 1–21. DOI: 10.3390/RS13010068

Pitoro, V. S. J., de Amorim, C. A., Franco, J. R., & Sánchez-Román, R. M. (2025). Application of low-cost soil moisture sensors for irrigation management in Brassica oleracea var. acephala cultivation. *Smart Agricultural Technology, 12*, 101596. DOI: 10.1016/J.ATECH.2025.101596

Rajak, P., Ganguly, A., Adhikary, S., & Bhattacharya, S. (2023). Internet of Things and smart sensors in agriculture: Scopes and challenges. *Journal of Agriculture and Food Research, 14*, 100776. DOI: 10.1016/J.JAFR.2023.100776

Rathor, A. S., Choudhury, S., Sharma, A., Nautiyal, P., & Shah, G. (2024). Empowering vertical farming through IoT and AI-Driven technologies: A comprehensive review. *Heliyon, 10*(15), e34998. DOI: 10.1016/J.HELIYON.2024.E34998

Reza, M. N., Lee, K. H., Karim, M. R., Haque, M. A., Bicamumakuba, E., Dey, P. K., Jang, Y. Y., & Chung, S. O. (2025). Trends of Soil and Solution Nutrient Sensing for Open Field and Hydroponic Cultivation in Facilitated Smart Agriculture. *Sensors 2025, Vol. 25, 25*(2). DOI: 10.3390/S25020453

Roccatello, E., Pagano, A., Levorato, N., & Rumor, M. (2025). State of the Art in Internet of Things Standards and Protocols for Precision Agriculture with an Approach to Semantic Interoperability. *Network 2025, Vol. 5, 5*(2). DOI: 10.3390/NETWORK5020014

Santaera, G., Zeni, V., Manduca, G., Canale, A., Mele, M., Benelli, G., Stefanini, C., & Romano, D. (2025). Development of an autonomous smart trap for precision monitoring of hematophagous flies on cattle. *Smart Agricultural Technology*, *10*, 100842. DOI: 10.1016/J.ATECH.2025.100842

Satoh, Y., & Kakiuchi, H. (2021). Calibration method to address influences of temperature and electrical conductivity for a low-cost soil water content sensor in the agricultural field. *Agricultural Water Management*, *255*. Advance online publication. DOI: 10.1016/J.AGWAT.2021.107015

Shahab, H., Naeem, M., Iqbal, M., Aqeel, M., & Ullah, S. S. (2025). IoT-driven smart agricultural technology for real-time soil and crop optimization. *Smart Agricultural Technology*, *10*, 100847. DOI: 10.1016/J.ATECH.2025.100847

Sharafat, S., Das Kabya, N., Emu, R. I., Ahmed, M. U., Onik, J. C., Islam, M. A., & Khan, R. (2025). An IoT-enabled AI system for real-time crop prediction using soil and weather data in precision agriculture. *Smart Agricultural Technology*, *12*, 101263. DOI: 10.1016/J.ATECH.2025.101263

Shu, M., Harfouche, A. L., Trtílek, M., Panzarová, K., Alasia, O. F., Lagergren, J. H., Labbé, A., Engle, N. L., Clark, M. M., Chen, J. G., Tuskan, G. A., & Tschaplinski, T. J. (2025). Leveraging hyperspectral phenotyping for accurate, non-destructive prediction of metabolite profiles in poplar under drought stress. *Environmental and Experimental Botany*, *237*, 106218. DOI: 10.1016/J.ENVEXPBOT.2025.106218

Subeesh, A., & Chauhan, N. (2025). Agricultural digital twin for smart farming: A review. *Green Technologies and Sustainability*, *100299*. Advance online publication. DOI: 10.1016/J.GRETS.2025.100299

Taye, M. M. (2023). Understanding of Machine Learning with Deep Learning: Architectures, Workflow, Applications and Future Directions. *Computers 2023, Vol. 12, 12*(5). DOI: 10.3390/COMPUTERS12050091

Thilakarathne, N. N., Abu Bakar, M. S., Abas, P. E., & Yassin, H. (2025). Internet of things enabled smart agriculture: Current status, latest advancements, challenges and countermeasures. *Heliyon*, *11*(3), e42136. DOI: 10.1016/J.HELIYON.2025.E42136

Vargas, C., Gamache, S., Henao, N., Agbossou, K., & Nagarsheth, S. (2024). A comprehensive cost mapping of digital technologies in greenhouses. *Smart Agricultural Technology*, *9*, 100634. DOI: 10.1016/J.ATECH.2024.100634

Vieira, G. H. S., & Ferrarezi, R. S. (2021). Use of Thermal Imaging to Assess Water Status in Citrus Plants in Greenhouses. *Horticulturae 2021, Vol. 7, 7*(8). DOI: 10.3390/HORTICULTURAE7080249

Zhang, Z., & Furman, A. (2021). Soil redox dynamics under dynamic hydrologic regimes - A review. *The Science of the Total Environment*, *763*. Advance online publication. DOI: 10.1016/J.SCITOTENV.2020.143026

Chapter 6
Transforming Irrigation Systems With IoT for Water Conservation and Efficiency

Amjid Khan
http://orcid.org/0009-0001-5475-5156
University of South Africa, South Africa

Malik Maaza
University of South Africa, South Africa

ABSTRACT

Efficient water management is crucial for sustainable agriculture amid climate variability and rising food demand. IoT-enabled irrigation enables precision farming through real-time monitoring, data analytics, and automation. This chapter reviews IoT applications, including system architecture, sensors, communication protocols, and data frameworks. Key sensors, soil moisture, weather, and plant response are discussed alongside low-cost and biodegradable options. AI, machine learning, and digital twins optimize water use, reduce waste, and improve yields. Emerging trends like AIoT, nano-sensors, and community IoT networks support climate-smart, resilient irrigation practices.

INTRODUCTION

Agriculture still is the single largest consumer of freshwater in the world, at the rate of almost seventy percent of withdrawals globally, putting immense pressure

DOI: 10.4018/979-8-3373-7257-0.ch006

on already strained water reserves (UNESCO, 2024). As climate change grows more rapid, many agricultural areas face water scarcity, erratic weather regimes, extended drought periods as well as declining groundwater tables. These adversities are compounded by the rapid increase in the world's population that requires an increase in food production from a shrinking land mass (Guiné, 2024). In this context, water efficiency cited goes from being a desirable characteristic to being an even more prerequisite to secure food security and maintain the ecological balance. Conventional irrigation practices - based on manual observation, regular schedule or intuition of the farmer - are not well equipped to respond effectively to the real time needs of crop water use (Bwambale, Abagale, & Anornu, 2022). Consequently, over-irrigation, waterlogging, nutrient leaching and excessive energy use are now deeply-rooted problems in many farming systems, at the expense of a resource efficient and good crop (Ingrao, Strippoli, Lagioia, & Huisingh, 2023).

Conventional irrigation approaches also do not consider temporal and spatial variations of soil characteristics, microclimatic conditions and plant water uptake dynamics. A field is rarely homogeneous: some areas may be water deficient, and others saturated and in this situation, uniform irrigation operations treat the field as a monolith (Kirkham, 2023c). Additionally, human error, inconsistent field monitoring, and the unavailability of precise soil moisture information are also common causes of lack of optimal decision-making. These limitations point out a critical need for advanced solutions that are able to continuously monitor the conditions of the field, read and interpret data collected, adjusting irrigation actions on-the-fly. The Internet of Things (IoT) has become a revolutionary technology with the ability to fill these gaps with interconnected sensor networks, automated control systems and analytics in the cloud (Choudhary, 2024).

IoT enabled irrigation systems use sensors for soil moisture, weather, track plant response and communication networks to provide real-time contextual insights that can help optimize irrigation schedules (Morchid et al., 2024). These systems help farmers use the right amount of water at the right time and place, improving water-use efficiency, minimizing water waste, helping crops grow stronger and also often saving money in operation costs. With the help of machine learning and data-driven models, IoT platforms can predict the irrigation needs, at the same time detect any anomalies and even turn on/off the pumps and valves automatically by using smart controllers (Waqas et al., 2025). The rising affordability of sensors, low-power wireless network developments, like LoRaWAN and NB- IoT and the availability of cloud-based farming tools make agricultural irrigation using IoT more accessible, even in developing regions. This technological transition is associated with both global goals of climate-smart agriculture and sustainable water management as well as environmentally responsible food production (García-Martín, Torralba, Hidalgo-Fort, Daza, & González-Carvajal, 2023).

This chapter is intended to give an overall picture of all aspects pertaining to IoT enabled irrigation systems and demonstrates their technological background, system architecture, sensor networks, communication protocols, and automaticity mechanisms. It synthesizes real world applications and challenges and opportunities for large scale implementation. The chapter also focuses on emerging innovations and provides concrete knowledge for researchers, policy makers and practitioners in the field of sustainable agricultural water management, based on data.

1. FOUNDATIONS OF IOT IN MODERN AGRICULTURE

1.1. Concepts and Components of IoT

The Internet of Things (IoT) refers to a collection of connected physical devices armed with sensors, communication functions and compute capability and therefore having the ability to collect, share and process data in an autonomous fashion with little or no human intervention (Kiran, 2019). In the context of agriculture, IoT extends the traditional paradigm of monitoring farms by introducing digital intelligence directly in the field environment. An IoT-based agricultural system is typically composed of four basic modules, namely: (1) sensing devices, which measure soil moisture, temperature, humidity and irrigation flow as well as crop stress, etc. (Kiran, 2019); (2) connecting infrastructure, which provides wireless communication via a Wi-Fi network, a LoRaWAN network, a ZigBee network, a NB-IoT network or a cellular network; (3) data processing units, in which raw sensor data is aggregated, analyzed and interpreted by clouds, edge devices or artificial intelligence algorithms (Krishnamurthi, Kumar, Gopinathan, Nayyar, & Qureshi, 2020); and (4) actuation mechanisms, which involve smart pumps, valves and controllers, which allow irrigation These components all work together to form a responsive data-driven irrigation ecosystem where decisions are based on actual conditions in real time as opposed to a preset (C. Wang, Zhao, Cheng, Ni, & Xiang, 2024). The modular nature of the IoT systems allows for customization in terms of different crop types, climatic conditions, and irrigation systems and therefore makes them suitable for either large-scale farm operations or smallholder settings.

1.2. Digital Agriculture Ecosystem

IoT is a building block in the large digital agriculture ecosystem, which includes the combination of cutting-edge technologies such as artificial intelligence (AI), remote sensing, unmanned aerial vehicles (drones), big data analytics, and decision support platforms (Ahvar, Ahvar, & Lee, 2024). Within this ecosystem IoT serves as the primary channel of data acquisition that continuously provides high resolution data, in real time, on conditions within the field back to digital platforms. Satellite imagery, as well as drone-based monitoring, add to the capabilities of the sensors operating at ground level to give multiple-scale insights into crop health, evapotranspiration and soil variability (Hafeez et al., 2023). Cloud platforms then perform synthesize of these heterogenous datasets to produce predictive goodnesses, optimization of resource-allocation and enabling evidence-based decision making. Mobile applications and farm management systems make these insights available to the farmers to help remotely monitor irrigation systems, receive alerts and adjust their operations using user-friendly interfaces. Consequently, the digital agriculture ecosystem is a perfect example of a seamless flow of information - from the field to the cloud and back - allowing an intelligent feedback loop for improved situational awareness, operational efficiency and sustainability. As digital infrastructure continues to spread worldwide, this integration seen for IoT is bringing about the shift from farming in the past to smart, connected and climate resilient farming in the future (Dong & Xu, 2025).

1.3. Benefits and Value Proposition for Irrigation Systems

IOT implementation in irrigation systems inspires revolutionary benefits that directly tackle water management challenges that have existed for long. First, real-time sensing allows farmers to water according to accurate soil and crop needs instead of fixed schedules so that water is not wasteful but instead water use is made 20-50% more efficient in many cases (Shahab, Naeem, Iqbal, Aqeel, & Ullah, 2025). Second, IoT facilitates spatially targeted irrigation so that each segment of a field will receive water based on its specific needs, thereby making the practice more uniform and improving the quality of the yield (Obaideen et al., 2022). Third, automated irrigation reduces the labor needs and eliminates human error, especially on large farms, or in areas with a poor availability of trained agricultural workers (OĞUztÜRk, 2025). Fourth, IoT systems help to increase energy efficiency, by preventing unnecessary pump operation and optimizing irrigation time in respect to the weather and the evapotranspiration models (Abdelmoneim et al., 2025). Additionally, these systems will give early warnings of leaks, blockages or equipment failures, as well as saving maintenance costs and crop stress. Overall, the value proposition of IoT in irrigation is based on the ability of IoT to make water management smart, efficient, adaptive and sustainable for economic benefits and environmental care.

2. ARCHITECTURE OF IOT-BASED IRRIGATION SYSTEMS

IoT-based irrigation systems are built on an architecture with multiple layers combined sensing, communication, data processing and automated decision-making. Each layer has a certain function that allows free flow of data from field to user and back to irrigation infrastructure (Kalpana et al., 2024). The layered architecture provides flexibility, scalability, and reliability to the system and thus makes IoT-based solutions applicable to a variety of agricultural settings, from small-scale farms to large commercial agricultural operations. The architecture is normally divided into four major layers: Sensor layer, Network layer, Cloud/Edge layer and User & Control layer. Collectively, these layers form an intelligent irrigation system that is capable of offering real-time monitoring, analytics-based decision making and automated water delivery.(Liu, Zhao, & Rezaeipanah, 2025). Sensor data and plant images are processed by the main controller via the cloud to intelligently actuate the relay-driven DC pump for precision drip irrigation as shown in *Figure 1*.

Figure 1. Architecture of an IoT-enabled smart irrigation system integrating environmental sensing, plant imaging, cloud connectivity, and automated pump control

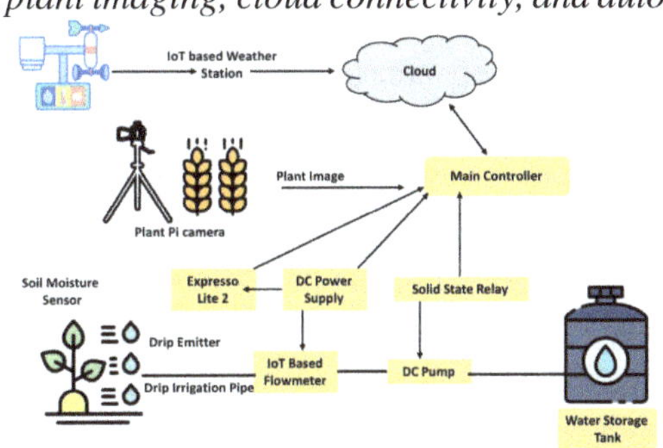

2.1. Sensor Layer: Field Data Acquisition

The sensor layer is the foundation of the IoT irrigation system and its job is to continuously and accurately gather the data from the field (García, Parra, Jimenez, Lloret, & Lorenz, 2020). This layer includes an array of sensors meant to collect important variables that affect irrigation decisions. Soil moisture sensors such as TDR, FDR, capacitive probes, and tensiometers gauge either water content or root zone water tension and give some thought to water availability at the root zone (X. Zhang, Feng, & Sun, 2024). Environmental sensors (e.g. air temperature, humidity, rainfall, wind speed, solar radiation) support the estimation of the evapotranspiration and the prediction of the irrigation needs. Plant - sequel beneath sensors, e.g., leaf wetness sensors, canopy temperature sensors, sap - flow meters provide exclusively direct assessments of water condition and stress reactions of crops. Additional sensors may include irrigation flow meters, pressure control and water quality probes. The accuracy and location of these sensors in the environment makes the decisions of irrigation safe. Sensor data is most often collected at a high temporal resolution and is transmitted either continuously or at set intervals to the network layer (Huynh, Tran, & Duong-Trung, 2023).

2.2. Network Layer: Connectivity and Communication

The network layer provides the ability to communicate between the sensors of the field, controllers, gateways, and cloud platforms. It acts as the important connecting link in transferring the data efficiently and reliably throughout the system. Depending on the size of the field, availability of infrastructure and power, various types of connectivity can be used. Short range communications technologies like Wi-Fi, ZigBee and Bluetooth Low Energy (BLE) are suitable for small or localized farms where there is dense sensor networks (Mastoi et al., 2022). Long range low power networks (LoRaWAN, NB-IoT, Sigfox, etc., and 4G/5G cellular networks) are preferred for large fields or remote areas where conventional networks are limited. These systems have a minimal power requirement for long fast communications, turning battery-operated sensors into long fast performing sensors for months or years. Data exchange protocols like MQTT, CoAP and HTTP are used for lightweight and energy-efficient transmission (Lalhriatpuii, Ruchi, & Wasson, 2025). A local or edge gateway may often aggregate data from the sensors before it is transmitted to the cloud, much to reduce the amount of bandwidth, and improve the scalability of the system. The functional robustness of the network layer is critical to guarantee the whole-time monitoring and decision taking of irrigation.

2.3. Cloud/Edge Layer: Data Processing and Storage

The cloud/edge layer at the top oversees deriving meaningful information from the raw data from the sensor. It is at this level that data is stored, processed, analyzed and fed into decision making models. Cloud computing platforms offer a large amount of memory and processing power, making them well suited for processed large data, implementing machine learning algorithms, producing predictive data, and even connecting with weather forecasts (Karami & Karami, 2025). Cloud dashboards enable users to visualize trends and receive alerts as well as view the historical performance. Simultaneously, edge computing plays a complementary role by processing data locally - close to the field - meaning that latency is reduced, and it is ensured that a rapid response will take place in situations where immediate action is necessary, such as the control of a valve, or detecting anomalies. Edge devices can also ensure the systems continue functioning at the time of network outages and therefore offer greater resilience. Hybrid cloud - edge is the trend and is becoming more the norm in today's irrigation system due to optimized speed, reliability and computational efforts. This layer forms the intelligence heart of IoT based irrigation and transforms the continuous flow of sensor data to meaningful decisions (Gong et al., 2022).

2.4. User & Control Layer: Decision Interfaces and Actuators

The user and control layer is responsible for the connection of the analytical capabilities of the system and the actual irrigation actions in the field. This layer encompasses user interfaces, like mobile applications and dashboards and farm management platforms, which allow farmers to track the soil moisture levels, weather, pump and irrigation schedules for real-time (Bleicher, Biermann, Drossel, Moehring, & Altintas, 2023). Alerts and recommendations - based on the intelligence of AI models or rule-based systems or thresholds they can set - prompt farmers on timely action. For sophisticated systems, the irrigation control is almost or totally automated using actuators such as smart valves, pumps, solenoid switches, and variable rate irrigation systems (Bleicher et al., 2023). These actuators can work based on the system generated commands thus allowing precision irrigation with very little manual representation. Integrated feedback loops are responsible for non-interruptive synchronization between the field conditions and irrigation operations. This layer is what adds value to the user experience, makes farm management easy, consistent and finally completes the cycle of data driven irrigation management (Hashemi, Darzi-Naftchali, Karandish, Ritzema, & Solaimani, 2024).

3. SENSOR TECHNOLOGIES FOR PRECISION IRRIGATION

Sensors are the building block of the Internet of Things-enabled irrigation system because they provide continuous and real-time measurements about soil, plants and environmental parameters (Samreen et al., 2022). These measurements allow agronomists to make decision evidence regarding the timing, spatial distribution and quantity of irrigation applied. Precision irrigation depends on the type of accuracy, reliability and site appropriate deployment of the sensors in order to achieve optimized water use efficiency. The main types of sensors used in irrigation industry are being delineated in below discussion with their major and minor details in term of their principle of operation, advantages, limitation and the upcoming innovations

(Aishwarya & Kumar, 2025). Field sensors and imaging devices transmit data via 4G/5G to the cloud for processing and mobile-based farm management as depicts in *Figure 2*.

Figure 2. Conceptual framework of a cloud-connected IoT smart agriculture system

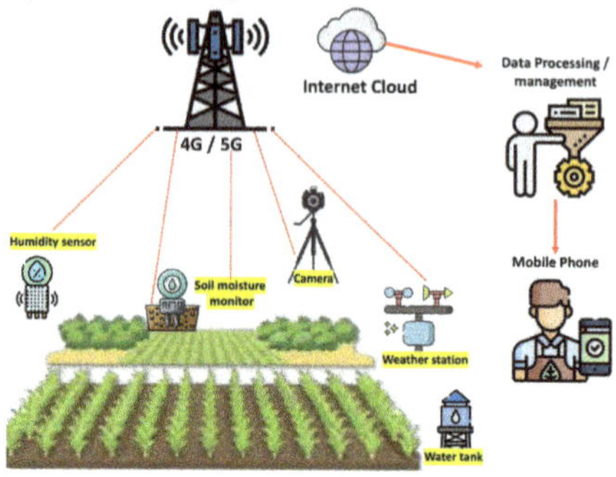

3.1. Soil Moisture Sensors

Soil moisture sensors are arguably the most important constituents of precision irrigation techniques in that a direct measure of the water availability in the root zone is obtained. Time Domain Reflectometry (TDR) sensors measure the soil moisture through the measurement of the propagation time of electromagnetic pulse passing through the soil matrix. They provide a high degree of accuracy and stability in varied soils making them fit for applications like research use or high value crops but at high cost (Aishwarya & Kumar, 2025).

Frequency Domain Reflectometry (FDR) It is a sensor that determines the dielectric permittivity at given frequencies and is a cost-effective alternative to TDR probes. FDR technology is very popular in commercial agriculture and provides reliable performance in variable field conditions. Capacitive sensors use the capacitive change between the electrodes embedded in the soil to estimate the soil moisture. These instruments are cheap, easy to install and they are compatible with IoT platforms because of their low power consumption. Nonetheless, they require regular calibration and are very sensitive to the soil salinity and temperature changes (Zhu, Irmak, Jhala, Vuran, & Diotto, 2019).

Tensiometers measure the tension of water in the soil, not the volumetric - or amount of water - in the soil. They are especially interesting for water sensitive crops, and they are used extensively in horticulture and orchard environments. Tensiometers provide direct information on plant available water but must be periodically maintained and will possibly not be as effective on arid soils (Kirkham,

2023a). Collectively, these soil moisture sensors form the data-driven scheme behind irrigation scheduling as they help to define the ideal soil moisture thresholds and alleviate potential under and over-irrigation of fields.

3.2. Weather and Environmental Sensors

Weather and environmental sensors help to capture microclimatic conditions affecting crops' water demand, evapotranspiration rate and best irrigation time. Temperature and humidity sensors help to calculate relative humidity and vapor-pressure deficiency, which are very important in the estimation of crop transpiration (Charles-Mary & Agbo, 2025). Rain gauges are used to detect the precipitation event allowing an automatic adjustment or a stop in irrigation after a precipitation event is detected. Wind speed sensors help to schedule irrigation by indicating possible loss of water due to drift in sprinkler irrigation. Solar radiation sensors measure the intensity of sunlight, and this is an input into evapotranspiration (ET) modelling. By integrating these sensors, they offer a whole-of-sensor system that looks at the overall atmosphere in ways that are directly related to water needs of the crop (Nava-Pintor et al., 2025). Incorporating weather information in IoT platforms improves the predictive modelling and correlates the irrigation decisions with the climatic variations to occur.

3.3. Plant-Response Sensors

Plant-response sensors provide direct measurement of the plant's water status, which is much more accurate in determining the irrigation requirement than using soil or environmental data on the water status. Such leaf wetness sensors measure the amount of water on a leaf, to help make irrigation changes and to manage diseases in crops susceptible to fungal infections (Kohzuma & Miyamoto, 2024). Sap flow sensors measure movement of water through plant stems and give direct information on rates of transpiration, as well as physiological stress. These sensors become particularly useful for perennial and high value crops (Kirkham, 2023b). Infrequent irrigation: Sensors improved by canopy temperature (water stress): Canopy temperature sensors Use infrared thermometry; Sensitive to small changes in leaf temperature which indicates a plant stress approaching because of poor availability of water (Siegfried et al., 2024). An increase in the temperature of leaves in the canopy relative to the air temperature signals the closing of stomata and the reduction in transpiration, and hence the requirement of irrigation. Plant - response sensors thus provide an advanced, crop-centered strategy of scheduling to improve accuracy and strengthen precision irrigation in heterogeneous fields.

3.4. Emerging Low-Cost and Biodegradable Sensors

Advances in material science, nanotechnology and printed electronics are promoting the development of low cost, disposable and biodegradable sensors for sustainable agriculture. Printed paper - based moisture sensor, biodegradable polymer sensors and low-cost graphene - based electrodes are a few examples of the new generation of field monitoring sensors that are the future (Sreejith, Leo Joseph, Kollem, Vijumon, & Ajayan, 2023). These devices reduce impact on the environment, save the trouble of retrieval or disposing of these products, and provide precision irrigation for smallholder farmers. Their ultralow power usage and compatibility with low range communication system make them perfect for resource constrained environments. In addition, self-powered sensors, using soil microbial fuel cells or solar nanogenerators, are now coming to the scene allowing sensors to run continuously without external power supplies (Abbas et al., 2022). These developments encourage the use of IoT systems in the agricultural sector on a broad scale as well as ecologically responsible sensor deployment.

Table 1. Key components and technologies of IoT-enabled irrigation systems

Layer / Component	Technologies / Examples	Function / Role	Benefits	References
Sensor Layer	Soil moisture sensors (TDR, FDR, Capacitive, Tensiometers), Plant-response sensors (Sap flow, Leaf wetness, Canopy temperature), Weather sensors (Temperature, Humidity, Rainfall, Solar radiation)	Data acquisition of soil, plant, and environmental conditions	Accurate irrigation decisions, crop stress monitoring, improved water-use efficiency	(Abbes, Jarray, & Farah, 2024; Gültekin, 2023)
Network / Connectivity Layer	Short-range: Wi-Fi, ZigBee, BLE; Long-range: LoRaWAN, NB-IoT, Sigfox	Transmission of sensor data from field to cloud / edge	Reliable communication, energy-efficient, scalable for small and large farms	(Ye, Yin, Cao, & Wang, 2024)
Data Processing Layer	Cloud computing, Edge computing, big data platforms, AI/ML algorithms	Data storage, analysis, predictive modeling, anomaly detection	Optimized irrigation scheduling, predictive insights, real-time decision support	(Ye et al., 2024)
User & Control Layer	Mobile apps, dashboards, automated valves and pumps, smart irrigation controllers	Interfaces for farmers; actuation of irrigation based on analytics	Reduced labor, precise water application, autonomous irrigation control	(Parra-López et al., 2025)

continued on following page

Table 1. Continued

Layer / Component	Technologies / Examples	Function / Role	Benefits	References
Emerging Technologies	AIoT, Digital twins, Nano-sensors, Biodegradable sensors, Community IoT networks	Advanced decision-making, sustainability, regional water management	Predictive irrigation, low environmental impact, climate-smart practices	(Muhammed et al., 2024)
Challenges / Barriers	Sensor drift, interoperability issues, connectivity gaps, high costs, digital literacy	Factors limiting adoption and efficiency	Helps in planning mitigation strategies and technology deployment	(Adli et al., 2023)

4. COMMUNICATION TECHNOLOGIES AND PROTOCOLS

Efficient communication is the focal point of IoT-enabled irrigation applications, where the field data can pass seamlessly from sensors to decision support platforms and irrigation commands are implemented in real time (Nsoh et al., 2024). Modern agricultural practice requires communication infrastructures that are energy-efficient, cost-effective, scale and are resistant to the more demanding field applications. A well-constructed communication strategy does not only improve the reliability of a system but determines the overall feasibility of IoT deployment in rural settings (Chirwa & Boikanyo, 2022). The following sections define the main types of communication available and their technical characteristics as well as their suitability for the precision irrigation applications.

4.1. Short-Range Wireless Systems

Short range wireless systems form the skeleton of many small scale and localized irrigation systems. Wi-Fi supports high data rate (often 10-100MBps) and thus can be used for applications that require real-time imaging, high sensor density deployment or on-farm remote monitoring applications (using smartphones). However, because of its high-power demand and short range (50-100 m) in the field it is limited for use in extensive farms (Ojo, Giordano, Procissi, & Seitanidis, 2018). ZigBee, based on the viable 802.15.4 standard, is favored for low power and low data rate agriculture applications where the use of the mesh networking capability to link sensors distributed over fields to relay information through each other, allowing for range limitations to be overcome (Pahlavan & Krishnamurthy, 2021). Bluetooth Low Energy (BLE) Bluetooth is suitable for ultra-low power, short range sensing applications and is being used in less bulky irrigation controllers, soil probes and handheld monitoring devices. Even though its range is limited, BLE is an excellent solution for energy-efficient point-to-point communication at the edge of the farm and in the greenhouse (Cäsar, Pawelke, Steffan, & Terhorst, 2022).

4.2. Long-Range Low-Power Networks

Long range communication technologies are critical for wide fields, isolated agricultural fields and areas without a conventional internet infrastructure. LoRaWAN provides a range of kilometers of communication with very low power consumption, which makes it ideal for battery operating soil sensors and weather sensors (de Carvalho Silva, Rodrigues, Alberti, Šolić, & Aquino, 2017). Its star-based network structure and data rate adaptive mechanism guarantee a strong network connection, even in the presence of interference. NB- IoT with Cellular network supports deep network coverage, high reliability, and secure data transmission. NB- is ideally suited to farms with existing cellular networks, and will support large numbers of devices with long battery life (Rastogi, Saxena, Roy, & Shin, 2020). Sigfox is another ultra-low-bandwidth technology that is being used for simple applications with low data rates, such as soil moisture sensing. Its global network availability, low energy consumption and subscription cost make it attractive to smallholder farmers, although, being uplink heavy, it limits the bidirectional controlling functions (Alqurashi, Bouabdallah, & Khairullah, 2023).

4.3. Data Exchange Protocols

Data exchange protocols are responsible for the exchange of data between IoT systems. MQTT (Message Queuing Telemetry Transport) is the most widely used protocol in the agricultural IoT by far, thanks to its lightweight publish-subscribe architecture, thanks to which it is easy to communicate even over low-bandwidth connections (LoRaWAN and NB-IoT) (El Alami, Sidna, Baina, & Najid, 2020). CoAP (Constrained Application Protocol) is optimized for resource limited IoT devices and uses UDP to function, so it can be used for rapid and low power devices into the sensor world (Yassein et al., 2020). HTTP, while being more bandwidth-intensive, is still valuable for applications that need web-based dashboards that need to be cloud APIs or are integrated with third-party platforms. Choosing a suitable protocol has an impact on energy use, system latency and scalability of irrigation networks (Kumari & Arowolo, 2024).

4.4. Connectivity Selection for Rurals and Remote Regions

Connectivity choices for irrigation systems based on IoT need to consider the terrain, network availability, power supply, and the farmer's affordability. In their efforts to build a more connected world, remote areas where cellular is spurious create LoRaWAN gateways as well as hybrid mesh networks as viable options (Chaoub et al., 2021). Solar powered communication node is often necessary to ensure continuous operation. Mixed-technology systems e.g., ZigBee mesh systems connected to an NB-IoT or LoRaWAN backhaul - have the benefit of providing both local reliability and long-range coverage. Ultimately, connectivity choice should take into consideration the balance between cost, data needs, field size, and environmental constraints for long rang sustainability and adoption of these systems by the farmer (Inthasuth, Kaewjumras, Somwong, & Boonsong, 2025).

5. DATA ANALYTICS, AI INTEGRATION, AND IRRIGATION DECISION-MAKING

The transformative potential of the Internet of Things (IoT) within irrigation systems is only pertained to once the raw data from the fields is converted by systems into actionable intelligence on a systematic basis (Obaideen et al., 2022). Using data analytics, artificial intelligence (AI) and latest decision support platforms agricultural practitioners are now moving away from rigid irrigation application times to dynamic demand driven applications. By utilizing soil moisture data trends, meteorological forecasts, phenology patterns of crops, and real-time variability in the field, using IoT analytics technology enables optimizing the use of water and maintaining crop productivity (OĞUztÜRk, 2025). This section explores this fascination of acknowledging and in the creation, processing, modeling and use of data for intelligent irrigation decision making with a specific focus on the increasing role of machine learning algorithms and digital twin technologies in predictive / autonomous water management.

5.1. Data Acquisition, Cleaning and Processing

Raw data coming from sensors, unmanned aerial vehicles, meteorological stations and satellite observation must undergo a systematic refinement before they could be considered useful. Data acquisition includes continuous data acquisition of soil moisture values, evapotranspiration parameters, microclimate data and plant physiological indicators (Z. Zhang & Zhu, 2023). Nonetheless, field data are often noisy, have missing data values, or drift, or outliers caused by environmental perturbations. Data cleaning - cleaning the data by filtering out the erroneous readings, calibrating the sensors and problems like applying a smoothing algorithm. Preparation beforehand using various pre-processing steps such as normalization or temporal alignment or feature extraction to transform given raw measurements into coherent datasets might be necessary. Later, the cleansed data are sent to the cloud or edge platforms for further analysis on analytical engines which aggregate, store and visualize the data (Edje, Abd Latiff, & Chan, 2023). Effective processing means that irrigation decisions are based on accurate and reliable field-specific information and are context-specific.

5.2. Irrigation Scheduling Models

Irrigation scheduling models are those that give scientific basis for determining the timing and volume of water application. Evapotranspiration (ET) models are based on the FAO-56 (Penman-Monteith) formulation and estimate the crop water demand according to climatic variables including temperature, radiation, humidity and wind speed (Pereira, Allen, Smith, & Raes, 2015; A. Singh & Haghverdi, 2023). Integrating ET models with the IoT sensor output helps in performing irrigation based on the growth stages and the prevailing weather. Threshold-based models are based on or governed by predefined soil moisture computed limits or plant stress-based indicators, and when sensor readings drop below the thresholds, irrigation events are automatically initiated. (Schweidtmann, Zhang, & von Stosch, 2024)

5.3. AI and Machine Learning Approaches

AI and machine learning methodologies raise the level of irrigation planning from rule-based decisions to being predictive and context aware. Supervised learning techniques, such as Random Forests, Support Vector Machines, and Gradient Boosting, are used to predict the dynamics of soil moisture, crop water needs and irrigation time using historical datasets and real time data (Bwambale, Abagale, & Anornu, 2023). Deep learning methods, e.g., long short-term memory (LSTMs) networks are especially powerful in prediction of future soil moisture behavior or water demand under different climatic conditions. Reinforcement learning (RL): With reinforcement learning, we can make irrigation systems learn the best watering strategies from the constant interaction with the surrounding environment which will improve water percentage and the crop quality (Lindemann, Müller, Vietz, Jazdi, & Weyrich, 2021). A further lending hand to these use cases is machine learning driven analytics that can also aid in detecting anomalies, so as to identify anomalies in a sensor (say a sensor failed) or a leak or an inefficient irrigation use. Collectively, AI promotes the creativeness, lowers the amount of manual intervention, and supports data-driven water resources management (Z. Wang & Li, 2025). Sensor and cloud data feed processing and modeling layers to guide expert-informed farm decisions as shown in *Figure 3.*

Figure 3. Framework of an IoT-driven decision support system for smart farming

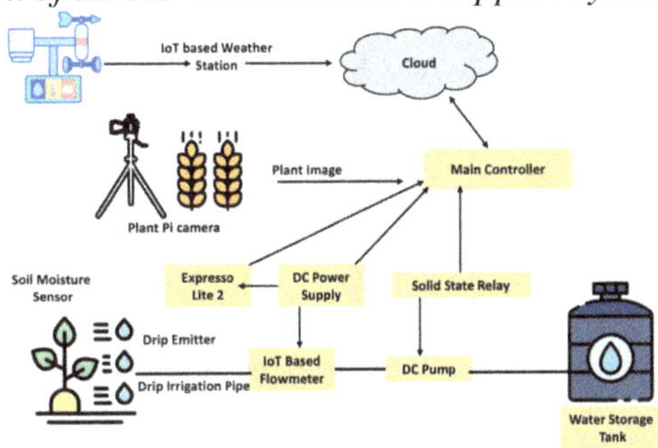

5.4. Digital Twins and Predictive Irrigation Systems

The representation of farm irrigating systems as virtual versions - digital twins - is the next step in smart agriculture. These digital models embrace sensor data, crop simulations, weather forecasts and irrigation infrastructure details to mimic real-time conditions that are found in the field (Purcell & Neubauer, 2023). Digital twins allow the assessment of "what-if" scenarios that can be applied based on, for example, nutrient uptake, ecological stress and crop response through predictive analytics such as higher crop yield due to reducing irrigation volume, adjusting irrigation interval or reacting to heatwaves, which facilitates better decision making (Boyes & Watson, 2022). Coupled with data streams from IoT environments, digital twins will enable near-autonomous irrigation environments in which water delivery systems automatically adapt water delivery on a preemptive basis based on predicted conditions. This makes for water sustaining, climatic and highly efficient irrigation management. Predictive irrigation systems with the support of digital twins are therefore a sign of evolvement from reactive water management to proactive water management, maximizing productivity and minimizing environmental impacts (Alves, Maia, & Lima, 2023).

6. REAL-WORLD IMPLEMENTATIONS AND CASE STUDIES

The practical effects of Internet of Things-enabled irrigation systems are probably best understood from analyses of real concrete applications. Across the world, various case studies show that incorporating IoT will help to improve the efficiency of water use, lower labor costs, and give growers a higher yield on their crops (Nižetić, Šolić, López-de-Ipiña González-de-Artaza, & Patrono, 2020). These implementations also point to the need for appropriate adaptation of technology to local environmental conditions, infrastructure availability and socio-economic context. This section provides reviews of notable examples in the developed and developing world and analyses some key lessons which have been drawn from their outcomes.

6.1. Success Stories from Developed Regions

In developed countries, precision irrigation with IoT has been widely applied in the high-value crop production and the technologically advanced farming system. For example, in California, USA, applications of vineyard development soil moisture sensors with weather stations and automatic drip irrigation systems connected through LoRaWAN using cloud platforms. These systems dynamically adjust water delivery based on evapotranspiration case studies as well as real-time soil information to reduce water consumption between 25 - 40% whilst maintaining grape quality (Liopa-Tsakalidi, Thomopoulos, Barouchas, Boursianis, & Goudos, 2024).

Likewise, in the Netherlands, greenhouses use sensor networks in high densities together with AI-driven irrigation controllers to optimize the supply of water and nutrients. Sensors check the leaf wetness, soil moisture and ambient parameters and at the same time the cloud-based platform simulates the crop growth and irrigation scenarios. This has given some great water savings and energy savings on pumps as well as plant health in the high value vegetable and flower production (Hosny, El-Hady, & Samy, 2025).

Australia is another good example large scale cotton farms take advantage of Internet of Things (IoT) systems combined with weather forecasts and soil moisture networks. Predictive irrigation scheduling helps farmers to manage the water efficiently during a long drought and consequently, or more high yield per unit of water used (Hardin, Barnes, Delhom, Wanjura, & Ward, 2022). These cases show the effects of mature technology, reliable infrastructure and the identification of well-trained personnel collectively resulting in the maximum benefit of IoT irrigation.

6.2. IoT Irrigation Adoption in Developing Countries

Developing countries face special problems: network coverage in developing countries is often low, the initial costs are high, technical know-how is low and cropping land is fragmented. Despite such barriers, pilot projects have had some promising results (Harahap, Kamp, & Ubacht, 2024). In India, the use of LoRaWAN for soil moisture monitoring of you see smallholder vegetable farms has helped farmers to reduce the frequency of irrigation activities while saving on the productivity of their crops. Solar-powered pumps with low cost sensors have been shown to be vital where electricity is not reliable (D. Singh, Sobti, Jain, Malik, & Le, 2022).

In Kenya, drip irrigation coupled with a mobile dashboard with alerts on soil moisture thresholds and weather forecasts has been adopted by small-scala maize farmers. As a result, water use has decreased by about 30 per cent, and maize yields have been improved because of reduced water stress during critical parts of the plant's growth stages.

Pakistan has also seen pilot projects using IoT enabled irrigation system in Punjab and Sindh provinces for wheat and vegetables. Low-cost capacitive soil sensors, LoRaWAN gateways and mobile notifications have enabled farmers to better irrigation and volume of water. Although adoption is so far limited, these pilots show the potential for the scale-up of IoT irrigation in resource limited agricultural environments (Majumder et al., 2023).

7. BARRIERS, RISK AND ENABLERS FOR LARGE-SCALE ADOPTION

Although the use of IoT-enabled irrigation systems can promise significant advances in terms of water-use efficiency and crop productivity, their large-scale deployment is inhibited by a raft of related challenges. Among the most significant technical barriers include sensor drift, inaccurate calibration, low interoperability between devices from disparate suppliers, vulnerability to cyberattacks, and all of these issues jeopardize the integrity of the data and increase system reliability (Thilakarathne, Yassin, Abu Bakar, & Abas, 2021). The interconnectedness of heterogeneous sensors, controllers and communications protocols also create operational challenges as they are integrated in a straightforward network and especially in a vast heterogeneous field. Economic and infrastructural constraints are additional barriers to adoption with initial costs associated with sensors, IoT controllers, connectivity gateways and automated pumps being prohibitive with smallholder or resource poor farmers. In addition, it has been found that intermittent electricity supply and poor internet also reduce system reliability and longevity, as does poor access to maintenance services (Pathmudi, Khatri, Kumar, Abdul-Qawy, & Vyas, 2023).

For more than the technological and economic factors, the policy and regulatory framework play a decisive role in the adoption, because absence of clear standards related to techniques of accuracy of sensors, security of data and to the interoperability and the lack of incentives, subsidies or technical assistance by governments negatively affect investments (Trapani & Longo, 2023). Social factors such as farmer acceptance, digital literacy, and the ability to interpret and act on data are similarly critical determinants. A significant number of farmers still lack knowledge of IoT systems and may be resistant or hesitant to adopt them because they may seem complex or uncertain of the benefits (Idoje, Dagiuklas, & Iqbal, 2021). Consequently, innovative training efforts, making mobile and web interface designs easier to use, and demonstrating real water and yield savings are necessary in building trust and engagement. Overcoming these barriers requires a multidimensional approach that combines resilient and cost-effective technologies, policy measures, sound infrastructure and mass initiatives that empower and develop capacity among farmers. When these elements converge, the potential for IoT driven irrigation can be fully realized, to bring sustainable, climate resilient and efficient water management in different agricultural contexts (Rajuroy, Alkhakani, & Navandar, 2025).

8. FUTURE DIRECTIONS, INNOVATIONS PATHWAYS

The prospective trajectory of the internet of things (IoT-enable) irrigation is expected to be influenced by the next-gen technologies, integration of irrigation systems into larger agricultural frameworks and research driven innovation that can take next-gen technologies to address current limitations and achieve efficiency and sustainability. Next-generation IoT innovated concepts include the convergence of Artificial Intelligence and IoT and are widely known as AIoT. This paradigm supports autonomous decision making of irrigation based on predictive analytics and real time learning based on distributed sensor networks. Concurrently the development of nanoscale, propellant-based sensors that can sense very small changes in nutrients, moisture content, and plant metabolites in soils is progressing, thus allowing very precise field monitoring while consuming very low amounts of power. Moreover, the development of community IoT networks, where several farms or agricultural cooperations share their sensor data, water resources and irrigation strategies, has the potential to optimize water use in the region, knowledge sharing and resilience in water strained areas.

Integration with climate-smart agriculture is another major source of critical path for innovation. IoT systems could be coordinated with weather forecasts, models for forecasting drought and simulating crop growth to account for climate change and implement strategies for adaptive irrigation practices that reduce water usage while reducing the risks of climate change. Coupled with Renewable Energy solutions (e.g. solar powered pumps and edge computing nodes), such systems could perform economically sustainably even in an off-grid or energy limited application.

Despite these advances, there continue to be a few research gaps and research opportunities. Sensor calibration at high precision under various soil conditions, the development of ultra low-cost and bio-degradable sensors, network protocols that are at scale for both remote and heterogeneous fields, cybersecurity, are topics that require urgent research. Furthermore, combining IoT irrigation with multi-dimensional farm management platforms which can include nutrient management, pest monitoring and yield prediction have huge potential in holistic digital agriculture. Research by sectors of engineers, agronomists, policy makers and farmers that identifies previous technological innovations and converts them into practical, adaptable and context specific technology innovations are important. By addressing these opportunities, future IoT enabled irrigation systems can be a new paradigm of precision, efficiency and sustainability and therefore contribute to global efforts towards water conservation, food security and climate resilient agriculture.

9. CONCLUSION

IoT-enabled irrigation is a new approach to sustainable water management and precision agriculture and therefore helps to solve some of the world's pressing problems concerning water scarcity, climatic variability, and rising food demands. From the basic understanding of the principles of the Internet of Things (IoT) and how it has been implemented in the digital agriculture ecosystem to the operational architecture that includes sensors, networks, cloud computing, and user interfaces, this chapter explains how connected technologies can help to make data-driven decisions related to irrigation. Advanced sensor modalities, including soil moisture probes, meteorological and environmental monitors, plant response sensors, etc., collect continuous measurements in real-time and emerging trends and less expensive solutions can lead to more sustainable and broadly accessible sensor configurations. Communication networks from the short-range of short-range protocols (ZigBee, BLE), up to long range low power communication networks (LoRaWAN, NB-IoT) allow efficient transmission of data from the field to the cloud and vice-versa, supporting reactive irrigation control. Data analytics, artificial intelligence, and machine learning can be used to transform raw data into actionable insights that can be used for predictive irrigation, anomaly detection, and decision making using digital twins, turning irrigation management from a reactive to a proactive and autonomous paradigm. Real world examples from both developed and developing contexts confirm measurable improvements in water use efficiency, in crop yields and resource optimization, while at the same time underlining the importance of context specific adaptation, train of mind and affordability. Nonetheless, the barriers to its widespread adoption include technical, economic, infrastructure, policy and social barriers, which will require integrated approaches based on resilient technology, supportive frameworks and capacity building at the level of farmers. Looking into the future, coming innovation trends like AIoT, nano sensors, community IoT networks and their combination with climate-smart agriculture promise to make it still more precise, sustainable, and resistant. Research opportunities abound in sensor development, scalable connect, cyber security and complete farm management integration. Overall, IoT enabled irrigation is a compelling statistically based approach to optimize an efficient use of water to augment agricultural productivity to contribute to climate resilient and sustainable agricultural systems around the world, signaling a paradigm shift in contemporary water resource management within modern agriculture.

REFERENCES

Abbas, S. Z., Wang, J.-Y., Wang, H., Wang, J.-X., Wang, Y.-T., & Yong, Y.-C. (2022). Recent advances in soil microbial fuel cells based self-powered biosensor. *Chemosphere*, *303*, 135036. DOI: 10.1016/j.chemosphere.2022.135036

Abbes, A. B., Jarray, N., & Farah, I. R. (2024). Advances in remote sensing based soil moisture retrieval: Applications, techniques, scales and challenges for combining machine learning and physical models. *Artificial Intelligence Review*, *57*(9), 224. DOI: 10.1007/s10462-024-10734-1

Abdelmoneim, A. A., Kimaita, H. N., Al Kalaany, C. M., Derardja, B., Dragonetti, G., & Khadra, R. (2025). IoT Sensing for Advanced Irrigation Management: A Systematic Review of Trends, Challenges, and Future Prospects. *Sensors (Basel)*, *25*(7), 2291. https://www.mdpi.com/1424-8220/25/7/2291

Adli, H. K., Remli, M. A., Wan Salihin Wong, K. N. S., Ismail, N. A., González-Briones, A., Corchado, J. M., & Mohamad, M. S. (2023). Recent Advancements and Challenges of AIoT Application in Smart Agriculture: A Review. *Sensors (Basel)*, *23*(7), 3752. https://www.mdpi.com/1424-8220/23/7/3752

Ahvar, E., Ahvar, S., & Lee, G. M. (2024). Artificial Intelligence of Things: Architectures, Applications, and Challenges. In Ziegler, S., Radócz, R., Quesada Rodriguez, A., & Matheu Garcia, S. N. (Eds.), *Springer Handbook of Internet of Things* (pp. 443–462). Springer International Publishing.

Aishwarya, K. A., & Kumar, P. (2025). Sensors-Based Irrigation for Increasing Crop and Water Productivity. In A. Kumar, V. K. Singh, Y. Singh, S. K. Singh, & P. Kumar (Eds.), *Modern Technology for Sustainable Agriculture* (pp. 101-113). Cham: Springer Nature Switzerland.

Alqurashi, H., Bouabdallah, F., & Khairullah, E. (2023). SCAP SigFox: A Scalable Communication Protocol for Low-Power Wide-Area IoT Networks. *Sensors (Basel)*, *23*(7). Advance online publication. DOI: 10.3390/s23073732

Alves, R. G., Maia, R. F., & Lima, F. (2023). Development of a Digital Twin for smart farming: Irrigation management system for water saving. *Journal of Cleaner Production*, *388*, 135920. DOI: 10.1016/j.jclepro.2023.135920

Bleicher, F., Biermann, D., Drossel, W. G., Moehring, H. C., & Altintas, Y. (2023). Sensor and actuator integrated tooling systems. *CIRP Annals*, *72*(2), 673–696. DOI: 10.1016/j.cirp.2023.05.009

Boyes, H., & Watson, T. (2022). Digital twins: An analysis framework and open issues. *Computers in Industry*, *143*, 103763. DOI: 10.1016/j.compind.2022.103763

Bwambale, E., Abagale, F. K., & Anornu, G. K. (2022). Smart irrigation monitoring and control strategies for improving water use efficiency in precision agriculture: A review. *Agricultural Water Management*, *260*, 107324. DOI: 10.1016/j.agwat.2021.107324

Bwambale, E., Abagale, F. K., & Anornu, G. K. (2023). Data-Driven Modelling of Soil Moisture Dynamics for Smart Irrigation Scheduling. *Smart Agricultural Technology*, *5*, 100251. DOI: 10.1016/j.atech.2023.100251

Cäsar, M., Pawelke, T., Steffan, J., & Terhorst, G. (2022). A survey on Bluetooth Low Energy security and privacy. *Computer Networks*, *205*, 108712. DOI: 10.1016/j.comnet.2021.108712

Chaoub, A., Lall, B., Bhatia, V., Kliks, A., Mendes, L., Rabie, K., & Zorzi, M. (2021). 6G for Bridging the Digital Divide: Wireless Connectivity to Remote Areas. *IEEE Wireless Communications*. Advance online publication. DOI: 10.1109/MWC.001.2100137

Charles-Mary, C., & Agbo, A. (2025). AN IOT-BASED ENVIRONMENTAL MONITORING SYSTEM USING TEMPERATURE, HUMIDITY, AIR QUALITY, LIGHT INTENSITY, AND RAIN DETECTION SENSORS WITH ESP 32 MICROCONTROLLER. *International Journal of Engineering*.

Chirwa, M., & Boikanyo, D. (2022). The role of effective communication in successful strategy implementation. *Acta Commercii*, *22*. Advance online publication. DOI: 10.4102/ac.v22i1.1020

Choudhary, A. (2024). Internet of Things: A comprehensive overview, architectures, applications, simulation tools, challenges and future directions. *Discover Internet of Things*, *4*(1), 31. DOI: 10.1007/s43926-024-00084-3

de Carvalho Silva, J., Rodrigues, J., Alberti, A., Šolić, P., & Aquino, A. (2017). *LoRaWAN - A Low Power WAN Protocol for Internet of Things: a Review and Opportunities*.

Dong, J., & Xu, J. (2025). Driving agricultural strength through digital transformation. *Scientific Reports*. Advance online publication. DOI: 10.1038/s41598-025-29878-3

Edje, A. E., Abd Latiff, M. S., & Chan, W. H. (2023). IoT data analytic algorithms on edge-cloud infrastructure: A review. *Digital Communications and Networks*, *9*(6), 1486–1515. DOI: 10.1016/j.dcan.2023.10.002

El Alami, H., Sidna, J., Baina, A., & Najid, A. (2020). *Analysis and evaluation of communication Protocols for IoT Applications.*

García, L., Parra, L., Jimenez, J. M., Lloret, J., & Lorenz, P. (2020). IoT-Based Smart Irrigation Systems: An Overview on the Recent Trends on Sensors and IoT Systems for Irrigation in Precision Agriculture. *Sensors (Basel), 20*(4), 1042. https://www.mdpi.com/1424-8220/20/4/1042

García-Martín, J. P., Torralba, A., Hidalgo-Fort, E., Daza, D., & González-Carvajal, R. (2023). IoT solution for smart water distribution networks based on a low-power wireless network, combined at the device-level: A case study. *Internet of Things : Engineering Cyber Physical Human Systems, 22*, 100746. DOI: 10.1016/j.iot.2023.100746

Gong, L., Yan, J., Chen, Y., An, J., He, L., Zheng, L., & Zou, Z. (2022). An IoT-based intelligent irrigation system with data fusion and a self-powered wide-area network. *Journal of Industrial Information Integration, 29*, 100367. DOI: 10.1016/j.jii.2022.100367

Guiné, R. P. F. (2024). The Challenges and Strategies of Food Security under Global Change. *Foods, 13*(13). Advance online publication. DOI: 10.3390/foods13132083

Gültekin, R. (2023). Future Agricultural Irrigation Technologies and Water Management. In (pp. 181-198).

Hafeez, A., Husain, M. A., Singh, S. P., Chauhan, A., Khan, M. T., Kumar, N., & Soni, S. K. (2023). Implementation of drone technology for farm monitoring & pesticide spraying: A review. *Information Processing in Agriculture, 10*(2), 192–203. DOI: 10.1016/j.inpa.2022.02.002

Harahap, V., Kamp, L. M., & Ubacht, J. (2024). A decision support scheme for solving the mobile coverage gap in rural areas in developing countries – Demonstrated with a case in Indonesia. *Technological Forecasting and Social Change, 207*, 123600. DOI: 10.1016/j.techfore.2024.123600

Hardin, R. G., Barnes, E. M., Delhom, C. D., Wanjura, J. D., & Ward, J. K. (2022). Internet of things: Cotton harvesting and processing. *Computers and Electronics in Agriculture, 202*, 107294. DOI: 10.1016/j.compag.2022.107294

Hashemi, S.-Z., Darzi-Naftchali, A., Karandish, F., Ritzema, H., & Solaimani, K. (2024). Enhancing agricultural sustainability with water and crop management strategies in modern irrigation and drainage networks. *Agricultural Water Management, 305*, 109110. DOI: 10.1016/j.agwat.2024.109110

Hosny, K. M., El-Hady, W. M., & Samy, F. M. (2025). Technologies, Protocols, and applications of Internet of Things in greenhouse Farming: A survey of recent advances. *Information Processing in Agriculture*, *12*(1), 91–111. DOI: 10.1016/j.inpa.2024.04.002

Huynh, H. X., Tran, L. N., & Duong-Trung, N. (2023). Smart greenhouse construction and irrigation control system for optimal Brassica Juncea development. *PLoS One*, *18*(10), e0292971. DOI: 10.1371/journal.pone.0292971

Idoje, G., Dagiuklas, T., & Iqbal, M. (2021). Survey for smart farming technologies: Challenges and issues. *Computers & Electrical Engineering*, *92*, 107104. DOI: 10.1016/j.compeleceng.2021.107104

Ingrao, C., Strippoli, R., Lagioia, G., & Huisingh, D. (2023). Water scarcity in agriculture: An overview of causes, impacts and approaches for reducing the risks. *Heliyon*, *9*(8), e18507. DOI: 10.1016/j.heliyon.2023.e18507

Inthasuth, T., Kaewjumras, Y., Somwong, S., & Boonsong, W. (2025). Comparative analysis of ZigBee, LoRa, and NB-IoT in a smart building: Advantages, limitations, and integration possibilities. [IJRES]. *International Journal of Reconfigurable and Embedded Systems*, *14*, 165–175. DOI: 10.11591/ijres.v14.i1.pp165-175

Kalpana, P., Smitha, L., Madhavi, D., Shaik, A., Kalpana, G., & Kodati, S. (2024). A Smart Irrigation System Using the IoT and Advanced Machine Learning Model: A Systematic Literature Review. *International Journal of Computational and Experimental Science and Engineering*, *10*. Advance online publication. DOI: 10.22399/ijcesen.526

Karami, A., & Karami, M. (2025). Edge computing in big data: Challenges and benefits. *International Journal of Data Science and Analytics*, *20*(7), 6183–6226. DOI: 10.1007/s41060-025-00855-3

Kiran, D. R. (2019). Internet of Things. In Kiran, D. R. (Ed.), *Production Planning and Control* (pp. 495–513). Butterworth-Heinemann.

Kirkham, M. B. (2023a). Tensiometers. In Kirkham, M. B. (Ed.), *Principles of Soil and Plant Water Relations* (3rd ed., pp. 65–81). Academic Press.

Kirkham, M. B. (2023b). Sap flow. In Kirkham, M. B. (Ed.), *Principles of Soil and Plant Water Relations* (3rd ed., pp. 411–429). Academic Press.

Kirkham, M. B. (2023c). Field capacity, wilting point, available water, and the nonlimiting water range. In (pp. 169-189).

Kohzuma, K., & Miyamoto, K.-i. (2024). Analysis of plant physiological responses based on leaf color changes through the development and application of a wireless plant sensor. *Sensing and Bio-Sensing Research*, *46*, 100688. DOI: 10.1016/j.sbsr.2024.100688

Krishnamurthi, R., Kumar, A., Gopinathan, D., Nayyar, A., & Qureshi, B. (2020). An Overview of IoT Sensor Data Processing, Fusion, and Analysis Techniques. *Sensors (Basel)*, *20*(21), 6076. https://www.mdpi.com/1424-8220/20/21/6076

Kumari, S., & Arowolo, M. O. (2024). Internet of things (iot): Concepts, protocols, and applications. In *Emerging Technologies and Security in Cloud Computing* (pp. 19–52). IGI Global Scientific Publishing.

Lalhriatpuii, Ruchi, & Wasson, V. (2025). *Comprehensive Exploration of IoT Communication Protocol: CoAP, MQTT, HTTP, LoRaWAN and AMQP.* Paper presented at the Machine Learning Algorithms, Cham.

Lindemann, B., Müller, T., Vietz, H., Jazdi, N., & Weyrich, M. (2021). A survey on long short-term memory networks for time series prediction. *Procedia CIRP*, *99*, 650–655. DOI: 10.1016/j.procir.2021.03.088

Liopa-Tsakalidi, A., Thomopoulos, V., Barouchas, P., Boursianis, A. D., & Goudos, S. K. (2024). A LoRaWAN-based IoT platform for smart irrigation in olive groves. *Smart Agricultural Technology*, *9*, 100673. DOI: 10.1016/j.atech.2024.100673

Liu, X., Zhao, Z., & Rezaeipanah, A. (2025). Intelligent and automatic irrigation system based on internet of things using fuzzy control technology. *Scientific Reports*, *15*(1), 14577. DOI: 10.1038/s41598-025-98137-2

Majumder, S., Kasirao, G., & Himavarsha, P., Himanshi, Chaudhary, S., Sekopo, K., . . . Verma, J. (2023). Assessing Low-cost Capacitive Soil Moisture Sensors: Accurate, Affordable, and IoT-ready Solutions for Soil Moisture Monitoring. *International Journal of Environment and Climate Change*, *13*, 2233–2242. DOI: 10.9734/ijecc/2023/v13i113386

Mastoi, M. S., Zhuang, S., Munir, H. M., Haris, M., Hassan, M., Usman, M., & Ro, J.-S. (2022). An in-depth analysis of electric vehicle charging station infrastructure, policy implications, and future trends. *Energy Reports*, *8*, 11504–11529. DOI: 10.1016/j.egyr.2022.09.011

Morchid, A., Jebabra, R., Khalid, H. M., El Alami, R., Qjidaa, H., & Ouazzani Jamil, M. (2024). IoT-based smart irrigation management system to enhance agricultural water security using embedded systems, telemetry data, and cloud computing. *Results in Engineering*, *23*, 102829. DOI: 10.1016/j.rineng.2024.102829

Muhammed, D., Ahvar, E., Ahvar, S., Trocan, M., Montpetit, M.-J., & Ehsani, R. (2024). Artificial Intelligence of Things (AIoT) for smart agriculture: A review of architectures, technologies and solutions. *Journal of Network and Computer Applications*, *228*, 103905. DOI: 10.1016/j.jnca.2024.103905

Nava-Pintor, J. A., Alcalá-Rodríguez, U. E., Guerrero-Osuna, H. A., Mata-Romero, M. E., Lopez-Neri, E., García-Vázquez, F., & Luque-Vega, L. F. (2025). Development and Evaluation of Solar Radiation Sensor Using Cost-Effective Light Sensors and Machine Learning Techniques. *Technologies*, *13*(5), 182. https://www.mdpi.com/2227-7080/13/5/182

Nižetić, S., & Šolić, P., López-de-Ipiña González-de-Artaza, D., & Patrono, L. (2020). Internet of Things (IoT): Opportunities, issues and challenges towards a smart and sustainable future. *Journal of Cleaner Production*, *274*, 122877. DOI: 10.1016/j.jclepro.2020.122877

Nsoh, B., Katimbo, A., Guo, H., Heeren, D. M., Nakabuye, H. N., Qiao, X., & Kiraga, S. (2024). Internet of Things-Based Automated Solutions Utilizing Machine Learning for Smart and Real-Time Irrigation Management: A Review. *Sensors (Basel)*, *24*(23). Advance online publication. DOI: 10.3390/s24237480

Obaideen, K., Yousef, B. A. A., AlMallahi, M. N., Tan, Y. C., Mahmoud, M., Jaber, H., & Ramadan, M. (2022). An overview of smart irrigation systems using IoT. *Energy Nexus*, *7*, 100124. DOI: 10.1016/j.nexus.2022.100124

OĞUztÜRk, G. E. (2025). AI-driven irrigation systems for sustainable water management: A systematic review and meta-analytical insights. *Smart Agricultural Technology*, *11*, 100982. DOI: 10.1016/j.atech.2025.100982

Ojo, M. O., Giordano, S., Procissi, G., & Seitanidis, I. N. (2018). A review of low-end, middle-end, and high-end iot devices. *IEEE Access : Practical Innovations, Open Solutions*, *6*, 70528–70554.

Pahlavan, K., & Krishnamurthy, P. (2021). Evolution and Impact of Wi-Fi Technology and Applications: A Historical Perspective. *International Journal of Wireless Information Networks*, *28*(1), 3–19. DOI: 10.1007/s10776-020-00501-8

Parra-López, C., Ben Abdallah, S., Garcia-Garcia, G., Hassoun, A., Trollman, H., Jagtap, S., & Carmona-Torres, C. (2025). Digital technologies for water use and management in agriculture: Recent applications and future outlook. *Agricultural Water Management*, *309*, 109347. DOI: 10.1016/j.agwat.2025.109347

Pathmudi, V. R., Khatri, N., Kumar, S., Abdul-Qawy, A. S. H., & Vyas, A. K. (2023). A systematic review of IoT technologies and their constituents for smart and sustainable agriculture applications. *Scientific African, 19*, e01577. DOI: 10.1016/j.sciaf.2023.e01577

Pereira, L. S., Allen, R. G., Smith, M., & Raes, D. (2015). Crop evapotranspiration estimation with FAO56: Past and future. *Agricultural Water Management, 147*, 4–20. DOI: 10.1016/j.agwat.2014.07.031

Purcell, W., & Neubauer, T. (2023). Digital Twins in Agriculture: A State-of-the-art review. *Smart Agricultural Technology, 3*, 100094. DOI: 10.1016/j.atech.2022.100094

Rajuroy, A., Alkhakani, A., & Navandar, P. (2025). *Advanced Smart Irrigation Systems: Leveraging AI*. IoT, and Nanotechnology for Sustainable Agriculture.

Rastogi, E., Saxena, N., Roy, A., & Shin, D. R. (2020). Narrowband Internet of Things: A Comprehensive Study. *Computer Networks, 173*, 107209. DOI: 10.1016/j.comnet.2020.107209

Samreen, T., Ahmad, M., Baig, M. T., Kanwal, S., Nazir, M. Z., & Sidra-Tul-Muntaha. (2022). Remote Sensing in Precision Agriculture for Irrigation Management. *Environmental Sciences Proceedings, 23*(1), 31. Retrieved from https://www.mdpi.com/2673-4931/23/1/31

Schweidtmann, A. M., Zhang, D., & von Stosch, M. (2024). A review and perspective on hybrid modeling methodologies. *Digital Chemical Engineering, 10*, 100136. DOI: 10.1016/j.dche.2023.100136

Shahab, H., Naeem, M., Iqbal, M., Aqeel, M., & Ullah, S. S. (2025). IoT-driven smart agricultural technology for real-time soil and crop optimization. *Smart Agricultural Technology, 10*, 100847. DOI: 10.1016/j.atech.2025.100847

Siegfried, J., Rajan, N., Adams, C. B., Neely, H., Hague, S., Hardin, R., & Thomasson, A. (2024). High-accuracy infrared thermography of cotton canopy temperature by unmanned aerial systems (UAS): Evaluating in-season prediction of yield. *Smart Agricultural Technology, 7*, 100393. DOI: 10.1016/j.atech.2023.100393

Singh, A., & Haghverdi, A. (2023). Development and evaluation of temperature-based deep learning models to estimate reference evapotranspiration. *Artificial Intelligence in Agriculture, 9*, 61–75. DOI: 10.1016/j.aiia.2023.08.003

Singh, D., Sobti, R., Jain, A., Malik, P., & Le, D.-N. (2022). LoRa based intelligent soil and weather condition monitoring with internet of things for precision agriculture in smart cities. *IET Communications, 16*. Advance online publication. DOI: 10.1049/cmu2.12352

Sreejith, S., Leo Joseph, L. M. I., Kollem, S., Vijumon, V. T., & Ajayan, J. (2023). Biodegradable sensors: A comprehensive review. *Measurement*, *219*, 113261. DOI: 10.1016/j.measurement.2023.113261

Thilakarathne, N., Yassin, H., Abu Bakar, M., & Abas, P. E. (2021). *Internet of Things in Smart Agriculture: Challenges*. Opportunities and Future Directions.

Trapani, N., & Longo, L. (2023). Fault Detection and Diagnosis Methods for Sensors Systems: A Scientific Literature Review. *IFAC-PapersOnLine*, *56*(2), 1253–1263. DOI: 10.1016/j.ifacol.2023.10.1749

Wang, C., Zhao, T., Cheng, W., Ni, Z., & Xiang, N. (2024). Microfluidic strategies in soft robotics: Actuators, control systems, and pumps. *Device*, *2*(9), 100551. DOI: 10.1016/j.device.2024.100551

Wang, Z., & Li, D. (2025). Big data and artificial intelligence-driven risk assessment and response for water resources: Model applications and future perspectives. *Advances in Resources Research*, *5*(1), 414–434.

Waqas, M., Naseem, A., Humphries, U. W., Hlaing, P. T., Dechpichai, P., & Wangwongchai, A. (2025). Applications of machine learning and deep learning in agriculture: A comprehensive review. *Green Technologies and Sustainability*, *3*(3), 100199. DOI: 10.1016/j.grets.2025.100199

Yassein, M. B., Hmeidi, I., Meqdadi, O., Alghazo, F., Odat, B., AlZoubi, O., & Smairat, A. (2020). *Challenges and techniques of constrained application protocol (CoAP) for efficient energy consumption*. Paper presented at the 2020 11th International Conference on Information and Communication Systems (ICICS).

Ye, Z., Yin, S., Cao, Y., & Wang, Y. (2024). AI-driven optimization of agricultural water management for enhanced sustainability. *Scientific Reports*, *14*(1), 25721. DOI: 10.1038/s41598-024-76915-8

Zhang, X., Feng, G., & Sun, X. (2024). Advanced technologies of soil moisture monitoring in precision agriculture: A Review. *Journal of Agriculture and Food Research*, *18*, 101473. DOI: 10.1016/j.jafr.2024.101473

Zhang, Z., & Zhu, L. (2023). A Review on Unmanned Aerial Vehicle Remote Sensing: Platforms, Sensors, Data Processing Methods, and Applications. *Drones (Basel)*, *7*(6), 398. https://www.mdpi.com/2504-446X/7/6/398

Zhu, Y., Irmak, S., Jhala, A., Vuran, M., & Diotto, A. (2019). Time-domain and Frequency-domain Reflectometry Type Soil Moisture Sensor Performance and Soil Temperature Effects in Fine- and Coarse-textured Soils. *Applied Engineering in Agriculture*, *35*, 117–134. DOI: 10.13031/aea.12908

Chapter 7
AI- and IoT-Driven Approaches for Crop Health Monitoring and Management

Rida Batool
http://orcid.org/0009-0007-8633-6320
Quaid-i-Azam University, Islamabad, Pakistan

Abdul Wahab
COMSATS University Islamabad, Pakistan

Mujtaba ul Hassan
COMSATS University Islamabad, Pakistan

Sana Azeem Kiani
Mirpur University of Science and Technology, Pakistan

Rashid Abbas Khan
http://orcid.org/0000-0002-5438-6274
College of Graduate Studies, University of South Africa, Pretoria, South Africa

ABSTRACT

The integration of Artificial Intelligence (AI) and the Internet of Things (IoT) is transforming crop health management by enabling predictive, precise, and sustainable agriculture. Traditional manual and reactive monitoring methods are increasingly inadequate for addressing climate variability, pest outbreaks, nutrient deficiencies, and abiotic stresses. AI–IoT systems combine high-resolution sensors, drones, edge computing, and advanced analytics to provide continuous real-time monitoring

DOI: 10.4018/979-8-3373-7257-0.ch007

of soil, plant, and environmental conditions. Machine learning, deep learning, and computer vision support early detection of diseases, pests, and stress while enabling data-driven decision making through predictive models and digital twins. Decision-support systems translate complex data into actionable recommendations for irrigation, nutrient management, and targeted pest control, while integration with automation and robotics improves operational efficiency. Despite significant economic and environmental benefits, adoption remains limited by connectivity gaps, costs, etc.

1. INTRODUCTION

Agriculture is going through a paradigm shift by the integration of digital technologies, mainly the Artificial Intelligence (AI) and the Internet of Things (IoT), that have redefined the monitoring, prediction and management of the crop health (Miller et al., 2025). This transformation is driven by keen realization of unprecedented challenges facing the world as a whole - including the impacts of climate variability and declining soil health, emergence of new pests and diseases, unpredictability of weather susceptibility, growing call of the need of sustainable solutions to meet the food demands - that require innovative solutions. Conventional agricultural practices, which for long time have relied on manual scouting, visual assessment and experiential knowledge, are becoming increasingly inadequate when it comes to tackling complex stress situations and enabling timely decision making in modern and current farming systems (Shamshiri et al., 2024). Such approaches often miss detecting early or latent symptoms of crop stress, have a limited spatial coverage and do not have the ability to produce data-driven insights in real time. Consequently, there are delays by farmers in the identification of threats leading to preventable yield losses, excessive use of resources, and increasing cost of production.

In this regard, digital and predictive agriculture has become a strong solution to overcome the long-standing limitations in crop health management. By combining sensor networks, drone surveillance, models for machine learning, and analytics offered on cloud, the AI-IoT systems are used to provide continuous, high resolution monitoring of soil, plant, and environment conditions (Mansoor et al., 2025). These technologies help to detect plant diseases early, predict pest outbreaks, detect nutrient deficiency, and monitor abiotic stresses of drought, heat and salinity with an incredible level of accuracy. Unlike the older ways, models built with artificial intelligence are able to analyze large datasets, spot patterns that we cannot see with our own eyes, and predict the upcoming risks before they turn into problems. IoT devices builds on this ecosystem further by real time, location specific data on moisture level, temperature, humidity, canopy condition, atmospheric changes hence enables precise and prompt intervention (Katie, 2024).

The integration of AI and IoT with Crop Health transcends the simple monitoring phase, but sets up the path towards predictive agriculture, which is a methodological framework to the problem which enables farmers to not only anticipate but also allocate the appropriate resource and come up with an informative decision based on sound data insights (Sharma & Shivandu, 2024). This is considered to support sustainable intensification including by reducing the need for applying undue amounts of chemicals to crops, improving efficiency of water and nutrient use, and supporting the development of healthier agro-ecosystems. Moreover, it empowers smallholders farmers through the access to digital tools and mobile based advisory platforms that provide expert-level recommendations instead of relying on the time consuming, offline learning (Kamal & Bablu, 2023). As agriculture struggles to support rising demands for global food supplies and understand mounting pressure to reduce environmental impact, AI - IoT systems are a pathway towards transformative farming by providing resilience to farming, productivity, and climate-smart farming. Their widening scope promises not just of the improved health of crops and stable yields but also of long-term sustainability and regional/and global food security(Kumar et al., 2024) .

The main goal of this chapter is to give a detailed understanding of the impact of the integrated use of AI and IoT technologies on the prediction, diagnosis, and management of crop health in the field of modern agriculture. It includes technological underpinnings, data ecosystems, applications, implementation frameworks, actual case studies and future innovation prospects. The scope lies between the base foundational concepts to advanced predictive modelling making sure that the chapter is a strategic reference for researchers, practitioners and policy makers who are concerned with digital and sustainable agriculture.

2. FOUNDATIONS OF ARTIFICIAL INTELLIGENCE IN AGRICULTURE

2.1. Machine Learning and Its Relevance to Crop Health

Machine learning (ML) is one of the key pillars of the latest digital agriculture, which makes agri technology systems to assimilate historical and real-time field data to prognosticate and manage the crop health with increased preciseness(Botero-Valencia et al., 2025). Its relevance to the vitality of crops lies in the ability to ask questions of complex datasets (comprising microclimatic logs, edaphic parameters, phenotypical metrics of leaves, and pest population dynamics) and reduce its information to actionable potentiality. Supervised learning formulates many classification and regression projects including the identification of disease from leaf images, yield prediction and nutrient deficiency prediction using tagged data. Unsupervised learning is used to complement the task of assessing the health of crops by grouping spatially contiguous areas in a field showing similar types of stress signatures causes anomalies in the sensor outputs to be detected and latent patterns to be understood in the crop images for which the number of classes are predefined. Reinforcement learning has played an increasingly prominent role in adaptive paradigms of decision making where it optimizes irrigation schedules or pesticide application strategies or the operation of robotic working fields through ongoing interactions of feedback and reward (Held et al., 2024). Collectively, these methodologies of ML enable a framework of prediction and response that brings accuracy to the crop health surveillance on a whole.

Figure 1. This flowchart illustrates the application of machine learning across various stages of crop cultivation, from nutrient and water management to post-harvest and yield management, ensuring optimal crop health and productivity

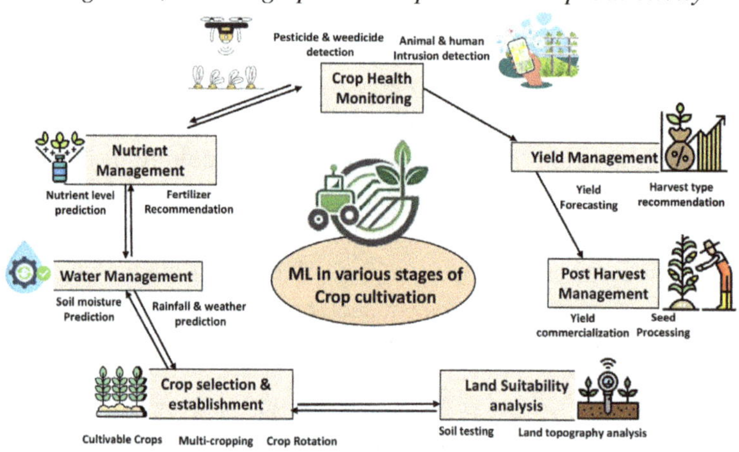

2.2. Deep Learning for Complex Agricultural Predictions

Deep learning is a specialized form of machine learning with a great potential in managing large agricultural data, especially those coming from imagery, time series from sensors and remote sensing platforms (Wang et al., 2022). Convolutional Neural Networks (CNNs) have shown proficiency in distinguishing plant diseases, canopy conditions and crops and weeds by extracting spatial information. Recurrent Neural Networks (RNNs) and their more advanced decoders, Long Short Term Memory (LSTMs) and Gated Recurrent Units (GRUs) are excellent at modeling information over time, and thus are very useful for predicting the growth stages of crops, fluctuations in soil moisture, and the impact of weather-induced stress patterns (Mienye et al., 2024). More recently, transformer architectures have taken center stage in agronomy because of their attention-based mechanisms that help integrate and amalgamate multimodal datasets to aid in high accuracy predictions of pest outburst, phenological development and interaction with the environment. Through the application of these sophisticated neural equivalents, deep learning provides a substantial enhancement of prediction capacities to the modern agricultural industry, as it cleavages real-time, strong insights into the dynamics of crop health (Choudhury & Banerjee, 2024).

Figure 2. This diagram highlights the diverse applications of deep learning in agriculture, spanning precision crop management, livestock farming, intelligent soil and water management, and emerging technologies like precision spraying and automated harvesting

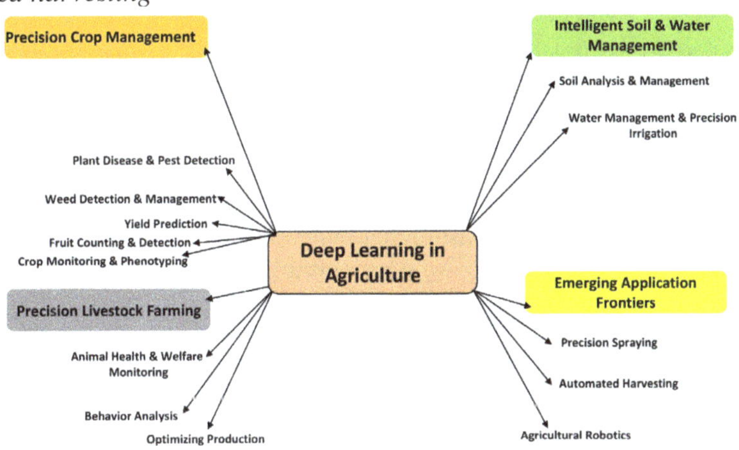

2.3. Computer Vision for Plant Stress and Disease Detection

Computer vision (CV) has become an unrivaled tool in AI-driven approaches to crop health management and visual analysis of plant conditions can also be done fully automated and at scale. By utilizing imaging modalities RGB, multispectral, hyperspectral and thermal infrared can be utilized to identify a broad array of stress indicators and in many cases prior to visualization to the human eye (Zinjurke et al., 2025). These technologies aid in early localization of the fungal, bacterial and viral ailment, nutrient deficiency signatures recognition, pest damage recognition, the delimitation of the water stress zones and evaluation of weed pressure. When integrated with unmanned aerial vehicles, satellite imagery, ground-based robotic platforms and smartphone applications, CV is providing constant surveillance, field-wide with high spatial and temporal resolution capability (Lausch et al., 2025). Such real time visual intelligence increases the ability of agronomists and growers to make diagnoses as soon as possible and make timely interventions that help mitigate the economic losses.

2.4. Predictive Analytics and Early-Warning Systems

Predictive analytics has an important role to play in converting raw agronomic data to foresight - driven knowledge that reduces the uncertainty in crop health management decisions. By integrating machine learning algorithms and data on environment, phenology and sensors, predictive models can help assess levels of risk in terms of diseases and pest outbreaks, drought and nutrient imbalance even before symptoms start to manifest themselves (Zarbakhsh et al., 2025). Early warning systems using predictive models help the farmer to have preventive measures, optimizing the use of inputs, and developing inherent resistance to climatic variability. Such systems normally combine real-time streams of Internet of Things (IoT) sensors, meteorological forecasts, satellite sequence, and historical time series, to generate alerts and recommendations to ensure that interventions are both concentrated and on-time (Dhanaraju et al., 2022). As the field of predictive analytics continues to advance toward becoming increasingly sophisticated in nature, it is facilitating the shift from reactive to proactive crop management practices in the process of enhancing productivity, sustainability and resource efficiency across agricultural landscapes (Titirmare et al., 2024).

3. INTERNET OF THINGS (IOT) IN CROP MONITORING

3.1. Architecture of Agricultural IoT Systems

The architecture of agricultural Internet of Things (IOT) system is the backbone of the digital crop monitoring system, the architecture enables free data collection, data transmitting and analysis in diverse field environments (Ahmed & Shakoor, 2025). A typical IoT architecture has several interconnection layers, starting from the sensing layer where distributed devices voice data in relation to the plant, soil and surrounding environment. This is then followed by the communication and network layers that transmit sensor readings either through a wired or wireless system to the preliminary cloud platforms or local servers. At the processing layer, advanced analytics, artificial intelligence models, and decision and support systems are used to ingest the data to gain some actionable insights from the data. Finally, this information is disseminated by the application layer, in dashboards, mobile applications or automated control systems, thus favoring real-time decision making in the fields of irrigation, fertilization, pest management or climate control (Quy et al., 2022).

3.2. Sensor Types for Plant, Soil, and Environmental Monitoring

IoT-enabled crop monitoring relies heavily on a whole range of sensors for measuring many different parameters which influence plant health and productivity. For soil assessment, sensors for moisture, nutrients, pH, salinity, and temperature help get insights on a continuous basis of soil condition and hence be informed for efficient irrigation and nutrient management (Faqir et al., 2024). Environmental sensors that measure temperature, humidity, light intensity, rainfall, CO2 concentrations, and wind patterns, help farmers to understand the microclimatic variations and their effect on the performance of their crops. Advanced optical sensors, such as hyperspectral and multispectral equipment, enable identification of the biochemical changes of plant tissues, to identify in advance signs of disease, and to monitor the level of plant health by taking into account certain wavelengths of the spectrum (Haq et al., 2025). These heterogeneous sensing technologies provide, together, a high resolution portrait of the dynamics in the field, allowing to trigger early diagnosis of stress and more precise allocation of resources.

3.3. Wireless Communication Technologies

Effective data transmission is essential to IoT systems that are deployed over large and variable atmospheric expanses in agriculture. Characteristic wireless communication technologies such as LoRaWAN, NB-IoT, 5G, and ZigBee are playing a pivotal role in the establishment of a reliable connection between sensors and edge devices and central platforms (Čolaković et al., 2021). LoRaWAN, which is famous for using the long-range/low-power internet, is largely used in terms of large-scale farms with scattered sensors. NB-IoT allows good indoor and underground penetration, i.e., it can be used for ground sensors and in green houses. Fifth - generation (5G) networks have ultra - low latency, high bandwidth, massive number of connected devices and therefore enable real - time monitoring and autonomous operations of machinery (Zreikat et al., 2025). ZigBee with its low power requirements and mesh technology is often found for field-level sensor clusters that require robust and short range communication (Ding et al., 2021). Collectively, these technologies provide for the overall reliability of data flow, the increased reliability of the system, and automation in a remote agricultural environment.

3.4. Edge Computing and Real-Time Data Processing

Edge computing has revolutionized the use of IoT in agriculture by allowing for the processing of data in real time directly at the point of collection which allows for less reliance on cloud based processing (Sathya et al., 2024). By acting on detector readings at the edge - by means of field controllers, gateways or embedded processing systems - systems are capable of immediately reacting to important events such as water stress, a sudden temperature change, pest invasion or disease threat. This localized processing not only helps keep the system's latency as low as possible but also reduces the bandwidth requirements while ensuring the operational continuity of the systems in the areas with limiting or unstable internet connectivity. (Peladarinos, 2023). Edge intelligence allows for further resilience of the system, including enabling on-site artificial intelligence algorithms that can make classifications on plant stress, predict irrigating requirements or autonomously regulate greenhouse conditions As edge computing becomes more integrated with IoT infrastructures it strengthens the reacted, dependable and efficient aspects of crop health management systems (Choudhary et al., 2025).

Figure 3. This diagram illustrates the process of agro-produce classification using deep convolutional neural networks (DCNN). It covers data processing, model selection, feature extraction, and stages of training, validation, and testing to classify agro-produce based on attributes like size, color, and damage detection

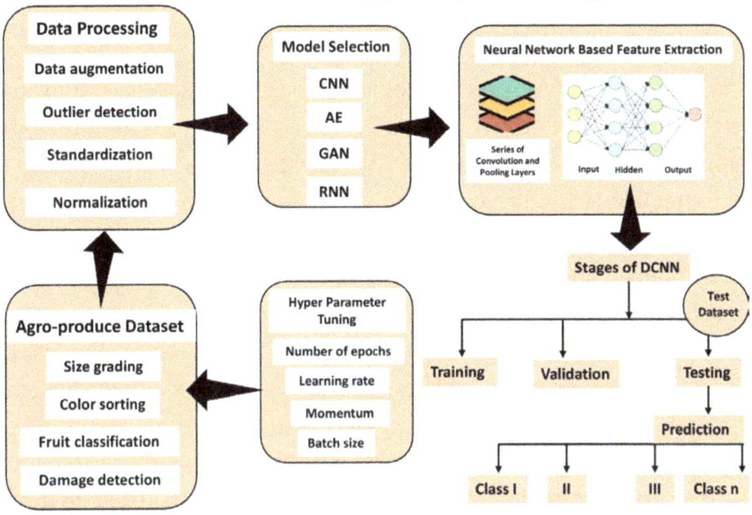

4. AI-IOT INTEGRATION FOR PREDICTING CROP HEALTH

4.1. Data Acquisition, Cleansing, and Fusion from IoT Devices

The fusion of AI and IoT in agriculture is based on strict data acquisition methods, where scattered sensors, unmanned aerial vehicles and field monitoring units also continuously compile high-frequency data sets related to soil properties, plant physiological state and environmental parameters (Rashid et al., 2025). Raw agricultural data usually contains noise, gaps and inconsistencies caused by the heterogeneity of the environment and drift of the sensors. For that reason, data cleansing operations, including filters, normalization, interpolation and outlier removal, are vital and necessary to ensure the accuracy and reliability of inputs used for AI models (Sharafat et al., 2025). Following data fusion techniques are further use cases that better increase the power of the prediction by aggregating heterogeneous details including information from soil sensors, canopy temperature maps, the spectral image map, and meteorological data into one of a kind. This multi source integration provides a thorough, high resolution picture of situation in the field, which in turn aids in the exact diagnosis of emerging threats and supports the development of a real-time capability for decisions (Ahmad et al., 2022).

4.2. AI Pipelines for Detecting Stress, Disease, and Nutrient Imbalance

Artificial intelligence (AI) pipeline focused on crop health has several consecutive steps to derive information from raw data i.e. observations to actionable information for recommendations (Yakkala et al., 2025). These processes are typically triggered by feature extraction which is based on the readings provided by the sensors or the imagery, and the model training using machine learning algorithms or deep learning techniques. Stress - detection systems use physiological signs of stress (e.g. chlorophyte fluorescence signal, thermal anomaly, soil moisture fluctuation) in order to identify early signs of drought, salinity or heat stress (Paul et al., 2025). Disease detection models place leaf lesions, discoloring or other abnormalities in leaf morphology by using advanced computer vision architectures and neural networks frameworks. Similarly, the forecasts of nutrient imbalance use parameters of spectral indices, soil elemental profiles and plant morphological design parameters to predict nutrient deficiencies or toxicities. After training, AI pipelines give real time outputs including flagging for anomalies, severity and recommending for remedial action. This end-to-end automation thus has the great ability to improve the speed and accuracy in crop health assessment (Wu et al., 2025).

4.3. Real-Time Crop Surveillance Using Drones, IoT, and AI Vision

The combination of unmanned aerial vehicles (UAVs), sensors with IoT devices and computer vision systems powered by artificial intelligence has made crop surveillance in real-time into a process that is highly precise and scalable (Fuentes-Peñailillo et al., 2024). UAVs with universally RGB, multispectral, hyperspectral and thermal cameras capture high resolution imagery at short intervals in large fields requiring little manual intervention. Ground-based IoT sensors are a useful addition to the aerial datasets and provide continuous and spatially specific data on the status of soil moisture, nutrient availability, microclimatic conditions and indicators of plant vigour. AI algorithms are used to process the images of UAVs to identify symptoms of disease, canopy stress, weed infestation, and spatial heterogeneity in crop productivity. When combined with IoT sensor streams, these interpretations derived using AI form a dynamic surveillance network that can be used to detect threats early and monitor the progression of stress in the system. The resulting system allows for targeted interventions and reduction of input wastage and increase the overall resilience of the crop (Naqvi et al., 2025).

4.4. Digital Twins for Crop Growth and Health Prediction

Digital twins are virtual representations of physical crop systems that have been created using the combination of information flow from IoT combined with simulations built on AI. These digital models are continually updated based on real-world sensor input which allows the growth trajectories, physiological states and environmental interactions of crops to be monitored on a real-time basis (Melesse, 2025). By simulating various scenarios - including different watering schedules or different nutrient application strategies or expected weather fluctuations, for example - digital twins help forecast how the crop is expected to react under different conditions. In addition to this, they enable early warning mechanisms through the prediction of risks through the proliferation of diseases, pest invasion, or abiotic stress (Subeesh & Chauhan, 2026). The live and evolving nature of digital twins also enables stakeholders to visualize field conditions and use them for predictive experimentation and optimizing management decisions without causing any physical changes in the field(Attaran & Celik, 2023). Their adoption adds to precision, reduces the uncertainty and promotes evidence based planning for agriculture.

4.5. Role of Cloud and Edge Analytics in Decision-Making

Cloud and edge analytics play complementary roles in the AI-IoT framework by each of them processing large amounts of data and facilitating decision-making processes. Cloud platforms based resources that offer scalable storage capacities, powerful computational resources and centralized environments for artificial intelligence model training make them fit to the manage of complex datasets from many different farms or longitudinal studies (Ficili et al., 2025). These platforms support superior forms of analytical techniques such as deep learning, predictive modelling, and remote-visualization dashboards for agricultural and agronomic planning on a long-term basis. Conversely, edge analytics deals with local data processing at gateways, UAVs or sensor locations, thus providing immediate reactions to important events economics (such as thermal spikes), water scarcity, pest incursions etc. By synergistically integrating cloud-based intelligence with edge-side low latency, the integrated provides periods of strategic and real time decision-making capability. This is a hybrid analytics framework that further increases the efficacy, accuracy and reliability of the crop health monitoring systems (Mushtaq et al., 2025).

Table 1. Overview of AI–IoT technologies for crop health management

Category	Technologies/Tools	Applications	Benefits	References
Sensors & IoT Devices	Soil moisture sensors, nutrient probes, temperature & humidity sensors, hyperspectral/ multispectral sensors, nano-sensors, biodegradable probes	Monitoring soil health, plant water and nutrient status, early stress detection	Real-time data collection, precise resource management, reduced input waste	(Tornese et al., 2024)
AI & Machine Learning	Supervised/unsupervised learning, reinforcement learning, CNNs, RNNs, transformers	Disease detection, stress assessment, pest forecasting, nutrient deficiency prediction	Early warning, accurate diagnosis, predictive analytics for proactive interventions	(Bilal et al., 2025)
Computer Vision & Imaging	RGB, multispectral, hyperspectral, thermal imaging, UAV/drone imaging	Leaf disease detection, canopy monitoring, weed identification, crop vigor assessment	Rapid, high-resolution monitoring, reduced manual labor, scalable field coverage	(Upadhyay et al., 2025)
Decision Support Systems (DSS)	Automated recommendation engines, mobile dashboards, cloud & edge analytics	Irrigation scheduling, fertilizer/pesticide management, risk assessment, yield-loss prediction	Improved operational efficiency, resource optimization, actionable insights for farmers	(R. Zhang et al., 2025)
Autonomous Systems & Robotics	Drones, robotic sprayers, autonomous tractors, UAV-based monitoring	Targeted interventions, variable-rate spraying, automated monitoring	Labor reduction, precise field management, seamless integration with AI predictions	(Nagaraja et al., 2024)
Integration & Predictive Models	Digital twins, AIoT platforms, data fusion, predictive modeling	Crop growth simulation, outbreak prediction, stress threshold modeling	Proactive crop management, enhanced resilience, climate-smart agriculture	(Awais et al., 2025)

5. APPLICATIONS IN DISEASES, PEST AND STRESS DETECTION

5.1. Early Detection of Plant Diseases through AI Vision and Integrated Sensors

Artificial intelligence - driven computer vision has become an exciting approach to early detection of plant diseases and can help agronomists detect symptomatic manifestations of plant diseases long before visual observation (Al-Bemani et al., 2025). Using High Res (RGB), Multi-Spectrum and Hyperspectral Imaging collected via Unmanned Aerial Vehicles, (UAV) and Stationary Field Cameras/Handheld Instrumentation, Artificial Intelligence (AI) models are able to interactively infer minute phenotypic characteristics such as the lesion morphology, chlorosis, discoloration and thermal energy gradients that are indicative of nascent fungal/bacterial/viral infections. (P. Zhang et al., 2025)If these imagery datasets are integrated with Internet-of-Things (IoT) inputs from sensors sensing the atmosphere, such as ambient humidity, canopy temperature, and leaf wetness, the accuracy of detecting the disease is increased and the contextual relevance is increased. As a result, these hybrid AI- IoT disease surveillance systems significantly reduce the diagnostic latency time, thereby allowing timely interventions, thereby reducing loss of yield and chemical usage. Presently, such systems are widely implemented in horticultural production, cereal production, orchard production and controlled environment agriculture (Morchid et al., 2024).

5.2. AI–IoT Systems for Pest Forecasting and Abiotic Stress Monitoring

Sensor networks that measure environmental variables are coupled to machine learning algorithms play a critical role in prognostication of pest outbreaks as well as real time monitoring of abiotic stress (Wang et al., 2024). Pest forecasting frameworks examine datasets that include temperature, soil moisture and wind velocity, as well as the number of insects collected by a pheromone trap, to represent the population dynamics of various important insect taxa, such as borers, fruit flies, whiteflies and aphids. Advanced forms of machine learning are also used to refine these predictive outputs by incorporating historical records of insect infestations and their correlation with variations in populations of climatic conditions (Chakrabarty et al., 2026). Analogously, abiotic stress detection platforms continuously monitor drought and stress via measurements of heat, salinity, and nutrient deficiency, as well as soil moisture detected by moisture probes in the soil.Analogously,stadt9 uv intelligence platform fixtures for the detection of abiotic stress to detect drought, heat, salinity, and nutrient deficiency continuously by increases taken from soil moisture probes, temperature loggers, electrical conductivity meters, and spectroscopic indices, producing the Normalized Difference Vegetation Index (NDVI) and Photochemical Reflectance Index (PRI) . By combining these stream of sensors with analytic pipelines (applying AI) the system outlines stress thresholds, severity and makes precise agronomic recommendations on how to act e.g. in strength regards irrigation, fertilizers and microclimate regulation (Chavanne & Frangi, 2024). This ongoing surveillance is considered to ensure that remedial actions occur before there is onset of irreversible stress responses.

5.3. Outbreak Prediction, Stress Threshold Modeling, and Global Case Studies

Being able to anticipate pest outbreaks and simulate stress levels is an advanced use case of AI - IoT platforms in the management of crop health. By analyzing multi-season datasets that use weather forecasts, edaphic attributes, maps of disease incidence and maps of pest movements AI models estimate the probability of pathogen or insect emergence and the temporal window of emergence (Nyakuri et al., 2025). Threshold models capture the set of environmental conditions from which dispersal of pathogens or reproduction of insects become favorable and thereby allows farmers to take precision preventative measures. Illustrative examples of global smart includes some innovative applications and innovations which demonstrate the transformative power of smart agriculture technologies: rice producers use Artificial Intelligence (AI) enabled unmanned aerial vehicles and Internet of Things (IoT) solutions to detect blast disease in the early stages of infection in Japan, producers of greenhouse crops use sensor networks and artificial intelligence machine learning computer algorithm to control humid conditions in real time for the control of powdery mildew in the Netherlands, and even pest forecasting platforms for predicting locust, fall armyworm and Whitefly invasion in India and Africa with smallholder farmers (Ali et al., 2025).

6. AI-BASED CROP HEALTH DECISION SUPPORT SYSTEM

6.1. Automated Recommendation Engines for Nutrient and Water Management

AI-fueled decision support systems (DSS) are one of the cornerstones in the optimization of nutrient and water management by analyzing raw data received from Internet of Things (IoT) sensors and converting it into actionable recommendations

(Eyasin et al., 2025). Automated engines study the soil moisture level, the rate of evapotranspiration, the concentration of nutrients, the weather and the phenology of crops to decide the precise schedule of irrigation and fertilization. Machine learning models constantly feed on past data to optimize predictive outputs, hence allowing systems to change over the years in response to the changing conditions in the field and the requirements of the crop. Such an intelligent recommendation engine helps reduce water wastage, improve efficiency in nutrients use, and also reduce the problems associated with excess application as well as stress caused due to deficient application (Waqas et al., 2025). Integration of real-time sensor feedback to ensure that interventions are spatially, and are efficient in terms of resources .

6.2. AI-Guided Pesticide and Fertilizer Application

AI guided application systems use predictive and spatial intelligence technologies for site specific and environmentally responsible pesticide and fertilizer applications (Chatrabhuj et al., 2025). These systems evaluate the risk of disease development, the risk of pests, the vigor and nutrient status of the canopy, and a far greater emphasis is placed on corrective applications of chemicals used on an as needed basis as opposed to the indiscriminate spraying of the crop. Computer vision algorithms, implemented on drones and robots on the ground, define the work areas, and artificial intelligence tools calculate the application volume and time of application depending on the condition of the crops and meteorological data (Ghazal et al., 2024). This approach not only contributes to the efficient usage of pest and disease control but also significantly reduces chemical inputs in order to promote sustainable agriculture and to minimize the ecological impact. AI guided application frameworks are becoming an increasing part of autonomous sprayers and precision boom systems to get an edge to being more accurate and as less work for the operator (Rajuroy, 2025).

6.3. Integration of DSS with Farm Machinery and Robotics

Modern DSS platforms are gradually being connected to high-tech farm machinery, driverless vehicles, and robotics to form a fully automated ecosystem for the management of crops (Mandal et al., 2024). Integration with tractors, smart sprayers, irrigation systems and robotic scouts allows AI-generated recommendations to be carried out with a minimal amount of human intervention. For example, autonomous irrigation systems make changes in irrigation delivery based on DSS information, and robotic sprayers provide variable - rate pesticide applications to heterogeneous areas of a field (Kalpana et al., 2024). Drones integrated with artificial intelligence (AI) vision go on scouting missions in the areas marked by the DSS alerts, which speeds up the process of the diagnosis and intervention (Bekmurzaeva et al., 2024). This ambition integration helps to create a seamless cycle of monitoring, analyzing and acting, therefore simplifying labor needs and improving operation precision at scale.

6.4. Risk Assessment Models and Yield-Loss Prediction

Risk assessment is an integral part of AI-enabled DSS by assigning a number to the threats from diseases, pests, climatic extremes or nutrient limitations (Pace et al., 2025). These models assess multidmensional datasets such as weather predictions, historic incidence of stress, soil profiles and crop growth indicators and calculate probabilities of risk and the potential impacts on yields. Predictive algorithms to identify how management choices would be divergent outcome in crops and help farmers implement priority management interventions according to economic and agronomic factors Yield-loss prediction tools also aid in planning, by helping predict results of delayed treatment, resources limitation or extreme environmental stress (Elbasi et al., 2023) . By providing forward-looking insights into it, these models make it possible to think ahead and not just react as a result of farming.

6.5. User-Friendly Dashboards and Mobile Applications

In order to ensure this adoption at a mass level, modern DSS platforms should focus on accessibility and offer intuitive dashboards and mobile apps. These interfaces interpret complex analytical results in potentially visually comprehensible formats, including heat maps, stress indicators, risk scores and recommendations for treatment (Krisnawijaya et al., 2022). Farmers can get real-time field updates, alerts and historical performance trends at the span of the smartphone or tablet, thus enabling quick decision making even in remote locations. Features like voice assistant integrated help, offline functionality, multi-language functionality and link with local advisory services, allows for improved usability. By combining advanced AI level analytics with farmer-friendly interfaces, the digital dashboards and applications can democratize the use of technology, as well as empower users to make informed decisions, in timely and necessary goals (Mendes et al., 2020).

7. ECONOMIC, ENVIRONMENTAL AND SOCIAL IMPACTS

7.1. Economic Benefits and Cost–Benefit Dynamics of AI–IoT Adoption

The economic effects of the integration of artificial intelligence and Internet of Things in the agricultural sector are huge, mainly due to improved productivity, less input costs and accurate crop health surveillance (Younas et al., 2025). Cost-and-benefit analyses always show that smart farming technologies allow farmers to optimize irrigation regimes, fertilizer application and pesticide deployment, thereby resulting in significant savings on the one hand and on the other hand increasing both quantity and quality of yield. Although the initial capital needs for sensors, connectivity options and artificial intelligence (AI) platforms may pose some challenges, especially for smallholder operations, the financial benefits from the reduction in labor requirements, minimized crop attrition and enhanced marketability of produce are often greater than the initial financial investment (Chad, 2025). Numerous governmental entities and private organizations now promote AI - IoT adoption with subsidies, credit protection, and public-private agreements which further promotes affordability and accessibility.

7.2. Environmental Sustainability and Resource-Use Efficiency

Artificial intelligence-Internet of Things systems plays a pivotal role in promoting ecological ways of agricultural practices by enabling accurate resource management and reducing the ecological footprint. Data-driven insights help farmers receive a better understanding of where and how to use fertilizers and pesticides, greatly reducing the amount of chemicals and preventing soil and water contamination (Azlan et al., 2025). Smart irrigation schemes based on real time data on soil moisture and evapotranspiration reduce water use efficiency by ensuring that crops receive only the required amount of water. Likewise, AI-driven nutrient management reduces Nitrogen and Phosphorus leaching which leads to healthier ecosystems and soil structure. By keeping track of abiotic stresses (heat, drought, salinity etc.) such systems foster adaptive agricultural techniques which reduce environmental degradation and contribute to long-term agro-ecosystem stability (Nurmalitasari et al., 2025).

7.3. Social Impacts, Digital Inclusion, and Climate-Smart Agriculture

The social ramifications of AI-IoT adoption are also seen in empowerment of the farmers, especially smallholders, through enabling access to on-time information, developing decision making capacity and also reducing dependence on traditional labor-intensive methods. Mobile-based dashboard and advisory services make sophisticated information about the condition of the crops available regardless of literacy, and create a structure of inclusivity and digital inclusivity (Omotayo et al., 2025). Access to predictive analytics provides enhanced resilience of farmers to climate induced risks with the ability to generate early warnings for droughts, pests and disease breakouts. This directly supports climate-smart agriculture through the facilitation of practices that improve productivity, reduce greenhouse gas emissions and support adaptation measures. Through better livelihoods, access and climate resilience, AI-IoT technologies support agricultural sustainability on a community, as well as national level (Nega, 2025).

8. LIMITATIONS, CHALLENGES, AND POLICY PERSPECTIVES

While artificial intelligence (AI) and the Internet of Things (IoT) show a lot of potential to revolutionize crop health monitoring and management, their adoption at a mass level is hindered by a convergence of technical, financial and policy-related issues. Technical barriers are the most immediate barriers; these barriers relate to connectivity gaps in rural areas, poor or inconsistent sensor performance, or interoperability between a wide range of different devices and platforms (Srivastav & Das, 2025). Many IoT devices may succumb to extreme field conditions (like high temperature, elevated levels of humidity, or dust presence) and hence, reduce the level of accuracy and reliability of data collected. Data quality problems, which include noise, missing data, and transmission latency, further add to the challenges of AI-based predictions and limit the power of decision-making systems that have been automated (Farooq et al., 2020).

Financial constraints, especially to smallholder and resource poor farmers, are a major constraint to deploying precision agriculture technologies. The costs of the smart-sensors, artificial intelligence, analytics platforms, unmanned aerial vehicles (UAVs) and other precision-farming equipment all add up to a potential cost that can be too expensive at the outset. Although the use of extended financing terms may potentially compensate for the initial costs with lower input costs and improved yields, access to credit, subsidies and supporting infrastructure are deficient in many developing regions (Touch et al., 2024). In addition, proper implementation of AI-IoT systems requires sufficient technical requirements and digital literacy. Oftentimes farmers, extension agents, and field technicians do not possess or possess adequate level to be able to interpret complex analytics, operate expensive machinery or to interact properly with integrated (AI-IoT) platform. As such, they are not able to harness such systems as they have already become part of the common harvesting and farming processes in modern times (Bhangar & Kashem, 2023).

Policy and regulatory factors are also critical factors in the direction of the digital agriculture adoption route. Lack of standardized protocols for data collection, storage and sharing is hindering interoperability and limiting innovation (Dibbern et al., 2024). Concerns about the privacy of farm level data combined with questions about who owns data can encourage farmers to not release data that is crucial for advanced predictive analytics. Additionally, inconsistent government policies and inconsistent support programs are some of the factors that further limit equitable adoption, as are low levels of investment in rural digital infrastructure (Khanna, 2021).

To overcome such hindrances, a concerted effort is a must. Such efforts should include the creation of standardized guidelines, offering financial incentives, the development of comprehensive training programs and the creation of open innovation ecosystems that build on collaboration between governments, research institutions and private entities (Becerra-Encinales et al., 2024). As such, by combating the interrelated technical, financial and regulatory challenges in a systematic way, it is possible to make AI-IoT systems more accessible, reliable and sustainable, and provide support to inclusive and climate-safe agriculture throughout the world (Yuan & Sun, 2024).

9. FUTURE DIRECTIONS AND INNOVATION ROADMAP

The future trajectory of AI knowledgeable IoT enabled crop health management seems to be prescribed by the next generation smart sensors, autonomous AI-O-T ecosystem and integrative frameworks that interlink genes, plant phenotyping and breeding. Advancements in the now more powerful sensor technology promise to allow for the implementation of nanosensors, biodegradable devices and multiparameter probing devices that will cheaply measure a wide range of physiological, biochemical and environmental parameters. These sensors are likely to produce real-time and continuous data streams on the soil nutrients, moisture, temperature, humidity and metabolic activity of plants while at the same time minimizing impact on the environment and decreasing the need for regular maintenance.

The combination of artificial intelligence and IoT in autonomous crops health ecosystems is expected to be another revolution in precision agriculture. Robotics, unmanned aerial vehicles and self-propelled field vehicles are likely to interact with edge AI systems to perform real-time surveillance of likely stress and disease hotspots, or take intervention actions with little human intervention. Self diffusive farms which will be able to optimize irrigation, nutrient application and pest control dynamically with persistence of data streams and dynamic adaptation of machine learning algorithms will ensure responsive and context aware management. Such autonomous systems are expected to work in concert to form completely connected field networks which would constantly monitor, predict and respond to problems related to the health of the crops.

Future predictive models for crop health will also take advantage of multidisciplinary integration, with the combination of AI - IoT data with plant genomics, phenotyping and breeding programs. This integration will make it possible to detect genetic traits for stress tolerance earlier, speed up the development of climate resilient varieties and provide useful information for the precision breeding. Models will include climate variability and extremes, and interactions of soil and plant, and increase in their predictive capacities and lead to increase their resilience against biotic and abiotic stresses.

In spite of these advances, there are a number of gaps in the research. Key deficiencies include the lack of standardized protocols for data fusion, insufficient and unrealistic analytics platforms to ease smallholder adoption of analytics, and a universal framework of ethical and sustainable deployment of AI-IoT technologies. Opportunities remain in integrating devices powered by renewable energy, global data-sharing discourses, and advisory services driven by AI to facilitate sharing of integrals towards promotions and its equitable access dynamical to technology. Projected to 2030, the future envisioned is a fully connected, predictive and sustainable agricultural ecosystem, in which AI - IoT ecologies will constantly monitor crops, anticipate threats, optimize the use of resources and assist in climate smart management strategies. This roadmap heralds a future of intelligent, autonomous and inclusive agriculture whereby production is not only improved but successfully environmental sustainability and global food security are simultaneously protected.

10. CONCLUSION

The steady convergence between Artificial Intelligence (AI) and Internet of Things (IoT) has had a significant effect on the management of crop health as it has made it possible to use predictive precision agriculture for sustainability purposes. Conventional methodologies - based on manual scouting and reactive interventions - are often insufficient to deal with problems such as climatic variability, emerging pest episodes, nutritional abnormalities and unpredictable weather occurrences. AI - IoT framework overcomes such limitations and uses high resolution sensor data machine learning, deep learning and computer vision techniques to detect disease, abiotic stresses and nutrient deficiencies in their early stages. Real-time monitoring (surveillance) through drone monitoring, ground-based sensors, edge computing: This creates coverage over long distances. Digital twins Predictive analytics: Predictive analytics is increasingly used to proactively take decision and optimize the use of resources. Decision support systems to translate the complex data sets and provide actionable recommendations for focused irrigation use, nutrient management, pest control etc. Decision support systems are being developed to work in tandem with the self-driving machines and robots for better operation and minimize human labor. Such developments bring a great number of economic, environmental, and social benefits such as improved productivity, reduced costs of inputs, reduced chemical applications, enhanced ecosystem sustainability, and increased resilience of small-holder farmers through available mobile platforms. Nevertheless, challenges glorious obstacles including restricted connectivity, sensor durability, rigorous preliminary capital expenditures, lack of digital literacy and much regulation frameworks, nonetheless continue to hinder expansive adoption. Over conquering these barriers requires the establishment of support policy measures, substantial training programs, and any standardization of data protocols that are crucial for removing barriers and maximizing the true potentials of AI - IoT systems. In conclusion, AI and IoT is a transformational pathway for evidence-based efficient and resilient management of crops health and thus increasing the productivity and sustainability of current agricultural systems.

REFERENCES

Ahmad, U., Nasirahmadi, A., Hensel, O., & Marino, S. (2022). Technology and Data Fusion Methods to Enhance Site-Specific Crop Monitoring. *Agronomy (Basel), 12*, 555. DOI: 10.3390/agronomy12030555

Ahmed, N., & Shakoor, N. (2025). Advancing agriculture through IoT, Big Data, and AI: A review of smart technologies enabling sustainability. *Smart Agricultural Technology*, *10*, 100848. 10.1016/j.atech.2025.100848

Al-Bemani, A., Al-Mahruqi, A., Balaji, G., & Suresh, U. (2025). A Relative Analysis for Plant Disease Detection with AI-Driven Techniques: Optimizing Precision Agriculture. *Procedia Computer Science*, *258*, 839–851.

Ali, Z., Muhammad, A., Lee, N., Waqar, M., & Lee, S. W. (2025). Artificial Intelligence for Sustainable Agriculture: A Comprehensive Review of AI-Driven Technologies in Crop Production. *Sustainability*, *17*(5), 2281. https://www.mdpi.com/2071-1050/17/5/2281

Attaran, M., & Celik, B. G. (2023). Digital Twin: Benefits, use cases, challenges, and opportunities. *Decision Analytics Journal, 6*, 100165. 10.1016/j.dajour.2023.100165

Awais, M., Wang, X., Hussain, S., Aziz, F., & Mahmood, M. Q. (2025). Advancing Precision Agriculture Through Digital Twins and Smart Farming Technologies: A Review. *AgriEngineering, 7*(5), 137. https://www.mdpi.com/2624-7402/7/5/137

Azlan, Z. H. Z., Junaini, S. N., Khan, A. S., & Mustafa, W. A. (2025). Data-Driven Agriculture: Unveiling the Power of Internet of Things and Data Analytics For Transformative Farming Practices. *Journal of Sensors, 2025*(1), 6205646. 10.1155/js/6205646

Becerra-Encinales, J. F., Bernal-Hernandez, P., Beltrán-Giraldo, J. A., Cooman, A. P., Reyes, L. H., & Cruz, J. C. (2024). Agricultural extension for adopting technological practices in developing countries: A scoping review of barriers and dimensions. *Sustainability*, *16*(9), 3555.

Bekmurzaeva, R., Kalimullin, R., & Aguzarova, F. (2024). Application of drones and artificial intelligence to monitor and protect natural ecosystems. *BIO Web of Conferences, 140*. https://doi.org/DOI: 10.1051/bioconf/202414001008

Bhangar, N., & Kashem, A. (2023). IoT and AI for Next-Generation Farming: Opportunities, Challenges, and Outlook. 8.

Bilal, M., Shah, A. A., Abbas, S., & Khan, M. A. (2025). High-Performance Deep Learning for Instant Pest and Disease Detection in Precision Agriculture. *Food Science & Nutrition, 13*(9), e70963. DOI: 10.1002/fsn3.70963 PMID: 40964151

Botero-Valencia, J., García-Pineda, V., Valencia-Arias, A., Valencia, J., Reyes-Vera, E., Mejia-Herrera, M., & Hernández-García, R. (2025). Machine Learning in Sustainable Agriculture: Systematic Review and Research Perspectives. *Agriculture, 15*(4), 377. https://www.mdpi.com/2077-0472/15/4/377

Chad, F. (2025). *Challenges of AI Adoption in Smallholder Farming: Infrastructure*. Cost, and Digital Literacy.

Chakrabarty, S., Deb, C. K., Marwaha, S., Haque, M. A., Kamil, D., Bheemanahalli, R., & Shashank, P. R. (2026). Application of artificial intelligence in insect pest identification - A review. *Artificial Intelligence in Agriculture, 16*(1), 44-61. 10.1016/j.aiia.2025.06.005

Chatrabhuj, M. K., Mishra, U., & Rathnayake, U. (2025). Application of artificial intelligence in agri-tech, environmental and biodiversity conservation. *Array, 26*, 100412. 10.1016/j.array.2025.100412

Chavanne, X., & Frangi, J.-P. (2024). A Sensor to Monitor Soil Moisture, Salinity, and Temperature Profiles for Wireless Networks. *Journal of Sensor and Actuator Networks, 13*(3), 32. https://www.mdpi.com/2224-2708/13/3/32

Choudhary, V., Guha, P., Pau, G., & Mishra, S. (2025). An overview of smart agriculture using internet of things (IoT) and web services. *Environmental and Sustainability Indicators, 26*, 100607. 10.1016/j.indic.2025.100607

Choudhury, S., & Banerjee, B. (2024). Deep Learning: A Catalyst for Sustainable Agriculture Transformation. In M. S. Raval, S. Chaudhary, J. Adinarayana, & W. Guo (Eds.), *Harnessing Data Science for Sustainable Agriculture and Natural Resource Management* (pp. 173-199). Springer Nature Singapore. https://doi.org/ DOI: 10.1007/978-981-97-7762-4_8

Čolaković, A., Haskovic Dzubur, A., & Karahodža, B. (2021). Wireless communication technologies for the Internet of Things. *Science Engineering and Technology, 1*, 1–14. DOI: 10.54327/set2021/v1.i1.3

Dhanaraju, M., Chenniappan, P., Ramalingam, K., Pazhanivelan, S., & Kaliaperumal, R. (2022). Smart Farming: Internet of Things (IoT)-Based Sustainable Agriculture. *Agriculture, 12*(10), 1745. https://www.mdpi.com/2077-0472/12/10/1745

Dibbern, T., Romani, L. A. S., & Massruhá, S. M. F. S. (2024). Main drivers and barriers to the adoption of Digital Agriculture technologies. *Smart Agricultural Technology*, *8*, 100459.

Ding, S., Liu, J., & Yue, M. (2021). The Use of ZigBee Wireless Communication Technology in Industrial Automation Control. *Wireless Communications and Mobile Computing*, *2021*(1), 8317862. 10.1155/2021/8317862

Eyasin, M. S. H., Sobhani, M. E., Nasrin, S., Al Rafi, A. S., & Muzahidul Islam, A. K. M. (2025). CropSynergy: Harnessing IoT Solutions for Smart and Efficient Crop Management. *Crop Design*, 100127. 10.1016/j.cropd.2025.100127

Faqir, Y., Qayoom, A., Erasmus, E., Schutte-Smith, M., & Visser, H. G. (2024). A review on the application of advanced soil and plant sensors in the agriculture sector. *Computers and Electronics in Agriculture*, *226*, 109385. 10.1016/j.compag.2024.109385

Farooq, M. S., Riaz, S., Abid, A., Umer, T., & Zikria, Y. B. (2020). Role of IoT Technology in Agriculture: A Systematic Literature Review. *Electronics (Basel)*, *9*(2), 319. https://www.mdpi.com/2079-9292/9/2/319

Ficili, I., Giacobbe, M., Tricomi, G., & Puliafito, A. (2025). From Sensors to Data Intelligence: Leveraging IoT, Cloud, and Edge Computing with AI. *Sensors (Basel)*, *25*(6), 1763. https://www.mdpi.com/1424-8220/25/6/1763 PMID: 40292910

Fuentes-Peñailillo, F., Gutter, K., Vega, R., & Silva, G. C. (2024). Transformative Technologies in Digital Agriculture: Leveraging Internet of Things, Remote Sensing, and Artificial Intelligence for Smart Crop Management. *Journal of Sensor and Actuator Networks*, *13*(4), 39. https://www.mdpi.com/2224-2708/13/4/39

Ghazal, S., Munir, A., & Qureshi, W. S. (2024). Computer vision in smart agriculture and precision farming: Techniques and applications. *Artificial Intelligence in Agriculture, 13*, 64-83. 10.1016/j.aiia.2024.06.004

Haq, S. I. U., Wang, G., Khan, S. N., Song, C., Ma, C., Zhang, X., & Lan, Y. (2025). Multi-dimensional optical remote sensing in agriculture: Spectral, angular, and spatial scaling for crop stress monitoring. *Smart Agricultural Technology*, *12*, 101583. 10.1016/j.atech.2025.101583

Held, L. K., Vermeylen, L., Dignath, D., Notebaert, W., Krebs, R. M., & Braem, S. (2024). Reinforcement learning of adaptive control strategies. *Communications Psychology*, *2*(1), 8. DOI: 10.1038/s44271-024-00055-y PMID: 39242891

Kalpana, P., Smitha, L., Madhavi, D., Shaik, A., Kalpana, G., & Kodati, S. (2024). A Smart Irrigation System Using the IoT and Advanced Machine Learning Model: A Systematic Literature Review. *International Journal of Computational and Experimental Science and Engineering*, *10*. Advance online publication. DOI: 10.22399/ijcesen.526

Kamal, M., & Bablu, T. (2023). Mobile Applications Empowering Smallholder Farmers. *An Analysis of the Impact on Agricultural Development.*, *8*, 36–52.

Katie, B. (2024). Internet of Things (IoT) for Environmental Monitoring. *International Journal of Computing and Engineering*, *6*, 29–42. DOI: 10.47941/ijce.2139

Khanna, M. (2021). Digital transformation of the agricultural sector: Pathways, drivers and policy implications. *Applied Economic Perspectives and Policy*, *43*(4), 1221–1242.

Krisnawijaya, N., Tekinerdogan, B., Catal, C., & Tol, P. P. J. (2022). Data analytics platforms for agricultural systems: A systematic literature review. *Computers and Electronics in Agriculture*, *195*, 106813. DOI: 10.1016/j.compag.2022.106813

Kumar, V., Sharma, K. V., Kedam, N., Patel, A., Kate, T. R., & Rathnayake, U. (2024). A comprehensive review on smart and sustainable agriculture using IoT technologies. *Smart Agricultural Technology*, *8*, 100487. 10.1016/j.atech.2024.100487

Lausch, A., Bumberger, J., Jung, A., Pause, M., Selsam, P., Zhou, T., & Herzog, F. (2025). Monitoring Agricultural Land Use Intensity with Remote Sensing and Traits. *Agriculture*, *15*(21), 2233. https://www.mdpi.com/2077-0472/15/21/2233

Mandal, S., Yadav, A., Panme, F. A., Devi, K. M., & Kumar, S. M. S. (2024). Adaption of smart applications in agriculture to enhance production. *Smart Agricultural Technology*, *7*, 100431. 10.1016/j.atech.2024.100431

Mansoor, S., Iqbal, S., Popescu, S. M., Kim, S. L., Chung, Y. S., & Baek, J.-H. (2025). Integration of smart sensors and IOT in precision agriculture: trends, challenges and future prospectives [Review]. *Frontiers in Plant Science, Volume 16 - 2025.* https://doi.org/DOI: 10.3389/fpls.2025.1587869

Melesse, T. Y. (2025). Digital twin-based applications in crop monitoring. *Heliyon*, *11*(2), e42137. 10.1016/j.heliyon.2025.e42137 PMID: 39911426

Mendes, J., Pinho, T. M., Neves Dos Santos, F., Sousa, J., Peres, E., Cunha, J., Cunha, M., & Morais, R. (2020). Smartphone Applications Targeting Precision Agriculture Practices—A Systematic Review. *Agronomy (Basel)*, *10*, 855. DOI: 10.3390/agronomy10060855

Mienye, I. D., Swart, T. G., & Obaido, G. (2024). Recurrent Neural Networks: A Comprehensive Review of Architectures, Variants, and Applications. *Information (Basel), 15*(9), 517. https://www.mdpi.com/2078-2489/15/9/517

Miller, T., Mikiciuk, G., Durlik, I., Mikiciuk, M., Łobodzińska, A., & Śnieg, M. (2025). The IoT and AI in Agriculture: The Time Is Now—A Systematic Review of Smart Sensing Technologies. *Sensors (Basel), 25*, 3583. DOI: 10.3390/s25123583 PMID: 40573470

Morchid, A., Marhoun, M., El Alami, R., & Boukili, B. (2024). Intelligent detection for sustainable agriculture: A review of IoT-based embedded systems, cloud platforms, DL, and ML for plant disease detection. *Multimedia Tools and Applications, 83*(28), 70961–71000. DOI: 10.1007/s11042-024-18392-9

Mushtaq, M. A., Ateeq, M., Ikram, M., Alam, S. M., Kaleem, M. M., Ashraf, M. A., Asim, M., Almutairi, K. F., Seleiman, M. F., & Shireen, F. (2025). Securing fruit trees future: AI-driven early warning and predictive systems for abiotic stress in changing climate. *Plant Stress, 17*, 100953. 10.1016/j.stress.2025.100953

Nagaraja, G., Shoba, H., Ms, S., & Pn, K. (2024). The impact of robotics and drones on agricultural efficiency and productivity. *International Journal of Research in Agronomy, 7*(9), 1001–1009.

Naqvi, S. M. Z. A., Tahir, M. N., Raghavan, V., Awais, M., Hu, J., Said, Y., Othman, N. A., Ashurov, M., & Khan, M. I. (2025). AI-enhanced IoT sensors for real-time crop monitoring: An era towards self-monitored agriculture. *Telecommunication Systems, 88*(3), 100. DOI: 10.1007/s11235-025-01326-7

Nega, A. (2025). Climate Change Impacts on Agriculture: A Review of Plant Diseases and Insect Pests in Ethiopia and East Africa, With Adaptation and Mitigation Strategies. *Advances in Agriculture, 2025*(1), 5606701. 10.1155/aia/5606701

Nurmalitasari, Nurchim, & Lestari, R. D. (2025). Artificial intelligence-driven solar smart irrigation for sustainable agriculture: Trends, challenges, and SDG implications – A systematic review. *Smart Agricultural Technology, 12*, 101665. 10.1016/j.atech.2025.101665

Nyakuri, J., Nkundineza, C., Gatera, O., Nkurikiyeyezu, K., & Mwitende, G. (2025). *AI-IoT driven system for agricultural pest outbreak risk prediction*. https://doi.org/ DOI: 10.21203/rs.3.rs-6881038/v1

Omotayo, A. O., Adediran, S. A., Omotoso, A. B., Olagunju, K. O., & Omotayo, O. P. (2025). Artificial intelligence in agriculture: Ethics, impact possibilities, and pathways for policy. *Computers and Electronics in Agriculture, 239*, 110927. 10.1016/j.compag.2025.110927

Pace, R., Schiano Di Cola, V., Monti, M. M., Affinito, A., Cuomo, S., Loreto, F., & Ruocco, M. (2025). Artificial intelligence in soil microbiome analysis: A potential application in predicting and enhancing soil health—a review. *Discover Applied Sciences, 7*(2), 85. DOI: 10.1007/s42452-024-06381-4

Paul, N., Sunil, G. C., Horvath, D., & Sun, X. (2025). Deep learning for plant stress detection: A comprehensive review of technologies, challenges, and future directions. *Computers and Electronics in Agriculture, 229*, 109734. 10.1016/j.compag.2024.109734

Peladarinos, N., Piromalis, D., Cheimaras, V., Tserepas, E., Munteanu, R. A., & Papageorgas, P. (2023). Enhancing Smart Agriculture by Implementing Digital Twins: A Comprehensive Review. *Sensors (Basel), 23*(16), 7128. DOI: 10.3390/s23167128 PMID: 37631663

Quy, V. K., Hau, N. V., Anh, D. V., Quy, N. M., Ban, N. T., Lanza, S., Randazzo, G., & Muzirafuti, A. (2022). IoT-Enabled Smart Agriculture: Architecture, Applications, and Challenges. *Applied Sciences (Basel, Switzerland), 12*(7), 3396. https://www.mdpi.com/2076-3417/12/7/3396

Rajuroy, A. (2025). Autonomous Sprayers and Precision Pesticide Application.

Rashid, A. B., Kausik, A. K., Khandoker, A., & Siddque, S. N. (2025). Integration of Artificial Intelligence and IoT with UAVs for Precision Agriculture. *Hybrid Advances, 10*, 100458. 10.1016/j.hybadv.2025.100458

Sathya, D., Thangamani, R., & Balaji, B. S. (2024). The Revolution of Edge Computing in Smart Farming. In Balasubramanian, S., Natarajan, G., & Chelliah, P. R. (Eds.), *Intelligent Robots and Drones for Precision Agriculture* (pp. 351–389). Springer Nature Switzerland., DOI: 10.1007/978-3-031-51195-0_17

Shamshiri, R. R., Sturm, B., Weltzien, C., Fulton, J., Khosla, R., Schirrmann, M., Raut, S., Basavegowda, D. H., Yamin, M., & Hameed, I. A. (2024). Digitalization of agriculture for sustainable crop production: a use-case review [Review]. *Frontiers in Environmental Science, Volume 12 - 2024*. https://doi.org/DOI: 10.3389/fenvs.2024.1375193

Sharafat, M. D. S., Kabya, N. D., Emu, R. I., Ahmed, M. U., Onik, J. C., Islam, M. A., & Khan, R. (2025). An IoT-enabled AI system for real-time crop prediction using soil and weather data in precision agriculture. *Smart Agricultural Technology, 12*, 101263. 10.1016/j.atech.2025.101263

Sharma, K., & Shivandu, S. K. (2024). Integrating artificial intelligence and Internet of Things (IoT) for enhanced crop monitoring and management in precision agriculture. *Sensors International, 5*, 100292. 10.1016/j.sintl.2024.100292

Srivastav, A. K., & Das, P. (2025). Challenges and Barriers to Adoption. In *Biotechnology and IoT in Agriculture and Food Production: Green Innovation* (pp. 161-168). Springer.

Subeesh, A., & Chauhan, N. (2026). Agricultural digital twin for smart farming: A review. *Green Technologies and Sustainability, 4*(2), 100299. 10.1016/j.grets.2025.100299

Titirmare, S., Margal, P., Gupta, S., & Kumar, D. (2024). AI-Powered Predictive Analytics for Crop Yield Optimization. In (pp. 89-110). https://doi.org/DOI: 10.1201/9781003570219-5

Tornese, I., Matera, A., Rashvand, M., & Genovese, F. (2024). Use of Probes and Sensors in Agriculture—Current Trends and Future Prospects on Intelligent Monitoring of Soil Moisture and Nutrients. *AgriEngineering, 6*(4), 4154-4181. https://www.mdpi.com/2624-7402/6/4/234

Touch, V., Tan, D. K. Y., Cook, B. R., Liu, D. L., Cross, R., Tran, T. A., Utomo, A., Yous, S., Grunbuhel, C., & Cowie, A. (2024). Smallholder farmers' challenges and opportunities: Implications for agricultural production, environment and food security. *Journal of Environmental Management, 370*, 122536. 10.1016/j.jenvman.2024.122536 PMID: 39299125

Upadhyay, A., Chandel, N. S., Singh, K. P., Chakraborty, S. K., Nandede, B. M., Kumar, M., Subeesh, A., Upendar, K., Salem, A., & Elbeltagi, A. (2025). Deep learning and computer vision in plant disease detection: A comprehensive review of techniques, models, and trends in precision agriculture. *Artificial Intelligence Review, 58*(3), 92. DOI: 10.1007/s10462-024-11100-x

Wang, J., Wang, Y., Li, G., & Qi, Z. (2024). Integration of Remote Sensing and Machine Learning for Precision Agriculture: A Comprehensive Perspective on Applications. *Agronomy (Basel), 14*(9), 1975. https://www.mdpi.com/2073-4395/14/9/1975

Wang, X., Zhang, J.-H., Xun, L., Wang, J., Wu, Z., Henchiri, M., Zhang, S., Zhang, S., Yun, B., Yang, S., Li, S., & Yu, X. (2022). Evaluating the Effectiveness of Machine Learning and Deep Learning Models Combined Time-Series Satellite Data for Multiple Crop Types Classification over a Large-Scale Region. *Remote Sensing*, *14*, 2341. DOI: 10.3390/rs14102341

1. Waqas, M., Naseem, A., Humphries, U. W., Hlaing, P. T., Dechpichai, P., & Wangwongchai, A. (2025). Applications of machine learning and deep learning in agriculture: A comprehensive review. *Green Technologies and Sustainability, 3*(3), 100199. 10.1016/j.grets.2025.100199

Wu, K., Ji, Z., Wang, H., Shao, X., Li, H., Zhang, W., Kong, W., Xia, J., & Bao, X. (2025). A Comprehensive Review of AI Methods in Agri-Food Engineering: Applications, Challenges, and Future Directions. *Electronics (Basel)*, *14*(20), 3994. https://www.mdpi.com/2079-9292/14/20/3994

Yakkala, V. S., Nusimala, K. V., Gayathri, B., Kanamarlapudi, S., Aravinth, S. S., Salau, A. O., & Srithar, S. (2025). Deep learning-based crop health enhancement through early disease prediction. *Cogent Food & Agriculture*, *11*(1), 2423244. DOI: 10.1080/23311932.2024.2423244

Younas, M., Akhtar, N., Batool, S., Owais, M., Sahar, S., & Anum, W. (2025). The Integration of Artificial Intelligence in Agriculture: Emerging Trends, Benefits and Challenges. *Journal of Asian Development Studies*, *14*, 1316–1333. DOI: 10.62345/jads.2025.14.1.105

Yuan, Y., & Sun, Y. (2024). Practices, Challenges, and Future of Digital Transformation in Smallholder Agriculture: Insights from a Literature Review. *Agriculture; Basel, 14*(12).

Zarbakhsh, S., Fakhrzad, F., Rajkovic, D., Niedbała, G., & Piekutowska, M. (2025). Approaches and challenges in machine learning for monitoring agricultural products and predicting plant physiological responses to biotic and abiotic stresses. *Current Plant Biology*, *43*, 100535. 10.1016/j.cpb.2025.100535

Zhang, P., Lu, B., Ge, J., Wang, X., Yang, Y., Shang, J., La, Z., Zang, H., & Zeng, Z. (2025). Using UAV-based multispectral and RGB imagery to monitor above-ground biomass of oat-based diversified cropping. *European Journal of Agronomy*, *162*, 127422. 10.1016/j.eja.2024.127422

Zhang, R., Zhu, H., Chang, Q., & Mao, Q. (2025). A Comprehensive Review of Digital Twins Technology in Agriculture. *Agriculture*, *15*(9), 903. https://www.mdpi.com/2077-0472/15/9/903

Zinjurke, S., Bhaltilak, O., Vidhate, R., & Nikumbh, J. (2025). Leveraging Computer Vision for Crop Health Monitoring, Automated Yield Estimation, and Enhanced Disease Detection in Precision Agriculture. *International Journal of Research and Analytical Reviews, 12*. Advance online publication. DOI: 10.56975/ijrar.v12i2.309776

Zreikat, A. I., AlArnaout, Z., Abadleh, A., Elbasi, E., & Mostafa, N. (2025). The Integration of the Internet of Things (IoT) Applications into 5G Networks: A Review and Analysis. *Computers, 14*(7), 250. https://www.mdpi.com/2073-431X/14/7/250

Chapter 8
Predicting and Managing Crop Health Through Artificial Intelligence and IoT

Tushar
http://orcid.org/0000-0001-7190-7687
United College of Engineering and Research, Prayagraj, India

Pooja Jaiswal
http://orcid.org/0009-0007-2507-1525
University of Allahabad, Prayagraj, India

Prabhat Chandra Shrivastava
http://orcid.org/0000-0003-0766-6422
University of Allahabad, Prayagraj, India

ABSTRACT

In this chapter, the author discusses the paradigm shift in modern agriculture that is motivated by the need to have sustainable food production and the use of resources efficiently through smart farming. Smart farming combines IoT, AI, ML, big data analytics, drones, and robotics and is used to increase productivity and environmental sustainability. Making sense of data is turning real-time sensor data into actionable insights in order to monitor the conditions of soil, crop development, weather conditions, and the health of livestock. Precision farming allows for the process of irrigating, fertilizing, pest control, and harvesting, which minimizes the amount of inputs wasted. Experiment results demonstrate machine learning algorithms as the best classifiers using the precision, recall, accuracy and F1-score

DOI: 10.4018/979-8-3373-7257-0.ch008

that play a key role in getting smarter farming to work. The model is stable, as the accuracy of disease detection is 98.26% and validation outcomes are more proficient. Agriculture 5.0 will encourage transparency, efficiency, and the use of less water, energy, and chemicals.

INTRODUCTION

Smart farming is a method of farming that is at the forefront of innovation. It is an innovative and technologically advanced type of agriculture, which monitors and regulates the agricultural processes with the assistance of automation, artificial intelligence (AI), and the Internet of Things (IoT). One such method of farming has evolved to be the most advanced and this is known as smart farming. In particular, it establishes a system of connected systems enabling gathering and processing of data of various sources. They possess sources to give useful information, which include sensors, drones, and satellites among other opportunities. Farmers with the utilization of such technology as yield monitors, variable rate applicators, and GPS guided tractors are able to apply these inputs to where they are required such as water, fertilizer and insecticides on a field rather than distributing them uniformly on a field. Without the concept of sustainable food supply networks, it is impossible to come up with these two concepts. Precision agriculture and smart farming help in cutting down pollution and preservation of natural resources as they maximize the production and reduce the waste. It is because food supply chain is more powerful since it starts with the farm and ends with the consumer since they tend to be more reliable, yielding high yields. It is shifting to old-fashioned approaches to fresh and highly technological approaches due to the emerging artificial intelligence. This will make the market size increase by moving it to 23.14 billion in 2021 as compared to the previous amount of 20.12 billion. This high increase indicates that individuals are willing to adopt new technology in agriculture, as (Haque et al., 2021) notes. The effect of AI on various aspects of the agriculture business is vast. In 2019 report by McKinsey and Company, the output in agriculture may also rise up by an incredible 70 per cent by the year 2050 through AI. Farmers are given the instruments to make improved decisions in the basis of genuine information via these advancements in technologies (Sharma et al., 2021, 2024). They also assist them in utilizing the fertilizer better, detecting diseases afflicting plants within the same day, and enhance the efficiency of the irrigation. The enhancements are consistent with the SDGs of the United Nations, i.e., Goal 2: Zero Hunger and Goal 12: Responsible Consumption and Production (Saetra, 2021).

In recent years, an application based on deep learning in agriculture, also referred to as smart agriculture, and has already become rather successful (Shin et al., 2021).

Deep learning, a cutting-edge technology, offers diverse techniques for both image processing and big data analysis (Tushar et al., 2026c). The next step in manufacturing and production is Industry 5.0, also known as the human-centered industry (Leng et al., 2022; Xu et al., 2021). It is a drastic difference to the Industry 4.0 that is marked by the introduction of extensive digital and communication technologies into the industrial processes that are concerned with full automation and very little human intervention (Wollschlaeger et al., 2017).

The researchers claimed that even the Industry 4.0 generation would require human participation in the key information streams and active role in the regulation of the output in the future (Turner et al., 2021). This notion is made in Industry 5.0, and one of the problems is how to collaborate best with technology and human beings (Huang et al., 2022) .It has enhanced smart farming by enabling farmers to control their farms remotely and provide them with convenient control systems. They gather material information. This information is then transmitted to every farmer by specialized sensor boards and centralised servers.

The mechanism of discovering and preventing Distributed Denial-of-Service (DDoS) attacks, particularly, HTTP based ones is also investigated a great deal. A method discussed in (Sivagnanapalani et al., 2021) also applies HTTP to defeat producer separated DDoS attacks with a multi-layer perceptron genetic algorithm (MLPGA). The methodology proposed by (OpokuDarko et al., 2020) is a combination of multiple symbol characteristics in DDoS validation. Their research developed an AI prototype which security investigators can utilize to acquire insights into the trends of danger. The data mining and artificial intelligence tool that we used to build the model was the weka tool, which is a Java-based tool. The supervised attribute filter of weka did the job of preprocessing the data and a decision tree based on C4.5 was taken to structure the preparation process.

To achieve a more sustainable world, the SDGs demand the cooperation of all the people of various walks of life including the governments, the civil society, and the corporations. The features agri-food supply chains AFSCs where the primary problems of food systems are pointed out and the larger influence of agri-food value chain activities (Camaréna, 2020). It is predicted that the capacity of AI to locate valuable information in large generated datasets and forecast the future trend will be one of the leading elements in the revitalization of agri-food supply chains (AFSCs). AI simplifies decision-making according to data, resource allocation, and being prepared to problems in AFSCs. AI is expected to play a significant role in offensive and defensive cyber security measures, requiring legal frameworks to adapt (Tushar et al., 2026a). Ultimately, AI (Tushar et al., 2026d) will also ensure that these crucial supply chains will be significantly efficient and sustainable.

The combination of Artificial Intelligence (AI), Internet of Things (IoT), and deep learning in smart agriculture has a massive impact on the productiveness, resource utilization, and sustainability of agriculture in comparison with the traditional farming practices. Developed systems assist in the monitoring of crops, detection of diseases, optimization of irrigation, and resilience of agri-food supply chains using AI-driven decision-making systems. In addition, safe AI-powered systems that have sophisticated DDoS detection systems enhance system resilience. Thus, the positive impacts of smart agriculture technologies on the environment sustainability, economic development, and SDG 2/ SDG 12 realization are observed.

SMART FARMING

Numerous studies have been conducted regarding the deployment of Artificial Intelligence in the detection and control of a disease. (Kethineni & Pradeepini, 2021) designed an AI system that uses computer vision to detect diseased plants. The varied AI solutions on smart farming used on Figure 1. As it has been stated, the timely identification of such problems can help farmers to reduce the possible outbreak and limit the damage to crops (Alpyssov et al., 2023; Sungheetha, 2022).

Figure 1. The smart farming agriculture has different approaches

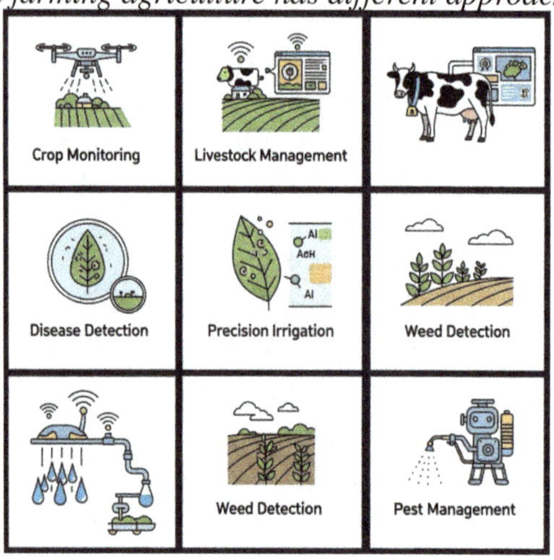

This fact revealed that AI is a highly receptive technology, whereby most sectors of agriculture have been improving tremendously as highlighted in this chapter (Alpyssov et al., 2023). The solutions presented to farmers are AI based, and using them, valuable insights can be offered, as well as aid in decision-making. It has facilitated a wide range of agricultural activities that includes crop farming, animal keeping, pest control and supply chain productivity optimization as shown in Table 1.

Table 1. Agriculture overview and smart farming classifications

Approach	Description
Crop Monitoring	Important data on crop health, growth and yield is gathered by satellite images and drones. This data is then processed through AI algorithms, which give information that is important in the optimization of crop management. The process, in its turn, can help farmers to optimize irrigation strategies, change the approach to fertilization, and to improve the practices in disease detection, which, in turn, will result in both the resource efficiency and increased yields.
Disease Recognition	Machine vision and AI systems analyze the photos of crops and animals to identify ailments and other issues. The technique accelerates disease diagnosis, treatment, and isolation of infection.
Controlled Irrigation	The strategy provides crops with water when they need it, and in the required quantity, minimizing water wastage and enhancing the efficiency of agricultural water. This is the exact irrigation control which minimizes agricultural water stress, enhances crop quality and yield, and makes sustainable management of agricultural water.
Weed Identification	The computer vision and AI are used to identify and classify farm pests. This is the method that farmers can use to differentiate between weeds and good crops. Precision pesticide application and mechanical weeding can be conducted by farmers with the help of the weed identification system that is based on AI. This lowers the amount of pesticides to be used and this lowers the adverse impacts of competition by weeds on crop production.
Pest Control	AI is used to identify and process pest movements by pest monitoring systems. This system is a tedious way of monitoring the number and activity of pests through sensors, traps and satellite photographs. The AI algorithms process the data to report the invasion of insects in real-time. This allows neighboring farmers to respond to prevent insect attacks. The use of AI-based pest management solutions also reduces the application of extensive use of pesticides, which renders pest management more sustainable and environmentally friendly.
Supply Chain Enhancement	AI algorithms are applied in the agricultural logistics, storage, and transportation business. This provides fresh and high-quality agricultural produce to the customers through monitoring, tracking and quality control of the entire process of farm to consumer. The use of Artificial Intelligence techniques enhances the work of the agricultural value chain, sustainability of the supply chain, and connection of farmers and markets.

The AI-based smart farming and agriculture can ensure optimal resource allocation, including pesticides, fertilizers, and irrigation water. This, consequently, causes low pollution rates and operation/establishment expenses, and it also raises the value of production, at the same time. The fact that the spread of plant/leaf diseases can be prevented in time with the assistance of artificial intelligence helps to reduce their spread; hence, fewer chemical solutions are applied, consequently, reducing pollution to a considerable extent (Li et al., 2021). The approach will make AI developed according to human values, ethical principles, and legal requirements as discussed in (Shneiderman, 2020). Nonetheless, fundamental general concepts and approaches can be employed to establish a future framework of HCAI, as reflected in Figure 2.

Figure 2. The overall guidelines and considerations of HCAI

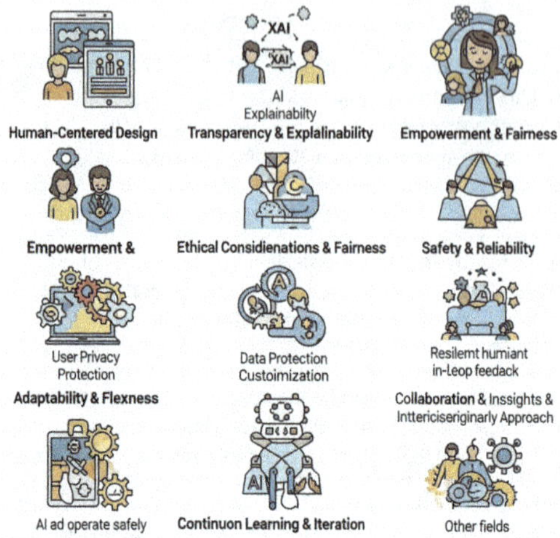

PRECISION AGRICULTURE

Multiple forms of cyber-attacks are capable of disrupting smart farming systems in precision agriculture, and an HTTP-deluge-based denial-of-service (DoS) attack, where the number of fake requests is too large to handle and paralyzes the processing of real-time agricultural information through the use of IoT-enabled farm servers, is one of the most significant threats. In an attempt to achieve its goal, it sends excessive number of HTTP GET or POST requests to a server which appears authentic. These requests are usually sent by a network of hijacked machines also known as a botnet. They consume excessive resources of the server, including CPU power and bandwidth. The attack causes the server to somehow be unavailable to actual users or extremely slow by consuming these resources. This eventually renders the service of the server useless. In order to execute the initial step of data collection, a CIDDS-001 dataset called Cobourg Intrusion Detection Dataset will be utilized. To accomplish the pre-processing of data, the irrelevant data and noisy data will be removed, and the pre-processed data will undergo data normalization (Kulanthaivelu et al., 2022). It is refined information that is inserted in the feature extraction phase where various operations are done to generate features on each event of network traffic in a numerical manner. All through the extraction process, pertinent features are discerned out of the processed dataset. Mathematical formulas such as mean, calculation of the entropy, median, and standard deviation are used to provide the numerical representation of each traffic flow in the dataset. The classification is the final step and it will include the use of a classifier that will be used to define the flows of input which could either be an attack or a normal occurrence.

Cybersecurity Risks in IoT-Enabled Farms

Web-based applications as a result of their constant growth and popularity are vulnerable to numerous security threats. These threats keep changing with the evolution of the internet in nature and magnitude. A number of methods that are aimed at mitigating such vulnerabilities, however, have not always been effective enough to provide the required web security and especially in the cloud environment. This is due to many reasons such as not changing the threat classification plans, lack of empirical evidence in design rules, and the inability to reason about new threat or response.

The HTTP protocol structure is expected to provide protocol integrity of the communication against protocol threats. This is a communication protocol that consists of messages such as the information of the HTTP, the HTTP requests and the HTTP responses. The short-form structure of an HTTP request and an HTTP response is observed in the following elements:

- **Start Line:** The start line has three facts; HTTP version,request method as well as Uniform Resource Identifier (URI). The communication line can receive messages such as POST, GET and PUT.
- **HTTP Request:** The HTTP request consists of headers,start line and a message body.
- **Headers:** The start line is reflected on the header messages. Other header is considered to be obligatory and must be present in every request, e.g. the host header. The detection system will also treat the request as malformed in case of the absence of this particular header. Headers are of four types namely Response, Request, General and Content.
- **Request Body:** It cannot be used including a GET message, still must be used with a POST message.
- **Response of HTTP:** The HTTP outcome gives a headers,status line and content to every HTTP request.
- **Status Line:** The responses are sent to target client with help of status ID.
- **Response Body:** This is the field with the content that is produced through the web-server interaction as the name implies.

HTTP Flooding Attack Threat

When considering general points of view, the constraints have been noted when using the features of the HTTP GET request method to identify systems of DDoS with the help of a destination when making an HTTP GET request, which usually leads to decreased accuracy because the threshold of detection may reach an uncontrolled amount. In particular, traditional systems are vulnerable to the advanced DDoS intrusions, which can flood infrastructures using a relatively small amount of HTTP requests. Figure 3 shows the steps involved in the activation of an HTTP flood attack on a cloud server.

The HTTP flood technique is normally used by the transfer of a large number of individually designed HTTP GET or POST requests to deplete a substantial amount of the resources of a server. When systems make such requests in great amounts, they constitute a botnet. Although the accuracy farming routing has a series of challenges, the initial steps were directed at the detection of Hypertext Transport Protocol (HTTP) threats in Industry 4.0-associated practices. Precision farming application was developed, and the machine learning algorithms and data mining were utilized to identify the HTTP threats in an appropriate way on the basis of the cloud. It is the contribution to the security issue encountered in precision farming experimental systems.

Figure 3.Threat of HTTP web services

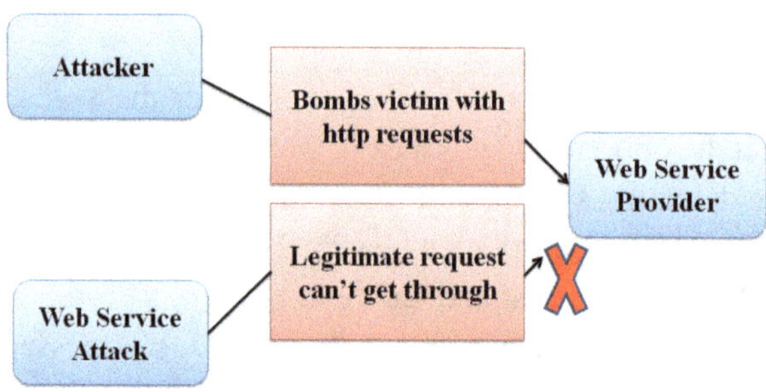

The key elements of precision farming as shown in Figure 4.

GPS-based precision agriculture

Precision farming is entirely based on GPS technology that aids in the acquisition of the right data in mapping irrigation systems, fields, and roads. It helps in variable rate application (VRA) to apply the appropriate quantity of seed and fertilizer, which ultimately results in maximum production efficiency.

GIS Technology in Precision Farming

Through GIS technology, farmers would be able to visualize and analyze the information on soil survey and crop conditions.

Drone-based spraying systems in precision farming

The standard type of crop dusters can be used in planting and watering, but the UAVs are more efficient, and they are supplied with a hi-tech instrument, i.e., a multispectral and thermal analysis.

Satellite Remote Sensing in Precision Farming

This technology provides real-time information on moisture, disease diagnosis, and levels of nutrient, which helps farmers to make improved decision-making.

Variable Rate Technology

Under the Variable rate Technology (VRT), pesticides fertilisers and seeds are sprayed on the sections of a field. The technique assists cultivators to have the highest possible utilization of resources, reduce the cost of inputs and is in a contribution to the sustainability of the environment.

Figure 4. Essential provisions of precision farming

The incorporation of the processing plans and the agriculture-related decisions was analyzed in detail (Kusumastuti et al., 2016). The other scrutinized analysis was on programming methods and kind of uncertainty models with the aim of dealing with uncertainties in systems of model of the AFSC. It was concluded that AFSC should have new decision models that take into account such aspects as seasonality, price, and market demand. The fruit supply chain (Soto-Silva et al., 2017), tomato industry, sugar industry, and mushroom production are examples of industries that have recorded production and logistic planning models in the agro-business. In this book chapter, the integration of various activities of the supply chain was proposed to achieve the holistic awareness of the trade-off that occur between the economic, social, and environmental objectives. This (Jonkman et al., 2019) chapter presented a strategic mathematical programming which applied harvesting, seasonality, processing, and perishability in the estimation of the most suitable supply chain architecture. To solve the problem of distribution and production within a closed-loop supply chain, the creation of a multi-objective model was used in the business of mushrooms (Banasik et al., 2017). The redesigning of the logistical structure of the mushroom supply chain led to an overall increase in the profitability and environmental performance of the supply chain as stipulated in the strategic model. The uncertainty of the same issue is tackled in a study which focuses on two stage stochastic programming model combining the production yields and the demands. A mixed-integer programming strategy was proposed to guarantee an increased overall profit and freshness of the supply chain of the fruit products (Grillo et al., 2017).

The model prepared is biased in support of the decisions concerning the order promising procedure to meet the orders of the clients. The fresh fruit would be purchased, transported and stored with the help of two models to the agro-industrial processing plants (Soto-Silva et al., 2017). Increasing the adoption of the CCUS technology can be of assistance in meeting the climate goals for the year 2050 (Tushar et al., 2025a). According to the studies, there are a series of factors, which are to be taken into consideration in case of making decisions related to the industry and agriculture at the same time. Figure 5 illustrates three stages of the supply chain namely agriculture, related to the text of transportation, and the industrial. To achieve sustainable crop rotation, it is attention to pay much consideration to choice of crop at initial stage. Farming decisions include crop planning and crop rotation, and the latter is the one that is being discussed now (Dury et al., 2011). To learn more concerning the crop rotation models, refer to (Li et al., 2015; Fikry et al., 2019).

The movement of the crops to the processing areas needs to be well coordinated because the products are in their optimum state when harvesting is taking place. Crop and process design is the basis of transportation of crops to the pre-processed or processed facilities. The production-distribution network is inclusive of transportation. Multiple forms of cyber-attacks are capable of disrupting smart farming systems in precision agriculture, and an HTTP-deluge-based denial-of-service (DoS) attack, where the number of fake requests is too large to handle and paralyzes the processing of real-time agricultural information through the use of IoT-enabled farm servers, is one of the most significant threats. Figure 5 depicts how inventory in Fresh crops and products is kept and conversion processes to achieve a predetermined demand occurs at the last stage. The production-logistics chain requirements dictate the coordination and the level of integration within the chain demanded by the successful management of the entire process.

Figure 5. A schematic order of the associated activities in the AFSC

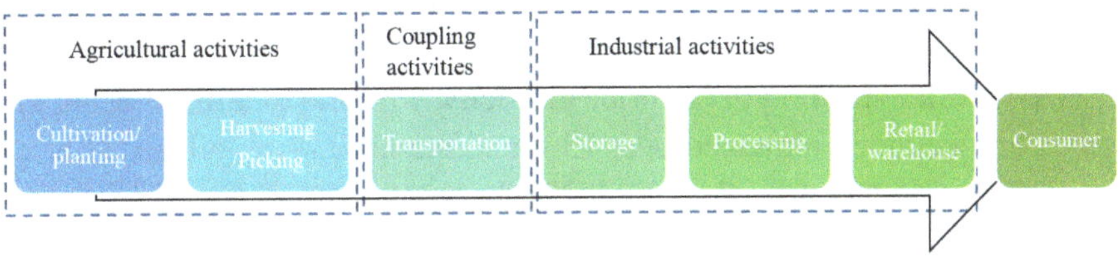

The digital technology is analyzed regarding SDGs 12 and 2, in particular, which can be impacted either directly or indirectly. Among the possible properties of digital technology that can support data analytics, one can single out the applications in demand forecasting. They can also prevent food waste, either in terms of as post-harvest losses, or in terms of consumable food (SDG objective 12.3). The other methods that the digital technologies help SDG 12 are the interactive data analysis of supply chain-wide data, the scenario analysis, and real-time monitoring. They are also capable of assisting in implementation of traceability systems and the AFSC to varying extents. All these can be fulfilled with the help of a computer program of automated dose delivery of natural fertilizer (SDG objective 12.5), which entails the judicious use of green chemicals as an agricultural input to farming (SDG objective 12.4). Rather, it is similar to the SDG 13 on climate change.

An Investigation: Global Agri-Food Trade Under the Ukraine Conflict

The invasion by the Russians exposed the largest number of civilian casualties ever in Ukraine, temporary occupation of territories and agricultural restrictions, destruction of infrastructure, a decline in GDP and trade turnover by 30 percent, and slower economic growth. Overall, it took the agricultural sector USD 40 billion to make losses to the economy. Ukraine suffered significant losses due to the armed invasion of Russia, in terms of lost agricultural firms of about 2,600 and a loss of 1.9 million hectares of planted areas. Having shipped more than 50000 tons of grain out of Ukraine to Russia and more than 20 percent of the livestock and poultry shell shocked. Adaptive system needs to be developed to minimize critical vulnerability among the agribusinesses. These consist of clearing farmland, repairing warehouse systems, developing importation logistics of seeds, feed, fertilizers, and veterinary provisions, increasing agricultural products, and providing sustenance to individuals in the frontlines and war-torn regions.

The revitalization of agriculture sector in Ukraine after the war should focus on sound export policies in order to enhance balance between trade and macro economy stability. The other global markets have been weak towards outside forces as compared to agriculture. It heals sooner (global exports of wheat, maize and sunflower oils increased by 20 percent in 2022 versus 2021). Different commodities are characterized by different strengths of price elasticity, with the wealthiest being grains (aggregate price index remained over 133) and the least being oilseeds (the price index of vegetable fats was 111.3 in 2022). In order to balance agri-food market, global social priorities should be emphasized by the international community. Establishing protectionist measures might have localised effects in the short term but create global price fluctuations in the long term. Sustainable development needs that a strong world-food security system should be established during hard time to fight hunger and poverty, as well as the environmental effect of military activities. The global hunger level stood at 18.1 and the percentage of undernourished individuals has increased to 9.2, quoting 735 million in 2022. Russia is persistent and the result of such aggression is unpredictable world responses and solutions.

METHODOLOGY

The SVM explanation is mathematically detailed but could be better linked to agricultural use cases (e.g., disease detection). SVM has a good reputation to perform classification activities and has a solid theoretical basis on the statistical learning theory. SVM models are very useful in high-dimensional data where decisions are susceptible. Another method that is like Support Vector Machines (SVMs) classifies important points of data by drawing an optimum hyperplane into high dimensions space. The equation below can be justified by a number of factors:

- SVM algorithm finds the hyperplane that separates the data points in two distinct groups. This decision plane enhances easier categorizing of recent data.

- It transfers data values to a higher dimension with a kernel function and produces a feasible separation curve. SVMs need not be in a position even to be able to linearly separate their data points in the feature space, and yet they may still produce a hyperplane to distinguish between classes.
- SVM is suitable with new data and it does not overfits. SVM works very well when the goal is to maximize the margin without any specific knowledge on the data involved and only limited information is known about it. Whereas, the goal function is maximized and regularized by the term.
- Regularization parameter C is equalized by training error and margin as illustrated in equation 1 and equation 2.The resulting formula can be obtained as follows:

$$minimize(\omega; b) = \left(\frac{1}{2}\left\|\omega\right\|^{2} + C\sum_{i-1}^{N}\xi i\right) \quad (1)$$

$$y_i(w^T x_i + b) \geq 1 - \xi i, \quad (2)$$

In the case of $\xi i>0$, i= 1, 2, and so on, N. w is the weight vector, b is the bias term, x_i is the input feature vector and y_i is the corresponding class and finally, slack variable is x_i. The use of support vector machines (SVMs) is especially common in the agricultural sector because their massive capabilities of processing complex and non-linear data are enormous. SVMs, such as, are very efficient in high dimensional processor data in sensors or imagery, and can be useful to know the initial symptoms of plant diseases (Kohut et al., 2024) and implement preventive measures against them. Moreover, SVMs are capable of real-time optimization of water usage on the accuracy of irrigation, thereby reducing the waste and increasing production. Nevertheless, it has been capable of forecasting the probability of illness using soil moisture and soil temperature which are instances of complex and non-linear relationships. It has been observed to be more appropriate in the case of binary classification which involves a predication of one of the following based on environmental data, whether acquisition of a crop disease is to be made or not. Another deep convolutional neural network called AlexNet is also used as an original network, having 8 layers. AlexNet network has 60 million parameters and 650,000 separate neurons. The data on which the model is operated contains the images of diverse species of pests like *Nilavarpata Lugens*, *Pomacea Canaliculata*, *Pyricularia Oryzae Neck Panicle* and *Stem borer larvae*.

Proposed in early 2021, Industry 5.0 paradigm can be interpreted to introduce new connections between the industrial practices and the society. It does not seem to be the direct continuation of Industry 4.0, and neither is it habituated as such but instead the existing abilities of Industry 4.0 are expanded by putting industry in the perspective of the future world i.e. Society 5.0 (Tushar et al., 2026b). In line with this vision, three major concepts of Industry 5.0 are outlined using in the Figure 6 as follows; Human-centric, Sustainable Resilient.

Figure 6. Industry 5.0 framework

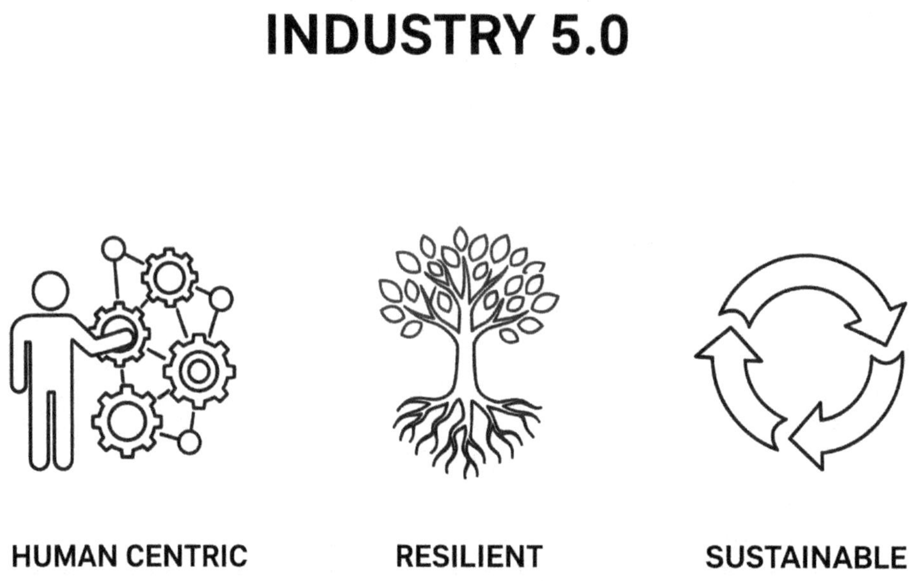

Figure 7 illustrates the hardwired control design that is used in this section and it consists of three parts; input, process and output. The system works with the use of ultrasonic sensor, TDS, pH and flow meter. These all sensors are linked with the ESP32 microcontroller where processing of data is done. After that, data obtained after processing sent to the valve of solenoid to actuate it and to the data store, where it is stored as information. At the same time, it is presented on applications for the Android platform.

Figure 7. Hardware connection diagram

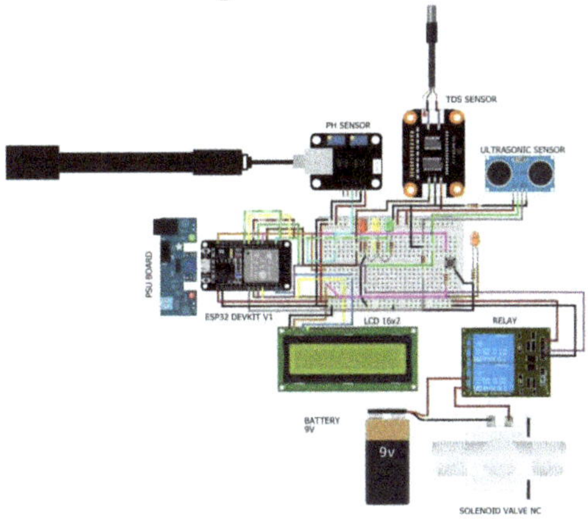

This system contains TDS sensor, pH sensor, a button and ultrasonic sensor. These four components are connected to an ESP32 microcontroller, where DevKit V1 that offers to support Wi-Fi and Bluetooth in-built can be used to support this Internet of Things project (Sayekti et al., 2024). The ESP32 also takes care of the internally received information of the sensors and the button. The processed data is then sent to an LED where it is indicated, a solenoid valve where it is sent to actuate and a database as the information which is in turn displayed in the Blynk application.

In trying to secure the sensitive information in the precision farming systems against cyber-attacks, the approach to identify flood attacks of HTTP in combination using systems of cloud storage is outlined in this module.

- Data acquisition using cloud services: in the Industry 4.0 case of the linked cloud storage architecture, an intrusion-tagged dataset, which is referred to as CIDDS-001, is ready (Kulanthaivelu et al., 2022).This data consists of regular normal and aggressive traffic, which is specifically created to serve Industry 4.0-facilitate apps connected to cloud-based computing.
- The pre-processing of data, Small practical networks and actual data reality data store or real-time documented internet traffic can be very sensitive to signal disturbances, absent values, or conflicting information thanks to their huge scale (many terabytes or more) and originating from many different and autonomous sources. When there is a very poor data quality then the next mining of data, results will also be degraded. Big practical world data storage and data distribution nodes often possess poor, outlier, and unequivocal data. There are multiple possible reasons that may cause inadequate information. Not all important characteristics can necessarily be easily accessible. Other details may be left out in case they were not understood to be important

in the collection process. Important information would also be avoided because of a misinterpretation or a system failure. Moreover, the information, which could not have been matched with other entries made, could be erased. Alternatively, we may also have the possibility that the data or even the process of collection in itself was not well thought out resulting in the introduction of missing data, especially the tuples that are missing some property. The data pre-processing process in data mining field is usually done in three steps:

- Data cleaning: It is aimed at solving discrepancies in dataset, even out noise and identifies anomaly and filling missing values. It would be beneficial to reference a sequence of data scheme tracking methods that involve identifying, removing, and rectifying errors and anomalies to live traffic using communication networks or benchmark data. In cases where there are many data, duplicate or missing data may yield wrong or false information and may give erroneous judgments. The steps are represented in Figure 8.
- Data Normalization: Data transformation can be carried out using one strategy that is normalization. Normalizing an attribute is often done by scaling the values of a dataset to a small fixed range often between 0 and 1. Normalization is of great help in the area of classification, clustering and nearest-neighbor classification among others. Normalization is a procedure that involves data conversion as well as assimilation.
- Non-Zero Analysis: An analysis of non-zero values is analyzed to identify the number of zero values which is large in a dataset, such as an offer market dataset. A high-percentage of zero values in the dataset might lead to an erroneous conclusion due to the analysis of the dataset. The values of an attribute are coded in a way that they have some numerical meaning in the event that more than a quarter of an attribute is coded with 0.

Figure 8. Algorithm 1: displays the process of suggested data cleaning procedure

```
Algorithm 1: Data Cleaning

Input
D: Raw network traffic
Output
C: Cleaned network traffic

C: Clean data vector initialization
For every record in dataset I ∈ D
    Extract the record I ∈ D
    Perform speculated data pattern cleaning
        keep a separate data combination for each column
    normalize evaluations of each column
    Store I in C
End for
Return (C)
```

Extraction of Features

The task of feature recognition consists in improving description of specimens of different categories by reducing the insignificance of the parameters or excluding the non-informative ones and choosing the best set of features. A feature extraction algorithm usually has two stages; the stage of building features and the stage of selecting features. Given some task like characterization or regression problems, feature generation is regarded as a crucial step in the representation of information. It is of great importance in determining the success of any later statistical analysis or modeling that would be done on raw data.

After the step of extracting features, a classifier is used to classify the flow according to whether there is an ordinary or anomalous flow. Definitely, the J48 Decision Tree (J48DT) has been used in this chapter. It uses a model which distinguishes the threats on the basis of the source IP, size of a packet, recipient IP, packet header, load and minimum value of traffic as features as demonstrated in Figure 9. Data have been separated into training and test set with ratios of 70/30 and 80/20 correspondingly. The Algorithm has a 99%precision, on the trained and test datasets when it is trained. According to these developments, the full process of identification (Kumar et al., 2023) of the vulnerabilities of HTTP within an Industry 4.0-related distributed computing environment can be demonstrated as shown in Algorithm 2 (Manwatkar et al., 2024).

Figure 9. Algorithm 2: Demonstrates AI-based attack detection

Algorithm 2: Attack Detection using AI

Input

D: Raw network traffic

Output

R: Classification results

1. Select data D
2. C: Perform data cleaning using algorithm 1
3. S: Apply feature selection on pre-processed data
4. Apply statistical computations as:
5. For each flow $f \in S$
6. $f1$: Extract entropy feature using Eq. (1)
7. $f2$: Extract mean feature using Eq. (2)
8. $f3$: Extract median feature using Eq. (3)
9. $f4$: Extract standard deviation feature using Eq. (3.4)
10. $v^f = \{f1, f2, f3, f4\}$
11. End for
12. Divide the v^f into $T1$: training (70 %) and $T2$: testing (30 %) samples
13. Apply classifier R: $J48DT\ (T1, T2)$
14. Return classification results R
15. Stop

RESULT AND DISCUSSION

To improve final results, the extremely suitable algorithm should be chosen using such metrics of performance as precision, recall, accuracy, and F1 Score by farmers. The comparison of the model performance on different tasks is outlined in the tables below, which can be used as reflects of the successful implementation of the model to smart farming. A closer examination of these measures of performance measures can be found in Table 2, which indicates the extent to which the performance of some tasks was accomplished. Table 2 indicates the results with the rest of the algorithms, such as Logistic Regression, and more various occupations, such as soil nutrient analysis. Table 3 indicates the performance of various algorithms with regard to various jobs and the contribution towards the goals of smart farming.

Table 2. Specific task performance comparison

Objective	Best Model	Accuracy	Recall	Precision	F1 Score
Disease of crops	RF	0.87	0.85	0.88	0.86
Analysis of soil nutrient	SVM	0.85	0.82	0.87	0.84
Identification of pests	DT	0.82	0.8	0.84	0.81
Prediction of crop yield	CNN	0.9	0.89	0.91	0.9

Table 3. Alignment with smart farming objectives

Assignment	Model(s)
Management of Enhanced Disease	Random Forest, SVM
Allocation of Optimal Resource	Neural Networks
Practices of Sustainable Farming	Decision Trees, SVM
Prediction of crop yield	Neural Networks
Analysis of Soil Nutrient	SVM

Figure 10. Accuracy and Loss curves of crop species detection

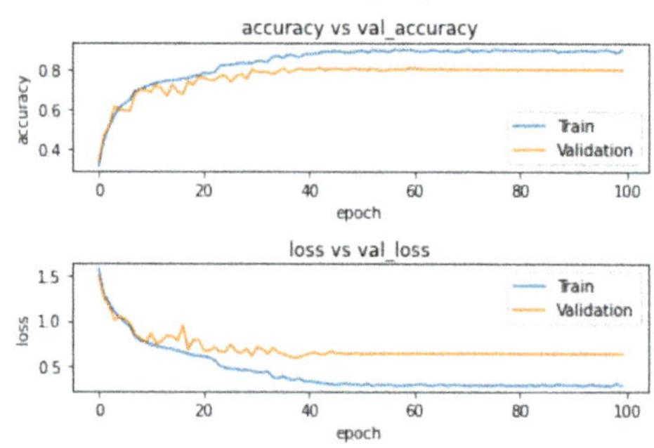

The process of finding crop categories is splintered into a number of stages. In order to control the time of training, there was a 20 epochs maximum, within which the algorithm would be trained. The model has been trained in an environment of TensorFlow, Python and Keras and has been used, and the graphs have been plotted later in Matplotlib. To evaluate the effectiveness, the model was trained on several datasets; Floria_dataset, Swedish_Leaf_Dataset, LeafSnap_dataset, and MalayaKew_dataset (Alfarisy et al., 2018). The model achieved an accuracy of 95.38% in terms of training. Figure 10 shows the training accuracy and loss validation. The training accuracy was established after the training process of the proposed architecture was done. The findings revealed that there was a total accuracy of 98.26% in the identification of diseases at the leaf level of the plants. The model correctness in terms of validation also stood at 96.77%, this indicated that significant correct outputs were obtained by the trained model in terms of classifying sick plants. The accuracy of the test as well was calculated using the evaluate method where the percentage showed 96.7738%. After that, the curves of training as well as validation was plotted, which is illustrated as Figure 11.

Figure 11. Detection of plant disease using training and validation curve

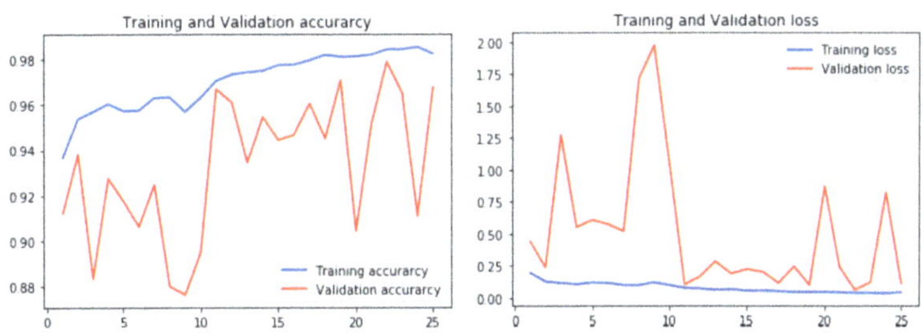

Within framework of human needs, human-centric industry and interests are prioritized in the middle of the harvest process. On Industry 5.0, workers are to be optimized in the production process, especially on the aspects of directions and training. In equation 3", agriculture 5.0 contains a representation (Atiku et al., 2024). For the framework:

$$\text{Agriculture 5.0} = \text{industry 5.0} + \text{Human-centered AI} \quad (3)$$

Each sensor node receives power supplied through a single battery. All the analog sensors should be calibrated before use in order to reduce the possibility of read error in data within a project. Figure 12 contains the pH sensor calibration program, and Figure 13 shows the results of TDS calibration program, which is associated with the Blynk application (Sayekti et al., 2024).

Figure 12. Code for pH calibration

PH.ino

```
int pH_Value = 0;
float Voltage = 0.0;

void setup (){
  Serial.begin(9600);
  pinMode(2, INPUT);
}

void loop (){
  pH_Value = 0;
  for(int test_cycle = 1 ; test_cycle <=10 ; test_cycle++)
  {
    pH_Value = pH_Value + analogRead(A0);
    delay (10);
  }
  float analog = (float)pH_Value / 10.0;
  Voltage = analog * (3.3 / 4095.00);
  Serial.print("Tegangan Analog = ");
  Serial.println(Voltage);
  delay (1000);

  float alibration_value = 21.34 - 1.53;
  float ph_act = -5.70 * Voltage + calibration_value;
  Serial.print("pH = ");
  Serial.println(ph_act);
  delay (2000);
}
```

Figure 13. Coding of TDS sensor

TDS-LCD-IOT.ino

```
#define BLYNK_PRINT Serial

#define BLYNK_TEMPLATE_ID "TMPL6OOmN8BI_"
#define BLYNK_TEMPLATE_NAME "hidroponik"
#define BLYNK_AUTH_TOKEN "c2A3JItIpeG0UHnxLAlgIUreyLfZxU_X"
#define LED_BUILTIN 2

#include <WiFi.h>
#include <BlynkSimpleEsp32.h>
#include <Wire.h>
#include <LiquidCrystal_I2C.h>

// Your WiFi credentials.
char ssid[] = "hihi";
char pass[] = "12345678";

LiquidCrystal_I2C lcd(0x27, 16, 2); // Alamat I2C LCD 0x27, ukuran 16x2

const int pH_pin = A0;
const int tdsSensorPin = A0; // Pin analog untuk sensor TDS
float tdsValue = 0;

void setup()
  {
  Serial.begin(9600);
  Blynk.begin(BLYNK_AUTH_TOKEN, ssid, pass);
  lcd.init();
  lcd.backlight();
  }
```

Figure 14 shows the comparative analysis of the different attack detection methods depending on samples test ratios of 20% and 30%. As with recall rates, accuracy and the precision rate of effective attack identification of the suggested method was higher as compared to any other available technique. Figure 15 investigated the performance of the F1-score and the trend was similar to the previously observed results.

Figure 14. Assessment of precision effectiveness

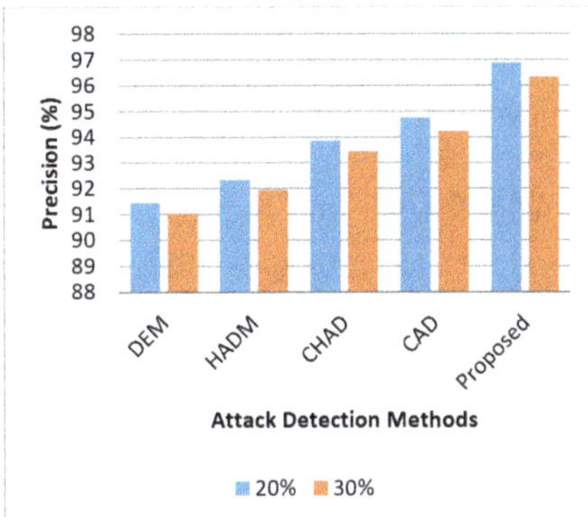

Figure 15. Assessment of F1-Score effectiveness

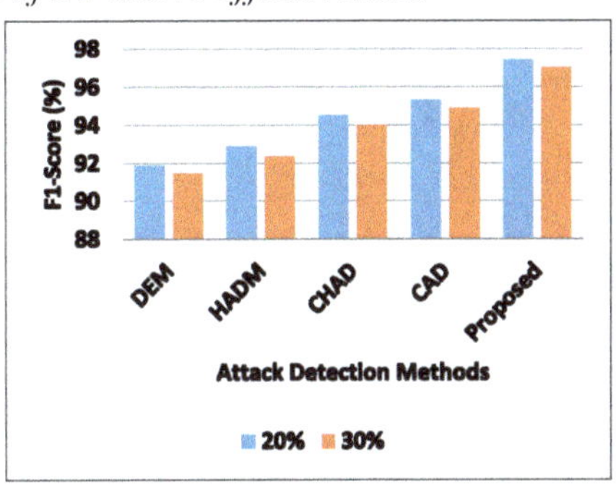

Outcomes and Achievements:

- Improved Efficiency and Optimization of Resources: This book chapter repeatedly shows that significant increases in the operational efficiency are attained with the help of the implementation of SFSCM practices. This involves streamlined logistics, lower energy and water usage, and massive improvements in the management of wastes at different levels in the farm to the consumer. As an example, the programs aimed at reducing food loss and wastes (FLW) are demonstrable to provide direct contributions to resources and economic savings, which imply that sustainability efforts are directly correlated with operational profits. Improved Environmental Performance: SFSCM results in and is primarily driven by the decrease in environmental footprints. The struggle of carbon emission reduction, switching to less polluting approaches of production examples: organic farming, and protection of nature's resources (examples: soil, biodiversity, water) are becoming more apparent. Although it is observed that the extent of impact differs, there is an evident shift in the direction of increased accountability on the environmental impacts, a fast and furious push into numerically assessment of GHG emissions and other ecological consequences along the chain-cycle.

- Social and economic benefits: Practices promoting sustainability can be also applied further than indicators of the environment to include social equity and economic feasibility. The findings suggest that SFSCM can support equitable distribution of food, corporate social responsibility, and livelihood, especially to smallholder farmers, through enhancing access to markets, and equitable remuneration.
- Growth in Transparency and Traceability: Sustainability has led to growth in transparency in supply chain. Such technologies as blockchain are becoming extremely important facilitators of effective traceability mechanisms enabling to monitor products more effectively in their farm-to-market supply chain journey.

CASE STUDIES

(i) A East Africa (cassava viral disease) smallholder offline AI: The performance of the smartphone AI application PlantVillage Nuru in cassava viral diseases and pest damage symptom recognition was compared to that of expert, extension agent, and farmer assessments through a field evaluation (Mrisho et al., 2020). This book states that the tool is able to achieve expert-level symptom recognition and outcompete non-expert users, which makes it helpful in enhancing the diagnostic capacity in areas with limited lab testing.

(ii) IoT adoption is high in viticulture since disease outbreaks (e.g., downy/powdery mildew) are closely related to the humidity, wetness of the leaves, and the canopy microclimate (Hnatiuc et al., 2023). This book indicates that IoT sensor networks with the aid of ML may aid in intelligent disease detection/ alerting in vineyards.

(iii) Individually, a wireless network need sub 6 GHz frequency (Tushar & Pradhan, 2024) and 2x2 MIMO antenna (Tushar et al., 2025b) of infrared sensors showed variability of vineyard within-field water stress, mapped canopy temperature-derived indices, and associated them with plant water status with a significant amount of correlation (Fuentes et al., 2024) .

(iv) Detection of HLB has prompted the search for remote sensing methods. One publication shows that visible–NIR and thermal imagery is able to provide meaningful classification results for symptomatic trees, and also corroborates bands/indices with high separability; there is also discussion on scaling up to larger areas using aerial platforms (Sankaran et al., 2013).

(v) Another study presented in-field HLB detection with vision sensor, showing a trend of operational monitoring instead of lab confirmation (Pourreza et al., 2016).

(vi) A deep learning method deployable on smartphone was developed, validated in a large labeled dataset of rice disease images and tested to separate image dataset, thereby demonstrating that fast diagnosis without special apparatus is possible (Deng et al., 2021).

(vii) Other works present a strong CNN/transfer learning pipeline for multi-class rice disease classification, indicating the maturity of deep learning in rice pathology (Latif et al., 2022).

(viii) A system for greenhouses used wireless visual sensor networks to identify a fungal presence on images and to activate other sensing/actuation (e.g., humidity measurement, control). Reported detection success in that controlled setting is high (Ali et al., 2019).

(ix) A review of satellite-based agricultural monitoring systems summarizes existing systems in place as well as barriers to adoption, and provides examples of how such systems form part of regional monitoring in Africa, important evidence that targeting macro-scale crop health requires different models, data governance, and validation standards compared with field-level disease detection (Nakalembe et al., 2021). Consortium framing: NASA Harvest is purposefully oriented toward actionable EO-based insights for resilience and decision-supports (policy support, risk assessments) in the U.S. and globally.

(x) The activities on seasonal time scales can be regionally dependent since planting dates, cloud coverage and revisit rates by the sensor along with sparsity in ground-truth data can vary widely from one place to another (Earth data NASA, 2022).

FUTURE CHALLENGES

Challenges in Smart Farming

The idea behind this stage was to foster collaboration as well as discovery of solutions in chapter in agricultural through the help of AI. To improve global food security, environmental issues, efficiency, Artificial Intelligence and advance farming should cooperate in the prospective. A Barket of "Agriculture 5.0" creates a balance amid the requirements of stakeholders, the conditions and technical power. Agriculture 5.0 needs to be successful with easy and available AI technologies that improve human efforts tremendously. Mixed cropping systems still have problems associated with the separation of species as well as detached measurements of nutrient supply, biomass, and health. Sensor based soil microbes as well as probabilistic models are also undergoing chapter in order to improve the health of plants as well as their nutrient absorption. Besides the fact that it is a technological improvement, this scheme is also a step towards a more efficient, responsible, and sustainable future of agriculture. There are various technical, ethical, social, legal and adoption complexities associated with the use of Artificial Intelligence in agriculture. Some of these are administrative hurdles, wasteful expenses, fears of unemployment, among the traditional farmers, as well as the low adoption rate. Concerning future opportunities, the advanced innovation, such as swarm-robot, blockchain, nano-technology and Integration of Web of Things devices, were examined. It is claimed that these inventions can transform considerably the agricultural industry and address global issues such as food security and sustainability.

Precision Agriculture Challenges

Large Initial Expenditure: The adoption of advanced agricultural technologies requires huge initial monetary commitment. They are the expenses relating to equipment acquisitions, software subscriptions, and some necessary training. Small farmers and those in less affluent areas are often lacking access to, and financing of these investments hence hindering their uptake of precision agriculture. Additionally, the constant need of regular maintenance, software changes, and technical support is an added value to the overall cost posing a big impediment to broad scale integration.

Technical Sophistication and Skills Gap: Precision Agriculture Technology Precision Agriculture technology requires the expertise and technical ability in different specialized fields such as: data analysis, GIS mapping, and equipment operation. These skills might not be acquired by many farmers, especially older generations or with less education, and it restrains their ability to make optimal use of precision technologies in agriculture.

Data confidentiality and security concerns: "The storage, transferring as well as gathering of agricultural dataset carry with them the implicit worry of privacy, security, and the ownership rights. There might be unwilling of the farmers to share sensitive data with the third-party service providers or the government, including field maps and yield data on the basis of how misused the data will be or the case of data being accessed unauthorized. Also, the centralization of the crop dataset in the cloud systems makes it more susceptible to cyber-attacks and data-breaches, thus, requiring the development of efficient data security practices and their extensive regulatory frameworks.

Cross-system integration Problems: Multiple industrial firms tend to use their data structures, proprietary software and this makes it difficult to share, collaborate between farmers and service providers. Additionally, due to the lack of interoperability, the adoption of precision agriculture technologies might have to involve implementing individual solutions to the problems and specific technical skills when utilizing them with a current farm management program or with outdated devices.

Climate Change ramifications and Adaptation: It is forecasted that more extreme weather conditions, including floods, droughts, severe storms and heat waves are bound to increase as well as affect cultivation, transit, and storage hence loss of food and volatility in prices. The changing climatic patterns are likely to require modifications in the choice of crops, mode of farming and even alteration in the location where farming activities are conducted. The increase in the water stress in many areas will have a direct effect on the irrigation based agriculture and will necessitate superior water management either by utilizing alternative methods of irrigation.

Resource Depletion and Environmental Deterioration: Continued erosion, loss of nutrients and salinization is likely to move the potential productivity of arable land and lead to threat of food security. Monoculture and intensive farming are what is adding to the depletion of essential pollinators and natural eliminators of pests, the growing use of external inputs. The agricultural chemicals are expected to find their way into water sources polluting them and have negative impacts on the ecosystem and human health.

Food Spoilage and Wastage: Large volumes of food in the whole food chain (farm to consumer) are likely to be lost or wasted with inefficient harvesting, poor storage and transportation of food, processing wastes, and consumer behavior. The minimization of this waste is taken to be of the utmost importance concerning the efficiency of resources and food security.

Technological Implementation and Digital Disparity: Technologies (AI, IoT, and blockchain) have enormous potential in terms of ensuring optimization and transparency, but their adoption is predicted to be uneven, specifically, in-developing areas, thus enhancing inequalities. The management of these technologies will cause challenges related to infrastructure, expertise, and cybersecurity as processing the large amounts of data given by these groundbreaking technologies will become harder to handle.

Economic Sustainability and Smallholder Farmer Resiliency: Smallholder farmers contribute much to the global food production and usually at the effect of price changes and are not always able to access markets in an equitable manner to invest in sustainable practices. The transition to sustainable methods of agriculture will involve higher initial expenses as a barrier to farmers with limited budgets. Sustainable foods also have a tendency to be pricier which acts as a hindrance to consumers whose prices are price sensitive.

CONCLUSION

The AI in agriculture is bound to revolutionize farming. This descriptive scout on the potential ways AI could help agricultural activities, payment levels, and resource distributions, as well as the agricultural performance, have clarified the role of AI in the agro-tech enterprise. The last section is the conclusion of the study that summarizes outcomes as well as Envisions its future significance in intelligent agriculture. This chapter has established that automation can cause labor costs to be cut by 50, agricultural yields can be boosted by 15 and irrigation water can be cut by 20 which results in an average 30 percent boost to productivity. This chapter encouraged the application of DL, Internet of Things, and other advanced technologies to the world of farm automation and smart agriculture (Arshad et al., 2022) and developed a solid group of tools, which may be implemented simultaneously. Deep learning and technologies seem to outperform in virtually any application. They include, but are not limited to: to promote sustainable Agriculture, our "Agriculture 5.0 framework" will combine Human-Centered as well as AI Industry 5.0. It encourages sustainability in the environment, health of the plants and soil, climate resilience, single health and consumer satisfaction. The Smart Farming system is a system comprised of two systems which include the controlling system and the monitoring system. The controlling system uses the Node MCU ESP32, which is the main microcontroller in connecting the PH meter, TDS sensor, speed sensor of water flow, and ultrasonic sensors. Smart Farming is implemented through the use of Node MCU ESP32 as it is utilized to inter-connect sensors as well as actuators. The system can work on the auto mode and transmit the data to the Blynk application and accept orders there. Precision agriculture is changing the modern agriculture by adopting the latest major technologies to enable efficient use of resources and maximize productivity, such as GPS, sensors, drones, and data analysis. Data-enabling approach allows the farmers to address the inconsistency of the fields, optimize the output, and minimize the footprints on the environmental grounds therefore streamline the effectiveness and self-sufficiency of agriculture. The strategic resource management, technological innovation and focus on quality in the development of the sustainable and economically resilient agri-food sector are indicated as incredibly important. The results of the study are deemed to be of importance both to the academic research and to the practical application, which is why it is important to have the comprehensive strategies used in the field of agriculture that involve combining the environmental, social and economic factors. Agri-food producers use it as a model where it advocates adoption of sustainable practices that would make the global agri-food supply chains more resilient and sustainable.

REFERENCES

Alfarisy, A. A., Chen, Q., & Guo, M. (2018). Deep learning based classification for Paddy pests & diseases recognition. *Proceedings of 2018 International Conference on Mathematics and Artificial Intelligence*, 21-25.

Ali, A., Chana, I., & Buyya, R. (2019). *A fungus detection system for greenhouses using wireless visual sensor networks and machine learning*

Alpyssov, A., Uzakkyzy, N., Talgatbek, A., Moldasheva, R., Bekmagambetova, G., Yessekeyeva, M., Kenzhaliev, D., Yerzhan, A., & Tolstoy, A. (2023). Assessment of plant disease detection by deep learning. *Eastern-European Journal of Enterprise Technologies*, *1*(2 (121)), 41–48.

Arshad, J., Rehman, A. U., & Abbas, H. (2022). Implementation of a LoRaWAN based smart agriculture decision support system for optimum crop yield. *Sustainability*, *14*(2), 827.

Atiku, S., Jeremiah, A., Semente, E., & Boateng, F. (2024). *Sustainable development in industry and society 5.0: Governance, management, and financial implications: Governance, management, and financial implications*. IGI Global.

Banasik, A., Kanellopoulos, A., Claassen, G., Bloemhof-Ruwaard, J. M., & Van der Vorst, J. G. (2017). Closing loops in agricultural supply chains using multi-objective optimization: A case study of an industrial mushroom supply chain. *International Journal of Production Economics*, *183*, 409–420.

Camaréna, S. (2020). Artificial intelligence in the design of the transitions to sustainable food systems. *Journal of Cleaner Production*, *271*, 122574.

Deng, R., Tao, M., Xing, H., Yang, X., Liu, C., Liao, K., & Qi, L. (2021). Automatic diagnosis of rice diseases using deep learning. *Frontiers in Plant Science*, *12*. Advance online publication. DOI: 10.3389/fpls.2021.701038

Dury, J., Schaller, N., Garcia, F., Reynaud, A., & Bergez, J. E. (2011). Models to support cropping plan and crop rotation decisions. A review. *Agronomy for Sustainable Development*, *32*(2), 567–580.

Earthdata (NASA). (2022, August 8). *The value of data in monitoring the health of crops*. https://www.earthdata.nasa.gov/dashboard/stories/sat-data-agriculture

Fikry, I., Gheith, M., & Eltawil, A. (2019). A new mathematical model for the deterministic crop rotation planning problem. *2019 IEEE International Conference on Industrial Engineering and Engineering Management (IEEM)*, 314-318.

Fuentes-Peñailillo, F., Ortega-Farías, S., Acevedo-Opazo, C., Rivera, M., & Araya-Alman, M. (2024). A Smart Crop Water Stress Index-Based IoT Solution for Precision Irrigation of Wine Grape. *Sensors (Basel)*, *24*(1), 25. DOI: 10.3390/s24010025

Grillo, H., Alemany, M., Ortiz, A., & Fuertes-Miquel, V. (2017). Mathematical modelling of the order-promising process for fruit supply chains considering the perishability and subtypes of products. *Applied Mathematical Modelling*, *49*, 255–278.

Haque, A., Islam, N., Samrat, N. H., Dey, S., & Ray, B. (2021). Smart farming through responsible leadership in Bangladesh: Possibilities, opportunities, and beyond. *Sustainability*, *13*(8), 4511.

Hnatiuc, M., Ghita, S., Alpetri, D., Ranca, A., Artem, V., Dina, I., Cosma, M., & Abed Mohammed, M. (2023). Intelligent grapevine disease detection using IoT sensor network. *Bioengineering (Basel, Switzerland)*, *10*(9), 1021. DOI: 10.3390/bioengineering10091021

Huang, S., Wang, B., Li, X., Zheng, P., Mourtzis, D., & Wang, L. (2022). Industry 5.0 and society 5.0-Comparison, complementation and Co-evolution. *Journal of Manufacturing Systems*, *64*, 424–428.

Jonkman, J., Barbosa-Póvoa, A. P., & Bloemhof, J. M. (2019). Integrating harvesting decisions in the design of agro-food supply chains. *European Journal of Operational Research*, *276*(1), 247–258.

Kohut, N.. (2024). Detecting plant diseases using machine learning models. *Sustainability*, *17*(1), 132.

Kulanthaivelu, R. K., Iyyanar, S., & Ramakrishnan, S. (2022). Climate change and agricultural losses in India. *American Journal of Economics and Sociology*, *81*(2), 339–358.

Kumar, A., Tiwari, R. G., & Trivedi, N. K. (2023). Smart farming: Design and implementation of an IoT-based automated irrigation system for precision agriculture. *2023 3rd International Conference on Innovative Sustainable Computational Technologies (CISCT)*, 1-6.

Kusumastuti, R. D., Donk, D. P., & Teunter, R. (2016). Crop-related harvesting and processing planning: A review. *International Journal of Production Economics*, *174*, 76–92.

Latif, G., Abdelhamid, S. E., Mallouhy, R. E., Alghazo, J., & Kazimi, Z. A. (2022). Deep learning utilization in agriculture: Detection of rice plant diseases using an improved CNN model. *Plants*, *11*(17), 2230. DOI: 10.3390/plants11172230

Leng, J., Sha, W., Wang, B., Zheng, P., Zhuang, C., Liu, Q., Wuest, T., Mourtzis, D., & Wang, L. (2022). Industry 5.0: Prospect and retrospect. *Journal of Manufacturing Systems*, *65*, 279–295.

Li, J., Rodriguez, D., Zhang, D., & Ma, K. (2015). Crop rotation model for contract farming with constraints on similar profits. *Computers and Electronics in Agriculture*, *119*, 12–18.

Li, L., Zhang, S., & Wang, B. (2021). Plant disease detection and classification by deep learning-A review. *IEEE Access : Practical Innovations, Open Solutions*, *9*, 56683–56698.

Manwatkar, A. B. KeshavraoDeshmukh, R., Junnarkar, A. A., JagdishKamble, S., Kachare, P. K., &PrashantLatkar, S. (2024). Securing industry 4.0-Based precision agriculture. *2024 IEEE 4th International Conference on ICT in Business Industry & Government (ICTBIG)*, 1-7.

Mrisho, L. M., Mbilinyi, N. A., Ndalahwa, M., Ramcharan, A. M., Kehs, A. K., McCloskey, P. C., Murithi, H., Hughes, D. P., & Legg, J. P. (2020). Accuracy of a smartphone-based object detection model, PlantVillage Nuru, in identifying the foliar symptoms of the viral diseases of Cassava–CMD and CBSD. *Frontiers in Plant Science*, *11*. Advance online publication. DOI: 10.3389/fpls.2020.590889

Nakalembe, C., Becker-Reshef, I., Bonifacio, R., Hu, G., Humber, M. L., Justice, C. J., Keniston, J., Mwangi, K., Rembold, F., Shukla, S., Urbano, F., Whitcraft, A. K., Li, Y., Zappacosta, M., Jarvis, I., & Sanchez, A. (2021). A review of satellite-based global agricultural monitoring systems available for Africa. *Global Food Security*, *29*, 100543. DOI: 10.1016/j.gfs.2021.100543

OpokuDarko, R., Liu, J., Yuan, S., K. Sam-Amoah, L., & Yan, H. (2020). Irrigated agriculture for food self-sufficiency in the sub-saharan African region. *International Journal of Agricultural and Biological Engineering*, *13*(3), 1–12.

Pourreza, A., Lee, W. S., Etxeberria, E., & Zhang, Y. (2016). Identification of citrus Huanglongbing disease at the pre-symptomatic stage using polarized imaging technique. *IFAC-PapersOnLine*, *49*(16), 110–115. DOI: 10.1016/j.ifacol.2016.10.021

Saetra, H. S. (2021). AI in context and the sustainable development goals: Factoring in the Unsustainability of the Sociotechnical system. *Sustainability*, *13*(4), 1738.

Sankaran, S., Maja, J., Buchanon, S., & Ehsani, R. (2013). Huanglongbing (Citrus Greening) detection using visible, near infrared and thermal imaging techniques. *Sensors (Basel)*, *13*(2), 2117–2130. DOI: 10.3390/s130202117

Sayekti, N. C., Fhaizal, M. A., Wijaya, W., Hendajany, N., Aprianti, I., &Kusmadi. (2024). Smart farming design with NodeMCU ESP32 for NFT hydroponic. *2024 18th International Conference on Telecommunication Systems, Services, and Applications (TSSA)*, 1-4.

Sharma, A., Jain, A., Gupta, P., & Chowdary, V. (2021). Machine learning applications for precision agriculture: A comprehensive review. *IEEE Access : Practical Innovations, Open Solutions*, *9*, 4843–4873.

Sharma, P., Kumar, P., Singh, P. K., & Rana, A. K. (2024). "Precision agriculture farming" enhancing farming efficiency through technology integration. *2024 1st International Conference on Advances in Computing, Communication and Networking (ICAC2N)*, 365-370.

Shin, J., Chang, Y. K., Heung, B., Nguyen-Quang, T., Price, G. W., & Al-Mallahi, A. (2021). A deep learning approach for RGB image-based powdery mildew disease detection on strawberry leaves. *Computers and Electronics in Agriculture*, *183*, 106042.

Shneiderman, B. (2020). Human-centered artificial intelligence: Reliable, safe & Trustworthy. *International Journal of Human-Computer Interaction*, *36*(6), 495–504.

Sivagnanapalani, P., Ansari, N. I., & Panda, P. (2021). Nd2Ti2O7 (NTO) with high Curie temperature (TC) for high temperature sensor applications. *Sensors International*, *2*, 100093.

Soto-Silva, W. E., González-Araya, M. C., Oliva-Fernández, M. A., & Plà-Aragonés, L. M. (2017). Optimizing fresh food logistics for processing: Application for a large Chilean Apple supply chain. *Computers and Electronics in Agriculture*, *136*, 42–57.

Sungheetha, A. (2022). State of art survey on plant leaf disease detection. *Journal of Innovative Image Processing*, *4*(2), 93–102.

Turner, C. J., Ma, R., Chen, J., & Oyekan, J. (2021). Human in the loop: Industry 4.0 technologies and scenarios for worker mediation of automated manufacturing. *IEEE Access : Practical Innovations, Open Solutions*, *9*, 103950–103966.

Tushar, D., Pradhan, N., & Singh, S. (2025a). A design of 2 × 2 element MIMO antenna with novel gain for 28/38 GHz 5G applications. *Lecture Notes in Electrical Engineering*, 257-270. https://doi.org/DOI: 10.1007/978-981-95-0203-5_22

Tushar, J. P., Shrivastava, P. C., & Pradhan, N. (2025b). *Carbon capture, utilization, and storage (CCUS) technologies: Net-zero strategies*. In *The future of green energy: Storage, materials, alternative fuels, and net-zero strategies* (pp. 251–281). IGI Global Scientific Publishing. https://doi.org/DOI: 10.4018/979-8-3373-3226-0

Tushar, P. N., Shrivastava, P. C., Jaiswal, P., Singh, S., & Mishra, V. (2026a). *Legal challenges in cloud computing*. In I. Oncioiu (Ed.), *Law and regulation of AI, blockchain, and digital rights* (pp. 63–102). IGI Global Scientific Publishing. https://doi.org/DOI: 10.4018/979-8-3373-3186-7.ch003

Tushar, P. N., Jaiswal, P., Shrivastava, P. C., Kumar, M., & Sundar, S. (2026b). *Precision farming and smart irrigation systems for energy management.* In B. Soufiene (Ed.), *Enhancing autonomous and adaptive systems with AI and IoT* (pp. 251–288). IGI Global Scientific Publishing. https://doi.org/DOI: 10.4018/979-8-3373-3146-1.ch009

Tushar, P. N., Shrivastava, P., Jaiswal, P., Mishra, S., & Yadav, A. (2026c). *AI regulation in K–12 and higher education institutions.* In K. Carlson (Ed.), Ethical, legal, and pedagogical perspectives on AI in education. IGI Global Scientific Publishing.

Tushar, P. N., Shrivastava, P. C., Jaiswal, P., Mishra, S., & Yadav, A. K. (2026d). AI Regulation in K-12 and Higher Education Institutions. In K. Carlson (Ed.), *Ethical, Legal, and Pedagogical Perspectives on AI in Education* (pp. 1-32). IGI Global Scientific Publishing. https://doi.org/DOI: 10.4018/979-8-3373-3000-6.ch001

Tushar, & Pradhan, N. (2024). Parametric analysis of DGS based sub-6 GHz antenna for WiFi communication. *2024 International Conference on Control, Computing, Communication and Materials (ICCCCM)*, 234-241. https://doi.org/DOI: 10.1109/icccm61016.2024.11039972

Wollschlaeger, M., Sauter, T., & Jasperneite, J. (2017). The future of industrial communication: Automation networks in the era of the Internet of things and industry 4.0. *IEEE Industrial Electronics Magazine*, *11*(1), 17–27.

Xu, X., Lu, Y., Vogel-Heuser, B., & Wang, L. (2021). Industry 4.0 and industry 5.0-Inception, conception and perception. *Journal of Manufacturing Systems*, *61*, 530–535.

Chapter 9
AI and Computer Vision for Plant Disease Detection in Sustainable Agriculture

Md. Shoeab Akhter
http://orcid.org/0009-0004-4582-2937
International University of Business, Agriculture, and Technology, Bangladesh

Sakibul Islam Ratul
http://orcid.org/0009-0000-7217-5190
Daffodil International University, Dhaka, Bangladesh

Dilsad Jahan Jui
http://orcid.org/0009-0003-5120-3924
Chattogram Veterinary and Animal Sciences University, Bangladesh

Farjana Sultana
International University of Business, Agriculture, and Technology, Bangladesh

Israt Jahan
http://orcid.org/0009-0005-9406-6223
East West University, Bangladesh

Md. Sakibul Hasan Dipto
International University of Business, Agriculture, and Technology, Dhaka, Bangladesh

ABSTRACT

Plant diseases significantly impact global food security, causing yield losses and reduced crop quality. Traditional detection methods are labor-intensive, slow, and hindering timely interventions. This chapter explores how AI, deep learning, and IoT technologies can revolutionize plant disease detection in sustainable agriculture. A review of recent research and experiments shows that convolutional neural networks (CNN), transfer learning models like ResNet and EfficientNet, and multispectral imaging can all be used to classify crops like rice, maize, and wheat with 90% to

DOI: 10.4018/979-8-3373-7257-0.ch009

98% accuracy. The integration of IoT sensors and cloud-based platforms enables real-time monitoring, early diagnosis, and optimized pesticide use, fostering sustainable practices. AI-driven disease detection helps make better predictions about crop yields, has less of an effect on the environment, and makes it easier to make decisions based on data. The chapter emphasizes the need for understandable AI systems and accessible tools for farmers to ensure equitable benefits and digital transformation in agriculture.

INTRODUCTION

Plant diseases have been identified to be a serious and continuous concern to food security worldwide, as they lower the yield of crops, alter quality, and also incur economic losses, which may account for the billions of dollars annually (Strange & Scott, 2005). The Food and Agriculture Organization (FAO) claims that with 20-40 percent losses to biotic stresses on staple crops like rice, wheat, and maize, and 50 percent losses in horticultural crops, uncontrolled cases result in losses of up to 50 percent of the potential yield of crops (Rathinapriya et al., 2025). The growing population pressure, climate change, and scarcity of agricultural resources worsen these difficulties and underline the necessity to find new solutions to disease management. Although these conventional procedures, including visual check-up, field scouting by experts, and laboratory-based identification of the pathogen, are effective in certain situations, they are usually labor-extensive, time-consuming, and may be subject to human error (Savary et al., 2012). This results in delay in diagnosis that may cause extra or improperly timed application of pesticides, which may increase the expenses, soil contamination, and health outcomes. The current technological developments have presented the opportunities of more effective, accurate, and environmentally friendly disease control methods.

Deep learning as a sub-field of Artificial Intelligence (AI) provides the means of disease detection automation based on the ability of deep learning algorithms to operate on large amounts of data, identifying complex patterns and symptoms that are inconspicuous and invisible (Prasad & Sharma, 2025). With the help of AI and computer vision, it is possible to analyze the image of a plant in real time to see the morphological and physiological alterations that indicate the presence of a disease. Moreover, the devices of the Internet of Things (IoT), soil moisture, temperature, humidity, nutrient, and light intensity sensors assist constant environmental control, which allows predicting the disease and providing an early response (Akhter et al., 2025; Chakraborty & Newton, 2011). With the combination of these technologies, one can develop a digital ecosystem that would be able to respond to adaptive, data-driven, and accurate plant health management.

Sustainable agriculture is based on the need to reduce the use of chemicals, maximize the utilization of water and nutrients, and lower the level of labor but still achieve high crop production. Plant health monitoring achieved with the help of AI can play a part in these targets since it allows taking practical decisions that align with the ecological and economic agendas (Ayorinde, 2025). Adaptive crop protection strategies are enhanced with the help of sensors, AI analytics, and automated systems, increasing resistance to new pathogens, environmental stress, and variability of climate. The chapter discusses the theoretical backgrounds, technological implementation, and functional implications of AI- and IoT-based plant disease detection, which provides a platform of sustainable, precision agriculture. It emphasizes the way in which digital technologies can revolutionize the conventional crop protection, increase the predictability of yields, and advance environmental friendly methods of farming, preconditioning resilient and data-driven agriculture globally (Thangamani et al., 2024).

Figure 1. Integrated framework of AI-driven plant disease detection for sustainable agriculture

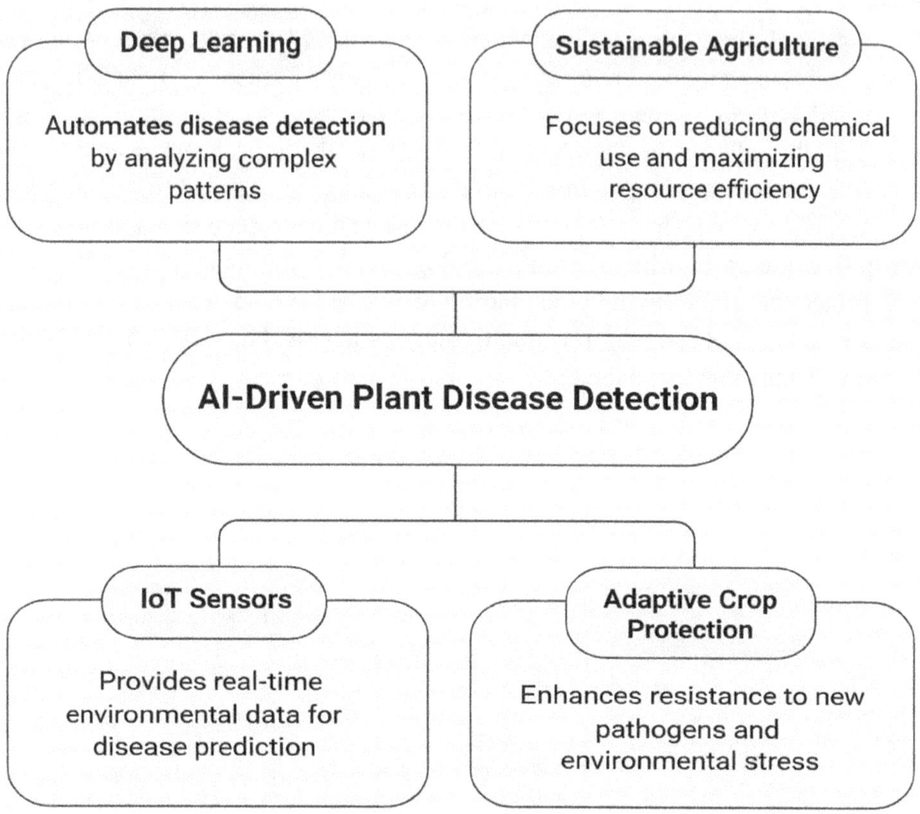

Figure 1 demonstrates how the use of modern artificial intelligence technologies can change the concept of plant disease detection as part of sustainable agriculture. The basic unit of this framework is AI-driven plant disease detection, which is the basic mechanism that connects deep learning, sustainability principles, integration of the IoT, and adaptive crop protection. The upper part brings out the importance of deep learning, especially convolutional neural networks, in the automation of disease diagnosis through the detection of complicated visual patterns in the leaves of plants. In addition sustainable agriculture is characterized by a decriminalization of chemical reliance and higher resource utilization, so that AI applications are used to engage in environmentally friendly farming. Two supporting components are also of critical importance, as shown in the lower part of the figure. IoT sensors provide real-time environmental measurements - humidity, soil conditions and temperature - which can be used to enables early prediction of diseases and monitoring of the field. Adaptive protection of crops relies on AI-enabled data to enhance the resilience of crops and assist plants to react more effectively to new pathogens and environmental factors.

All these interrelated aspects show a whole-body, technological approach to disease management. The figure supports the main theme of the chapter in which it is stated that the combination of AI, sensor technologies, and sustainable practices can help to achieve great progress in terms of diagnostic accuracy, environmental footprint, and more robust agricultural ecosystems.

Objectives of the Chapter

1. To explore the AI, deep learning, and computer vision in the automation of the precise detection of plant disease.
2. To examine how IoT sensors and cloud/edge computing can be integrated in order to perform real-time monitoring as well as early disease detection of crops.
3. To determine the way AI-IoT-vision systems allow sustaining agriculture by cutting down the use of chemicals, labor, and environmental effects.
4. To investigate real-life examples and applications of smart technologies used in agriculture that prove to be effective in various crops and regions.

2. THEORETICAL OVERVIEW

2.1 The concept of AI and machine learning in the fields of agriculture

Artificial Intelligence (AI) is a term used to describe the computational systems that may assume the tasks that had been performed by human intelligences, including pattern recognition, decision-making, and problem-solving (Akhter, et al., 2025). Artificial intelligence in the agricultural industry is absolutely diverse: crop yield, pests and diseases, fine-tuning of irrigation, optimal use of fertilizers, weather adjustment, etc. Machine Learning (ML) is a sub-field of AI that enables the systems to be trained on previous and current data, identify patterns, and make predictions without any particular code (Jafar et al., 2024). Supervised learning has been used extensively in the detection of plant diseases, in which models are trained on labeled data sets comprising of plant images labeled with a known disease or healthy condition (Wang et al., 2025). The most common algorithms used to classify diseases with either visual or sensor-derived features are the Convolutional Neural Networks (CNNs), Support Vector Machines (SVMs), and Random Forests (Majdalawieh et al., 2025). A wide range of unsupervised learning methods, such as clustering and anomaly detection, enable the models to discover a pattern in unlabeled data, which can be used to detect signs of diseases earlier than before. Reinforcement learning can then optimize resources allocation, irrigation and fertilizer application, through learning of optimal action by providing more information on the reward feedback loop. The AI-based systems offer a number of benefits compared to the conventional methods in the agricultural sector. They have the ability to handle multi-modal sensor, satellite, UAV, and historical crop records, and thus, conduct timely and predictive interventions. Continuous learning algorithms are enhanced as time goes by to suit new diseases, crop varieties and environmental conditions (Mishra & Mishra, 2023). AI decreases labor dependence and reduces human error and improves decision-making by facilitating automated and scalable, as well as precise monitoring. The application of AI to precision agriculture schemes is the basis of smart farming, which enables greater productivity, sustainability and resiliency in contemporary farming (Ahmad & Nabi, 2021).

2.2 Significant Perspectives of Image Processing and Computer Vision

Automated plant disease detection requires computer vision and image processing. Image processing is the conversion of raw visual information into supervisory format enhancing the manipulation of relevant data and reducing noise. The main preprocessing methods are resizing, normalization, demising, color correction, and augmentation techniques, which are flipping, rotation, and scaling and contrast adjustment. The concept of data augmentation is especially important to deep learning, because it enhances the diversity of the used data and alleviates overfitting in the used model. Using feature extraction images are transformed into numerical data which display disease related features, such as color, texture, shape and size. Classical approaches include manual features, including color histograms, edge detection techniques and Gray Level Co-occurrence Matrix (GLCM) texture analysis measures (Iqbal et al., 2021; Wiranata et al., 2025). The feature extraction process is automated by advanced computer vision methods, especially CNNs and deep learning networks, which learn hierarchical representations using low level features (edges, gradients) up to complex (lesion shapes, patterns of discoloration) ones. Multispectral and hyperspectral imaging also increase the detection abilities by capturing the information that is not visible and that identifies physiological stress, chlorophyll deficits and water stress before the visible symptoms are detected. Thermal imaging reveals temperature changes related to changes in the metabolism, water stress or hotspots of infections. High-resolution spatial coverage of large field's scalable Monitoring plant health UAVs and drones offer fast, highly-scaled spatial coverage of the fields. Combination of AI models with computer vision enables life-time disease identification, early diagnosis, and accurate control, which is the foundation of accuracy in sustainable agriculture (Wiley & Lucas, 2018).

2.3 IoT, Sensor, and AI integration to provide crop monitoring services

Internet of Things (IoT) is a type of continuity data-driven monitoring ecosystem which links field-deployed sensors, devices and compute platforms, and produces emerging ecosystems of data monitoring and analytics. Water content, pH and soil moisture are measured using soil probes, temperature, humidity and light intensity are measured using environmental sensors. Supplementary visual data would be offered by imaging devices such as RGB, multispectral, hyperspectral and thermal cameras. LoRaWAN, ZigBee and 5G wireless communication protocols allow stable communication of the data even in rural areas of agriculture (Majumdar et al., 2025). Sensor fusion integrates the heterogeneous sources of data enhancing reliability and minimizing noise in detecting diseases. The approaches such as the Kalman filtering, Bayesian inference, and machine learning-based fusion, combine both environmental and visual data to identify the stress patterns in plants correctly. Edge computing devices can analyze data on-site to give instantaneous data collection and notifications where cloud computing facilitates mass data analysis, model training, and storage. Integration of historical data increases predictive power, which allows early detection of diseases outbreaks. AI tools use combined sensor and imaging data to produce insights to be used to irrigate, fertilize, and disease control. Cloud dashboards and mobile applications are accessible at a distance and give farmers the ability to make decisions based on data. Modern agriculture, with its combination of IoT, AI, and computer vision, can be adaptive, resilient, and sustainable in the management of plant health to maximize the use of resources and minimize the environmental impact (Sharma & Shivandu, 2024).

3. TECHNIQUES OF DETECTING DISEASES IN PLANTS

The detection of plant diseases is based on the complex of imaging technologies and data preprocessing, feature selection, and deep learning models that allow obtaining accurate and scalable results. The contemporary methods make use of ground and remote image recording through the use of UAVs, RGB, hyperspectral, multispectral, and thermal imaging systems to provide a rich visual data of a vast field. RGB cameras are easily available and easy to implement and can capture the visible symptoms (spots on leaves, chlorosis, and necrotic lesions) (Mukhtar et al., 2025). Multispectral and hyperspectral sensors offer extra spectral categories that can be subjected to detect physiological stress prior to the appearance of its symptoms in detecting diseases such as rice blast, wheat rust, or maize leaf blight early. The thermo imaging process records temperature changes related to the state of water stress or infection and makes the detection of the affected regions possible based on the metabolic and transpiration differences (Fevgas et al., 2025). Large-scale agricultural fields also have a quick and high-resolution coverage, minimizing the time spent on scouting with manual methods and allowing timely action to be taken by UAVs with such sensors (Sankaran et al., 2010).

3.1 Image Acquisition Methods

The first step in AI-based disease detection is image acquisition which is crucial. Ground-based imaging offers close-up images of parts of the plants that are at close range and thus early signs of disease can be captured in high resolution. On the contrary, aerial imaging performed by UAV composes larger zones and captures canopy-level patterns on several plots (Islam et al., 2025; Majumdar et al., 2025). RGB cameras are color cameras that are high-resolution and would be suitable in visible disease symptoms. Hyperspectral cameras and multispectral cameras also capture reflectance patterns outside the visible spectrum like near-infrared cameras and display plant stress and nutrient deficiencies. The thermal cameras can identify the temperature change in the canopy, and this can be used to identify low transpiration areas or areas of infection. The main considerations in the acquisition of images are the light conditions, sensor resolution, flight altitude in the case of UAVs, the angle of capture, and variability of nature. These factors should be optimized to guarantee the quality of data and model performance (Nguyen et al., 2021).

3.2 Preprocessing and Feature Extraction of Data

The images taken in the field tend to be noisy, have inconsistencies in illumination, and contain irrelevant background information. The preprocessing methods that include resizing, normalization, color correction, and denoising make the dataset standard and boost feature extraction. Image augmentation methods, such as rotation, flipping, scaling, and contrast changes, make deep learning models less susceptible to overfitting and makes them more diverse. The process of extracting features is a transformation of images into numerical form which brings out the disease-relevant features in the images. The classic techniques are based on manually designed characteristics in the form of color histograms, texture analysis with the help of Gray Level Co-occurrence Matrix (GLCM), edge detection, and shape descriptors (Setiawan & Budiman, 2025). Current methods embrace convolutional neural networks (CNNs) to learn hierarchical features, starting with edges and gradients at the low-levels and ending with high-level patterns reflecting qualifications of disease severity, in an automatic manner. Interpolated multispectral indices are NDVI (Normalized Difference Vegetation Index), GNDVI, and PRI (Photochemical Reflectance Index), which are used to increase detection accuracy by emphasizing patterns of physiological stress (Islam et al., 2025). Using the visual features together with the data of environmental sensors leads to a more detailed image of the health of plants, enhancing the predictions of models (Upadhyay et al., 2025) (Hall et al., 1971).

3.3 Disease Identification Model architectures

Learning complex patterns has been the order of the day in making deep learning models the standard in terms of automated disease detection. The CNN architectures have wide use in image classification tasks to retrieve hierarchical features on diverse layers. A pre-trained classifier, such as Resnet and Efficient, makes use of transfer learning; hence, it is able to be trained on a plant disease dataset with small sample sizes using a pre-trained model, such as ImageNet. Model optimization techniques of pruning and quantization and ensembles help to improve computational efficiency and strength. The classification of these models represents a correct classification of 90-98 percent of crops like rice, wheat, maize and tomato. Models that are edge deployed can provide real-time detection at farm and models that are cloud deployed can provide large-scale analysis, trend detection and model retraining. A system of internet of things sensor data and these architectures can be used to develop a holistic monitoring system more effective in the detection of disease early on and resorting to precision interventions (Dargan et al., 2020) (Table 1).

Table 1. Comparison of imaging techniques and artificial intelligence to detect disease in plants

Imaging Method	Sensor Type	Features Captured	AI Model Used	Crop Examples	Accuracy (%)
RGB Camera (Dolatabadian et al., 2025)	Ground / Drone	Color, leaf spots, lesions	CNN, ResNet	Wheat, Rice, Tomato	90–95
Multispectral Imaging (H. Sharma et al., 2025)	Drone / UAV	Physiological stress, NDVI, GNDVI	CNN, EfficientNet	Maize, Vineyards	92–97
Hyperspectral Imaging (Chandel et al., 2022)	UAV	Chlorophyll, water stress, NPK	ResNet, DenseNet	Rice, Wheat	93–98
Thermal Imaging (Guaragnella, 2024)	Ground / Drone	Canopy temperature, water stress	CNN, ResNet	Tomato, Wheat	90–94

In this part, the very basis of the AI-based detection of plant diseases is established, and the priorities are put on the bundle of the image acquisition, its preprocessing, feature extraction, and deep learning structures. Modern systems have provided a solution for precision disease monitoring with scalability and accuracy in a timely manner due to the use of advanced imaging, computational models, and IoT integration.

4. AI AND DEEP LEARNING STRUCTURES IN DISEASE DIAGNOSIS

Artificial intelligence and deep learning models are critical in the process of automating the detection of plant diseases using the images and sensor data to learn the complex images. Those models combine various computational schemes, such as supervised and unsupervised learning, transfer learning, model optimization, edge/cloud deployment, etc., to deliver precise, scalabler and real-time disease monitoring. Through synergizing these methods, precision agriculture systems will be able to identify the diseases at an earlier stage, optimize actions, and decrease the effect on the environment (Sharma et al., 2022).

4.1 The supervised and unsupervised learning models are explained below

Supervised learning has been widely applied in the detection of plant diseases, and in the process of training the models, the labeled data sets are utilized, and these data sets include pictures of both healthy and diseased plants. Most of the supervised models rely on the convolutional neural networks (CNNs) because the convolutional neural networks can learn hierarchical features by themselves. Studies to detect rice blast, based on CNN, had classification accuracies of more than 92 percent, and classification accuracies of models to detect maize leaf blight were at 95 percent. Other supervised classification models, such as Support Vector Machines (SVM) and Random Forests, are used to do the smaller datasets or feature-based classification and can be interpreted and have relatively low computational power requirements. The unsupervised learning may be advantageous in case of a limitation in the number of labeled data and in situations when it is required to distinguish a new or emerging disease. The clustering methods, such as k-means or hierarchical clustering, can identify the aberrant tendencies in the plant images or sensor data, and this is an indication that there is a possible disease outbreak. Autoencoders and self-organizing maps (SOMs) are other tools that assist in the detection of anomalies through learning the typical patterns of plants and alerting of anomalies. By incorporating both unsupervised and supervised models, hybrid structures will be capable of improving the detection and handling of disease cases that have not been observed before (Rajoub, 2020).

4.2 Transfer Learning and Model Optimization

This section provides an overview of this area aimed at applying the insights gained from this course to apply to this domain by implementing the knowledge of transfer learning and model optimization.

Transfer learning uses the training of large datasets like ImageNet, like ResNet, EfficientNet, or DenseNet, to learn faster on plant disease datasets. This method is especially useful on crops that have few labeled images, as it will save on training time and enhance accuracy. Experimenting with fine-tuning of pre-trained models trained on crop-specific data enables the network to be adjusted to plant disease classification tasks using model optimization techniques, which also improve the performance and efficiency. The overfitting is avoided by regularization techniques such as dropout and batch normalization whereas pruning and quantization techniques reduce the size of the model and computational costs to enable its deployment to edge devices (Alahe et al., 2025). Ensemble techniques, that is, the combination of several models, enhance stability and precision, particularly in different environmental conditions or with different crops. Transfer learning is complemented by data augmentation (to enhance generalization) to augment data diversity, e.g., rotation, scaling, and color jittering. Transfer learning A report of tomato leaf disease classification studies with up to 98 accuracy showed the usefulness of combining pre-trained models with optimized architectures (Jiang et al., 2018).

4.3 Detection of disease in real-time with the help of edge and cloud computing

Detection of diseases in real-time demands the smooth combination of AI models and IoT devices, edge computing platforms, and cloud architecture. Edge computing enables the processing of data on the sensor or on the mobile device and AI models, minimizing the latency and allowing timely notifications to farmers. Minimal CNN networks can be put on drones or smartphones to identify leaf images in a couple of seconds and assist in the quick detection of infected zones.

Major benefits of cloud computing include massive storage of data, retraining of models, and cross-field analysis. Cloud platforms combine data of various farms, allowing for prediction of disease outbreaks, trend analysis, and risk mapping. The edge and cloud computing combination makes it scalable, precise and responsive in real-time. Southeast Asian rice farms that are connected to the IoT can be used as an example: edge-deployed CNNs are used to detect cases of the blast disease immediately, whereas cloud dashboards can be used to see patterns of the infection spread across the regions, making AI models have a direct effect on the disease management strategies by minimizing the use of chemicals and labor. Automated systems, including irrigation valves and drone-based spraying, can be connected to the IoT to ensure that AI models can directly affect the disease management procedures, using fewer chemicals and requiring less labor. Through the integration of edge computing, cloud computing, and the Internet of Things, real-time disease identification systems improve precision agriculture to guarantee on-time, effective, and sustainable crop security (Badidi, 2023).

5. SMART SYSTEMS WITH IOTS TO MONITOR THE HEALTH OF PLANTS

IoT combined with AI and computer vision has transformed the health monitoring of plants, making it possible to acquire data on a continuous basis in real time and make reasonable decisions. IoT-based systems are systems constituted of sensors, communication networks, and computation platforms that collect and analyze the environmental and plant data to enhance the management of the disease, resource efficiency, and sustainable agriculture (Ataei Kachouei et al., 2023).

5.1 Data collection and Fusion techniques based on sensors

IoT sensor networks would be important in recording important environmental and physiological parameters. Typical sensors are soil moisture, nutrient (NPK), temperature/humidity, pH, light, and gas sensors to detect volatile organic compounds that are released by stressed plants. The use of imaging devices (RGB, multispectral, hyperspectral, and thermal cameras) offers supplementary visual data, and AI models are able to identify disease symptoms and patterns of plant stress using the data fusion techniques. The fusion of multi-modal data can identify disease onset and progression using methods including Kalman filtering, Bayesian inference, and deep learning-based approaches. Combined soil moisture, temperature data and leaf reflectance index can be used to detect fungi in rice and wheat early so that the interventions can be focused. Sensor calibration, redundancy and error-checking procedures are vital to provide accurate and reliable data in field condition (Izadi et al., 2015).

5.2 Intelligent Decision-Support Systems and Automation

IoT-driven, AI-driven decision support systems (DSS) are used to deliver actionable insights to manage diseases. Through these systems sensor and imaging data are analyzed to produce recommendations on irrigation, fertilization, application of pesticides and crop rotation. It makes use of mobile and cloud-based dashboards to display the health of plants, identify hotspots of infection, and predict the risk of disease using predictive models. Interventions are more efficient through automation. As an example, smart irrigation systems regulate the amount of water distributed according to the current soil moisture rates to avoid stress, which may predispose the crops to illnesses (Sahu et al., 2025). Likewise, pesticides can be sprayed by the use of drone-based or robotic spraying systems, which can spray pesticides only in the needed areas, lessening the use of chemicals and lowering work expenses. The combination of predictive analytics and automation will guarantee timely reaction to the emergent threats to keep crops productive and to allow sustainable farming (Patel & Kamrani, 1996).

5.3 Case Studies: IoT usages in Real-Life Ag scenarios

The efficacy of the IoT-enabled plant health monitoring is presented by a number of actual applications:

5.3.1 Rice Blast Detection in Southeast Asia:

A case study of rice blast detection in Southeast Asia was conducted to investigate the following questions. The following are questions that were investigated in a case study of rice blast detection in Southeast Asia.

The IoT-based monitoring systems in Southeast Asian rice fields comprised soil moisture, humidity and temperature sensors that were combined with the multispectral imaging of UAVs. Analysis of data was based on Convolutional Neural Network (CNN) models as a means to detect rice blast disease in an early stage. The farmers got automated alerts through mobile applications, which thus helped them to apply fungicides accurately and on time. This has cut down chemical use to almost 30% maintained soil health and ensured a steady production even during varying climatic conditions. The case demonstrates the use of the IoT technologies to protect crops in the developing countries, to make real-time decisions, and to also conserve resources (Saleh et al., 2014).

5.3.2 Wheat Rust Management in Europe:

Wheat rust is a disease affecting the wheat species in Europe caused by a fungus named *Blumeriella W. epiphytica.*

Growth: Wheat rust is a fungal disease that affects the wheat species in Europe due to a fungus *Blumeriella W. epiphytica*. IoT-based networks with thermal imaging and leaf wetness sensors on European wheat farms made it possible to detect signs of rust infection in real time. Information was sent to a cloud-based dashboard upon which AI algorithms forecasted patterns of disease spread. Farmers would have the ability to monitor fields remotely and apply site-specific fungicides, which would decrease the use of chemicals and limit the impact on the environment. IoT built into predictive analytics was found to be more accurate and improved precision agriculture by managing disease better and enhancing the quality of yield. The case proves that the use of advanced IoT infrastructure can facilitate the implementation of data-driven and sustainable crop protection initiatives (Jørgensen et al., 2018).

5.3.3 Tomato Greenhouses in North America:

The North American Tomato Greenhouses were established in 1971. The production of tomatoes in North America has incorporated the IoT-based greenhouses with temperature, humidity, and CO_2 sensors connected to automated irrigation and ventilation systems. Prediction models based on AI were used to prevent the outbreak of fungi and bacteria, based on real-time data. The system has ensured the best microclimatic conditions, and the disease incidence was reduced by 25%, thereby the quality of the fruits was enhanced. Growers were able to remotely adjust irrigation and nutrient supply through the central monitoring of their farms. The case underlines the importance of IoT-based smart greenhouses, which will improve productivity, minimize the number of manual workers, and foster sustainable horticulture under controlled conditions (Peet & Welles, 2005).

5.3.4 Vineyard Disease Surveillance in Mediterranean areas:

Some vineyards in the Mediterranean are also reported to have a disease known as Vineyard Disease Surveillance.

Mediterranean vineyards involved the use of IoT-based disease surveillance involving multispectral UAV sensors as well as soil moisture and weather sensors. The field-level variations were handled by the data analytics platforms to identify early symptoms of powdery mildew. This technology would allow farmers to localize areas of attack, selective pruning and use fungicides only when absolutely needed, which would cut down the farm chemical use and labor expenses. The system boosted the quality of the yields and sustainability of the environment through the advancement of precision viticulture. The current case study is an indication of how IoT-driven decision-support systems can transform the management of diseases in high-value perennial crops (Portela et al., 2024).(Table 2).

Table 2. Applications of IoT and AI models in plant health monitoring and disease management

Application Area	IoT Devices / Sensors	Data Collected	AI Models Used	Benefits
Rice Blast Detection (Zhou et al., 2021)	Soil moisture, temperature, UAV	Moisture, canopy images	CNN, ResNet	Early detection, reduced fungicide use
Wheat Rust Management (R. Singh et al., 2023)	Thermal cameras, humidity, soil	Temp, humidity, leaf stress	CNN, SVM	Targeted pesticide application, cost reduction
Tomato Greenhouses (Akbar et al., 2024)	pH, humidity, temperature sensors	Microclimate, nutrient levels	EfficientNet, CNN	Reduced fungal outbreaks, optimized irrigation
Vineyard Disease Monitoring (Nkwocha & Chandel, 2025)	Multispectral UAV, weather sensors	NDVI, canopy reflectance, temp	ResNet, DenseNet	Timely pruning, localized treatment, yield improvement

The intelligent systems within the article with the IoT features indicate how sensor networks, data fusion, AI analytics, and automation can be used to improve monitoring of plant health. These systems are not only efficient in the detection and use of resources needed in the disease but also reduce environmental impact and reduce the operational cost so as to promote sustainable and precision agriculture. The farmers can adopt proactive actions, maintain crop production, and react to evolving environmental behavior using real-time data and automated decision-making (Verdouw et al., 2019).

6. CONCEPTUAL FRAMEWORK

The AI analytics, IoT-based sensing, and computer vision are all unified into a single system that allows adaptive and resilient crop protection based on the conceptual framework of sustainable plant disease management. It is grounded on a digital plant health monitoring framework under which multi-mode data, including soil parameters, microclimatic conditions, spectral signatures of the leaves, and canopy temperature, are constantly collected by environmental sensors, images, and UAVs (Garg et al., 2023).

Figure 2. Conceptual framework of AI-based plant disease detection integrating image processing and IoT systems

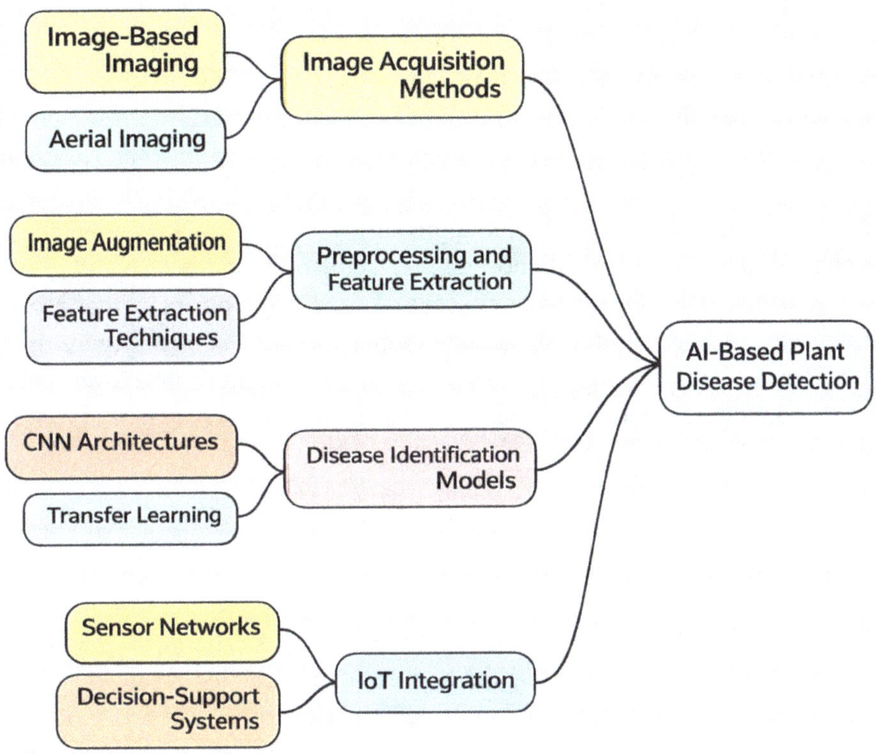

Wireless IoT networks transmit this information to cloud and edge computing platforms, and AI algorithmsin the cloud recognize patterns and categorize disease symptoms and future outbreaks. Deep learning (such as CNNs, ResNet, and EfficientNet) is combined to enable automatic feature detection and proper disease identification, but transfer learning ensures that it can be used inter-crop and inter-geography. The model aims at the relations between AI analytics and IoT sensing and sustainability goals. This system provides an alternative to using chemical pesticides, introducing predictive modelling of the diseases and real-time monitoring of the environment, thus improving the water and nutrient management and minimizing the losses of the yield, thus improving the ecological balance and economic efficiency.

The IoT-based automation will enable timely interventions, e.g., smart irrigation or accurate application of fungicides, and minimize the amount of resources wasted and the effects on the environment (Sharma et al., 2025). In addition, the system will have feedback loops, which will allow learning to be continuous so that the AI models can update and respond to new patterns of the disease and climate variability. In addition, the framework supports resiliency based on adaptive crop protection strategies. The predictive devices of diseases in particular areas are made possible through historical and geospatial integration, whereas decision-making is assisted by cloud-based dashboards for farmers and agronomists. This framework can foster an active and evidence-based sustainable approach in crop health management through the integration of technological knowledge, sustainability indicators, and practical practices in the field. Overall, the conceptual model demonstrates that artificial intelligence, the Internet of Things, and computer vision can reconfigure the traditional disease detection approach and create a holistic and adaptable system that is more productive, resource-efficient, and sustainable in the modern agricultural environment (Garg et al., 2025).

7. FINDINGS

The review of the existing chapter on the issues of AI and IoT-based plant disease detection demonstrates that the high-technological and practical successes are accomplished in the sphere of sustainable agriculture. AI-based models can achieve high classification rates on numerous crops, including rice, wheat, maize, and tomatoes, as well as vineyards, with some of the algorithms achieving above 90 percent, and in some studies the models have reached above 98 percent accuracy with CNNs, ResNet, and EfficientNet (Mohyuddin et al., 2024). Combining multispectral with hyperspectral imaging and deep learning will allow them to detect physiological stress and disease symptoms that cannot be seen by the naked eye in time to institute relevant interventions and minimize crop losses. Combining the IoT sensors goes an extra mile to improve the model performance with real-time environmental context to make the disease detection algorithms more predictive. The empirical comparison of AI models shows that the transfer learning methods achieve higher performance than the models that are trained directly, particularly when working on the datasets where the number of labeled images is small. The ensemble-based methods that integrate two or more architectures enhance crop and environmental robustness and generalization. Models deployed on edges are used to provide quick on-field detection, and cloud-based models are used to support large-scale monitoring, trend analysis and forecasting. The chapter point to the fact that multi-modal data integration, i.e. integration of imaging, soil and microclimate sensors, improves the accuracy and operational efficiency. There are socioeconomic and environmental advantages (Sharma et al., 2024). According to farmers, systems applied with the support of AI-IoT display a decrease in the number of pesticides applied by 20-35 percent, a decrease in labor expenses by automation, and a greater level of stability of the harvest. Environmentally, the accurate management of the disease reduces the release of chemicals and soil and water pollution and leads to sustainable farming activities. These results indicate that digital plant health monitoring can improve crop production as well as ecological stewardship, cost-efficient business, and data-driven decisions, which support the claims about the potential of AI and IoT systems as revolutionary means of contemporary agriculture (Pintus et al., 2025). The chapter managed to describe how CNNs, ResNet, EfficientNet, hyperspectral imaging, and transfer learning can be used to identify patterns of complex diseases with 90-98 percent accuracy. It also demonstrates the superiority of these technologies in comparison to the manual diagnosis. The document shows the clear contribution of temperature, humidity, soil, multispectral, and thermal sensors to real-time environmental information, which can detect the environment early on, predictive analytics and automated decision support systems. The chapter is a good source of evidence that AI-IoT systems minimize the use of pesticides by 20-35%, increase the efficiency of resources, reduce soil contamination, and sustain the protection of the environment in the management of crops. The chapter contains effective real-life examples such as rice blast in Southeast Asia, wheat rust in Europe, tomato greenhouses in North America and vineyard monitoring, where there was an improved stability of yield, lower costs, and precision agriculture (Akhter et al., 2025).

8. DISCUSSION

In spite of the prospective development of AI and IoT-based plant disease detection, a number of limitations and difficulties remain. One highly important question is the quality and accessibility of data; training deep learning models needs large, high-quality, and correctly labeled datasets, which are sometimes challenging to access regarding a particular crop, rare disease, or even a wide range of environmental conditions. Lighting variability, sensor calibration and field variability can cause noise, which impacts model reliability and generalization. As well, interpretability of models is also an issue since sophisticated deep learning models are usually black-box models, and farmers and agronomists are unable to comprehend the reasoning behind prediction. This is not very transparent, and this could make its adoption difficult, particularly for smallholder farmers, because they need to be guided through actionable guidelines. There are also ethical concerns in digitized agriculture. Farm data collection and transmission should be done in a way that assures privacy, security and fair access. The implementation of AI and IoT technologies must not distort digital inequalities, so small-scale farmers are not left out. Connection with the traditional farming knowledge is essential to functional efficiency. Although AI offers predictive information, the recommendations are contextually appropriate because field testing by qualified agronomists and farmers is carried out. Integrating the novel AI-based systems with the traditional methods, e.g., crop rotation, intercropping, and manual scouting, will enhance resilience, reduce the bias of the models, and foster trust in the system (among farmers). In general, the discussion indicates that the AI and IoT frameworks can create transformative opportunities; however, the implementation process should focus on data quality, ethical considerations, model transparency, and alignment with field-level practices, which will guarantee sustainable, reliable, and user-friendly plant disease management (Jafar et al., 2024).

9. RECOMMENDATIONS

A number of recommendations are offered to derive the most optimal value of the AI and IoT technologies to manage the plant diseases in a sustainable manner. Policy and Research Sp recommendations Governments and agricultural agencies ought to come up with policies, which will encourage use of digital agriculture such as financing AI and IoT research, standardization of data format, and open-source databanks of plant diseases. The subsidies can assist in making the introduction of incentives to use smart sensors or cloud services faster. The partnership between agritech business and farmers with the university can aim at coming up with crop-specific AI models, multispectral imaging plans and low cost IoT that fit the local conditions. Capacity Building and Training: In order to reap maximum benefits of AI-based disease detection, a requirement would be to train farmers. The application of smart sensors, mobile apps, and decision-support dashboard should have hands-on workshops offered in the extension programs. The training of the knowledge concerning the output of the AI models, disease alert interpretation, and recommendations integration into the current farm management practice should also be trained. The knowledge gap can be closed by the use of digital literacy courses so that farmers are educated on technology application. Increase the Accessibility to Smallholder Farmers: The smallholder farmers constitute the largest number of the agricultural workforce in most localities, but the cost, infrastructure, and knowledge barriers have mostly been associated with the system of the digital solutions adoption. To enhance accessibility, the development of devices, applications as mobile apps, sensor networks that can be shared among communities can be developed that are low-cost and easy to use. The cooperation with cooperatives, local governments may also lead to the accumulation and analysis of the group data, decreasing the cost of each investment. Having provided equal opportunity to take advantage of AI-based disease detection, not only the farm productivity will be promoted, but the food security and sustainability of the environment will be enhanced, too. These suggestions can enable policy-makers, researchers, and farmers to build an ecosystem in which AI and IoT technologies will help farmers to control the health of their plants in a sustainable, accurate, and non-discriminatory manner (A. Ahmad et al., 2024).

10. CONCLUSION

The IoT-enabled and AI-enabled systems of plant disease detection will become a revolutionary solution to the contemporary agricultural industry. These systems can be used to monitor the diseases in different crops and the conditions in real time by real and precise methods with the aid of computer vision, deep learning networks, and sensor networks. The recent research has also depicted high classification rates, early detection, and optimization of resources, and this proves the technological appropriateness and viability of the methods. The introduction of AI-based disease detectors that make the use of pesticides limited, environmental impact, and predictability all allow sustainable agriculture. Connecting IoT-based sensors with edge and cloud computing, it is possible to monitor continuously, intervene within a short time, and make decisions based on the data. These systems are a reflection and complement of the traditional agricultural expertise to provide implementable outcomes that enhance the capability to respond to new pathogens and variability in climate change. Also, the technologies of AI and IoT do not only result in socioeconomic benefits, including the reduction in labor, savings, and profitability, but also promote ecological custodianship. Precision farming plans encourage farmers to utilize their resources in the best manner by utilizing predictive modelling, real-time surveillance, and automatic treatment to achieve sustainability in the long term. In conclusion, AI, IoT, and computer vision convergence is an all-inclusive and adaptable plant health management system. The adoption and impact in the future will be based on additional research, data quality improvement, fair and equal access to technology, and capacity building of farmers. As digital agriculture evolves, the technologies will be able to reshape the concept of crop protection, enhance food security, and promote the use of environmentally friendly methods of growing crops worldwide.

11. FUTURE STUDIES

The future research of AI- and IoT-regulated plant diseases must focus on the enhancement of the system's adaptability, scalability, and explainability. One of the prospects that would seem to be promising is the integration of the multimodal AI systems, which would involve the integration of computer vision, genomic data, environmental sensors, and weather information. The systems can provide a full picture of the health of plants and, therefore, early detection of an illness, understanding of the interactions between pathogens, and anticipation of a disease outbreak under various climatic conditions (Atapattu et al., 2024). The development of understandable and adaptive AI models that will result in farmer trust and adoption is important. The manners of how models predict can be described by the approaches of explainable AI (XAI), which suggests the features or environmental influences that affect the categorization of the diseases. Perhaps, we should be able to make the crops, geographical regions, and new strains of pathogens more resilient by learning adaptive models, which learn depending on the new information. The other frontier is the smart farming autonomous disease monitoring development. Continuous observation and specific interventions may be supported with the help of autonomous drones, robotic sprayers, and mobile edge devices to react to the stress in the plants in real-time. Such systems, in conjunction with a system of IoT sensors, can generate optimization of resources to be used, decreased environmental impact, and an intervention that can be scaled to large-scale farms and small hold farms. Moreover, the AI and IoT combined with the decision-support systems could facilitate the predictive analytics, risk mapping, and early-warning systems and provide actionable information to farmers and policymakers. The cooperative research, readily available databases, and cheap technology software will play an essential role in attaining equitable profitability in a broad spectrum of agricultural environments (Mu et al., 2024) In line with these suggestions, new studies can be conducted later to be more accurate, sustainable, and resilient in controlling the plant diseases and inform the next generation of smart, information-focused, and greener agricultural technologies (Melhim & Isaifan, 2025).

REFERENCES

Ahmad, A., Liew, A. X. W., Venturini, F., Kalogeras, A., Candiani, A., Di Benedetto, G., Ajibola, S., Cartujo, P., Romero, P., Lykoudi, A., De Grandis, M. M., Xouris, C., Lo Bianco, R., Doddy, I., Elegbede, I., D'Urso Labate, G. F., García del Moral, L. F., & Martos, V. (2024). AI can empower agriculture for global food security: Challenges and prospects in developing nations. *Frontiers in Artificial Intelligence*, *7*. Advance online publication. DOI: 10.3389/frai.2024.1328530

Ahmad, L., & Nabi, F. (2021). *Agriculture 5.0: Artificial Intelligence, IoT and Machine Learning*. CRC Press.

Akbar, J. U. M., Kamarulzaman, S. F., Muzahid, A. J. M., & Rahman, Md. A., & Uddin, M. (2024). A Comprehensive Review on Deep Learning Assisted Computer Vision Techniques for Smart Greenhouse Agriculture. *IEEE Access : Practical Innovations, Open Solutions*, *12*, 4485–4522. DOI: 10.1109/ACCESS.2024.3349418

Akhter, Md. S., Jahan, I., Tasnim, M., Miah, M. S., & Karim, A. J. M. S. (2025). AI-Driven Collaborative Robotics in Agriculture: Optimizing Soil and Crop Yield. In K. K. Reddy C., B. Saju, A. Hulus, T. Sibalija, & H. Nakouri (Eds.), *Advances in Computational Intelligence and Robotics* (pp. 57–88). IGI Global Scientific Publishing. https://doi.org/DOI: 10.4018/979-8-3373-4207-8.ch003

Alahe, M. A., Chang, Y., Kemeshi, J., Won, K., Yang, X., & Wei, L. (2025). Real-Time agricultural image encryption algorithm using AES on edge computing devices. *Computers and Electronics in Agriculture*, *237*, 110594. DOI: 10.1016/j.compag.2025.110594

Ataei Kachouei, M., Kaushik, A., & Ali, Md. A. (2023). Internet of Things-Enabled Food and Plant Sensors to Empower Sustainability. *Advanced Intelligent Systems*, *5*(12), 2300321. DOI: 10.1002/aisy.202300321

Atapattu, A. J., Perera, L. K., Nuwarapaksha, T. D., Udumann, S. S., & Dissanayaka, N. S. (2024). Challenges in Achieving Artificial Intelligence in Agriculture. In Chouhan, S. S., Saxena, A., Singh, U. P., & Jain, S. (Eds.), *Artificial Intelligence Techniques in Smart Agriculture* (pp. 7–34). Springer Nature., DOI: 10.1007/978-981-97-5878-4_2

Ayorinde, A. (2025). Sustainable Farming Through AI-IoT Synergy: Mitigating Ecological Damage and Financial Resource Leakage. *International Journal of Research Publication and Reviews*, *6*, 7845–7862. DOI: 10.55248/gengpi.6.0625.2273

Badidi, E. (2023). Edge AI for Early Detection of Chronic Diseases and the Spread of Infectious Diseases: Opportunities, Challenges, and Future Directions. *Future Internet, 15*(11), 370. DOI: 10.3390/fi15110370

Chakraborty, S., & Newton, A. C. (2011). Climate change, plant diseases and food security: An overview. *Plant Pathology, 60*(1), 2–14. DOI: 10.1111/j.1365-3059.2010.02411.x

Chandel, N. S., Rajwade, Y. A., Dubey, K., Chandel, A. K., Subeesh, A., & Tiwari, M. K. (2022). Water Stress Identification of Winter Wheat Crop with State-of-the-Art AI Techniques and High-Resolution Thermal-RGB Imagery. *Plants, 11*(23), 3344. DOI: 10.3390/plants11233344

Dargan, S., Kumar, M., Ayyagari, M. R., & Kumar, G. (2020). A Survey of Deep Learning and Its Applications: A New Paradigm to Machine Learning. *Archives of Computational Methods in Engineering, 27*(4), 1071–1092. DOI: 10.1007/s11831-019-09344-w

Dolatabadian, A., Neik, T. X., Danilevicz, M. F., Upadhyaya, S. R., Batley, J., & Edwards, D. (2025). Image-based crop disease detection using machine learning. *Plant Pathology, 74*(1), 18–38. DOI: 10.1111/ppa.14006

Fevgas, G., Lagkas, T., Papadopoulos, P., Sarigiannidis, P., & Argyriou, V. (2025). Integrating thermal infrared and RGB imaging for early detection of water stress in lettuces with comparative analysis of IoT sensors. *Smart Agricultural Technology, 11*, 100881. DOI: 10.1016/j.atech.2025.100881

Garg, G., Gupta, S., Mishra, P., Vidyarthi, A., Singh, A., & Ali, A. (2023). CROP-CARE: An Intelligent Real-Time Sustainable IoT System for Crop Disease Detection Using Mobile Vision. *IEEE Internet of Things Journal, 10*(4), 2840–2851. DOI: 10.1109/JIOT.2021.3109019

Garg, T., Kumar, U., Yadav, H., & Jain, A. (2025). Fostering Sustainable Growth via the Fusion of Artificial Intelligence and Computer Vision. In Whig, P., Silva, N., Elngar, A. A., Aneja, N., & Sharma, P. (Eds.), *Sustainable Development through Machine Learning, AI and IoT* (pp. 102–109). Springer Nature Switzerland., DOI: 10.1007/978-3-031-71729-1_10

Guaragnella, G. (2024). *RGB and Thermal Camera Integration for Advanced Perception of an Agricultural Robot* [Laurea, Politecnico di Torino]. https://webthesis.biblio.polito.it/31892/

Hall, E. L., Kruger, R. P., Dwyer, S. J., Hall, D. L., Mclaren, R. W., & Lodwick, G. S. (1971). A Survey of Preprocessing and Feature Extraction Techniques for Radiographic Images. *IEEE Transactions on Computers*, *C–20*(9), 1032–1044. DOI: 10.1109/T-C.1971.223399

Iqbal, N., Mumtaz, R., Shafi, U., & Zaidi, S. M. H. (2021). Gray level co-occurrence matrix (GLCM) texture based crop classification using low altitude remote sensing platforms. *PeerJ. Computer Science*, *7*, e536. DOI: 10.7717/peerj-cs.536

Islam, Md. R., Abdullah, H. M., Rahman, M. F., Islam, M., Tuhin, A. K., Ashiquzzaman, M., Islam, K. S., & Geisseler, D. (2025). Mitigation of Water-Deficit Stress in Soybean by Seaweed Extract: The Integrated Approaches of UAV-Based Remote Sensing and a Field Trial. *Drones (Basel)*, *9*(7), 487. DOI: 10.3390/drones9070487

Izadi, D., Abawajy, J. H., Ghanavati, S., & Herawan, T. (2015). A Data Fusion Method in Wireless Sensor Networks. *Sensors (Basel)*, *15*(2), 2964–2979. DOI: 10.3390/s150202964

Jafar, A., Bibi, N., Naqvi, R. A., Sadeghi-Niaraki, A., & Jeong, D. (2024). Revolutionizing agriculture with artificial intelligence: Plant disease detection methods, applications, and their limitations. *Frontiers in Plant Science*, *15*. Advance online publication. DOI: 10.3389/fpls.2024.1356260

Jiang, M., Huang, Z., Qiu, L., Huang, W., & Yen, G. G. (2018). Transfer Learning-Based Dynamic Multiobjective Optimization Algorithms. *IEEE Transactions on Evolutionary Computation*, *22*(4), 501–514. DOI: 10.1109/TEVC.2017.2771451

Jørgensen, L. N., Matzen, N., Hansen, J. G., Semaskiene, R., Korbas, M., Danielewicz, J., Glazek, M., Maumene, C., Rodemann, B., Weigand, S., Hess, M., Blake, J., Clark, B., Kildea, S., Batailles, C., Ban, R., Havis, N., & Treikale, O. (2018). Four azoles' profile in the control of Septoria, yellow rust and brown rust in wheat across Europe. *Crop Protection (Guildford, Surrey)*, *105*, 16–27. DOI: 10.1016/j.cropro.2017.10.018

Majdalawieh, M., Martins, C., Radi, M., Alaraj, M., & Khan, S. (2025). Precision agriculture in the age of AI: A systematic review of machine learning methods for crop disease detection. *Smart Agricultural Technology*, *12*, 101491. DOI: 10.1016/j.atech.2025.101491

Majumdar, P., Mitra, S., Bhattacharya, D., & Bhushan, B. (2025). Enhancing sustainable 5G powered agriculture 4.0: Summary of low power connectivity, internet of UAV things, AI solutions and research trends. *Multimedia Tools and Applications*, *84*(17), 17389–17433. DOI: 10.1007/s11042-024-19728-1

Melhim, S. H., & Isaifan, R. J. (2025). The Energy-Economy Nexus of Advanced Air Pollution Control Technologies: Pathways to Sustainable Development. *Energies*, *18*(9), 2378. DOI: 10.3390/en18092378

Mishra, H., & Mishra, D. (2023). Artificial Intelligence and Machine Learning in Agriculture: Transforming Farming Systems (pp. 1–16). https://doi.org/DOI: 10.5281/zenodo.15157536

Mohyuddin, G., Khan, M. A., Haseeb, A., Mahpara, S., Waseem, M., & Saleh, A. M. (2024). Evaluation of Machine Learning Approaches for Precision Farming in Smart Agriculture System: A Comprehensive Review. *IEEE Access : Practical Innovations, Open Solutions*, *12*, 60155–60184. DOI: 10.1109/ACCESS.2024.3390581

Mu, W., Kleter, G. A., Bouzembrak, Y., Dupouy, E., Frewer, L. J., Radwan Al Natour, F. N., & Marvin, H. J. P. (2024). Making food systems more resilient to food safety risks by including artificial intelligence, big data, and internet of things into food safety early warning and emerging risk identification tools. *Comprehensive Reviews in Food Science and Food Safety*, *23*(1), e13296. DOI: 10.1111/1541-4337.13296

Mukhtar, S., Arbabi, A., & Viegas, J. (2025). Advances in Spectral Imaging: A Review of Techniques and Technologies. *IEEE Access : Practical Innovations, Open Solutions*, *13*, 35848–35902. DOI: 10.1109/ACCESS.2025.3544476

Nguyen, C., Sagan, V., Maimaitiyiming, M., Maimaitijiang, M., Bhadra, S., & Kwasniewski, M. T. (2021). Early Detection of Plant Viral Disease Using Hyperspectral Imaging and Deep Learning. *Sensors (Basel)*, *21*(3), 742. DOI: 10.3390/s21030742

Nkwocha, C. L., & Chandel, A. K. (2025). Towards an End-to-End Digital Framework for Precision Crop Disease Diagnosis and Management Based on Emerging Sensing and Computing Technologies: State over Past Decade and Prospects. *Computers*, *14*(10), 443. DOI: 10.3390/computers14100443

Patel, S. A., & Kamrani, A. K. (1996). Intelligent decision support system for diagnosis and maintenance of automated systems. *Computers & Industrial Engineering*, *30*(2), 297–319. DOI: 10.1016/0360-8352(95)00173-5

Peet, M. M., & Welles, G. (2005). Greenhouse tomato production. In *Tomatoes* (pp. 257–304). https://doi.org/DOI: 10.1079/9780851993966.0257

Pintus, M., Colucci, F., & Maggio, F. (2025). Emerging Developments in Real-Time Edge AIoT for Agricultural Image Classification. *IoT*, *6*(1), 13. DOI: 10.3390/iot6010013

Portela, F., Sousa, J. J., Araújo-Paredes, C., Peres, E., Morais, R., & Pádua, L. (2024). A Systematic Review on the Advancements in Remote Sensing and Proximity Tools for Grapevine Disease Detection. *Sensors (Basel)*, *24*(24), 8172. DOI: 10.3390/s24248172

Prasad, A., & Sharma, A. (2025). From lesion detection to outcome prediction: Artificial intelligence and deep learning applications in multiple sclerosis. *Neurological Sciences*, *47*(1), 46. DOI: 10.1007/s10072-025-08661-2

Rajoub, B. (2020). Supervised and unsupervised learning. In Zgallai, W. (Ed.), *Biomedical Signal Processing and Artificial Intelligence in Healthcare* (pp. 51–89). Academic Press., DOI: 10.1016/B978-0-12-818946-7.00003-2

Rathinapriya, P., Maharajan, T., Jothi, R., Prabakaran, M., Lee, I.-B., Yi, P.-H., & Jeong, S. T. (2025). Unlocking biochar impacts on abiotic stress dynamics: A systematic review of soil quality and crop improvement. *Frontiers in Plant Science*, *15*. Advance online publication. DOI: 10.3389/fpls.2024.1479925

Sahu, A., Acharya, B., & Sahoo, P. S. (2025). Agricultural farming decision support system using artificial intelligence: A comparative analysis. In *Optimizing Smart and Sustainable Agriculture for Sustainability*. CRC Press.

Saleh, D., Milazzo, J., Adreit, H., Fournier, E., & Tharreau, D. (2014). South-East Asia is the center of origin, diversity and dispersion of the rice blast fungus, Magnaporthe oryzae. *The New Phytologist*, *201*(4), 1440–1456. DOI: 10.1111/nph.12627

Sankaran, S., Mishra, A., Ehsani, R., & Davis, C. (2010). A review of advanced techniques for detecting plant diseases. *Computers and Electronics in Agriculture*, *72*(1), 1–13. DOI: 10.1016/j.compag.2010.02.007

Savary, S., Ficke, A., Aubertot, J.-N., & Hollier, C. (2012). Crop losses due to diseases and their implications for global food production losses and food security. *Food Security*, *4*(4), 519–537. DOI: 10.1007/s12571-012-0200-5

Setiawan, A., & Budiman, A. A. (2025). Optimization of Gray Level Co-occurrence Matrix (GLCM) Texture Feature Parameters in Determining Rice Seed Quality. *EMITTER International Journal of Engineering Technology*, *13*(1), 110–123. DOI: 10.24003/emitter.v13i1.928

Sharma, D. K., Chatterjee, M., Kaur, G., & Vavilala, S. (2022). 3—Deep learning applications for disease diagnosis. In Gupta, D., Kose, U., Khanna, A., & Balas, V. E. (Eds.), *Deep Learning for Medical Applications with Unique Data* (pp. 31–51). Academic Press., DOI: 10.1016/B978-0-12-824145-5.00005-8

Sharma, K., & Shivandu, S. K. (2024). Integrating artificial intelligence and Internet of Things (IoT) for enhanced crop monitoring and management in precision agriculture. *Sensors International*, *5*, 100292. DOI: 10.1016/j.sintl.2024.100292

Sharma, N. K., Anand, A., Budhlakoti, N., Mishra, D. C., & Jha, G. K. (2024). Artificial Intelligence and Machine Learning for Rice Improvement. In Singh, A., Singh, S. K., & Shrestha, J. (Eds.), *Climate-Smart Rice Breeding* (pp. 273–300). Springer Nature., DOI: 10.1007/978-981-97-7098-4_11

Sharma, R., Kumar, N., & Sharma, V. (2025). Computer Vision for Disease Detection—An Overview of How Computer Vision Techniques Can Be Used to Detect Diseases in Medical Images, Such as X-Rays and MRIs. In *AI in Disease Detection* (pp. 77–98). John Wiley & Sons, Ltd., DOI: 10.1002/9781394278695.ch4

Singh, R., Krishnan, P., Singh, V. K., & Das, B. (2023). Estimation of yellow rust severity in wheat using visible and thermal imaging coupled with machine learning models. *Geocarto International*, *38*(1), 2160831. DOI: 10.1080/10106049.2022.2160831

Strange, R. N., & Scott, P. R. (2005). Plant Disease: A Threat to Global Food Security. *Annual Review of Phytopathology, 43*(Volume 43, 2005), 83–116. https://doi.org/DOI: 10.1146/annurev.phyto.43.113004.133839

Thangamani, R., Sathya, D., Kamalam, G. K., & Lyer, G. N. (2024). AI Green Revolution: Reshaping Agriculture's Future. In Balasubramanian, S., Natarajan, G., & Chelliah, P. R. (Eds.), *Intelligent Robots and Drones for Precision Agriculture* (pp. 421–461). Springer Nature Switzerland., DOI: 10.1007/978-3-031-51195-0_19

Upadhyay, A., Chandel, N. S., Singh, K. P., Chakraborty, S. K., Nandede, B. M., Kumar, M., Subeesh, A., Upendar, K., Salem, A., & Elbeltagi, A. (2025). Deep learning and computer vision in plant disease detection: A comprehensive review of techniques, models, and trends in precision agriculture. *Artificial Intelligence Review*, *58*(3), 92. DOI: 10.1007/s10462-024-11100-x

Verdouw, C., Sundmaeker, H., Tekinerdogan, B., Conzon, D., & Montanaro, T. (2019). Architecture framework of IoT-based food and farm systems: A multiple case study. *Computers and Electronics in Agriculture*, *165*, 104939. DOI: 10.1016/j.compag.2019.104939

Wang, S., Xu, D., Liang, H., Bai, Y., Li, X., Zhou, J., Su, C., & Wei, W. (2025). Advances in Deep Learning Applications for Plant Disease and Pest Detection: A Review. *Remote Sensing*, *17*(4), 698. DOI: 10.3390/rs17040698

Wiley, V., & Lucas, T. (2018). Computer Vision and Image Processing: A Paper Review. *International Journal of Artificial Intelligence Research*, *2*(1), 29–36. DOI: 10.29099/ijair.v2i1.42

Wiranata, A. D., Syahri, R., Jusman, Y., Saifuddin, S., & Borman, R. I. (2025). Hybrid Deep Learning Model with GLCM and CIELAB Feature Fusion for Tea Leaf Disease Classification. *BIO Web of Conferences, 199*, 01008. https://doi.org/DOI: 10.1051/bioconf/202519901008

Zhou, X.-G., Zhang, D., Lin, F., Zhou, X.-G., Zhang, D., & Lin, F. (2021). UAV Remote Sensing: An Innovative Tool for Detection and Management of Rice Diseases. In *Diagnostics of Plant Diseases*. IntechOpen. https://doi.org/DOI: 10.5772/intechopen.95535

Chapter 10
Farm Automation Empowered by Artificial Intelligence for Enhanced Efficiency

Zeeshan Ahmad
http://orcid.org/0009-0009-9459-659X
COMSATS University Islamabad, Pakistan

Aneeqa Nadeem
Comsats University Islamabad, Pakistan

Urooj Farid
COMSATS University Islamabad, Pakistan

Sana Azeem Kiani
COMSATS University Islamabad, Pakistan

Muhammad Zakria Iqbal Khan
COMSATS University Islamabad, Pakistan

Mujtaba Ul Hassan
COMSATS University Islamabad, Pakistan

Abdul Wahab
COMSATS University Islamabad, Pakistan

ABSTRACT

Artificial intelligence (AI) is transforming agriculture by enabling automation that improves efficiency, sustainability, and resilience. This chapter reviews the transition from conventional mechanization to AI-driven autonomous farming systems, driven by labor scarcity, climate variability, and cost pressures. Key enabling technologies such as sensors, IoT connectivity, cloud and edge computing, and AI/machine learning support real-time data perception and decision-making. Applications include autonomous ground robots and unmanned aerial systems for planting,

DOI: 10.4018/979-8-3373-7257-0.ch010

irrigation, fertilization, pest management, harvesting, and crop monitoring. The chapter highlights digital twins, decision support systems, and closed-loop control architectures that integrate multisource data to optimize resource use while minimizing environmental impact. Challenges related to technical complexity, economic feasibility, interoperability, cybersecurity, and regulation are discussed, along with future directions for scalable, climate-resilient AI-based farm automation.

INTRODUCTION

1.1. Background and Global Need for Automation

Agricultural systems around the world are in the process of rapid change as food production is under more unprecedented pressure from population growth, resource scarcity, and climatic variability (Seppelt et al., 2022). Traditional farming methods, highly dependent on manual labor and decision making based on experience, are no longer, however, sufficient to meet the requirements in order to achieve high productivity, monitoring year-round and the sustainable management of inputs. Labor paucities arising from rural-urban migration, demographic changes and the declining interest towards agriculture among the younger generation have only added more boon to the rush for automated solutions (Selod & Shilpi, 2021). Climate change is adding further complexity as it results in changes in pest and diseases, an irregular precipitation pattern, and more extreme weather events (Skendžić et al., 2021).

1.2. Defining AI-Enabled Farm Automation

AI-enabled farm automation is the interconnection of artificial intelligence and mechanical, robotic and digital farm automation in which agricultural activities are carried out autonomously with a minimal human touch (Sharma & Shivandu, 2024). It includes machine learning, computer vision, sensor fusion, and autonomous decision-making, which allow machines to sense the conditions in the field, analyses the data received and take precise actions. This approach is therefore significantly different to previous forms of automation which have been based on fixed rules, manual calibration or limited sensing capabilities. Instead, the AI driven solutions constantly learn from the field variability, past trends, and real-time feedback to become more accurate and optimize the operation at any agricultural seasons (Rashid & Kausik, 2024).

Figure 1. Automation in agriculture

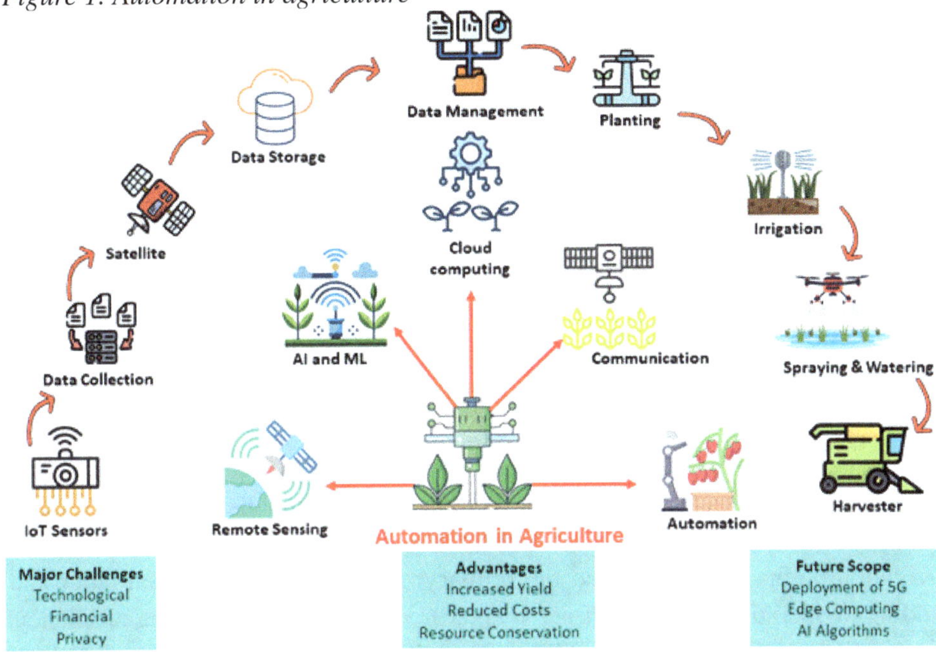

1.3. Key Drivers: Labor, Climate, Cost, Sustainability

A number of worldwide factors are leading to the acceleration of the use of AI enabled automation. Labor shortages are still a dominant issue with some areas having ageing farming populations and high dependency on seasonal workers. Automated systems, the need for manual labor is reduced and the operations such as planting, weeding and harvesting can be done with a greater degree of precision (Maró et al., 2025). Climate variability makes up the uncertainties and unpredictability of growing conditions, which demand real time monitoring and adaptive responses which AI systems are well suited to do thanks to their ability to analyses huge and complex data sets to inform timely interventions. Rising production costs related to fertilizers, pesticides and water put economic pressure on farmers, thus motivating the use of technologies that optimize the use of resources (Tudi et al., 2021).

This chapter is a comprehensive examination of how modern agriculture is changing through AI powered automation. It discusses the historical development of systems in agricultural automation, the tech behind smart systems, and integrating robotics, sensors, and AI algorithms into the automation process of any of several agricultural processes. The chapter also talks about the role of digital twins, decision support systems and closed loop automation in enhancing the level of decision-making in the farms. Moreover, it addresses accuracy in resource management solutions such as irrigation, nutrient administration, and automated pest control and its role in sustainability and production (Dhirani et al., 2023). The chapter concludes by discussing what future research is required on the problem to enable future innovation policies and trends that would enable the scaling of AI-enabled farming automation practices across the diverse agricultural landscapes.

2. EVOLUTION OF FARM AUTOMATION AND CONCEPTUAL FOUNDATIONS

2.1. From Mechanization to Autonomous Intelligence

The development of farm automation represents an evolution of mechanization focused on labor savings and one of data-driven autonomous intelligence. Early mechanization in the twentieth century (characterized by tractors, harvesters and irrigation pumps), however, had the primary aim of replacing manual labor and adding to operational scale (Kumar et al., 2023). While this increasing efficiency, these machines required constant human control and judgement. The second phase was introducing electromechanics into automation, including GPS-guided tractors and variable rate applicators, which provided a great degree of precision but were highly dependent on human supervision. The beginning of the third phase was the advent of digital agriculture, through which sensors, remote sensing and management information systems allowed to generate and analyses large amounts of farm data (Abiri et al., 2023).

Figure 2.

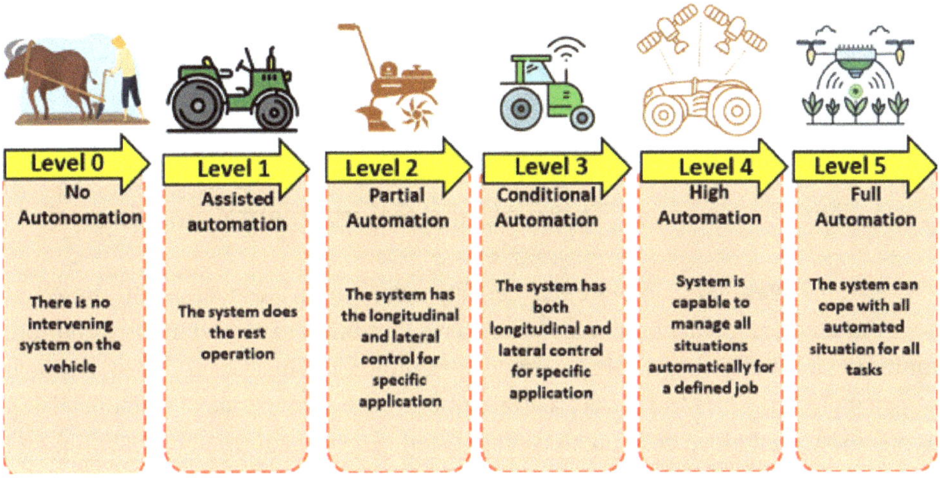

2.2. Core Principles of AI-Driven Automation

AI driven automation is based on various underlying principles to support intelligent behavior in agriculture systems. The principle of perception focuses on the ability of machines to perceive data from their environment by using cameras, spectral sensors, LiDAR and IoT devices (Jin et al., 2025). The principle of prediction makes use of machine learning models to analyses the patterns and predict crop growth, soil movements, pest pressures and yield results. The principle of reasoning is the basis for decision-making through algorithms that make in-depth evaluations, optimize resource allocation and derive actionable insights (Zhao et al., 2024).

2.3. Conceptual Models: Perception–Decision–Action Loop

At the heart of agriculture farm automation enabled by AI is the perception - decision - action (PDA) loop - a conceptual model identifying the working principle of intelligent systems within the agricultural environment. During the perception stage, the sensors collect real time data about the crop health, soil moisture and weeds, and environmental parameters (Mana et al., 2024). This information gets fed into the decision stage where AI algorithms analyze the data and make their predictions and determine what the best actions. Following this, in the action stage, these decisions are made using autonomous machinery e.g sprayers, irrigation values, robotics manipulators or drones (Ghraizi et al., 2023). The loop is continuous and allows the system to learn from the feedback, improve the predictions, and adjust up the performance previously given.

3. ENABLING TECHNOLOGIES IN AI-BASED AUTOMATION

3.1. Sensors and Perception Systems

3.1.1. Optical, Spectral, Thermal, and Environmental Sensors

Artificial intelligence-based enhanced farm automation is fundamentally based on sophisticated sensing technologies which have the ability to collect high-resolution, multimodal data from dynamic changing agriculture (Padhiary et al., 2024). Optical sensor including RGB and hyperspectral camera is the cornerstone of machine vision applications including crop phenotyping, weed discrimination, pest detection and autonomous navigation. Hyperspectral and multispectral sensors take this ability to the next level by acquiring reflectance measurements in dozens to hundreds of wavelengths and, as a result, are able to accurately measure plant stress, nutrient deficiency, onset of disease and soil variability long before it can be visually detected by the human eye (Dian et al., 2021).

3.1.2. Calibration, Accuracy, and Field Conditions Challenges

Although there have been huge advances in technology, global sensory systems that are deployed within agricultural landscape, have major challenges. Inconsistent spectral signatures can be generated by the variability of sunlight, shadows, dust, precipitation, fog, temperature variation and heterogeneity of the canopy and, consequently, the accuracy of the model can be lowered (Pande & Moharir, 2023). Routine calibration (radiometric and geometric) is therefore not an option if reliable data maintenance is to be maintained over seasons and under widely varying environmental conditions. Low-cost sensors, although they are ubiquitous, tend to drift, have a low dynamic range, and be very prone to noise(H. Zheng et al., 2025)

3.2. Connectivity and Computing Infrastructure

3.2.1. IoT Networks, 5G, and Satellite Connectivity

The principle of connectivity is the foundation of modern farm automation, that incorporates the free flow of information among sensors, machinery, and cloud computing and decision support systems. With protocols such as LoRaWAN, NB-IoT and Zigbee, Internet of ThingsS (IoT) networks - use low power long range protocols to manage distributed environmental sensors and smart devices There is an emergence of the 5G networks enabling application of artificial intelligence to control autonomous tractors, drones and robot harvesters, and high-resolution image streaming to edges Use Cases Internet of Things (IoT) networks The 5G networks opens up negative ultra-low latency and high bandwidth, which can be used to control autonomous tractors, drones (Khalifeh et al., 2025).

3.2.2. Edge vs Cloud Computing in Farm Automation

Both the cloud and the edge have complementary roles to play in AI enabled automation. Cloud platforms provide a big data storage, long term data analytics, predictive modelling, fleet level optimization for tractors, drones and irrigation systems (Gong et al., 2025). However, they are sometimes constrained by bandwidth limitations and latency if they are time-critical, however, decisions are to be made. Edge computing - processing data on premises such as robots, UAVs and field sensors - helps to create an instant reaction for navigation, obstacle avoidance, targeted spraying and anomaly detection (Karami & Karami, 2025).

3.3. AI/ML Frameworks for Automation

3.3.1. Deep Learning and Reinforcement Learning

Artificial intelligence frameworks are the basis of the decision-making layer of automated farming systems. Deep learning (especially convolution neural network (CNN) and transformers based) enables object detection, crop disease classification, yield estimation, and real-time machine vision (robotic crop harvester, self-driving cars etc.) (Taherdoost, 2023). The predictive analytics for weather, irrigation scheduling and crop growth modelling are supported by recurrent neural networks (RNN) and temporal models (Lu et al., 2025).

Figure 3. AI and agriculture

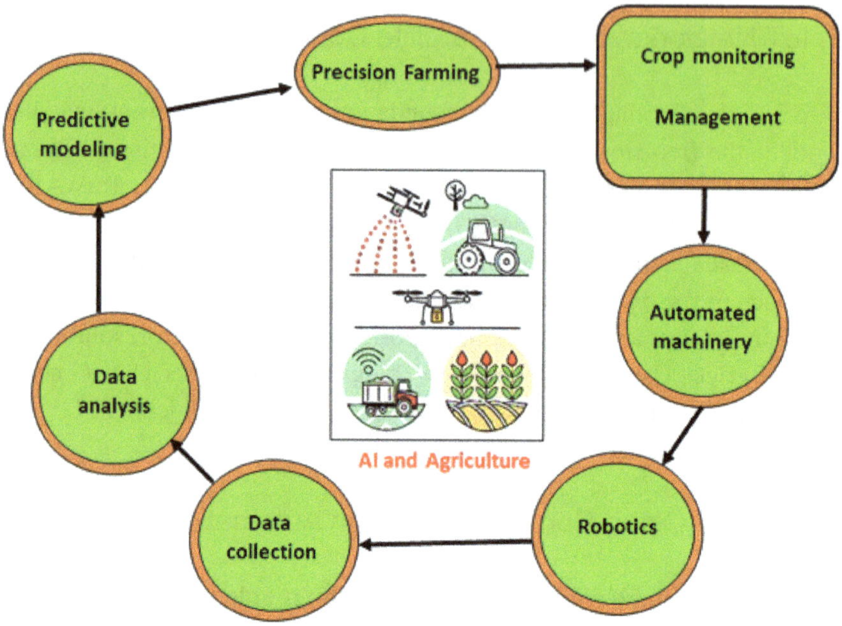

3.3.2. Explainable and Trustworthy AI

In the growing inclination toward automation decisions in food production, transparency and dependability are on the list of priorities. Explainable AI (XAI) guarantees that model predictions-not as earth shaking as detecting disease or providing farmers and agronomists with input suggestions, but still (Longo et al., 2024), such techniques as saliency mapping, SHAP values and a simplification of a model assist the user to see a logic behind the decisions made by the system. Privacy protection, adversarial attack proof, field condition diverse fairness and agricultural data governance policy compliance are also the properties of trustful AI frameworks. Those are the values, which are central to foster user trust and ethical AI-in-agriculture application (Nikolinakos, 2023).

3.4. Data Standards, Interoperability, and Digital Integration

The success in the efficacy of AI driven automation relies on seamless integration across sensors, machinery, platforms and software systems. Data standards, such as ISOBUS, AgGateway and ISO 11783, strive to standardize machine communication protocols, and in turn with this, enable tractors, implements, drones and robots built by different manufacturers to work cooperatively (Olliver et al., 2022). Interoperability frameworks are helping to ensure consistent data formats, metadata structures and API-based connectivity to counter digital agriculture ecosystem fragmentation (Ahmed & Shakoor, 2025).

4. AUTONOMOUS GROUND ROBOTICS AND MOBILE PLATFORMS

4.1. Types of Robotic Platforms

4.1.1. Autonomous Tractors

Autonomous tractors form base technology of contemporary robotic agriculture and consist of various elements like GPS controlled steering systems, several sensor perceptions, and AI assisted decision making algorithms to do the field operations without constant human involvement (Srivastav & Das, 2025). These vehicles are equipped with LiDAR, radars and multiple camera suites for real-time sensing of the environment to maintain exact control over their trajectory, sense and avoid obstacles in their path and adapt to varying soil/crop conditions. (Dutta, 2024). As a result, they mechanize labor intensive labor requiring operations such as ploughing, tilling, sowing, hauling etc. This reduces considerably the cost of labor, and enhances the productivity of the field manifold (Georgiadis et al., 2019).

4.1.2. Small Field Robots

Small, low-power, lightweight field robots that are generally called “agri-bots” perform specialized tasks with high agility and minimum soil compaction. These platforms include scouting robots for disease and nutrient monitoring, robotic weeders for mechanical removal of weeds and micro-sprayers for delivery of pesticides (Shinde et al., 2024). They usually use electric drivetrains, machine vision-based navigation and machine tool attachments. Their small dimensions enable to operate between narrow crop rows and in sensitive fields, hence supporting continuous monitoring and micro - management of the crop health. Swarm robotics is an emerging concept in this category, where multiple cheap units perform weeding, mapping or sampling the soil zones collaboratively thereby improving the resilience and at the same reducing the cost of operation (Albiero et al., 2022).

4.1.3. Manipulators and Robotic Arms

Robotic arms and manipulators are finding ever more application in horticulture and greenhouse operations as well as post-harvest management. Equipped with grippers, suction pads or soft robotic end effectors, these systems are intended to work with delicate produce, without causing mechanical damage to it (Barbosa Júnior et al., 2024). They combine force sensors, machine vision systems, and dexterous control algorithms for performing precision tasks like picking fruits, pruning and transplanting and packing. The advent of soft robotics has further increased their ability to manipulate irregular shapes and fragile commodities and therefore increasing the accuracy and reducing post post-harvest losses (Wang et al., 2024).

4.2. Navigation, Perception, and Field Mobility

The functionality of autonomous robotic platforms that operate in unstructured and highly variant outdoor environments is based on navigation and mobility. The AI-assisted navigation algorithms utilize global positioning (global navigation system, GNSS), real time correction (RTK), inertial navigation devices, LiDAR and vision-based localization to maintain the precision of centimeter (Taleb et al., 2025). Simultaneous localization and mapping (SLAM) algorithms are algorithms which enable robots to develop a detailed map of the field as they move around the field and will assist in the adaptive route planning and avoidance of obstacles (Ugenti et al., 2023).

4.3. Robotics for Key Field Operations

4.3.1. Planting, Weeding, and Spraying

Robotics is transforming work in mainstream areas through precision, consistency and chemical depravity. Machine vision and geospatial analytics can be deployed by autonomous planters to optimally position seeds based on their spacing and depth irrespective of the skill of the operator (Javaid et al., 2021). Robotized weeders, equipped with AI Sung at least weed models, which can selectively weed using mechanical blades, electric pulse or focused lasers and reducing the quantity used of herbicides and promoting regenerative agriculture. The autonomous ground robots are used in spraying to deliver pesticides and fertilizers with site specific patterns and variable rate technology (Ghafar et al., 2023).

4.3.2. Elective Harvesting and Fruit Picking

Selective harvesting requires fine motor control and precise perception-features that are available through the use of A.I. enabled manipulators. Robotic harvesters are able to identify the ripe produce by the color, shape, spectral signatures, and maturity indices based on machine learning algorithms, ensuring that only the market-ready fruits are collected (Montoya-Cavero et al., 2022). These systems are particularly important in labor-intensive industries, such as in strawberries, apples, citrus and tomatoes. Integrated robotic arms carefully remove fruits without bruising, whilst parallel sorting modules sort produces by quality measures. In high value crops, selective harvesting is enhancing profitability, through maximizing the yield quality, as well as through labor shortages in peak seasons (Arikapudi & Vougioukas, 2023).

4.4. Performance Metrics, Economics, and Deployment Challenges

The assessment of autonomous robotic system should be based on a multi-dimensional framework of performance metrics that include technical efficiency, operational reliability, economic return and environmental impact. The key performance indicators include field cover rate, navigational accuracy, task completion accuracy, energy consumption, and system uptime (Asih et al., 2020). Economic assessments include capital investment, maintenance expenditures, labor replacement, life span financial costs, input optimization opportunities and the reduction of down time. Although such systems have high promises, there are different obstacles which prohibit large-scale deployment (Stasiak-Betlejewska & Potkány, 2015).

Table 1. Key technologies, applications, and benefits of AI-driven farm automation

Technology	Applications in Farming	Key Benefits	Challenges	References
Autonomous Tractors	Planting, tilling, harvesting	Labor reduction, precise field operations, increased efficiency	High initial cost, terrain adaptability, maintenance	(S. Zheng et al., 2025)
Small Field Robots	Weeding, targeted spraying, monitoring	Reduced chemical use, selective intervention, labor saving	Battery life, field mobility, crop type limitations	(Pané et al., 2021)
Robotic Manipulators	Pruning, transplanting, selective harvesting	Precision handling, reduced crop damage	Complex control algorithms, cost, operator training	(Zhang & Wei, 2017)

continued on following page

Table 1. Continued

Technology	Applications in Farming	Key Benefits	Challenges	References
UAVs / Drones	Crop monitoring, disease detection, yield estimation	Rapid area coverage, high-resolution imagery, early problem detection	Flight regulations, weather dependency, payload limits	(Mohsan et al., 2023)
Satellite Imaging	Large-scale crop monitoring, soil health, climate impact	Long-term trend analysis, integration with precision agriculture	Low temporal resolution, cloud cover interference	(Milrad, 2018)
IoT Sensor Networks	Soil moisture, nutrient content, temperature, humidity	Real-time monitoring, data-driven interventions, resource optimization	Sensor calibration, data management, connectivity issues	(De & Moessner, 2023)
AI / ML Frameworks	Disease prediction, yield forecasting, decision support	Improved accuracy, predictive capabilities, autonomous decision-making	Data quality, explainability, model generalization	(Ofosu-Ampong, 2024)
Digital Twins	Virtual farm simulation, scenario testing, closed-loop control	Predictive planning, resource optimization, risk reduction	High computational requirement, model validation	(Verdouw et al., 2021)
Controlled-Environment Systems	Greenhouses, vertical farms, hydroponics	Year-round production, precise environmental control, resource efficiency	High energy use, setup cost, technical expertise	(Sharma et al., 2024)
Precision Resource Management	AI-driven irrigation, fertigation, pest control	Reduced waste, optimized inputs, environmental sustainability	Sensor reliability, integration complexity, maintenance	(Raj et al., 2025)

5. AERIAL SYSTEMS, REMOTE SENSING, AND IMAGING PIPELINE

5.1. UAV Platforms, Payload Technologies, and Satellite Monitoring

Unmanned Aerial Vehicles (UAVs) represent essential tools in the current agronomic practice, especially in those that make use of Artificial Intelligence; it is the instrument that allows to acquire high-resolution data in heterogeneous field environments. UAV platforms vary on a scale from small-scale quad opting units suitable for scouting in close proximity to fixed wing platforms placed permanently that can cover an area of up to hundreds of hectares with a single sortie (Mohsan et al., 2023). These systems include a diversified payload architecture where RGB cameras are used for general mapping purposes, multispectral and hyperspectral sensors for diagnosing crop health, thermal probes for irrigation management and LiDAR units are for producing high-density three-dimensional field models (Zhao et al., 2022).

5.2. AI-Powered Image Processing and Agricultural Intelligence

Artificial - intelligence driven image analytics take the raw imagery of an aerial and produce actionable agriculture insights. Deep learning algorithms such as convolutional neural network (CNN) and vision models based on transformers are used for pixel level classifications, anomaly detection and detection of minor stress indicators, which are invisible to the human eye (Indolia et al., 2018). Such models provide the basis for high accuracy monitoring of crops by identifying changes in canopy reflectance, chromatic changes and thermal signatures. Disease detection systems trained using annotated data sets can detect the initial symptoms of fungal, bacterial or viral infections, and allow for the timely intervention before these outbreaks expand. Similarly, weed-recognition models speak of differentiating between crop and weed species under complex canopy and, therefore, are in favor of site-specific weed management (Adhinata et al., 2024).

5.3. Operational Constraints, Regulatory Frameworks, and Data Integration with Ground Systems

Although aerial systems have the potential to bring change, though, there are several operational and clean air regulation challenges to their operation. Strict regulations regarding flight altitude, line of sight/line of-sight operation, maximum payload mass, entr reminder to a specific airspace are in place in order to maintain safety and privacy in the aviation arena. Satellites despite being hardy have issues concerning cloud interference and spatial resolution this can imply that they are not able to capture small scale features. Another challenge is that of management of data, as high volume imagery demands not only effective data storage but processing and standardization of data. The ability to actually interpret is of much importance by ensuring that it is cross-platform interoperable. (Theodorakopoulos et al., 2024).

Integrating aerial data with ground-based automation systems is the future in precision agriculture. UAV and satellite outputs are consumed by decision support platforms that are coordinated with autonomous tractors, robots that deal with weeds, irrigation systems, and soil sensors (Rashid et al., 2025). For example, hot spots of disease, which are detected through the use of aerial imaging systems, can be sent on to ground robots for specific spraying or mechanical removal. Likewise, moisture maps using thermal or multispectral data can be used to fuel variable rate irrigation systems utilizing automated pivots or drip irrigation systems (Meshram et al., 2022). This vertically integrated approach - from air to ground - enables closed-loops of automation where sensing and analysis also occur and action is implemented. Such interconnected systems are more efficient with respect to resources, reduce labor requirements and allow for real-life, adaptive management across heterogeneous landscapes (R. Liu et al., 2025).

6. DIGITAL TWINS, DECISION SUPPORT SYSTEMS, AND CLOSED-LOOP CONTROL

6.1. Digital Twin Farms and Modelling Architectures

Digital twins in agriculture represent dynamic virtual duplicates of currently existing farming systems that constantly ingest data coming from sensors, in-field equipment, aerial photography and environmental models. A digital twin farm combines geospatial repositories, soil factors, crop physiological parameters, machinery performance parameters as well as climatic factors to create a synchronized and real-time simulation environment (Peladarinos et al., 2023). The architecture of such systems typically includes three levels: (1) the physical farm subsystem with associated Internet of Things (IoT) sensors, robotics and imaging platforms; (2) a data integration layer responsible for harmonizing data from multiple sources with the aid of standardized ontologies as well as cloud-based infrastructures; and (3) a computational modelling level used to support real time simulation, prediction and scenario testing (Zhang et al., 2025).

6.2. Modelling Paradigms: Process-Based and Data-Driven Frameworks

Two prevalent types of modelling are in control of digital twin simulation environments-process-based models and data-driven models. Process based models rely on biological and physical principles to model crop growth, nutrient cycling, evaporation and transpiration, soil - water interaction, and plant - climate interaction. (Nzimande et al., 2025)

Conversely, data drivenness uses machine learning algorithms, like deep neural networks, Gaussian processes, ensemble learners, etc. to understand complex and non-linear relationships reflected from historical datasets (Kufel et al., 2023). They exceed process-based models when high resolution data sets are available in great amounts and in cases where real-time adaptation is necessary. Increasingly, hybrid models are being developed combining mechanistic knowledge and inference based on data, which makes them more accurate and robust predictors. (Wu et al., 2023).

6.3. Decision Support Systems and Closed-Loop Automation

The insights of the digital twins are put into practice by Decision Support Systems (DSSes) which send recommendations to farmers, agronomists and automated farm implements. DSS platforms play a role in integrating descriptive analytics (a summary of real-time conditions on the field), predictive analytics (those that forecast crop performance, disease or irrigation demand prevalence in a given area and) and prescriptive analytics (what interventions should be deployed, taking into account the constraints, resources and desired outcome). (Zhang et al., 2025). The systems operate on multiple temporal scales that affect has a decision that is more operational in a day-to-day scope or a long-range strategic level. AI-controlled DSS Can Direct Digital-Twin Implications to Field Objectives and Environmental Specifications The variable-rate seeding, nutrient rate, pests, and harvest logistics may be managed by AI-controlled DSS. (Barbone et al., 2026).

6.4. Uncertainty, Validation, and Human-in-the-Loop Systems

While digital twins and DSS enable a level of decision accuracy that was previously unattainable, agriculture systems are generally uncertain by their nature because of their variable weather and soil heterogeneity, and biological complexity (David et al., 2025). Models therefore need to be validated on an ongoing basis using ground truth data from sensors, field trials and machine feedback.

Despite the progress in autonomy, human-in-the-loop means that it is essential to have hand in Evaluation of reliability, interpretability and ethical deployment. Farmers and agronomists question the recommendations of the systems, confirm the predictions of models, and intervene in the face of unexpected situations (Wu et al., 2022). Human expertise is further required in making decisions that can be very context-specific, and may not be fully accounted for by AI, e.g. economic considerations, crop preferences, and market uncertainties. The mixture of human intelligence and AI-driven automation builds the collaborative intelligence paradigm, which leads to safer, adaptive, and trustworthy farm management systems (Tarafdar, 2025).

7. PRECISION RESOURCE MANAGEMENT VIA AI AUTOMATION

7.1. Precision Irrigation and Water-Saving Automation

AI-driven irrigation systems are among the most important developments in precision agriculture that help optimize how water is distributed throughout heterogeneous fields. These systems contain soil moisture sensors, thermal imaging, weather prediction, and evaporation/transpiration models that can be used to estimate the crop water demand in real time. Machine learning algorithms are used to analyses temporal trends, identify water stress signatures and predict the optimal time to irrigate the field, a process where minimal water can be wasted and optimum performance of crops can be ensured (Zhang et al., 2024). Smart drip irrigation systems, variable rate pivot systems and autonomous irrigation robots help water be delivered to only areas of need and prevent over-irrigation and the leaching of nutrients (Hargrove et al., 2023).

7.2. AI-Enabled Nutrient Management

Since many different factors influence the plant's efficiency in different soils, nutrient optimization through AI automation can overcome the challenges that have faced agriculture for decades: inefficient fertilizer use, the variability in soil fertility and the rising input costs. AI enabled nutrient systems integrate the soil nutrient sensors, multispectral imagery, yield map and the plant physiological models to estimate nutrient deficiencies at fine spatial scales (Chen et al., 2025). Machine-learning frameworks to predict the nutrient uptake of crops, simulate fertilizations scenarios and create variable rate nutrient application maps for different crop growth stages Autonomous tractors and robotic applicators then follow these recommendations to a sub centimeter accuracy (Ennaji et al., 2023).

7.3. Automated Pest and Disease Control

AI-Driven management of pests and disease diseases is the coming together of technologies that encompass automated detection and precision intervention systems, thereby achieving very efficient crop protection. Deep learning and computer vision algorithms, which have been trained using thousands of annotated images, can be used to detect early stage infections, pest infestations and physiological disorders that are challenging for human scouts to identify (Chai et al., 2021). UAV imagery, ground-based cameras and hyperspectral sensors take these subtle signatures

- chlorosis, necrotic lesions and anomalous reflectance patterns - into sight. (X. Liu et al., 2025).

Once the detection is complete, automated systems are deployed to take control measures that are specific. Sprayers that are mobile robots with nozzles regulated by artificial intelligence (AI) devices adjust the volume of the spray depending on the density of the canopies, severity of the pathogens, and micro-climate conditions (Vijayakumar et al., 2023).

7.4. Ecosystem Monitoring and Sustainability Indicators

AI aids in holistic managing an ecosystem by processing ecological variables affecting long-term farm sustainability. Sensors and imaging systems are used to capture soil organic matter, indicators of biodiversity, carbon sequestration potential, pollinator activity and water quality. Machine-learning models help farmers spot trends and when the ecology has crossed the point of no return to help them jump to regenerative farming (Waqas et al., 2025).

7.5. Key Performance Indicators: Resource Efficiency and Environmental Benefits

The effect of precision management of resources is measured using well-defined Key Performance Indicators which are KPIs. Technical KPIs are water-use efficiency, nutrient-use efficiency, pesticide reduction, energy savings and carbon footprint (Ferrero et al., 2024). Economic KPIs represent a reduction in input costs and the return on investment and yield improvements associated with optimized allocation of resources. Environmental KPIs include reduced chemical run off, improved soil health, improved biodiversity and reduced emissions. (Attaran & Celik, 2023).

8. FARM AUTOMATION IN CONTROLLED-ENVIRONMENT AGRICULTURE

8.1. Importance of Controlled-Environment Agriculture (CEA)

Controlled-Environment Agriculture (CEA) is a group of agricultural production systems in which environmental variables are carefully controlled for optimum plant growth, including temperature, humidity, amount of light, concentration of carbon dioxide in the atmosphere, and provision of nutrients to the plants (Wang et al., 2023). Exemplars of CEA include greenhouses, vertical farms, plant factories and hydroponic or aeroponic installations, which all enable continuous production, more yields per unit area, and less reliance on the external climatic conditions. (Zhang et al., 2022).

8.2. Sensor Networks and Environmental Control Systems

Automation in CEA is based on the provision of dense networks of sensors, which continuously measure microclimatic parameters, nutrient concentration, pH, electrical conductivity of the growth medium (Ojo & Zahid, 2022). Sensor outputs go to AI-based controllers, which control heating, cooling and ventilation systems, supplemental lighting and irrigation, and nutrient delivery systems. Advanced control algorithms (many based on model-predictive control, or MPC, or reinforcement learning), which offer continuous adjustment depending on plant developmental stages, energy optimization goals and instant feedback from the environment (Nagarsheth et al., 2025).

8.3. Automation of Core CEA Operations

Beyond environmental control, with CEA, automation also applies to planting, pruning, harvesting, and post-harvest handling. Robotic manipulators perform accurate transplanting, shape management and selection harvesting with a minimal degree of damage (Cowan et al., 2022). Conveyor-based systems and automated trays and vertically oriented moving racks for multi-layered vertical farming provide optimal use of space and labor. Integrated imaging systems in combination with machine learning algorithms identify earliest signs of nutrient deficiencies, diseases or pest outbreaks paving the way for initiated interventions (Sanchez-Cubillo et al., 2024).

8.4. Resource Optimization and Sustainability in CEA

CEA automation provides an unrivalled resource economy compared to traditional field agriculture. Water -for instance Water -use efficiency is maximized through recirculating hydroponic and aeroponic systems that are governed by AI models that prevent over -irrigation (Dsouza et al., 2023). Nutrient-use efficiency is increased through precision fertigation using in real time plant and media readings. Energy efficiency is enhanced with new smart lighting and climate control in conjunction with renewable energy sources as much as is practical. The enclosed nature of CEA gives the option for this precision in carbon footprint assessment, reducing agrochemical run off, and increasing biosecurity. Sustainability metrics, such as water savings, reduction of energy consumption, minimization of waste, and enhanced productivity units per square meter is continuously monitored and optimized using AI, and digital twin frameworks (Vatistas et al., 2022).

9. CHALLENGES, ADOPTION BARRIERS, AND POLICY IMPLICATIONS

9.1. Technical, Economic, and Social Challenges

Despite significant advancements in the use of artificial intelligence for automation in agriculture, there are several challenges that must be overcome to ensure its use in agricultural application becomes common (Salomi & Samsudeen, 2024). Technically, the integration of heterogeneous systems is as follows; autonomous tractor, unmanned aerial vehicle, robots, sensors, and artificial intelligence analytics platforms; this integration is rather a formidable task. Reliable nature of operation under fluctuating weather condition, soil type and crop state requires sturdy hardware and precise calibration and maintenance of software. Emerging threats in cybersecurity while they may deal with unauthorized access to data streams or manipulation of the automated controls can pose significant threats to safety and operational integrity of the farm (Hider & Zafer, 2024).

Economic barriers do exist in constraining the uptake of technology. The massive capital needed upfront for new technologies such as autonomous machinery, drones, sensors and computational infrastructure is prohibitive especially for smallholders and resource limited-enterprises (Krishna et al., 2024). Continuous maintenance, repair and specialized technical support further add to the cost. Although long term benefits, including saving in labor and inputs once a crop is planted, as well as yield improvement, are clear, the longtime taken to realize a payback on investment, might dissuade early adopters (Curry et al., 2021).

9.2. Policy, Regulatory, and Strategic Considerations

Policy frameworks and regulations given the importance of the responsible deployment of farm automation powered with AI. Aviation regulations govern the operation of drones by specifying altitude, line of sight and airspace access. Machinery and robotics standards promote safety, interoperability and compatibility among devices from different manufacturers (Burns et al., 2019). Orne said that "Data governance policies, such as ownership, privacy, sharing and cybersecurity, are essential to protect the digital assets of the farmers and ensure the ethical use of AI analytics."

Strategic policy interventions such as financial incentives, subsidies or loans with low interest rates to facilitate the purchase of high-tech equipment that can speed up the adoption process. Investment in digital infrastructure, such as high-speed internet, Internet of Things network and cloud platforms is critical to support real-time automation in rural areas (Choudhary, 2024).

10. FUTURE DIRECTIONS AND STRATEGIC ROADMAP

The potential of farm automation through AI is great; however, there are significant hurdles - technological, scalability and policy-wise. A multimodal strategy for data fusion that takes the unmanned aerial vehicle imagery (UAV), sensors imagery on the ground, satellite, and environment measurements to synthesize data is very important for high precision and Realtime PDS. Moreover, reinforcement learning (RL) architectures will need to be improved so that they can handle the dynamic, stochastic conditions experienced during field operations, and domain-adaptation techniques will need to be used so that model generalizability can be ensured across different crop species, edaphic conditions and climatic regimes. Progress in robotics and artificial intelligence provide a base for climate resilient agriculture. Autonomous platforms, continuously outfitted with advanced perception systems, complacent actuators and coordination based swarms activities, are ready to carry out precision-mechanics tasks with higher efficacy, like harvesting, pruning, and insect control. In addition, AI models that have a climate forecast synthesizing with physiological stress indicators of crops allow for proactive interventions with the aim of strengthening resilience against abiotic stresses, such as drought, heat or pathogen outbreaks.

11. CONCLUSION

The integration of artificial intelligence and farm automation is a paradigm shift in the modern world of farming with never-before-seen possibilities for enhancing efficiency, sustainability, and resiliency. The advancement of agricultural technology from initial mechanization, to total autonomy has directed the way towards Artificial Intelligence enables (AI) decision making, precision resource management and also real-time adaptive operations. Foundational enabling technologies - such as sensors, internet of things (IoT), cloud and edge computing, and advanced artificial intelligence (AI) frameworks - lay the foundation for an accurate perception of the farm environment and predictive analytics and autonomous interventions under varying conditions on the farm. Autonomous ground robots, manipulators and aerial platforms (UAV have redefined the essence of the agronomic process - from planting to irrigation, fertilization, pest management, selective harvesting. The integration of aerial imagery, ground based sensors in combination with digital twins will create closed loop systems and continuous monitoring of the vitality of crops, soil parameters and environmental variables, which will aid in data driven and site-specific intervention. Controlled- environment agriculture is another example of the possibility's automation brings in terms of amplifying productivity, maximizing the use of resources and reducing environmental impact within a high-value or urban growing environment.

REFERENCES

Abiri, R., Rizan, N., Balasundram, S. K., Shahbazi, A. B., & Abdul-Hamid, H. (2023). Application of digital technologies for ensuring agricultural productivity. *Heliyon, 9*(12), e22601. 10.1016/j.heliyon.2023.e22601 PMID: 38125472

Adhinata, F. D. Wahyono, & Sumiharto, R. (2024). A comprehensive survey on weed and crop classification using machine learning and deep learning. *Artificial Intelligence in Agriculture, 13*, 45-63. 10.1016/j.aiia.2024.06.005

Ahmed, N., & Shakoor, N. (2025). Advancing agriculture through IoT, Big Data, and AI: A review of smart technologies enabling sustainability. *Smart Agricultural Technology*, *10*, 100848. 10.1016/j.atech.2025.100848

Albiero, D., Pontin Garcia, A., Kiyoshi Umezu, C., & Leme de Paulo, R. (2022). Swarm robots in mechanized agricultural operations: A review about challenges for research. *Computers and Electronics in Agriculture*, *193*, 106608. 10.1016/j.compag.2021.106608

Arikapudi, R., & Vougioukas, S. G. (2023). Robotic Tree-fruit harvesting with arrays of Cartesian Arms: A study of fruit pick cycle times. *Computers and Electronics in Agriculture*, *211*, 108023. 10.1016/j.compag.2023.108023

Asih, I., Purba, H., & Sitorus, T. (2020). KEY PERFORMANCE INDICATORS: A SYSTEMATIC LITERATURE REVIEW. *8*, 142-155.

Attaran, M., & Celik, B. G. (2023). Digital Twin: Benefits, use cases, challenges, and opportunities. *Decision Analytics Journal, 6*, 100165. 10.1016/j.dajour.2023.100165

Barbone, A., Bicocchi, N., Martinelli, M., Morandi, R., & Picone, M. (2026). On-device AI and digital twins: A synergistic approach to intelligent cyber-physical systems. *Future Generation Computer Systems*, *175*, 108068. 10.1016/j.future.2025.108068

Barbosa Júnior, M. R., Santos, R. G. D., Sales, L. A., & Oliveira, L. P. (2024). Advancements in Agricultural Ground Robots for Specialty Crops: An Overview of Innovations, Challenges, and Prospects. *Plants, 13*(23). Advance online publication. DOI: 10.3390/plants13233372 PMID: 39683165

Burns, T., Cosgrove, J., & Doyle, F. (2019). A Review of Interoperability Standards for Industry 4.0. *Procedia Manufacturing*, *38*, 646–653. 10.1016/j.promfg.2020.01.083

Chai, J., Zeng, H., Li, A., & Ngai, E. W. T. (2021). Deep learning in computer vision: A critical review of emerging techniques and application scenarios. *Machine Learning with Applications*, *6*, 100134. 10.1016/j.mlwa.2021.100134

Chen, X., Zhang, H., & Wong, C. U. I. (2025). Dynamic Monitoring and Precision Fertilization Decision System for Agricultural Soil Nutrients Using UAV Remote Sensing and GIS. *Agriculture*, *15*(15), 1627. https://www.mdpi.com/2077-0472/15/15/1627

Choudhary, A. (2024). Internet of Things: A comprehensive overview, architectures, applications, simulation tools, challenges and future directions. *Discover Internet of Things*, *4*(1), 31. DOI: 10.1007/s43926-024-00084-3

Cowan, N., Ferrier, L., Spears, B., Drewer, J., Reay, D., & Skiba, U. (2022). CEA Systems: the Means to Achieve Future Food Security and Environmental Sustainability? [Review]. *Frontiers in Sustainable Food Systems, Volume 6 - 2022*. https://doi.org/DOI: 10.3389/fsufs.2022.891256

Curry, G. N., Nake, S., Koczberski, G., Oswald, M., Rafflegeau, S., Lummani, J., Peter, E., & Nailina, R. (2021). Disruptive innovation in agriculture: Socio-cultural factors in technology adoption in the developing world. *Journal of Rural Studies*, *88*, 422–431. 10.1016/j.jrurstud.2021.07.022

David, I., Shao, G., Gomes, C., Tilbury, D., & Zarkout, B. (2025, 2025//). Interoperability of Digital Twins: Challenges, Success Factors, and Future Research Directions. Leveraging Applications of Formal Methods, Verification and Validation. Application Areas, Cham.

De, S., & Moessner, K. (2023). Sensor Networks: Physical and Social Sensing in the IoT. *Sensors (Basel)*, *23*(3). Advance online publication. DOI: 10.3390/s23031451 PMID: 36772505

Dhirani, L. L., Mukhtiar, N., Chowdhry, B. S., & Newe, T. (2023). Ethical Dilemmas and Privacy Issues in Emerging Technologies: A Review. *Sensors (Basel)*, *23*(3). Advance online publication. DOI: 10.3390/s23031151 PMID: 36772190

Dian, R., Li, S., Sun, B., & Guo, A. (2021). Recent advances and new guidelines on hyperspectral and multispectral image fusion. *Information Fusion*, *69*, 40–51. 10.1016/j.inffus.2020.11.001

Dsouza, A., Newman, L., Graham, T., & Fraser, E. D. G. (2023). Exploring the landscape of controlled environment agriculture research: A systematic scoping review of trends and topics. *Agricultural Systems*, *209*, 103673. 10.1016/j.agsy.2023.103673

Dutta, S. (2024). Lean Manufacturing and Process Optimization: Enhancing Efficiency in Modern Production. *International Journal of Scientific Research in Computer Science. Engineering and Information Technology, 10*, 265–273. DOI: 10.32628/CSEIT241051024

Ennaji, O., Vergütz, L., & El Allali, A. (2023). Machine learning in nutrient management: A review. *Artificial Intelligence in Agriculture, 9*, 1-11. 10.1016/j.aiia.2023.06.001

Ferrero, F., Rolando, G., Tabacco, E., & Borreani, G. (2024). Water Use Efficiency in Agriculture: From the Water Footprint Concept to Water Management Policies and Practices. In Minella, M., Bianco Prevot, A., & Maurino, V. (Eds.), *Water Reuse and Unconventional Water Resources: A Multidisciplinary Perspective* (pp. 483–510). Springer Nature Switzerland., DOI: 10.1007/978-3-031-67739-7_20

Georgiadis, G. P., Elekidis, A. P., & Georgiadis, M. C. (2019). Optimization-Based Scheduling for the Process Industries: From Theory to Real-Life Industrial Applications. *Processes (Basel, Switzerland), 7*(7), 438. https://www.mdpi.com/2227-9717/7/7/438

Ghafar, A. S. A., Hajjaj, S. S. H., Gsangaya, K. R., Sultan, M. T. H., Mail, M. F., & Hua, L. S. (2023). Design and development of a robot for spraying fertilizers and pesticides for agriculture. *Materials Today: Proceedings, 81*, 242–248. 10.1016/j.matpr.2021.03.174

Ghraizi, D., Talj, R., & Francis, C. (2023). An Overview of Decision-Making in Autonomous Vehicles. *IFAC-PapersOnLine, 56*(2), 10971–10983. 10.1016/j.ifacol.2023.10.793

Gong, R., Zhang, H., Li, G., & He, J. (2025). Edge Computing-Enabled Smart Agriculture: Technical Architectures, Practical Evolution, and Bottleneck Breakthroughs. *Sensors (Basel), 25*(17), 5302. https://www.mdpi.com/1424-8220/25/17/5302 PMID: 40942729

Hargrove, W. L., Heyman, J. M., Mayer, A., Mirchi, A., Granados-Olivas, A., Ganjegunte, G., Gutzler, D., Pennington, D. D., Ward, F. A., Chavira, L. G., Sheng, Z., Kumar, S., Villanueva-Rosales, N., & Walker, W. S. (2023). The future of water in a desert river basin facing climate change and competing demands: A holistic approach to water sustainability in arid and semi-arid regions. *Journal of Hydrology. Regional Studies, 46*, 101336. 10.1016/j.ejrh.2023.101336

Hider, U., & Zafer, B. (2024). Emerging Threats in Cybersecurity: A Comprehensive Analysis of Modern Attack Vectors.

Indolia, S., Goswami, A. K., Mishra, S. P., & Asopa, P. (2018). Conceptual Understanding of Convolutional Neural Network- A Deep Learning Approach. *Procedia Computer Science*, *132*, 679–688. 10.1016/j.procs.2018.05.069

Javaid, M., Haleem, A., Singh, R. P., & Suman, R. (2021). Substantial capabilities of robotics in enhancing industry 4.0 implementation. *Cognitive Robotics, 1*, 58-75. 10.1016/j.cogr.2021.06.001

Jin, W., Wang, N., Zhang, L., Tian, X., Shi, B., & Zhao, B. (2025). A Review of AI-Driven Automation Technologies: Latest Taxonomies, Existing Challenges, and Future Prospects. *Computers, Materials & Continua*, *84*(3), 3961–4018. 10.32604/cmc.2025.067857

Karami, A., & Karami, M. (2025). Edge computing in big data: Challenges and benefits. *International Journal of Data Science and Analytics*, *20*(7), 6183–6226. DOI: 10.1007/s41060-025-00855-3

Khalifeh, A., Al-Qaisi, A., Aldahdouh, K., Al-Sit, W. T., Olaimat, A. A. L., Alouneh, S., & Darabkh, K. A. (2025). Low power wide area network (LPWAN) protocols: Enablers for future wireless networks. *Results in Engineering*, *27*, 105866. 10.1016/j.rineng.2025.105866

Krishna, V., Akhila, S., & Rakesh, K. (2024). Technology Adoption Barriers and Economic Solutions. In (pp. 212-227).

Kufel, J., Bargieł-Łączek, K., Kocot, S., Koźlik, M., Bartnikowska, W., Janik, M., Czogalik, Ł., Dudek, P., Magiera, M., Lis, A., Paszkiewicz, I., Nawrat, Z., Cebula, M., & Gruszczyńska, K. (2023). What Is Machine Learning, Artificial Neural Networks and Deep Learning?-Examples of Practical Applications in Medicine. *Diagnostics (Basel)*, *13*(15). Advance online publication. DOI: 10.3390/diagnostics13152582 PMID: 37568945

Kumar, M., Malik, V., Tomar, B., Tomar, S., Kumari, M., & Gupta, S. (2023). *Agricultural Mechanization Development.* In (pp. 1-11).

Liu, R., Luan, T. H., Qu, Y., Xiang, Y., Gao, L., & Zhao, D. (2025). Internet of Digital Twin: Framework, Applications and Enabling Technologies. *IEEE Communications Surveys and Tutorials*.

Liu, X., Zhang, Q., Min, W., Geng, G., & Jiang, S. (2025). Solutions and challenges in AI-based pest and disease recognition. *Computers and Electronics in Agriculture*, *238*, 110775. 10.1016/j.compag.2025.110775

Longo, L., Brcic, M., Cabitza, F., Choi, J., Confalonieri, R., Ser, J. D., Guidotti, R., Hayashi, Y., Herrera, F., Holzinger, A., Jiang, R., Khosravi, H., Lecue, F., Malgieri, G., Páez, A., Samek, W., Schneider, J., Speith, T., & Stumpf, S. (2024). Explainable Artificial Intelligence (XAI) 2.0: A manifesto of open challenges and interdisciplinary research directions. *Information Fusion*, *106*, 102301. 10.1016/j.inffus.2024.102301

Lu, Y., Hegyi, A., Maria Salomons, A., & Wang, H. (2025). Reference RL: Reinforcement learning with reference mechanism and its application in traffic signal control. *Information Sciences*, *689*, 121485. 10.1016/j.ins.2024.121485

Mana, A. A., Allouhi, A., Hamrani, A., Rehman, S., el Jamaoui, I., & Jayachandran, K. (2024). Sustainable AI-based production agriculture: Exploring AI applications and implications in agricultural practices. *Smart Agricultural Technology*, *7*, 100416. 10.1016/j.atech.2024.100416

Maró, Z. M., Nagy, J., Molnár, E. M., & Mizik, T. (2025). Challenges and potential solutions to employment issues in the agri-food sector of developed countries - A systematic literature review. *Sustainable Futures : An Applied Journal of Technology, Environment and Society*, *10*, 100895. 10.1016/j.sftr.2025.100895

Meshram, A. T., Vanalkar, A. V., Kalambe, K. B., & Badar, A. M. (2022). Pesticide spraying robot for precision agriculture: A categorical literature review and future trends. *Journal of Field Robotics*, *39*(2), 153–171.

Milrad, S. (2018). 11 - Satellite Imagery. In Milrad, S. (Ed.), *Synoptic Analysis and Forecasting* (pp. 145–161). Elsevier., 10.1016/B978-0-12-809247-7.00011-9

Mohsan, S. A. H., Othman, N. Q. H., Li, Y., Alsharif, M. H., & Khan, M. A. (2023). Unmanned aerial vehicles (UAVs): Practical aspects, applications, open challenges, security issues, and future trends. *Intelligent Service Robotics*, *16*(1), 109–137. DOI: 10.1007/s11370-022-00452-4 PMID: 36687780

Montoya-Cavero, L.-E., Díaz de León Torres, R., Gómez-Espinosa, A., & Escobedo Cabello, J. A. (2022). Vision systems for harvesting robots: Produce detection and localization. *Computers and Electronics in Agriculture*, *192*, 106562. 10.1016/j.compag.2021.106562

Nagarsheth, S., Agbossou, K., Henao, N., & Bendouma, M. (2025). The Advancements in Agricultural Greenhouse Technologies: An Energy Management Perspective. *Sustainability*, *17*(8), 3407.

Nikolinakos, N. T. (2023). Ethical Principles for Trustworthy AI. In Nikolinakos, N. T. (Ed.), *EU Policy and Legal Framework for Artificial Intelligence, Robotics and Related Technologies - The AI Act* (pp. 101–166). Springer International Publishing., DOI: 10.1007/978-3-031-27953-9_3

Nzimande, T. N. M., Chimonyo, V. G. P., Wimalasiri, E. M., & Mabhaudhi, T. (2025). Using AquaCrop, DSSAT and the SIMPLE to Estimate Water Use of Underutilised Cereal in South Africa. In Mabhaudhi, T., Chimonyo, V. G. P., Senzanje, A., & Chivenge, P. P. (Eds.), *Enhancing Water and Food Security Through Improved Agricultural Water Productivity: New Knowledge, Innovations and Applications* (pp. 201–221). Springer Nature Singapore., DOI: 10.1007/978-981-96-1848-4_9

Ofosu-Ampong, K. (2024). Artificial intelligence research: A review on dominant themes, methods, frameworks and future research directions. *Telematics and Informatics Reports, 14*, 100127. 10.1016/j.teler.2024.100127

Ojo, M. O., & Zahid, A. (2022). Deep Learning in Controlled Environment Agriculture: A Review of Recent Advancements, Challenges and Prospects. *Sensors (Basel), 22*(20), 7965. https://www.mdpi.com/1424-8220/22/20/7965 PMID: 36298316

Olliver, A., Schlingmann, N., & van der Vlugt, P. (2022). ISOBUS Technologies – The Standard for Smart Agriculture. In Zhang, Q. (Ed.), *Encyclopedia of Smart Agriculture Technologies* (pp. 1–18). Springer International Publishing., DOI: 10.1007/978-3-030-89123-7_117-2

Padhiary, M., Saha, D., Kumar, R., Sethi, L. N., & Kumar, A. (2024). Enhancing precision agriculture: A comprehensive review of machine learning and AI vision applications in all-terrain vehicle for farm automation. *Smart Agricultural Technology, 8*, 100483. 10.1016/j.atech.2024.100483

Pande, C. B., & Moharir, K. N. (2023). Application of Hyperspectral Remote Sensing Role in Precision Farming and Sustainable Agriculture Under Climate Change: A Review. In Pande, C. B., Moharir, K. N., Singh, S. K., Pham, Q. B., & Elbeltagi, A. (Eds.), *Climate Change Impacts on Natural Resources, Ecosystems and Agricultural Systems* (pp. 503–520). Springer International Publishing., DOI: 10.1007/978-3-031-19059-9_21

Pané, S., Wendel-Garcia, P., Belce, Y., Chen, X. Z., & Puigmartí-Luis, J. (2021). Powering and Fabrication of Small-Scale Robotics Systems. *Current Robotics Reports, 2*(4), 427–440. DOI: 10.1007/s43154-021-00066-1 PMID: 35036926

Peladarinos, N., Piromalis, D., Cheimaras, V., Tserepas, E., Munteanu, R. A., & Papageorgas, P. (2023). Enhancing Smart Agriculture by Implementing Digital Twins: A Comprehensive Review. *Sensors (Basel), 23*(16). Advance online publication. DOI: 10.3390/s23167128 PMID: 37631663

Raj, P., Sharma, S., Kapri, A., & Gulaiya, S. (2025). *Precision Agriculture: A Strategic Approach to Resource Efficiency and Sustainable Farming* (Vol. 31). https://doi.org/DOI: 10.9734/jsrr/2025/v31i83381

Rashid, A. B., Kausik, A. K., Khandoker, A., & Siddque, S. N. (2025). Integration of Artificial Intelligence and IoT with UAVs for Precision Agriculture. *Hybrid Advances, 10*, 100458. 10.1016/j.hybadv.2025.100458

Rashid, A. B., & Kausik, M. D. A. K. (2024). AI revolutionizing industries worldwide: A comprehensive overview of its diverse applications. *Hybrid Advances, 7*, 100277. 10.1016/j.hybadv.2024.100277

Salomi, M., & Samsudeen, S. (2024). Agricultural Artificial Intelligence: Obstacles and Opportunities. In Chouhan, S. S., Saxena, A., Singh, U. P., & Jain, S. (Eds.), *Artificial Intelligence Techniques in Smart Agriculture* (pp. 55–66). Springer Nature Singapore., DOI: 10.1007/978-981-97-5878-4_4

Sanchez-Cubillo, J., Del Ser, J., & Martin, J. L. (2024). Toward Fully Automated Inspection of Critical Assets Supported by Autonomous Mobile Robots, Vision Sensors, and Artificial Intelligence. *Sensors (Basel), 24*(12). Advance online publication. DOI: 10.3390/s24123721 PMID: 38931502

Selod, H., & Shilpi, F. (2021). Rural-urban migration in developing countries: Lessons from the literature. *Regional Science and Urban Economics, 91*, 103713. 10.1016/j.regsciurbeco.2021.103713

Seppelt, R., Klotz, S., Peiter, E., & Volk, M. (2022). Agriculture and food security under a changing climate: An underestimated challenge. *iScience, 25*(12), 105551. 10.1016/j.isci.2022.105551 PMID: 36458255

Sharma, A., Hazarika, M., Heisnam, P., Pandey, H., Devadas, V. S., & Wangsu, M. (2024). Controlled Environment Ecosystem: A plant growth system to combat climate change through soilless culture. *Crop Design, 3*(1), 100044. 10.1016/j.cropd.2023.100044

Sharma, K., & Shivandu, S. K. (2024). Integrating artificial intelligence and Internet of Things (IoT) for enhanced crop monitoring and management in precision agriculture. *Sensors International, 5*, 100292.

Shinde, G. U., Agrawal, R. C., Mani, I., Agrawal, A., Khodke, U. M., Muley, S. V., Tekale, D. D., Bhalerao, S. R., & Kakade, O. D. (2024, 2024//). Digital Farming Solution by Automation Using Agri-Bot, Agri-Drone, and Agri-AGV for Organic Farming Practices. Agriculture Digitalization and Organic Production, Singapore.

Skendžić, S., Zovko, M., Živković, I. P., Lešić, V., & Lemić, D. (2021). The Impact of Climate Change on Agricultural Insect Pests. *Insects*, *12*(5). Advance online publication. DOI: 10.3390/insects12050440 PMID: 34066138

Srivastav, A. K., & Das, P. (2025). Agricultural Robotics and Autonomous Systems. In Srivastav, D. A. K., & Das, D. P. (Eds.), *Biotechnology and IoT in Agriculture and Food Production: Green Innovation* (pp. 201–225). Apress., DOI: 10.1007/979-8-8688-1469-3_22

Stasiak-Betlejewska, R., & Potkány, M. (2015). Construction Costs Analysis and its Importance to the Economy. *Procedia Economics and Finance*, *34*, 35–42. 10.1016/S2212-5671(15)01598-1

Taherdoost, H. (2023). Deep Learning and Neural Networks: Decision-Making Implications. *Symmetry*, *15*(9), 1723. https://www.mdpi.com/2073-8994/15/9/1723

Taleb, M. A., Korsoveczki, G., & Husi, G. (2025). Automotive navigation for mobile robots: Comprehensive review. *Results in Engineering*, *27*, 105837. 10.1016/j.rineng.2025.105837

Tarafdar, R. (2025). *Human-AI Collaboration in Workflow Optimization: A Framework for Hybrid Decision Systems in Automation-Heavy Industries* (Vol. 11). https://doi.org/DOI: 10.32628/CSEIT251112384

Theodorakopoulos, L., Theodoropoulou, A., & Stamatiou, Y. (2024). A State-of-the-Art Review in Big Data Management Engineering: Real-Life Case Studies, Challenges, and Future Research Directions. *Eng*, *5*(3), 1266-1297. https://www.mdpi.com/2673-4117/5/3/68

Tudi, M., Daniel Ruan, H., Wang, L., Lyu, J., Sadler, R., Connell, D., Chu, C., & Phung, D. T. (2021). Agriculture Development, Pesticide Application and Its Impact on the Environment. *International Journal of Environmental Research and Public Health*, *18*(3). Advance online publication. DOI: 10.3390/ijerph18031112 PMID: 33513796

Ugenti, A., Galati, R., Mantriota, G., & Reina, G. (2023). Analysis of an all-terrain tracked robot with innovative suspension system. *Mechanism and Machine Theory*, *182*, 105237. 10.1016/j.mechmachtheory.2023.105237

Vatistas, C., Avgoustaki, D. D., & Bartzanas, T. (2022). A Systematic Literature Review on Controlled-Environment Agriculture: How Vertical Farms and Greenhouses Can Influence the Sustainability and Footprint of Urban Microclimate with Local Food Production. *Atmosphere, 13*(8), 1258. https://www.mdpi.com/2073-4433/13/8/1258

Verdouw, C., Tekinerdogan, B., Beulens, A., & Wolfert, S. (2021). Digital twins in smart farming. *Agricultural Systems, 189*, 103046. 10.1016/j.agsy.2020.103046

Vijayakumar, V., Ampatzidis, Y., Schueller, J. K., & Burks, T. (2023). Smart spraying technologies for precision weed management: A review. *Smart Agricultural Technology, 6*, 100337. 10.1016/j.atech.2023.100337

Wang, L., Lian, G., Harris, Z., Horler, M., Wang, Y., & Chen, T. (2023). The controlled environment agriculture: a sustainable agrifood production paradigm empowered by systems engineering. In Kokossis, A. C., Georgiadis, M. C., & Pistikopoulos, E. (Eds.), *Computer Aided Chemical Engineering* (Vol. 52, pp. 2167–2172). Elsevier., 10.1016/B978-0-443-15274-0.50345-0

Wang, Y., Wang, Y., Mushtaq, R. T., & Wei, Q. (2024). Advancements in Soft Robotics: A Comprehensive Review on Actuation Methods, Materials, and Applications. *Polymers, 16*(8). Advance online publication. DOI: 10.3390/polym16081087 PMID: 38675005

Waqas, M., Naseem, A., Humphries, U. W., Hlaing, P. T., Dechpichai, P., & Wangwongchai, A. (2025). Applications of machine learning and deep learning in agriculture: A comprehensive review. *Green Technologies and Sustainability, 3*(3), 100199. 10.1016/j.grets.2025.100199

Wu, H., Ji, P., Ma, H., & Xing, L. (2023). A Comprehensive Review of Digital Twin from the Perspective of Total Process: Data, Models, Networks and Applications. *Sensors (Basel), 23*(19). Advance online publication. DOI: 10.3390/s23198306 PMID: 37837136

Wu, X., Xiao, L., Sun, Y., Zhang, J., Ma, T., & He, L. (2022). A survey of human-in-the-loop for machine learning. *Future Generation Computer Systems, 135*, 364–381. 10.1016/j.future.2022.05.014

Zhang, D., & Wei, B. (2017). A review on model reference adaptive control of robotic manipulators. *Annual Reviews in Control, 43*, 188–198. 10.1016/j.arcontrol.2017.02.002

Zhang, R., Zhu, H., Chang, Q., & Mao, Q. (2025). A Comprehensive Review of Digital Twins Technology in Agriculture. *Agriculture*, *15*(9), 903. https://www.mdpi.com/2077-0472/15/9/903

Zhang, X., Feng, G., & Sun, X. (2024). Advanced technologies of soil moisture monitoring in precision agriculture: A Review. *Journal of Agriculture and Food Research*, *18*, 101473. 10.1016/j.jafr.2024.101473

Zhang, Y., Yasutake, D., Hidaka, K., Okayasu, T., Kitano, M., & Hirota, T. (2022). Crop-localised CO2 enrichment improves the microclimate, photosynthetic distribution and energy utilisation efficiency in a greenhouse. *Journal of Cleaner Production*, *371*, 133465. 10.1016/j.jclepro.2022.133465

Zhao, J., Zhao, W., Deng, B., Wang, Z., Zhang, F., Zheng, W., Cao, W., Nan, J., Lian, Y., & Burke, A. F. (2024). Autonomous driving system: A comprehensive survey. *Expert Systems with Applications*, *242*, 122836. 10.1016/j.eswa.2023.122836

Zhao, Q., Yu, L., Du, Z., Peng, D., Hao, P., Zhang, Y., & Gong, P. (2022). An Overview of the Applications of Earth Observation Satellite Data: Impacts and Future Trends. *Remote Sensing*, *14*(8), 1863. https://www.mdpi.com/2072-4292/14/8/1863

Zheng, H., Shen, L., Tang, A., Luo, Y., Hu, H., Du, B., Wen, Y., & Tao, D. (2025). Learning from models beyond fine-tuning. *Nature Machine Intelligence*, *7*(1), 6–17. DOI: 10.1038/s42256-024-00961-0

Zheng, S., Xu, S., & Rai, R. (2025). Trailer-referenced autonomous navigation of agricultural tractor–trailer systems. *Computers and Electronics in Agriculture*, *237*, 110514. 10.1016/j.compag.2025.110514

Chapter 11
Robotics and Drones: Shaping the Future of Agriculture With Automation

Amjid Khan
http://orcid.org/0009-0001-5475-5156
University of South Africa, South Africa

Aisha Israr
http://orcid.org/0009-0005-4204-8007
University of Houston, USA

Afnan Khan Shinwari
http://orcid.org/0009-0008-7448-2942
University of Houston, USA

ABSTRACT

Agricultural automation is redefining modern agriculture through the integration of robotics, AIoT ecosystems, smart sensors, autonomous machinery, and cloud-based analytics to enhance productivity, efficiency, and sustainability. This chapter per the authors synthesizes how autonomous platforms, including drones, robotic harvesting, precision spraying, automated irrigation, and AI-driven monitoring tools, are transforming crop and livestock management. Emphasis is placed on real-time data acquisition, site-specific interventions, and intelligent decision support to optimize input use, reduce agrochemical dependency, and improve soil, water, and climate-smart outcomes. The chapter evaluates economic benefits, highlighting improved profitability and long-term cost efficiency despite high initial investments, while critically reviewing social implications such as labor displacement, skill requirements, and technological divides. Key challenges and future pathways toward next-generation autonomous farming systems are also discussed.

DOI: 10.4018/979-8-3373-7257-0.ch011

INTRODUCTION

Agriculture is moving towards an era where automation, intelligent machines and unmanned aerial systems are redefining agriculture's cultivation, monitoring and harvesting crops. Global food systems are growing under extruding pressure because of demographic boom, climate change, resource depletion and uncertain pressure in the environment. Consequence of this is increased demand for technologies which increase people's productivity, are less wasteful of inputs, and that will remain resilient to changing climatic conditions (Awan & Yaqoob, 2023). Within this context, robotics and drones have become revolutionary instruments that can execute complex operation in the agricultural field with precision, efficiency, and with limited human involvement. Their integration is the next paradigm shift in the way conventional labor-intensive practices are operating toward an intelligent data driven agricultural framework.

The development of agricultural mechanization can be marked in several different phases starting from the development of rudimentary hand tools and culminating in animal-drawn tools that increased labor productivity. The Industrial Revolution was a defining or turning point with the introduction of steam-driven machines, then internal combustion engines that made tractors, sprayers, and threshers an essential part of farming in the modern world (Ahmed et al., 2025). By the end of the twentieth century, mechanization became relatively mature, as was demonstrated by the mechanization of large-scale cultivation, which has become common place in this industry. Nevertheless, for all these advancements, conventional mechanization was very much dependent on human decision-making, manual controls and homogeneous field management (Groumpos, 2021). The limitations inherent in this way of doing things, especially the non-ability to address issues brought about by within-field variability, paved the way for digital and automated solutions.

The advent of precision agriculture in the 1990s brought about the availability of instruments to farmers, such as GPS-guided tractors, soil mapping techniques and variable rate technologies, which allowed farmers to apply inputs with more accuracy and efficiency (Mamabolo et al., 2025). Precision agriculture was a major advance towards site specific management, but the human operator was still needed to interpret the data, control the machinery and carry out most operations in the field. In the last 10 years, advances in artificial intelligence, robotics, computer vision and machine learning have supported the shift from precision agriculture toward total automation, where machines can sense, analyses, decide and take actions with very little human supervision (Maurya et al., 2024). Autonomous tractors, robotic harvesters, intelligent weeders and AI-driven drones are now performing a variety of tasks from planting to spraying to crop monitoring and yield estimation. This move marks the arrival of the "autonomous farm," in which sensing and real-time decision-making will lead constantly optimized agricultural operations.

Several related factors have contributed to the increasing use of robotics and drones in agriculture. Labor shortages - worsened by the migration of people out of the countryside, an aging farming population, and the seasonal nature of agricultural work - have farmers turning to automated solutions that can take as long as possible to work, and do so with consistency (Liu et al., 2023). Climate stress and greater environmental unpredictability demand systems that are capable of fast acting, precise and data-driven responses to disease outbreaks, droughts, nutrient deficiencies and pest infestations. Moreover, the need for efficient use of resources has been strengthened, due to increased costs of inputs and global sustainability goals. Robotics and drones enable the management of crops at the micro level and therefore help in decreasing the wastage of water, fertilizers, and pesticides (Cheng et al., 2024). Thus, the convergence between sensor technologies, artificial intelligence (AI), cloud computing and geo-positioning systems has made a considerable contribution to making automation accessible, cost-efficient and scalable to a degree where robotics and unmanned aerial vehicles have become building stones in modern agriculture practice. This chapter takes an in-depth examination of the technical basis of new solutions, potential applications, and practical implementations of robotics and UAV technology in modern-day agronomy. It is critically concerned with innovations where operations are concerned, outlines operational frameworks, and examining mechanisms of integration that are compatible with a variety of farming systems. Furthermore, it ponders the environmental, economic, and social consequences, outlines current problems and potential opportunities for the future to develop a full roadmap for the large-scale introduction of agricultural automation.

1. FOUNDATIONS OF AGRICULTURAL ROBOTICS AND DRONE TECHNOLOGIES

1.1. Core Components: Sensors, Actuators, Controllers, and AI

Agricultural robotics and drone systems are based on an architecture, which is integrated with sensing modules, mechanical actuators, computational controllers, and artificial intelligence (AI) algorithms. The interplaying of these elements allows the autonomous and semi-autonomous operation, according to the complex agronomic environments (Hernández et al., 2025). Sensors form the main interface between robotic platforms and the agricultural environment, making continuous measurements of real time data about crop health, soil characteristics, surrounding environmental conditions and physical obstacles. Contemporary systems routinely employ the use of multispectral and hyperspectral imaging modalities to quantify the physiological status of plants as well as identify stress signals well in advance of visual symptoms, observable for humans (Yan et al., 2025). Thermal imaging helps in evaluation of the canopy temperature and water - stress indicators, while RGB camera helps in fruit detection, weed mapping and phenotyping. These optical systems are often combined with LiDAR for high resolution 3D reconstruction of the orchards, vineyards and greenhouse with ultrasonic and proximity sensors for obstacle detection on short range (Wen et al., 2023). Moreover, soil probes which measure soil moisture, pH and concentration levels of nutrients provide key information for precision irrigation and fertilization schemes.

Actuators are the mechanical body of agricultural robots, which can perform mechanical work in the environment such as harvesting, cutting, spraying, roughing the ground, etc. Electric motors, hydraulic assemblies and pneumatic drives are designed to work under tough agronomic conditions with dust, moisture, heavy load application, and constant vibration (Pustavrh et al., 2023). Advanced robotic harvesters combine actuators with force and torque sensors to achieve the delicate handling of fruits and vegetables, and soft robotic grippers and suction-based end-effector minimize mechanical damage during harvesting. Embedded controllers coordinate the inputs from sensors and outputs from actuators by means of real-time processing platforms, microcontrollers (ARM-based, edge computing, AI optimized processors designed to support in-situ computer vision and machine learning tasks) (Kure, 2022).

Artificial intelligence serves as the brain of today's agricultural systems and is responsible for absorbing information from sensors, recognizing complex patterns, and participating in automated decision making in the fields (Jan et al., 2025). Machine learning based models are helping in high precision weed - crop discrimination, in-the-fly disease detection, predicting yield and in planning adaptive paths. Convolutional neural networks are used in the image based diagnostic accuracy and reinforcement learning algorithms improving the robotic paths in dynamic and heterogeneous terrain (Coleman et al., 2022). Collectively, these parts of technology bring both computational intelligence and mechanical functionality together to promote fully automated practices in agriculture.

1.2. Navigation and Localization: GPS, RTK, LiDAR, and Visual SLAM

Navigation and localization systems allow the robots used in agriculture and unmanned aircraft to acquire the spatial position at high accuracy and to create maps of the environment, being able to carry out the precise movement through agricultural fields in variable and complex environments. Global Navigation Satellite Systems (GNSS), such as GPS, Galileo, GLONASS, and BeiDou, have wide coverages in the spatial domain suitable for outside operations for agriculture, such as boundary mapping, guiding agricultural tractors and others on their tracks (Jin et al., 2024). However, standard GPS accuracy is often not high enough to be used for high-precision agronomic applications. As a result, Real Time Kinematic (RTK) correction systems are used for increasing the accuracy of the position up to cm level (Sirish Kumar & Srilatha Indira Dutt, 2020). RTK-enabled navigation plays a significant role in applications where precise row spacing during seeding operations is needed, or in autonomous spraying systems where precise positioning is needed, and for high resolution field agronomics maps during seeding operations when the use of drones is needed. As shown in Figure 1, precision technologies in navigation form the basis for all areas of the crop production continuum.

The localization is further improved with the use of LiDAR technology and the generation of dense three-dimensional representations of agricultural landscapes. Such a type of spatial datasets is especially useful in an orchard, a vineyard, an undulating region, where robotic systems need to detect the structures of canopies, avoid barriers, and control the paths without collisions (Shi et al., 2023). LiDAR guided navigation is particularly relevant in spatially sensitive type tasks including automated pruning, canopy management and inter row weeding. In environments where GNSS signals are poor or exclusive like inside greenhouses or tunnels, dense orchards and under the tree canopy, robots use Visual Simultaneous Localization and Mapping (SLM). Visual SLAM uses features extracted from the camera information to build the real-time map of the surrounding environment and the relative movement of the robot (Theodorou et al., 2022). This capability allows for high-precision indoor operations, i.e., fruit harvesting and scavenging in greenhouses, to be operated without relying on satellite. The unification of some of the best available positioning technologies, GNSS, RTK, LiDAR, and Visual SLAM in united control software frameworks guarantees robust positioning, environmental cognizance, and stability in otherwise diverse agricultural environments (Diao et al., 2025).

Figure 1. Role of precision navigational technologies in agriculture, highlighting satellite positioning, IMU-based machinery estimation, environmental sensing, and 3D point-cloud processing to support tillage, sowing, crop growth management, and harvesting across the production chain

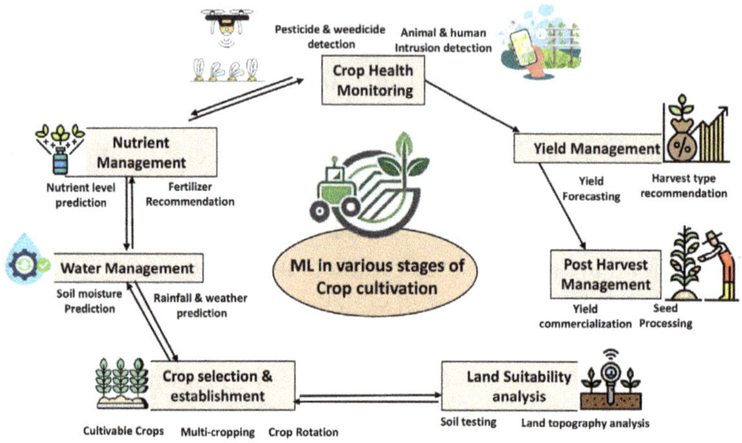

2. GROUND-BASED AGRICULTURAL ROBOTS

Ground based agricultural robots are one of the most disruptive technological advancements in modern agriculture and is the subsequent use of robotics in the farm industry. These systems allow the precise operations to be performed in a more efficient and less manpower steps during the entire crop production cycle (Barbosa Júnior et al., 2024). Designed to role in complex environments on Earth, they interact directly with the soil or plant tissue, and can perform sanitary tasks, that is, repeating or completing routine tasks with incredible accuracy. Their integration to farming systems marks a fundamental shift towards an ever more automated approach to farming, where robots on the farm do not just complement the work of a human operating the machinery, but often also outperform them in terms of precision, consistency, and operational lifetime. As labor shortages continue to rise worldwide and calls for sustainable food production rise, robots with the ability to function on the ground become increasingly important parts of technologically advanced farming ecosystems (Heydari, 2024).

2.1. Field Robots for Planting, Seeding, and Soil Preparation

Ground based robots designed to put the seed down, plant, seed and till the ground, are the foundation for automated crop establishment. These systems combine high precise navigation with advanced soil interaction technologies (i.e. irrigation system, channels) and automation components, to obtain homogenous seed positioning, optimal distance between each seed and optimal germination conditions. Robotic seeders based on real-time Kinematic (RTK) Global Positioning System (GPS) guidance, keep seeding rate and positioning accuracy at centimeter level; thus ensuring uniformity of earth rows and little wastage of seeds (Zhang et al., 2020). Soil preparation robots use smart tillage modules which can adjust weight, pressure and speed in real-time according to measured soil moisture, soil compaction and soil texture. This type of tillage is adaptive in nature and therefore reduces soil disturbance, conserves energy and promotes regenerative practices including the maintenance of optimal soil structure (Yang et al., 2023).

Autonomous unmanned planting robots do use systems of seed metering machines that can vary seeding rates based on spatial variability determined by on-board sensors or by combining it with field maps. In addition, this variable rate seeding strategy helps to improve plant spacing, optimize the use of resources, and improve early stage vigor of the crop (Du et al., 2022). In the case of conservation agriculture, robots referred to as direct-seeders executive the least possible soil inversion and maintain organic content, combining automation with the best land management practices (Naim et al., 2025). Collectively, these technologies are to allow more uniform crop stand establishment with less than manual labor and large tractor-based operations.

2.2. Autonomous Weeding Robots and Laser-Guided Weed Control

Weed control is one of the most labor demanding and expensive parts of crop production, and the development of autonomous weeding robots is a precise and chemical-free solution to the age-old challenge. In some cases, these robots have made use of sophisticated machine vision algorithms, multispectral imaging and deep learning-based plant classification models to differentiate between weeds and crops with immense accuracy (Shahria et al., 2022). Once the weed is detected the robots deploy a targeted mechanical, thermal or laser-based intervention to destroy the weed without disturbing the rest of the plants nearby. Mechanical weeding systems use rotating blades, hoes, or micro - weeding equipment to dig up the weeds while thermal or flame-based systems use controlled heat to dry out unwanted weeds (Xiang et al., 2024).

Laser guided control of weeds turns out to be the latest cutting edge in autonomous weed control. These robots employ high energy AI-aligned laser beams that produce precise hits on the meristem of weeds which impairs their growth without the need for the herbicides (Elhamali et al., 2025). The use of lasers allows for no chemical runoff, minimum development of herbicide resistance, and is in line with organic farming practices. Continuous autonomous operation allows weeding robots to workday and night while always keeping the fields weed free throughout the crop cycle.

Many of the modern weeding robots also include real-time data logging which produces detailed weed maps for predictive analytics and long-term weeds pressure modelling. As such, by eliminating manual weeding labor, as well as being able to reduce the need for chemical herbicides, autonomous weeding robots are part of more sustainable, environmentally responsible weed management systems (Khadatkar et al., 2025).

2.3. Harvesting and Picking Robots for Fruits, Vegetables, and Grains

Harvesting represents one of the most technically demanding tasks in agricultural automation, due to the large variability in crop dimension, morphology, ripeness and mechanical sensitivity of the crops that need to be harvested. Robotic harvesting platforms incorporate leading computer vision and three-dimensional imaging technologies and machine learning algorithms in order to accurately assess quality, dimensional and chromatic analysis, and selective picking (Tang et al., 2020). In fruit orchards, robotic harvesters come with compliant grippers or suction-based end-effectors that helps to gently detach the individual fruit which reduces bruising and postharvest damages. These systems generally involve the use of robotic arms with multi-joints and a large degree of freedom, assisted by real time visual feedback to ensure that components/objects are accurately and controlled. Vegetable-harvesting robots are designed to be used in crop processing and harvesting of produce such as lettuce, tomato, strawberry, cucumber and peppers. For example, another use involves strawberry-picking robots that use stereoscopic vision and deep neural networks to determine berries at the optimum stage of ripening and cut them without damaging the quality. Lettuce harvesting systems use laser cutters or automated blades, which ensure the clean separation of heads from the base of the stems, while ensuring very high sanitary standards for humidity and hygiene (Tituaña et al., 2024). In the level of production, in cereal based systems autonomous combine harvesters by incorporating sensor networks, positioning-based guidance, and real-time yield monitoring with onboard subsystems that allow for real-time dynamic optimization of cutting height and threshing performance and grain separation efficiency in real time. (Droukas et al., 2023).

2.4. Swarm Robotics for Coordinated Field Operations

Swarm robotics is a new paradigm of agricultural automation that involves the use of small, cooperative robotic groups that communicate with each other in order to perform agricultural tasks (Albiero et al., 2022). Taking inspiration from collective behaviors exhibited by biological systems, such as ant colonies and bee swarms, and extending these characteristics to robots, these distributed agents exchange information over wireless networks and use AI-based coordination algorithms to make it possible to share data in real-time and to make collaborative decisions. Swarm based architectures come with benefits in terms of increase in scaling, redundancy, and resiliency of the system as the operational activities are spread across many operational units and not concentrated in one big machine (Abbass & Mostaghim, 2025).

Moreover, swarm intelligence is the basis of adaptive learning and self-organization, which allows a robotic agent to alter and fine tune their operational strategy according to shared experiential data, and information gathered about the environment, accumulated over time. This collective learning ability makes swarm robotics especially promising for other applications, including distributed environmental sensing, large-scale scouting in the field and precision micro-management of the crops (Nitti et al., 2025).

3. AERIAL DRONES IN AGRICULTURE

Aerial drones, also known widely as unmanned aerial vehicles, or UAVs, have become essential in the modern agricultural system because they provide high resolution datasets within a short period of time which have an important impact on effective decision making and field management practices (Mohsan et al., 2023). In contrast to ground-based robotic platforms requiring direct physical interaction with crops and soil, UAVs are specialized for acquisition of spatially rich information, implementation of expeditious field surveys and implementation of precision tasks across extensive landscapes, all whilst minimizing the human element of involvement. Their ability to get through Crops rows, Orchards, Forested areas and Pasture lands, the ability to monitor Health Indices, Biotic and Abiotic Stress, Canopy Vigor and inputs with exceptional accuracy (Csikós & Tóth, 2023). As the technologies in photonic sensors, battery chemistries, and autonomous flight algorithms have continued to be perfected and refined, drones are moving from primitive cartographic technologies to advanced ones with the ability to perform real time analytics, site specific spraying, autonomous pollination and ecological restoration. The following sections define the main types of Drones in agriculture and explain their fast-growing uses in precision- and automation-based agriculture.

3.1. Drone Types: Fixed-Wing, Rotary-Wing, and Hybrid VTOL

Agricultural UAVs can be broadly divided into three main configurations; fixed-wing, rotary-wing and hybrid vertical-take-off and landing (VTOL) platforms, depending on the operational requirements for which they are built. Fixed-wing UAVs mimic small planes, with their strong aerodynamic efficiency which makes them ideal for large coverage area crop mapping, watershed mapping, landscape-scale ecosystem mapping, and topographic mapping. Their ability to fly without using much energy allows for extended missions, i.e. to gather a lot of information over large fields. However, fixed-wing platforms require runway equivalent runways for both take-off and landing, thus forbidding sustained hovering ability (Imran & Li, 2025).

Rotary-wing UAVs, which include quadcopters, hexacopters and octocopters, offer better maneuverability, the ability to take-off and land vertically (VTOL), and accurate loitering. These characteristics allow one to gather high-resolution plant health data, and to scout areas of problems, or to apply agrochemicals precisely and with high spatial resolution. For this reason, rotary platforms are preferred for orchard, vineyard and fragmented land field configurations where agility of operation overrides requirements for endurance (Li, 2025).

Hybrid VTOL UAVs are a combination of the features of fixed wing and rotary wing. By taking off using vertical flight and later transitioning to efficient forward flight regimes, hybrid platforms are both enabled to provide the long-range coverage and precise, low altitude, operations. This dual mode capability makes them especially useful for large area mapping where it is necessary to cover enormous distances but still achieve pinpoint takeoff precision within often cramped spaces and variegated terrain or remote locations that lack any form of runway (Osman et al., 2025). Collectively, these typologies of UAVs enable a wide range of applications in the agriculture industry spanning from immediate operation (diagnosis) to commercial-scale applications (input delivery). Agricultural unmanned aerial vehicles can be categorized into several platform types, including multirotor, fixed-wing, hybrid VTOL, and spray drones, as illustrated in Figure 2.

Figure 2. Classification of agricultural unmanned aerial vehicles (UAVs).

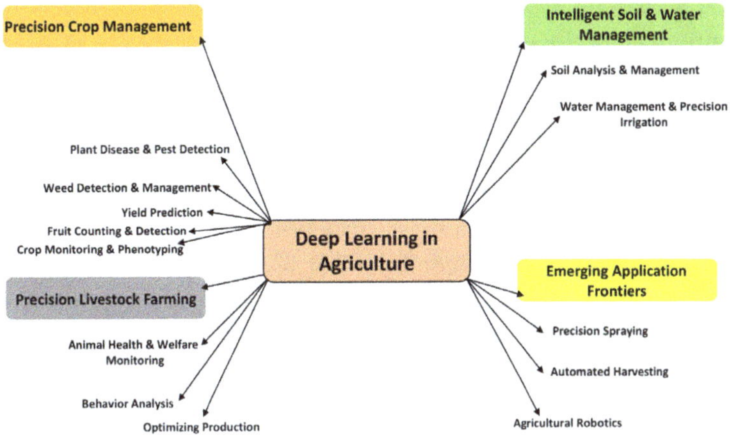

The diagram presents major categories of agricultural UAV platforms, including multirotor drones, fixed-wing drones, hybrid VTOL systems, observation drones, heavy-duty spray drones, and nano/micro drones, highlighting their diverse structural designs and application-specific roles in precision agriculture

3.2. Multispectral, Hyperspectral, and Thermal Imaging for Crop Monitoring

The combination of advanced imaging sensors with UAVs represents a revolutionary development in the field of precision agriculture, allowing for the extraction and information about plant and soil level which is beyond the perception range of eyesight. Multispectral sensors, which record reflectance at a limited number of murals of prominent energy, including near-infrared (NIR), red-edge, red, green and blue, provide the possibility to calculate vegetation indices such as the Normalized Difference Vegetation Index (NDVI), Normalized Difference Red-edge (NDRE), Soil Adjusted Vegetation Index (SAVI), as well as Green Normalized Difference Vegetation Index (GNDVI) (Zeng et al., 2022). These indices are essential for the assessment of the crop vigor, chlorophyll status, nutrient deficiency and water stress, as well as for early deviation disease signs. By obtaining data through regular intervals, UAVs empower growers to track crop development and take regulations before irreversibility of stress manifestations.

Hyperspectral imaging goes even deeper by gathering densities of spectral bands (often numbering in the hundreds), in order to deliver granular biochemical signatures UAV-borne hyperspectral platform is capable of detecting fine differences in the concentration of pigments, nutrient uptake, pathogen activity and physiological stress conditions and helps diagnose diseases and detect pests even at the early stages of occurrence, even though no visible symptoms are present (Bhargava et al., 2024). The resultant high resolution spectral data sets are the basis of site-specific management strategies and predictive development of agronomic management models.

Thermal Imaging is complementary with spectral modalities in being able to record temperature differences in the surface of the canopy. Deviations from leaf or soil temperature can indicate conditions of water stress, water inefficiency of irrigation water, buildup of salinity and drainage anomalies (Gerhards et al., 2016; Simbeye et al., 2023). Thermal land covers are particularly beneficial in areas of orchards, vineyards and high value vegetable systems where uniform irrigation is essential to yield and quality production. When combined with artificial intelligence-driven analytics, such imaging technology shares their insights to optimize crop health monitoring, fertilization timings and resources allocation. Repetitive UAV missions produce time- series data sets since they serve as early warning systems and for long-term farm planning.

3.3. Variable-Rate Spraying and Targeted Agrochemical Delivery

Recent advancement in the field of UAV engineering has led to the development of such platforms (i.e., specialized spraying platforms) that can spray fertilizers, pesticides, herbicides and biological agents with high spatial accuracy. Agricultural Spraying UAVs have controllable droplet applicators, atomizers, smart nozzles, and flow regulators that control spray volume based on parameters such as flight altitude, velocity, canopy density and even environmental parameters (Talaeizadeh et al., 2025). Variable-rate spraying - augmented by onboard sensors and artificial intelligence algorithms ensures that agrochemicals are applied only to areas where they are needed, therefore reducing wastage and limiting the contamination of the environment.

UAVs can cover tougher gradients; flooded fields; terraced tops and risky areas where conventional equipment cannot be safely used. For example, rice paddies, tea plantations and montane orchards benefit from the UAV based spraying especially due to improved accessibility and lower crop damage. Autonomous flight paths while Real-Time Kinematic (RTK) Global Positioning System integration provide centimeter level accuracy for defining accurate boundaries and effective coverage and coverage without overspray or drift (Hafeez et al., 2023).

Contemporary innovations also include UAVs exclusively designed to deliver biological agents in a highly targeted way like beneficial microbes, natural predators, and biopesticides. Such kind of precision minimizes the need for chemical additives and fits in sustainable and organic farming concepts (Mgendi, 2024). In large-scale commercial operations, synchronized fleets of UAVs enable large acreages to be dealt with in a short period of time with high accuracy, which enhances both the temporal responsiveness and efficacy of crop protection interventions.

4. SENSOR NETWORKS AND DATA INTELLIGENCE FOR AUTOMATED FARMING

Sensor networks and data intelligence are the cores of digital basis of automated agriculture activities which enable real time monitoring, predictive analysis and autonomous decision making in different agronomic scenarios. Contemporary sensor ecosystems encompass the Internet of Things (IoT), Artificial Intelligence of Things (AIoT) and advanced communication frameworks to build intelligence from the raw data in the field. These technologies facilitate robotic, aerial and smart farm machinery to work with a very minimal intervention from humans while maintaining high levels of accuracy, efficiency and sustainability (Abdulhussain et al., 2025). As shown in Figure 3. throughput increases from traditional to automated deep learning methods.

Figure 3. Workflow for plant disease detection using image-based approaches. Images from RGB, IR, MS, HS, CF, and 3D sensors undergo pre-processing, annotation, and feature extraction (hand-crafted or deep learning), followed by machine learning for disease identification and model evaluation. Throughput improves from traditional to automated deep learning methods

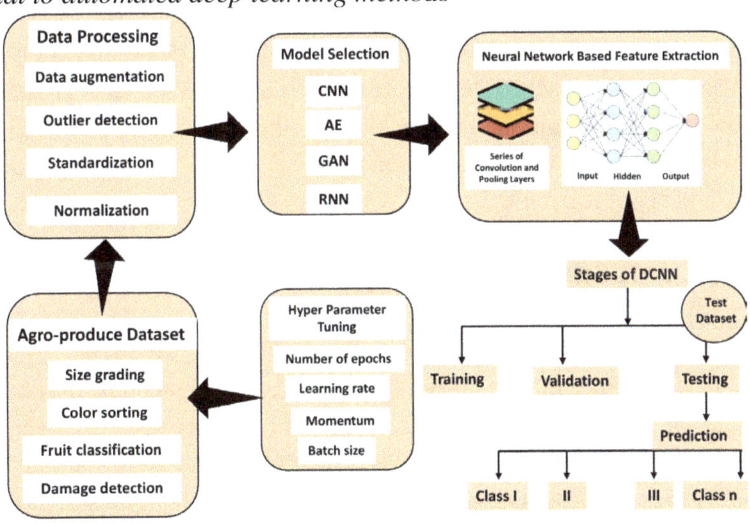

4.1. In-field IoT and AIoT Sensor Ecosystems

In field sensor networks consist of distributed devices continuously monitoring biophysical variables such as soil moisture, nutrient concentrations, canopy temperature, plant water potential as well as atmospheric variables. IoT based systems use wireless communication technologies such as LoRaWAN, NB IOT, Zigbee, 5G, and WiFi Mesh to send large amounts of data from the remote fields to the centralized servers or edge devices (Mekki et al., 2019). AI enabled sensor arrays can use this data to detect the early signs of nutrient imbalances, pest outbreaks, irrigation shortfalls or microenvironmental stressors, in order to enable autonomous robots and drones to respond in a timely way (El khediri et al., 2024). The convergence of robotics and IOT sensors creates a synchronized ecosystem in which machines should adjust the way that they operate depending on the dynamics of the field in real time.

4.2. Real-Time Crop, Soil, and Microclimate Sensing

Real than sensing technologies provide the opportunity for high resolution information on spatial and temporal variability in fields for the purpose of site-specific management and autonomous system response. Soil-sensing platforms measure soils moisture gradients, electrical conductivity, organic matter and pH to get support irrigation and nutrient optimization (Zhang et al., 2024). Crop level sensing using spectral cameras and chlorophyll meters as well as fluorescence sensors provide information on physiological stress, signs of diseases, growing rates and canopy structure. The addition of microclimate sensors further improves the performance by manipulating and recording humidity, wind speed, leaf wetness and temperature profiles across various canopy height (Bhujel et al., 2020). When these sensing modalities are combined with robotics and unmanned aerial vehicles (UAVs), dynamic maps of the field can be created, which are used to target and spray, to weed only in specific areas, to schedule irrigation optimally and to automate harvesting, among other tasks. Continuous feedback loops allow autonomous machineries to modify their actions on the go including aiding in accurately executing the achieved operations and economizing the use of resources.

4.3. Edge Computing and Cloud-Based Analytics

Edge computing is a vital part of automated farming, as it allows data processing closer to the source thereby limiting latency and reliance on high bandwidth communication. Robots, drones, sensor hubs and smart gateways act as edge nodes, taking sensor data and running machine learning models and making local decisions. This capability is essential for time critical functions such as obstacle detection, variable rate spraying and auto-nav (Beniwal & Singhrova, 2022). Cloud platforms balance the advantages of edge computing with big-storage and advanced modelling environments and long-term trend analysis. Big data analytics, digital twins and predictive algorithms running on cloud servers help to find patterns on crop growth, disease spread, soil health and weather condition. Hybrid edge-cloud frameworks keep farm automation systems efficient, robust, and scalable, regardless of the terrain and size of the farm.

4.4. AI-Driven Decision Support Systems (DSS)

AI powered decision support systems pull in information from sensors, records and information over the years, satellite imagery and environmental models to provide evidence-based recommendations for crop and farm management conditions. Through machine learning, deep learning and statistical modelling, DSS platforms produce insights relating to irrigation scheduling, nutrient application, pest control, yield forecasting and crop harvest date (Jabed & Azmi Murad, 2024). These systems increase autonomy by allowing robots and drones to perform actions that follow prediction, rather than reaction from the presence of dosages and observations. Progressive DSS platforms use reinforcement learning and digital twin simulations to test different decision situations and provide the farmers and automatic machines with the ability to execute optimal strategies in real-time. By turning complex data into useful insights, AI-driven DSS enables fully automated intelligent self-adjusting agricultural systems to optimize production while reducing its impact on the environment (Soori et al., 2026).

5. INTEGRATION OF ROBOTICS, DRONES AND SMART MACHINERY

The combination of robotics, unmanned aerial vehicles, and smart machinery has been a revolutionary shift toward invigorating fully automated, interconnected agricultural systems that can operate as a whole ecosystem rather than as a tool of seclusion (Pal et al., 2023). Contemporary agribusinesses are increasingly relying on multi-platform coordination, where robots operate on the ground and in the sky, autonomous tractors are making their place and artificial intelligence-driven management platforms are working together to perform complex processes with increased precision and efficiency. This convergence is driven by the development of sensing, communications networks, edge - cloud computing and digital twin technologies which oversee that there is a continual interplay of data and cooperative operation (Zhang et al., 2023). As agriculture is moving towards extensive automation there are integrated systems that allow continuous and coupled flow production schemes (soil preparation, planting, crop monitoring, spray application, harvesting, etc.) which optimize utilization of resources, improve accuracy of decision-making processes and save labor dependency.

5.1. Robotic–Drone Coordination for Full-Field Coverage

Robotic diviner coordination is possible because it allows for both broad monitoring of the fields and performing the tasks, by taking advantage of the complementing abilities of the terrestrial and the aerial platform. Drones provide quick and high-resolution aerial imagery that is used to delineate variations, stress areas, weeds, and pest outbreaks in the crop in broad areas (Behera, 2025). These insights are then passed on to the robots on the ground, who then conduct targeted interventions such as precision weeding, targeted spraying or site-specific fertilization. Coordination is provided by the same shared data lakes, AI-store mission planning software, protocols for real-time communication protocols (such as 5G, Wi-Fi mesh, and ultra-wideband positioning technology) to enable autonomous fleets to perform complex missions - such as in parallel, mapping fields, navigating in and around fields, and treating crops (such as simultaneously treating and accessing the same field); while minimizing overlaps, reducing energy consumption, and improving the operational speed (Rovira-Sugranes et al., 2022). The result is a co-operative robotic ecosystem that has superior levels of efficiency and accuracy compared with single platform systems.

5.2. Digital Twins for Simulation and Predictive Agriculture

Digital twin technology has become a central force behind the integrated farm automation that gives access to virtual versions of fields, machinery, and crops, as well as environmental conditions on district-wide scales. These virtual environments mimic real-time workings of agricultural systems and allow the digital testing of scenarios and in turn optimize the workflow and forecast the crops performance even before implementing them in real world (Purcell & Neubauer, 2023).). In the areas of robotics and drone operations, digital twins enable the best possible use of routes, help avoid collisions, and optimize the use of energy, payload and mission duration. They further combine meteorological models, soil data, historical proceedings and real time sensors data streams to provide predictive information related to disease breakouts, irrigation needs, biomass build up or pest dynamics. By matching real life operations with simulation, digital twins help create a basis of proactive and precision-based decision making, aiding in greatly improving the reliability and safety of autonomous farm equipment in the field (Sharafat et al., 2025).

5.3. Autonomous Tractors, Sprayers, and Harvesters

Autonomous tractors, sprayers and harvesters are the backbone of large-scale mechanized automation doing high power field operations with limited human supervision. Equipped with multi sensor navigation systems like RTK-GPS, LiDAR and AI based perception modules, these machines offer accuracy levels of centimeter range over different terrain and climatic conditions (Zhuang et al., 2023). Autonomous tractors for soil preparation, planting and towing tools, automated sprayers for preventing applications of pesticides, herbicides and fertilizers based on variable rates in the field based on real time information. Autonomous harvesters use machine vision to identify the plants reach maturation and to align the cutting mechanisms and optimize the throughput. Their integration with drones and ground robots can guarantee synchronized flows of work - such as drone observation of the moisture in the grain and adjustment of harvesting mechanisms of the workers or automatic robots for transporting the collected produce (Islam et al., 2022). These intelligent machines significantly improve high productivity levels, reduce wastage of chemicals, ensure uniform field operations, especially in regions of labor shortage.

5.4. Interoperability and Farm Management Platforms

Interoperability is essential to fully integrated agricultural automation which ensures communication of heterogeneous machines, sensors and software systems. Farm management platforms provide digital platforms that provide consolidated dashboards where data from drones, robots, autonomous tractors, IoT sensors, and weather stations come together into a digital farm management platform (Reinking et al., 2020). These platforms use standardized data formats and open API and communication protocols like ISOBUS, MQTT, and OPC- encara as well as OPC UA to achieve interoperability of the brands and technologies involved. Advanced systems have incorporated AI-driven analytics, work flow automation, algorithms for resource optimization, as well as automated scheduling of tasks, and that foster farms to become integrated digital ecosystems (Domínguez-Bolaño et al., 2022). Interoperability also promotes traceability, scalability and long-term maintainability where agronomists can add new robotic tools or sensor systems without having to revamp their digital infrastructure. Ultimately, farm-management platforms are the central nervous system of automated agriculture, driving the violent coordination of machinery and analyzing data and making decision operation.

6. APPLICATIONS ACROSS MAJOR AGRICULTURAL SECTORS

The use of robotics and unmanned flying devices across the largest agricultural industries has transformed crop, orchard, greenhouse and livestock management with unprecedented gains in precision, productivity and sustainability (Chouhan et al., 2025). Automation makes it easy to constantly monitor, by data, agricultural operations that in the past have been dominated by labor intensity and high variability. Its adoption in different sectors, from both staple field crops to protected horticulture and livestock, attests to the low-cost representativeness and adaptability of human-intelligent machines in different environmental circumstances, crop architectures, and fields of operation. Within the framework of an international agricultural landscape facing challenges in climate change, labor scarcity and the need for sustainable intensification, robotics and drone technology work as transformative tools that allow the creation of highly efficient farming systems that work with the specific called for on different crops and livestock (Prajapati et al., 2024).

6.1. Field Crops: Automation Pathways

Automating major field crops saw a rapid growth, from the use of robots, autonomous tractors and drones to manage precise management of wide expanses of arable land. Autonomous planters that use the principles of RTK-controlled navigation have good uniformity in terms of seed placement thus optimizing plant density and resource allocation (Pareek et al., 2025). Drones equipped with multispectral sensors outline spatial heterogeneity intrepidity, the nutrients deficit stress, etc. Site specific interventions. Robotic sprayers with real time imaging using AI-based weed recognition that can selectively spray the weeds reducing herbicide use and increased environmental sustainability. Harvest Automation Systems, including self-driving combinations and grain carts synchronize operations with machine-to-machine communication allowing for an efficient harvest with minimal loss of throughput (Ehrampoosh et al., 2025). Moreover, advanced analytics generated through drone and sensor data provide predictive modelling of yield, scheduling of irrigation application and forecasting of diseases.

6.2. Horticulture and Orchards: Precision Picking and Canopy Mapping

Horticultural systems including orchard and vineyards are excellent examples of such demanding applications due to their unique characteristics such as irregular tree architecture, spatial differences of fruit positions, and the inherent requirement of the application for delicacy (Freitas & Silva, 2022). The solutions to robotics deployed within application areas leverage advanced computer vision, three-dimensional modelling and dexterous end-effector of the robot to achieve precision harvesting. Fruit-picking robots equipped with soft grippers or ones that use suction-based technology identify ripe fruit using color, size, firmness, and spectral signatures and pick only ripe fruit without any damage. Drones have an instrumental role in orchard canopy mapping by using LiDAR & multi spectra imaging to create exactly mapped tree structure images, leaf density and crop loads distribution (Popescu et al., 2023). These insights are responsible for precision pruning; focused agrochemical use and automatic thinning.

6.3. Protected Agriculture: Greenhouse Robots and Environmental Automation

Environments that are controlled in some sense such as greenhouses, vertical farms, and polyhouses are an ideal setting for robotics, thanks to how the parameters of the environment are precisely controlled, as well as how organized the layouts are. In agricultural greenhouse environments, robots have cameras, thermographic sensors and artificial intelligence algorithms that monitor the growth of plants, identify signs of disease, and perform a variety of other tasks such as pruning, pollination and nutrient delivery (Cañadas-Aránega et al., 2024). Mobile robots use visual simultaneous localization and mapping (SLAM) techniques or rail guided navigation systems to inspect plants at different phases of growth to continuously gather data on growth rates and variation in microclimatic conditions. Environmental Automation Systems Putting all the sensors together, such as CO2, Humidity, pH, Electrical Conductivity, Temperature sensors etc., it becomes possible to optimize the use of resources, such as Irrigation, Ventilation, Lighting etc. (Narayana et al., 2024). Within vertical farms, robotic arms are used for seeding, transplanting, and harvesting with high preciseness, and thus, all operations can be carried out with much higher productivity levels, without requiring much human labor, and that allows for maintaining the uniformity of environmental control.

6.4. Livestock Automation: Robotic Feeding, Health Monitoring, and Drone Surveillance

Livestock systems were also increasingly incorporating robotics and drone technology to improve animal welfare, optimize feeding regimens, and also supplement surveillance on the farm. Robotic feeders cook exact amounts of feed primarily based on the nutrient needs of individual animals and reduce waste and encourage consistent growth. Automated milking systems use sensors and machine vision for monitoring the general health of cows and milk from animals, animal behavior, etc. Automatic systems help detect the early stage of mastitis or metabolic disorders in cows (Cogato et al., 2021). Wearable sensors and smart collars monitor the movements of the animals, the rumination patterns, the temperature, and the physiologic indicators, producing health warnings based on the results by artificial intelligence-powered analytics (Cogato et al., 2021). Drones complement these systems by performing aerial surveillance of grazing fields by checking the condition of pastures, tracking the movement of herds, and identifying the location of distressed and wounded animals. Overall, livestock automation improves productivity, as well as decreasing the human labor needed and increasing animal welfare, through constant monitoring and targeted interventions.

Table 1. Overview of robotics, drones, and automation technologies in agriculture

Technology	Primary Applications	Key Features	Advantages	Challenges	References
Ground Robots (Field & Harvesting)	Planting, seeding, soil prep, harvesting, weeding	Sensors, actuators, AI, GPS/RTK, LiDAR	Labor saving, precision, continuous operation	Terrain complexity, battery life, cost	(Saggi & Jain, 2022)
Autonomous Tractors & Sprayers	Tillage, fertilization, pesticide application	RTK GPS, AI navigation, variable-rate systems	High efficiency, reduced chemical use	High capital cost, maintenance, regulatory approvals	(Bechar & Vigneault, 2017)
Aerial Drones (Fixed-wing, Rotary, Hybrid VTOL)	Crop monitoring, spraying, pollination, reseeding	Multispectral & thermal imaging, GPS, AI flight control	Rapid surveillance, targeted application, accessibility	Flight time limitations, regulatory restrictions	(Tsouros et al., 2019)
Sensor Networks & IoT/AIoT Platforms	Soil, crop, microclimate monitoring	Soil moisture, nutrient sensors, edge computing, cloud integration	Real-time data, precision interventions	Data security, connectivity, integration complexity	(Rajak et al., 2023)
Digital Twin & Decision Support Systems	Predictive modeling, simulation, autonomous decision-making	AI algorithms, cloud computing, field simulation	Optimized resource use, proactive management	High computational requirements, data accuracy	(Kritzinger et al., 2018)
Livestock Robotics	Feeding, health monitoring, milking, drone surveillance	Smart collars, sensors, robotic feeders	Animal welfare, labor reduction, real-time monitoring	High cost, adoption resistance, sensor reliability	(Tullo et al., 2019)

7. CHALLENGES AND FUTURE DIRECTION

Agricultural automation is at a critical point today, where technological advancements happen very quickly but at the same time, there are great barriers preventing wider implementation. Persistent technical limitations, such as inadequate battery life in terms of endurance and reduced reliability of sensors in extreme environmental conditions combined with difficulty of navigating through heterogenous terrains, always affect the stability and scale of operations. Concurrently, cybersecurity vulnerabilities, lack of clarity in the ownership of data as well as regulatory issues (especially for unmanned aerial vehicles and autonomous machinery) introduce further institutional and governance issues. Human capital constraints are also as predominant; many smallholder farmers in low-income regions, for example, commonly face constraints in accessing credit, having the digital infrastructure, or being technically trained. Sociocultural and ethical aspects also influence adoption processes because concerns about the loss of jobs, lack of appropriate technical skills, and/or the breakdown of traditional agrarian societies may create resistance towards the more advanced integration of technologies. Despite these impediments, the future course of agricultural automation holds a great deal of opportunity for sustainable intensification and climate-smart agricultural. Emerging technologies such as nanosensors, compliant (soft) robotics, artificial intelligence (AI)-based IoT, smart fields and cloud-based robotic platforms are expected to contribute to better precision, more flexibility and energy efficiency. The integration of collaborative robot - drone ecosystems and fully automated field operations combined with digital - twin based predictive modelling frameworks have transformative potentials for the creation of continuously optimized self-regulating production systems. Realizing this vision requires the provision of coherent policy support, harmonization of international standards on the issue and targeted capacity building programs and inclusive financing mechanisms to enable equitable deployment to large scale and smallholder farming systems.

8. CONCLUSION

Agricultural robotics and drone technologies are intrinsically changing the agricultural system around the use of labor-intensive and reactive farming applications and replacing them with precision-oriented, data-driven, and autonomous production models. The integration of ground robots, unmanned aerial vehicles, sensor networks and intelligent machinery, in addition to AI-based analytics, has created an entire technological ecosystem that can optimize the entire continuum of crop and livestock management. Automated planting, weeding, harvesting, canopy mapping, targeted spraying and microclimate monitoring have a sum total of offering greater resource-use efficiency, agrochemical dependency, carbon footprints, and overall system resilience. These advancements are further reinforced with digital twins, cloud robotics, Artificial Intelligence of things and coordinated robot - drone workflows allowing predictive, scenario-based and increasingly autonomous farm management. Nevertheless, the path to the large-scale adoption is still held back by technical challenges, cybersecurity threats, capital intensive investments, regulatory uncertainty and socio-ethical considerations especially in developing economies. Overcomes these barriers to global increases in the safe, inclusive and responsible adoption of automation, including strategic investments in enabling policy, technical capacity development, equitable financing and globally harmonized standards. Looking toward the future, next multigeneration innovations such as soft robotics, nanosensors arrays, next-generation autonomous intelligence and multi-robot collaborative ecosystems are likely to fundamentally improve the elements of precision, adaptability and sustainability. Ultimately, agricultural automation is not only a technology evolution but is considered a system change that is necessary to ensure global food security and environmental stewardship, combined with providing resiliency for farmers during the terrible challenges of the 21st century.

REFERENCES

Abbass, H., & Mostaghim, S. (2025). The road forward with swarm systems. *Philosophical Transactions. Series A, Mathematical, Physical, and Engineering Sciences*, *383*(2289), 20240135. DOI: 10.1098/rsta.2024.0135 PMID: 39880030

Abdulhussain, S. H., Mahmmod, B. M., Alwhelat, A., Shehada, D., Shihab, Z. I., Mohammed, H. J., Abdulameer, T. H., Alsabah, M., Fadel, M. H., Ali, S. K., Abbood, G. H., Asker, Z. A., & Hussain, A. (2025). A Comprehensive Review of Sensor Technologies in IoT: Technical Aspects, Challenges, and Future Directions. *Computers*, *14*(8), 342.

Ahmed, A., Saleem, S. R., Tahir, M. N., Manzoor, S. H., Zaman, Q., Zhang, Z., & Ahmed, R. (2025). Design and development of industrial prototype of spot specific orchard sprayer using ultrasonic sensing system and advanced pressure control mechanism. *Computers and Electronics in Agriculture*, *237*, 110690. 10.1016/j.compag.2025.110690

Albiero, D., Pontin Garcia, A., Kiyoshi Umezu, C., & Leme de Paulo, R. (2022). Swarm robots in mechanized agricultural operations: A review about challenges for research. *Computers and Electronics in Agriculture*, *193*, 106608. 10.1016/j.compag.2021.106608

Awan, A. G., & Yaqoob, R. (2023). Economic value of introducing technology to improve productivity: An ARDL approach. *Innovation and Green Development*, *2*(3), 100069. 10.1016/j.igd.2023.100069

Barbosa Júnior, M. R., Santos, R. G. D., Sales, L. A., & Oliveira, L. P. (2024). Advancements in Agricultural Ground Robots for Specialty Crops: An Overview of Innovations, Challenges, and Prospects. *Plants*, *13*(23). Advance online publication. DOI: 10.3390/plants13233372 PMID: 39683165

Bechar, A., & Vigneault, C. (2017). Agricultural robots for field operations. Part 2: Operations and systems. *Biosystems Engineering*, *153*, 110–128. 10.1016/j.biosystemseng.2016.11.004

Behera, S. (2025). *Drones and Aerial Imaging for Crop Health and Soil Analysis*. In (pp. 64–88).

Beniwal, G., & Singhrova, A. (2022). A systematic literature review on IoT gateways. *Journal of King Saud University. Computer and Information Sciences*, *34*(10, Part B), 9541–9563. 10.1016/j.jksuci.2021.11.007

Bhargava, A., Sachdeva, A., Sharma, K., Alsharif, M. H., Uthansakul, P., & Uthansakul, M. (2024). Hyperspectral imaging and its applications: A review. *Heliyon*, *10*(12), e33208. 10.1016/j.heliyon.2024.e33208 PMID: 39021975

Bhujel, A., Basak, J., Khan, F., & Kim, H. (2020). Sensor Systems for Greenhouse Microclimate Monitoring and Control: A Review. *Journal of Biosystems Engineering*, *45*. Advance online publication. DOI: 10.1007/s42853-020-00075-6

Cañadas-Aránega, F., Moreno, J. C., Blanco-Claraco, J. L., Giménez, A., Rodríguez, F., & Sánchez-Hermosilla, J. (2024). Autonomous collaborative mobile robot for greenhouses: Design, development, and validation tests. *Smart Agricultural Technology*, *9*, 100606. 10.1016/j.atech.2024.100606

Cheng, H.-C., Shu, M.-H., & Huang, J.-C. (2024). Economic strategies for efficient use of natural resources: The impact of carbon taxation and fiscal policy. *Resources Policy*, *92*, 104927. 10.1016/j.resourpol.2024.104927

Chouhan, S. S., Patel, R. K., Singh, U. P., & Tejani, G. G. (2025). Integrating drone in Agriculture: Addressing technology, challenges, solutions, and applications to drive economic growth. *Remote Sensing Applications: Society and Environment*, *38*, 101576. 10.1016/j.rsase.2025.101576

Cogato, A., Brščić, M., Guo, H., Marinello, F., & Pezzuolo, A. (2021). Challenges and Tendencies of Automatic Milking Systems (AMS): A 20-Years Systematic Review of Literature and Patents. *Animals (Basel)*, *11*(2). Advance online publication. DOI: 10.3390/ani11020356 PMID: 33572673

Coleman, S., Kerr, D., & Zhang, Y. (2022). Image Sensing and Processing with Convolutional Neural Networks. *Sensors (Basel)*, *22*(10). Advance online publication. DOI: 10.3390/s22103612 PMID: 35632021

Csikós, N., & Tóth, G. (2023). Concepts of agricultural marginal lands and their utilisation: A review. *Agricultural Systems*, *204*, 103560. 10.1016/j.agsy.2022.103560

Diao, Z., Chen, L., Yang, Y., Liu, Y., Yan, J., He, S., & Zhang, B. (2025). Localization technologies for smart agriculture and precision farming: A review. *Computers and Electronics in Agriculture*, *236*, 110464. 10.1016/j.compag.2025.110464

Domínguez-Bolaño, T., Campos, O., Barral, V., Escudero, C. J., & García-Naya, J. A. (2022). An overview of IoT architectures, technologies, and existing open-source projects. *Internet of Things : Engineering Cyber Physical Human Systems*, *20*, 100626. 10.1016/j.iot.2022.100626

Droukas, L., Doulgeri, Z., Tsakiridis, N. L., Triantafyllou, D., Kleitsiotis, I., Mariolis, I., Giakoumis, D., Tzovaras, D., Kateris, D., & Bochtis, D. (2023). A Survey of Robotic Harvesting Systems and Enabling Technologies. *Journal of Intelligent & Robotic Systems, 107*(2), 21. DOI: 10.1007/s10846-022-01793-z PMID: 36721646

Du, Z., Yang, L., Zhang, D., Cui, T., He, X., Xiao, T., Xie, C., & Li, H. (2022). Corn variable-rate seeding decision based on gradient boosting decision tree model. *Computers and Electronics in Agriculture, 198*, 107025. 10.1016/j.compag.2022.107025

Ehrampoosh, A., Hettiarachchi, P., Koirala, A., Hassan, J., Islam, N., Ray, B., Nabi, M. N., Tolba, M., Mazid, A. M., Xu, C.-Y., Ashwath, N., Dzitac, P., & Moore, S. (2025). Intelligent weed management using aerial image processing and precision herbicide spraying: An overview. *Crop Protection (Guildford, Surrey), 194*, 107206. 10.1016/j.cropro.2025.107206

El khediri, S., Benfradj, A., Thaljaoui, A., Moulahi, T., Ullah Khan, R., Alabdulatif, A., & Lorenz, P. (2024). Integration of artificial intelligence (AI) with sensor networks: Trends, challenges, and future directions. *Journal of King Saud University - Computer and Information Sciences, 36*(1), 101892. 10.1016/j.jksuci.2023.101892

Elhamali, S., Musbah, H., Zawi, L., Shuwehdi, A., Faris, H., & Mahdawe, A. (2025). Artificial intelligence meets laser technology: A review of recent advances. *Results in Surfaces and Interfaces, 19*, 100484. 10.1016/j.rsurfi.2025.100484

Freitas, J., & Silva, P. (2022). Sustainable Agricultural Systems for Fruit Orchards: The Influence of Plant Growth Promoting Bacteria on the Soil Biodiversity and Nutrient Management. *Sustainability, 14*(21), 13952.

Gerhards, M., Rock, G., Schlerf, M., & Udelhoven, T. (2016). Water stress detection in potato plants using leaf temperature, emissivity, and reflectance. *International Journal of Applied Earth Observation and Geoinformation, 53*, 27–39. 10.1016/j.jag.2016.08.004

Groumpos, P. P. (2021). A Critical Historical and Scientific Overview of all Industrial Revolutions. *IFAC-PapersOnLine, 54*(13), 464–471. 10.1016/j.ifacol.2021.10.492 PMID: 38620687

Hafeez, A., Husain, M. A., Singh, S. P., Chauhan, A., Khan, M. T., Kumar, N., Chauhan, A., & Soni, S. K. (2023). Implementation of drone technology for farm monitoring & pesticide spraying: A review. *Information Processing in Agriculture, 10*(2), 192–203. 10.1016/j.inpa.2022.02.002

Hernández, H. A., Mondragón, I. F., González, S. R., & Pedraza, L. F. (2025). Reconfigurable agricultural robotics: Control strategies, communication, and applications. *Computers and Electronics in Agriculture*, *234*, 110161. 10.1016/j.compag.2025.110161

Heydari, M. (2024). Cultivating sustainable global food supply chains: A multifaceted approach to mitigating food loss and waste for climate resilience. *Journal of Cleaner Production*, *442*, 141037. 10.1016/j.jclepro.2024.141037

Imran, & Li, J. (2025). UAV Design for Agricultural Applications. In Imran & J. Li (Eds.), *UAV Aerodynamics and Crop Interaction: Revolutionizing Modern Agriculture with Drone* (pp. 169–202). Springer Nature Singapore. DOI: 10.1007/978-981-96-8402-1_6

Islam, S. F., Uddin, M. S., & Bansal, J. (2022). *Harvesting Robots for Smart Agriculture.* In (pp. 1–13). DOI: 10.1007/978-981-16-9991-7_1

Jabed, M. A., & Azmi Murad, M. A. (2024). Crop yield prediction in agriculture: A comprehensive review of machine learning and deep learning approaches, with insights for future research and sustainability. *Heliyon*, *10*(24), e40836. 10.1016/j.heliyon.2024.e40836 PMID: 39720079

Jan, B., Mohapatra, S., Samal, M., Choudhary, B., & Basist, P. (2025). Role of artificial intelligence in food and agriculture industries: An updated review. *Nutrire*, *50*(2), 96. DOI: 10.1186/s41110-025-00392-w

Jin, S., Meng, X., Dardanelli, G., & Zhu, Y. (2024). Multi-Global Navigation Satellite System for Earth Observation: Recent Developments and New Progress. *Remote Sensing*, *16*(24), 4800.

Khadatkar, A., Sujit, P. B., Agrawal, R., Vishwanath, K., Sawant, C. P., Magar, A. P., & Chaudhary, V. P. (2025). WeeRo: Design, development and application of a remotely controlled robotic weeder for mechanical weeding in row crops for sustainable crop production. *Results in Engineering*, *26*, 105202. 10.1016/j.rineng.2025.105202

Kritzinger, W., Karner, M., Traar, G., Henjes, J., & Sihn, W. (2018). Digital Twin in manufacturing: A categorical literature review and classification. *IFAC-PapersOnLine*, *51*(11), 1016–1022. 10.1016/j.ifacol.2018.08.474

Kure, S. (2022). Sensors, Actuators and Microcontrollers. In (pp. 9).

Li, M. (2025). Beyond Conventional Drones: A Review of Unconventional Rotary-Wing UAV Design. *Drones (Basel)*, *9*(5), 323.

Liu, J., Fang, Y., Wang, G., Liu, B., & Wang, R. (2023). The aging of farmers and its challenges for labor-intensive agriculture in China: A perspective on farmland transfer plans for farmers' retirement. *Journal of Rural Studies*, *100*, 103013. 10.1016/j.jrurstud.2023.103013

Mamabolo, E., Mashala, M. J., Mugari, E., Mogale, T. E., Mathebula, N., Mabitsela, K., & Ayisi, K. K. (2025). Application of precision agriculture technologies for crop protection and soil health. *Smart Agricultural Technology*, *12*, 101270. 10.1016/j.atech.2025.101270

Maurya, D., Maurya, S. K., Kumar, M., Chaubey, C., Gupta, D., Patel, K., Mehta, A., & Yadav, R. (2024). A Review on Precision Agriculture: An Evolution and Prospect for the Future. *International Journal of Plant and Soil Science*, *36*, 363–374. DOI: 10.9734/IJPSS/2024/v36i54534

Mekki, K., Bajic, E., Chaxel, F., & Meyer, F. (2019). A comparative study of LPWAN technologies for large-scale IoT deployment. *ICT Express*, *5*(1), 1–7. 10.1016/j.icte.2017.12.005

Mgendi, G. (2024). Unlocking the potential of precision agriculture for sustainable farming. *Discover Agriculture*, *2*(1), 87. DOI: 10.1007/s44279-024-00078-3

Mohsan, S. A. H., Othman, N. Q. H., Li, Y., Alsharif, M. H., & Khan, M. A. (2023). Unmanned aerial vehicles (UAVs): Practical aspects, applications, open challenges, security issues, and future trends. *Intelligent Service Robotics*, *16*(1), 109–137. DOI: 10.1007/s11370-022-00452-4 PMID: 36687780

Naim, M., Rizzo, D., Sauvée, L., & Medici, M. (2025). Advancing agroecology and sustainability with agricultural robots at field level: A scoping review. *Computers and Electronics in Agriculture*, *237*, 110650. 10.1016/j.compag.2025.110650

Narayana, T. L., Venkatesh, C., & Kiran, A. J, C. B., Kumar, A., Khan, S. B., Al-musharraf, A., & Quasim, M. T. (2024). Advances in real time smart monitoring of environmental parameters using IoT and sensors. *Heliyon, 10*(7), e28195. 10.1016/j.heliyon.2024.e28195

Nitti, A., de Tullio, M. D., Federico, I., & Carbone, G. (2025). A collective intelligence model for swarm robotics applications. *Nature Communications*, *16*(1), 6572. DOI: 10.1038/s41467-025-61985-7 PMID: 40670391

Osman, M., Xia, Y., Mahdi, M., & Ahmed, A. (2025). Hybrid VTOL UAV technologies: Efficiency, customization, and sector-specific applications. *Alexandria Engineering Journal*, *120*, 13–49. 10.1016/j.aej.2024.12.087

Pal, O., Shovon, M., Ph, D. M., & Shin, J. (2023). *A Comprehensive Review of AI-enabled Unmanned Aerial Vehicle: Trends, Vision, and Challenges*. /arXiv.2310.16360DOI: 10.48550

Pareek, C. M., Tewari, V. K., & Nare, B. (2025). A mechatronic seed metering control system for improving sowing uniformity of planters. *Journal of Engineering Research, 13*(2), 808–819. 10.1016/j.jer.2023.10.041

Popescu, D., Ichim, L., & Stoican, F. (2023). Orchard monitoring based on unmanned aerial vehicles and image processing by artificial neural networks: A systematic review. *Front Plant Sci, 14*, 1237695. DOI: 10.3389/fpls.2023.1237695 PMID: 38089793

Prajapati, H., Yadav, K., Hanamasagar, Y., Kumar, M., Khan, T., Belagalla, N., Thomas, V., Jabeen, A. G. G., & Malathi, G. (2024). Impact of Climate Change on Global Agriculture: Challenges and Adaptation. *International Journal of Environment and Climate Change, 14*, 372–379. DOI: 10.9734/ijecc/2024/v14i44123

Purcell, W., & Neubauer, T. (2023). Digital Twins in Agriculture: A State-of-the-art review. *Smart Agricultural Technology, 3*, 100094. 10.1016/j.atech.2022.100094

Pustavrh, J., Hočevar, M., Podržaj, P., Trajkovski, A., & Majdic, F. (2023). *Comparative study of a hydraulic, pneumatic and electric linear actuator system*. DOI: 10.21203/rs.3.rs-3184855/v1

Rajak, P., Ganguly, A., Adhikary, S., & Bhattacharya, S. (2023). Internet of Things and smart sensors in agriculture: Scopes and challenges. *Journal of Agriculture and Food Research, 14*, 100776. 10.1016/j.jafr.2023.100776

Reinking, J., Arnold, V., & Sutton, S. G. (2020). Synthesizing enterprise data through digital dashboards to strategically align performance: Why do operational managers use dashboards? *International Journal of Accounting Information Systems, 37*, 100452. 10.1016/j.accinf.2020.100452

Rovira-Sugranes, A., Razi, A., Afghah, F., & Chakareski, J. (2022). A review of AI-enabled routing protocols for UAV networks: Trends, challenges, and future outlook. *Ad Hoc Networks, 130*, 102790. 10.1016/j.adhoc.2022.102790

Saggi, M. K., & Jain, S. (2022). A Survey Towards Decision Support System on Smart Irrigation Scheduling Using Machine Learning approaches. *Archives of Computational Methods in Engineering, 29*(6), 4455–4478. DOI: 10.1007/s11831-022-09746-3 PMID: 35573028

Shahria, M. T., Sunny, M. S. H., Zarif, M. I. I., Ghommam, J., Ahamed, S. I., & Rahman, M. H. (2022). A Comprehensive Review of Vision-Based Robotic Applications: Current State, Components, Approaches, Barriers, and Potential Solutions. *Robotics (Basel, Switzerland)*, *11*(6), 139.

Sharafat, M. D. S., Kabya, N. D., Emu, R. I., Ahmed, M. U., Onik, J. C., Islam, M. A., & Khan, R. (2025). An IoT-enabled AI system for real-time crop prediction using soil and weather data in precision agriculture. *Smart Agricultural Technology*, *12*, 101263. 10.1016/j.atech.2025.101263

Shi, P., Li, J., & Zhang, Y. (2023). A fast LiDAR place recognition and localization method by fusing local and global search. *ISPRS Journal of Photogrammetry and Remote Sensing*, *202*, 637–651. 10.1016/j.isprsjprs.2023.07.008

Simbeye, D. S., Mkiramweni, M. E., Karaman, B., & Taskin, S. (2023). Plant water stress monitoring and control system. *Smart Agricultural Technology*, *3*, 100066. 10.1016/j.atech.2022.100066

Sirish Kumar, P., & Srilatha Indira Dutt, V. B. S. (2020). The global positioning system: Popular accuracy measures. *Materials Today: Proceedings*, *33*, 4797–4801. 10.1016/j.matpr.2020.08.380

Soori, M., Jough, F. K. G., Dastres, R., & Arezoo, B. (2026). AI-based decision support systems in Industry 4.0, a review. *Journal of Economy and Technology*, *4*, 206–225. 10.1016/j.ject.2024.08.005

Talaeizadeh, A., Sharifi, I., Alasty, A., & Ghatrehsamani, S. (2025). Agricultural spraying drones: A comprehensive review. *Smart Agricultural Technology*, *12*, 101519. 10.1016/j.atech.2025.101519

Tang, Y., Chen, M., Wang, C., Luo, L., Li, J., Lian, G., & Zou, X. (2020). Recognition and Localization Methods for Vision-Based Fruit Picking Robots: A Review. *Front Plant Sci*, *11*, 510. DOI: 10.3389/fpls.2020.00510 PMID: 32508853

Theodorou, C., Velisavljevic, V., Dyo, V., & Nonyelu, F. (2022). Visual SLAM algorithms and their application for AR, mapping, localization and wayfinding. *Array (New York, N.Y.)*, *15*, 100222. 10.1016/j.array.2022.100222

Tituaña, L., Gholami, A., He, Z., Xu, Y., Karkee, M., & Ehsani, R. (2024). A small autonomous field robot for strawberry harvesting. *Smart Agricultural Technology*, *8*, 100454. 10.1016/j.atech.2024.100454

Tsouros, D., Bibi, S., & Sarigiannidis, P. (2019). A Review on UAV-Based Applications for Precision Agriculture. *Information (Basel)*, *10*, 349. DOI: 10.3390/info10110349

Tullo, E., Finzi, A., & Guarino, M. (2019). Review: Environmental impact of livestock farming and Precision Livestock Farming as a mitigation strategy. *The Science of the Total Environment*, *650*, 2751–2760. 10.1016/j.scitotenv.2018.10.018 PMID: 30373053

Wen, T., Li, J.-H., Wang, Q., Gao, Y.-Y., Hao, G.-F., & Song, B.-A. (2023). Thermal imaging: The digital eye facilitates high-throughput phenotyping traits of plant growth and stress responses. *The Science of the Total Environment*, *899*, 165626. 10.1016/j.scitotenv.2023.165626 PMID: 37481085

Xiang, M., Qu, M., Wang, G., Ma, Z., Chen, X., Zhou, Z., Qi, J., Gao, X., Li, H., & Jia, H. (2024). Crop detection technologies, mechanical weeding executive parts and working performance of intelligent mechanical weeding: a review [Review]. *Frontiers in Plant Science, Volume 15 - 2024*. DOI: 10.3389/fpls.2024.1361002

Yan, H.-F., Zhao, Y.-Q., Chan, J. C.-W., Kong, S. G., Ei-Bendary, N., & Reda, M. (2025). Hyperspectral and multispectral image fusion: When model-driven meet data-driven strategies. *Information Fusion*, *116*, 102803. 10.1016/j.inffus.2024.102803

Yang, Q., Du, X., Wang, Z., Meng, Z., Ma, Z., & Zhang, Q. (2023). A review of core agricultural robot technologies for crop productions. *Computers and Electronics in Agriculture*, *206*, 107701. 10.1016/j.compag.2023.107701

Zeng, Y., Hao, D., Huete, A., Dechant, B., Berry, J., Chen, J. M., Joiner, J., Frankenberg, C., Bond-Lamberty, B., Ryu, Y., Xiao, J., Asrar, G. R., & Chen, M. (2022). Optical vegetation indices for monitoring terrestrial ecosystems globally. *Nature Reviews. Earth & Environment*, *3*(7), 477–493. DOI: 10.1038/s43017-022-00298-5

Zhang, L., Du, Q., Lu, L., & Zhang, S. (2023). Overview of the Integration of Communications, Sensing, Computing, and Storage as Enabling Technologies for the Metaverse over 6G Networks. *Electronics (Basel)*, *12*(17), 3651.

Zhang, X., Feng, G., & Sun, X. (2024). Advanced technologies of soil moisture monitoring in precision agriculture: A Review. *Journal of Agriculture and Food Research*, *18*, 101473. 10.1016/j.jafr.2024.101473

Zhang, Z., Luo, J., & Chen, B. (2020). Spatially explicit quantification of total soil erosion by RTK GPS in wind and water eroded croplands. *The Science of the Total Environment*, *702*, 134716. 10.1016/j.scitotenv.2019.134716 PMID: 31726344

Zhuang, Y., Sun, X., Li, Y., Huai, J., Hua, L., Yang, X., Cao, X., Zhang, P., Cao, Y., Qi, L., Yang, J., El-Bendary, N., El-Sheimy, N., Thompson, J., & Chen, R. (2023). Multi-sensor integrated navigation/positioning systems using data fusion: From analytics-based to learning-based approaches. *Information Fusion*, *95*, 62–90. 10.1016/j.inffus.2023.01.025

Chapter 12

Ethical Considerations of Artificial Intelligence and Automation in Agricultural Decision Making

Afshan Ahmad
http://orcid.org/0009-0002-6335-5602
NeuraVue Ltd., Canada

ABSTRACT

The use of AI and automation is changing how farmers make decisions according to their fields and farms. Irrigation timing, outbreak prediction and crop grade are only some of the decisions that AI can help with. These technologies do provide real benefits such as more output, efficiency, resource management and climate adaptation. But they simultaneously pose ethical questions that merit more than a theoretical solution. Who actually owns the massive datasets that are generated on the farm and who benefits? How clear should automated suggestions be for the people who rely on them for money? When algorithms start replacing human judgement, What happens to rural areas?

DOI: 10.4018/979-8-3373-7257-0.ch012

PART A — ETHICAL CONSIDERATIONS OF ARTIFICIAL INTELLIGENCE

1. Introduction

Artificial intelligence (AI) and automation are becoming embedded across agricultural value chains in ways that are increasingly visible to those working on the ground. Satellite-supported crop diagnostics, automated grading systems, irrigation algorithms, and livestock monitoring tools now inform decisions that were once made through direct observation and experience. Their rapid uptake reflects a convergence of pressures, including climate variability, labour shortages, rising input costs, and the growing expectation that productivity gains must be achieved under tightening ecological constraints (FAO, 2023). As these systems move from experimental tools to routine infrastructure, they begin to shape not only outcomes, but the terms on which agricultural decisions are made.

Agricultural knowledge has historically been developed through sustained engagement with land, crops, animals, and seasonal rhythms. When decision authority shifts toward algorithmic systems, that knowledge is not simply replaced; it is reconfigured. In commercial Canadian greenhouse operations, for instance, climate and fertigation platforms now generate automated recommendations that previously depended on growers' close interpretation of plant response over time. In parallel, smallholder farmers in Kenya, India, and Brazil increasingly receive mobile-based advisory alerts linked to AI-driven weather forecasting and crop disease classification (Barrett, 2022; Wright & Chan, 2022). These tools influence planting choices, input use, and risk tolerance, even when the reasoning behind specific recommendations remains only partially visible to their users.

The ethical challenges raised by these developments cannot be resolved through technical performance metrics alone. Questions of who controls agricultural data, whose knowledge is reflected in algorithmic models, how labour roles are reshaped, and how ecological consequences unfold over time emerge gradually, across the lifespan of AI systems. Treating ethics as a checkpoint at the moment of deployment risks obscuring how authority, responsibility, and risk distribution evolve as systems are trained, adapted, and institutionalized.

For this reason, the chapter adopts an ethical lifecycle perspective, illustrated in *Figure 1*, which traces agricultural AI from data collection and model development through human oversight, regulatory response, and longer-term sustainability outcomes. This framing emphasizes continuity rather than intervention, highlighting how ethical considerations surface repeatedly as systems interact with social, institutional, and ecological contexts.

The chapter is organized in two parts. *Part A* examines core ethical considerations of artificial intelligence in agriculture, including data governance, interpretive authority, labour transformation, and trust across diverse production systems. *Part B* turns to artificial intelligence and automation in agricultural decision-making, drawing on comparative case examples from Canada and the Global South to explore how governance choices shape resilience, human agency, and equitable outcomes in practice.

These questions are best understood as part of a *continuous ethical lifecycle*, rather than a one-time evaluation at the moment of deployment. This lifecycle is illustrated in *Figure 1*, which maps the movement from data collection to model development, through human oversight and policy response, and finally to long-term sustainability outcomes.

Figure 1. Ethical lifecycle of agricultural artificial intelligence

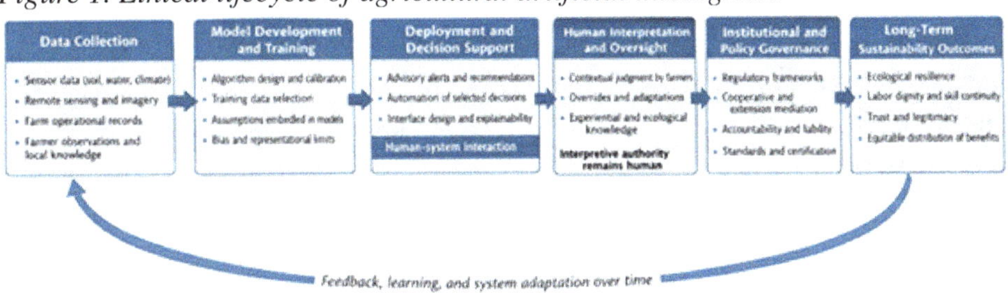

This lifecycle model provides the conceptual foundation for *Part A*, where the focus is on understanding how ethical questions arise, what they look like in practice, and how they can be addressed thoughtfully and collaboratively.

2. Data Ownership and Digital Sovereignty in Agriculture

The deployment of AI depends on continuous streams of agricultural data: remote-sensing imagery, soil sensor measurements, yield histories, supply chain throughput, and livestock behavior patterns. These datasets hold commercial value, ecological insight, and strategic importance. The question of who owns this data, and under what terms it may be accessed or monetized, is therefore central to ethical agricultural AI governance (FAO, 2022).

In many commercial agricultural software platforms, farmers are required to upload operational data that is subsequently stored in proprietary cloud environments. While users typically retain nominal rights to their data, platforms often reserve broad permissions to analyze, aggregate, or sell insights derived from it (Beerbaum

& Otto, 2023). *Scholars have noted that such platform-based arrangements can consolidate informational power upstream, limiting farmers' practical ability to negotiate terms or exit systems once data dependency is established (*Carbonell, 2016*;* Klerkx, Jakku, & Labarthe, 2019*)*. The imbalance of bargaining power makes meaningful consent difficult—particularly in regions where digital infrastructure is scarce and alternatives are limited.

In field deployments across Canadian grain cooperatives and Kenyan smallholder seed networks, concerns over data misuse and external dependency consistently arose. Farmers voiced hesitation not about digital tools themselves, but about losing control over how their knowledge and production patterns would be used downstream—especially if used to inform pricing, land forecasting, or credit decisioning systems. To address this, emerging modes of data governance are emphasized in *Table 1* below:

Table 1. Ethical data governance principles and corresponding practical mechanisms observed in agricultural AI systems

Ethical Principle	Practical Mechanism (Observed in Real Systems)
Data remains locally governed	Cooperative or community-based data custodianship
Consent is revocable and specific	Layered permissions, not blanket service agreements
Value created from data is shared	Revenue-sharing, insight-sharing, or federated analytics
Data processing can occur locally when possible	Edge computing or on-farm storage nodes

These approaches align with international recommendations that emphasize data sovereignty, particularly in the Global South, *where data extraction without reciprocal governance has been linked to contemporary forms of digital and knowledge colonialism (*Couldry & Mejias, 2019*;* Birhane, 2021*;* Wright & Chan, 2022*)*.

The governance relationships linking *technical safeguards, organizational agreements, and national regulation* are depicted in *Figure 2*.

Figure 2. Multi-level governance framework for responsible AI in agriculture

3. Algorithmic Bias, Context, and Fairness

AI models do not operate in abstraction; they reflect the assumptions embedded in their training data. When datasets are drawn from limited geographies or production conditions, model outputs may reproduce systematic advantage for some farming systems while disadvantaging others (Garske et al., 2021).

As illustrated in Figure 3, these biases do not arise solely at the level of model design but emerge through the interaction between algorithmic logic, human interpretation, and the social and ecological contexts in which systems are deployed.

Examples observed across regions include:

- Pest detection tools trained on monocrop imagery misidentifying disease symptoms in intercropped fields.
- Yield prediction models calibrated for high-input fertilizer regimes underestimating regenerative or low-input soil systems.
- Irrigation advisory platforms failing in regions where water access is variable and decision-making is socio-politically negotiated rather than purely agronomic.

This form of error is not merely technical; it redistributes economic and ecological vulnerability.

Smallholder networks in India and Kenya have begun addressing bias by participatory dataset annotation—extension workers and farmer groups classify local crop health images to retrain models. This practice increases both model accuracy and legitimacy, because system reasoning becomes visibly tied to lived field conditions.

Recent work on participatory and community-generated datasets shows that such co-produced data practices improve contextual validity while reducing structural bias in agricultural AI systems (Barrett & Rose, 2022; Birhane et al., 2022*).*

This dynamic—where technical fairness depends on contextual co-interpretation—is represented visually in Figure 3, which depicts ethical layers in agricultural automation as overlapping rather than hierarchical.

Figure 3. Ethical layers in agricultural automation

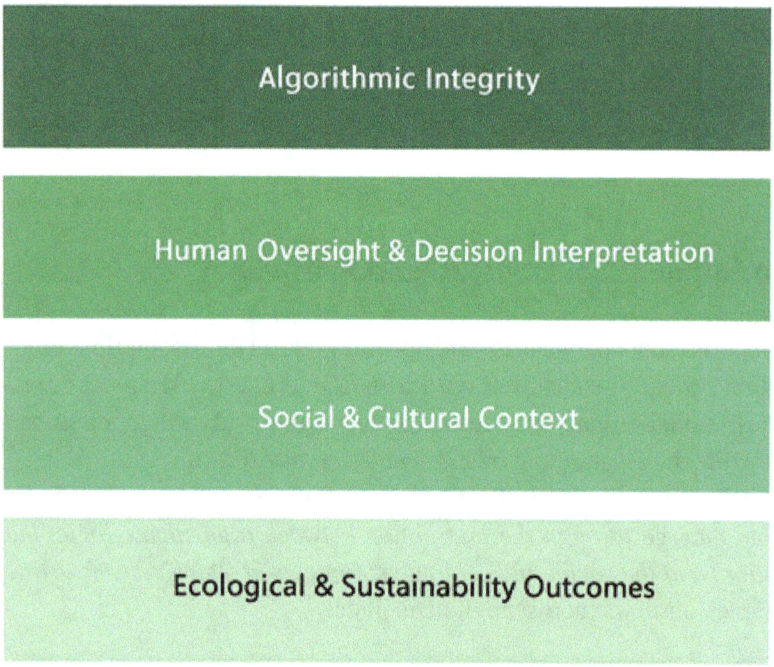

This figure shows how algorithmic integrity, human judgment, socio-cultural context, and ecological responsibility interdepend to shape ethical outcomes.

4. Human Judgment, Interpretive Authority, and Accountability

Even when agricultural systems are automated, accountability does not disappear. Decisions still carry consequences—financial, ecological, and social—that must be carried by someone. The core ethical concern is therefore not whether intelligent systems make recommendations, but who is ultimately responsible when those recommendations produce harm.

In commercial greenhouse operations in Ontario, irrigation-control systems learn environmental response patterns and adjust watering schedules automatically. These systems generally perform well—until a threshold condition shifts (such as a sudden humidity spike after a storm). When this happens, growers report that they must "step in to feel the plants," re-engaging sensory knowledge to re-stabilize the system. Their human judgment is activated precisely at the moment where automated logic loses contextual grounding. This dynamic is neither accidental nor temporary—it is structural (Liang & Fernandez, 2021).

Similarly, in Kenyan smallholder advisory networks, farmers use mobile-based alert systems to plan interventions. Yet, when model outputs conflict with local cues—such as wind direction, soil softness underfoot, insect sound, or communal intuition—farmers routinely choose not to follow the algorithmic recommendation. They do so not because they reject technology, but because risk in agriculture is shared and relational, and decisions are evaluated with regard to harm tolerance and community resilience, not accuracy alone (Mwangi et al., 2023).

As established earlier in Section 4, interpretive authority in agricultural AI systems does not reside solely in model outputs but remains situated with human actors who contextualize, negotiate, and bear responsibility for outcomes. This layered relationship is reflected in the model shown earlier (Figure 3). Algorithmic systems may offer pattern recognition at scale, but interpretive authority—the ability to make meaning of those patterns—remains grounded in human judgment. Ethical design therefore requires systems that anticipate farmer overrides and treat them not as error, but as expertise.

Three forms of responsibility therefore need to be explicitly retained in agricultural AI deployment, as summarized in Table 2 below.

Table 2. Distribution of ethical responsibilities across agricultural AI systems, illustrating how operational, design, and system-level duties are shared among farmers, developers, and institutions

Responsibility	Held By	Description
Operational responsibility	Farmers, growers, extension workers	Deciding whether to act on a recommendation
Design responsibility	Developers, data scientists	Ensuring the model learns appropriately from diverse conditions
System responsibility	Institutions and regulators	Ensuring risk is not unfairly transferred to those with the least resilience

Without maintaining these distinct layers of accountability, the introduction of AI into agriculture risks a subtle but significant transfer of liability: systems make decisions, but farmers absorb losses. This dynamic is already visible in several regions where crop insurance frameworks do not yet recognize algorithmic influence on decision-making (Hess & Tadesse, 2022).

As established earlier in Section 4, preventing this transfer requires treating human judgment as an ethical safeguard rather than a system inefficiency. In practice, this means designing agricultural AI systems not as replacements for expertise, but as partners in distributed decision-making:

- AI models should explain why a recommendation is made—using language or visual cues, not abstract indices.
- Farmers should be able to challenge or reverse system guidance without penalty.
- Institutions should recognize interpretive judgment as a form of responsibility, not deviation.

In practice, this means designing agricultural AI systems where risk is shared rather than shifted, where trust is built through transparency rather than dependency, and where human interpretation remains central rather than residual.

5. Trust, Legitimacy, and Cultural Compatibility in AI-Supported Agriculture

Trust in agricultural AI does not emerge from model accuracy alone. It is built relationally—through communication, shared interpretation, and alignment with the cultural logics of farming communities (Villalba & Alarcón, 2022). In many regions, agriculture is not simply a technical activity but a social and ecological practice shaped by seasonal rhythms, kinship networks, and collective memory. When AI systems enter this landscape, they also enter these meaning systems.

5.1 Canada: Trust Built Through Co-Interpretation

In commercial greenhouse operations in Ontario and British Columbia, growers described trusting AI-based climate control systems only when they could trace how decisions were made. Transparency mattered most during non-routine conditions—cold snaps, heat waves, or disease-pressure surges—when growers needed to verify whether the model's assumptions still matched reality. Systems that provided explanatory feedback loops, such as natural language rationale summaries ("humidity increased due to transpiration spike; adjusting venting"), were considered more legitimate than those that simply provided a numeric directive.

Trust formed over time through repeated co-interpretation, rather than at the moment of installation. Growers described this relationship as one of collaboration rather than delegation, where both system and operator adapted to evolving conditions (Harper et al., 2023).

5.2 India: Contextual Knowledge and Irrigation Risk

In semi-arid regions of Maharashtra and Karnataka, AI-driven irrigation advisories are used alongside local water-sharing norms. Here, irrigation decisions are collectively negotiated, shaped by rainfall expectations, canal availability, and social reciprocity obligations. Farmers routinely adjust or reject AI irrigation alerts when they conflict with community signals such as the sound of groundwater pumps in neighboring fields or cloud movement patterns historically associated with rainfall (Patil & Deshmukh, 2021).

Trust in this context depends less on optimization accuracy and more on whether AI systems acknowledge irrigation as a shared social responsibility rather than an individual decision variable.

5.3 Kenya: Cooperative Interpretation Networks

In Kenyan farmer cooperatives, digital advisory platforms are often interpreted collectively rather than individually. Disease alerts may be reviewed by a cooperative lead, who contextualizes the recommendation based on seed variety, informal agronomic knowledge, and market timing. In this way, algorithmic outputs are filtered through an existing social interpretation layer, where expertise is distributed rather than centralized (Mwangi et al., 2023).

This mediation strengthens trust because the technology is anchored within already legitimate cooperative structures, rather than positioned as an external authority.

5.4 Brazil: Legitimacy and Ecological Worldviews

In agroforestry regions of Pará and Mato Grosso, AI tools are increasingly used to monitor land-use change, disease incidence, and biodiversity indicators. However, systems are accepted only when they align with local ecological worldviews, where forest health is understood as relational—a balance among soil organisms, canopy moisture, and seasonal winds (Junior & Silva, 2022). Tools that frame land management as extractive optimization often face resistance, while those supporting ecological stewardship gain legitimacy.

Here, trust is tied not to technological sophistication, but to whether AI reinforces or disrupts existing ecological ethics.

Across all four contexts—Canada, India, Kenya, and Brazil—trust emerges where human interpretation remains central, as illustrated earlier in Figure 3. Systems that recognize farmers as interpreters rather than endpoints maintain legitimacy; those that obscure their logic risk rejection, even when technically accurate. This underscores that ethical agricultural AI cannot be culturally neutral—it must be situated, mediated, and co-developed rather than simply deployed.

6. Labor and Skill Transformation in AI-Supported Agriculture

The introduction of artificial intelligence and automation into agriculture is not only a technological transition; it is also a redefinition of agricultural labor, skill, and economic agency. While automation is often framed in terms of efficiency—reduced manual effort, improved prediction accuracy, and optimization of inputs—the social and ethical consequences of these changes are unevenly distributed. Impacts differ substantially across production scales, institutional environments, and cultural contexts.

In many Canadian greenhouse and controlled-environment farming operations, climate-control systems, autonomous fertigation units, and digital monitoring platforms have reduced routine manual observation while increasing the importance of sensor interpretation, system calibration, and cross-disciplinary reasoning. Workers who were once valued primarily for experiential pattern recognition—such as detecting subtle plant stress through leaf texture or color—are now expected to translate that knowledge into parameter adjustments and data-informed decisions. Expertise is therefore not eliminated, but reshaped.

Field observations suggest that this transition can lead to skill deepening rather than displacement, but only where institutional support is present. When workers are given time and training to compare their own interpretations with system outputs, automation is experienced as a collaborative tool rather than an external authority (Harper et al., 2023). Where such support is lacking, automation instead produces dependency, with workers following system recommendations without fully understanding the reasoning behind them.

In smallholder contexts in Kenya and India, agricultural skill is rarely individual. Knowledge is embedded within kinship networks, cooperative labor arrangements, and shared seasonal decision-making. AI-based advisory systems enter these environments as an additional interpretive voice rather than a replacement for existing practices. In Indian irrigation communities, for example, decisions about when to release water remain shaped by social reciprocity and negotiated fairness. Farmers routinely adjust or disregard algorithmic schedules when they conflict with communal obligations or locally observed environmental cues (Patil & Deshmukh, 2021; Mwangi et al., 2023).

These dynamics highlight the risk of skill hollowing, which occurs when individuals are expected to act on automated recommendations without understanding their logic. This risk is heightened when AI systems are introduced rapidly, training is minimal, or model reasoning is opaque (Liang & Fernandez, 2021). Under such conditions, authority shifts upward toward platform providers and system designers, while responsibility for outcomes remains with those working on the land.

Preventing skill hollowing requires treating human interpretation as an ethical safeguard rather than a system inefficiency. In practice, this means designing AI deployments that explicitly preserve interpretive authority and support skill continuity.

Table 3. Ethical priorities that safeguard human expertise, shared authority, and ecological knowledge in agricultural AI systems, with corresponding mechanisms

Ethical Priority	Mechanism	Example
Preserve interpretive skill	Co-interpretation sessions; joint alert validation	Growers calibrating sensor thresholds collectively in BC greenhouses
Share decision-making authority	Override pathways recognized as expertise, not error	Cooperative advisory review groups in Kenya
Ensure continuity of indigenous/ ecological knowledge	Local dataset annotation + participatory model retraining	Brazilian agroforestry biodiversity mapping
Support transition rather than displacement	Progressive role evolution, not abrupt automation	Shared diagnostic roles in Canadian greenhouse teams

Table 3 summarizes the ethical priorities and governance mechanisms observed across different agricultural contexts that help safeguard human expertise, distribute decision-making authority, and maintain ecological knowledge.

Where such mechanisms are in place, automation tends to support adaptive capacity and resilience. Where they are absent, ethical risk increases, particularly for workers and communities with limited ability to absorb system failure or market shocks.

Section summary.

Labor transformation in AI-supported agriculture is shaped less by automation itself than by the governance choices surrounding training, transparency, and authority. Preserving human interpretive capacity remains central to ethical deployment, particularly as systems become more deeply embedded in everyday agricultural practice.

7. Institutional Governance, Policy Alignment, and Shared Responsibility

If AI and automation are to support sustainable agricultural futures, governance structures must ensure that benefits are distributed fairly and that risks are not concentrated among those least able to bear them. In practice, this means designing institutions capable of absorbing technological complexity, translating it into socially meaningful guidance, and maintaining oversight even as systems evolve.

7.1 Governing Complexity, Not Just Compliance

Traditional agricultural regulation focuses on inputs (fertilizers, pesticides), products (food safety standards), and environmental impacts (water quality, emissions). AI complicates this framework because it is:

- Adaptive (models update over time)
- Relational (systems interact with climate, markets, and social norms)
- Distributed (decision influence moves through networks)

Therefore, governance cannot be limited to pre-deployment certification. It must include continuous evaluation, similar to post-market surveillance used in pharmacology (OECD, 2021). Institutions must be structured to:

1. Monitor performance drift in changing environmental contexts
2. Assess fairness impacts across socio-economic groups

3. Respond to emergent, unintended consequences
4. Update standards iteratively rather than reactively

This is an adaptive governance model, not a static regulatory one, emphasizing learning-oriented oversight and institutional responsiveness as systems evolve (FAO, 2023; OECD, 2022).

7.2 The Role of Cooperatives and Intermediary Institutions

One of the strongest determinants of ethical AI integration observed across Canada, Kenya, India, and Brazil was the presence of intermediary institutions — cooperatives, growers' associations, extension agencies. These organizations act as interpretive bridges between:

- Farmers and technical specialists
- Local knowledge and algorithmic models
- Individual decision-making and policy environments

In Kenya, for instance, cooperative agronomists routinely translate AI advisories into place-specific action plans, allowing for feasibility adjustments based on seed access, market conditions, or labor availability (Mwangi et al., 2023). The cooperative becomes not only a distributor of information, but a guardian of contextual appropriateness.

In Canada, provincial extension specialists play a similar interpretive role, though increasingly under pressure due to staffing reductions (Harper et al., 2023). This suggests that sustaining advisory bodies is not ancillary — it is a core ethical requirement of digital agriculture.

Where advisory intermediaries weaken, power shifts toward platform providers; where they are strengthened, interpretive authority remains local.

7.3 Aligning Incentives and Accountability

Ethical governance also requires aligning responsibilities with the capacity to act on them. A system is unjust if:

- Farmers are liable for outcomes they did not fully choose
- Developers face no obligation to communicate uncertainty
- Institutions mandate technology adoption without shared support systems

To prevent asymmetry, three linked policy commitments are necessary, as reflected in Table 4 below.

Table 4. Governance goals for ethical agricultural AI, paired with policy approaches and practical examples from irrigation management

Governance Goal	Policy Approach	Practical Example
Maintain interpretive human authority	Require **explainability** in agricultural AI	Natural-language rationales in irrigation control dashboards
Prevent risk concentration	**Shared liability frameworks**	Crop insurance recognizing algorithm-influenced decisions
Distribute benefit fairly	**Value-sharing or cooperative data trusts**	Community-run data repositories in India and Brazil

This model frames AI not as a *transfer of control* from humans to systems, but as a *distributed collaboration* where accountability is shared, transparent, and legally grounded.

The movement from *algorithmic output human interpretation institutional support ecological outcome*, mapped earlier in *Figure 3*, becomes an organizing principle for ethical governance. The multi-level governance model shown in *Figure 2* aligns with this, reinforcing the need to treat technology as *embedded within social, cultural, and ecological systems*, not external to them.

8. Measuring Ethical Impact and Accountability

Measuring the ethical performance of AI systems in agriculture requires more than evaluating model accuracy or yield outcomes. Ethical impact emerges over time, through interaction between technology, people, institutions, and ecosystems. Thus, evaluation must consider both technical indicators (e.g., prediction precision, resource efficiency) and social indicators (e.g., autonomy, fairness, cultural legitimacy) (Barrett, 2022; Mwangi et al., 2023).

One of the limitations frequently noted in deployments across Canada, Kenya, India, and Brazil is that evaluation frameworks often focus on system outputs rather than system relationships. Yet relationships are where meaning and trust form—especially in agriculture, where knowledge is historically communal, context-dependent, and ecologically embedded.

8.1 Dimensions of Ethical Impact

To support meaningful assessment, ethical impact can be evaluated along five interlinked dimensions. *Table 5 shows* the core ethical dimensions used to evaluate agricultural AI systems, alongside guiding questions and example non-technical indicators drawn from real deployments across diverse farming contexts.

Table 5. Core ethical dimensions for evaluating agricultural AI, with guiding questions and example non-technical indicators

Ethical Dimension	Key Question	Example Indicator (Non-Technical)
Transparency	Can users understand how decisions were made?	Clarity of model explanation interfaces
Autonomy	Do farmers retain authority to override or adapt recommendations?	Documented override pathways
Fairness	Are impacts distributed equitably across groups and scales?	Bias audits across geographies and production systems
Collective Benefit	Are cooperatives, communities, and ecosystems strengthened?	Shared data governance structures
Sustainability Outcomes	Does the system reinforce ecological resilience?	Soil health, biodiversity, water-use stability over time

These dimensions mirror the ethical layering illustrated in Figure 3, reflecting how outcomes depend on interactions among algorithmic logic, human judgment, institutional context, and ecological conditions.

8.2 The Ethical Performance Cycle

To make assessment ongoing rather than retrospective, we adopt a Continuous Ethical Performance Cycle, where evaluation is integrated before, during, and after deployment, not only at procurement.

The performance cycle should involve farmers, agronomists, cooperative leads, data scientists, and institutional stewards as co-evaluators—not passive recipients of system reports. *As illustrated in Figure 4, ethical oversight operates as a contin uous loop rather than a linear checkpoint, with learning and adjustment occurring across all stages of system use.*

Figure 4. Ethical lifecycle of agricultural AI

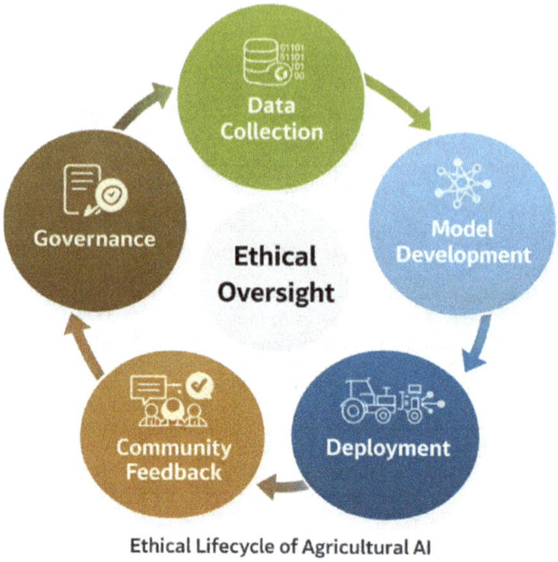

This diagram maps the cyclical stages of ethical oversight in agricultural AI, showing how data collection, model development, deployment, community feedback, and governance continuously inform one another.

8.3 Practical Evaluation Methods (Blended Model)

To ensure both rigor and usability, evaluation must rely on mixed-method approaches. *Table 6 outlines* the primary methods used to evaluate ethical impact in agricultural AI systems, combining quantitative, qualitative, participatory, and ecological measures.

Table 6. Methods for evaluating ethical impact in agricultural AI

Method	Measures	Example Use Case
Quantitative monitoring	Yield variance, input efficiency, detection sensitivity	Greenhouse sensor performance validation (Canada)
Qualitative field diaries	Interpretive decision reasoning, trust, experiential fit	Smallholder irrigation advisory reflections (India)
Participatory validation sessions	Community review of recommendations	Cooperative disease alert review groups (Kenya)
Ecological longitudinal metrics	Soil carbon trends, biodiversity indices	Agroforestry monitoring in Pará (Brazil)

These methods reinforce that ethical performance is not a checklist, but a co-evolving social–technical practice. This shift reframes ethical assessment from monitoring compliance to sustaining relationships. The transition mirrors broader trends in adaptive governance (OECD, 2022), where responsibility is distributed and learning is continuous.

9. Comparative Case Analysis Across Agricultural Contexts

Although AI and automation are often discussed in global terms, their meaning and impact vary significantly depending on *agricultural scale, ecological conditions, labor structure, cultural norms, and institutional support*. The same algorithm may reinforce autonomy in one region while eroding it in another. This section compares four contexts—Canada, Kenya, India, and Brazil—to illustrate how ethical outcomes are shaped not by technology alone, but by *relationships, governance, and local knowledge systems*.

This analysis builds on the layered ethical framework introduced earlier in *Figure 3*, where decision-making outcomes arise through interactions among system design, human interpretation, and institutional governance. It also aligns with the *Continuous Ethical Performance Cycle* shown in *Figure 4*, emphasizing that trust and accountability develop through iteration rather than mere deployment.

9.1 Canada – Data-Rich Greenhouse Systems and the Negotiation of Expertise

Vignette

While auditing AI-enabled climate-control systems in Canadian greenhouses, I observed a quiet shift unfolding on the shop floor. Workers who had historically been valued for their tactile intuition—recognizing humidity shifts through leaf texture—were now being asked to interpret dashboard analytics and adjust nutrient solutions based on models rather than sense perception. One technician described the transition as "learning to see plants through the computer's eyes first, and my own second." He did not dismiss the system—he negotiated with it.

Analysis

Canadian greenhouse systems demonstrate how *automation reshapes the definition of expertise*. Workers are not displaced; they are re-skilled into *data interpreters and system calibrators*. These role changes, however, depend on institutional culture as identified in Table 7.

Table 7. Conditions influencing skill outcomes in automated agricultural workflows

Support Condition	Outcome
Training + collaborative interpretation	Skill deepening, shared decision authority
Automation without dialogue	Skill hollowing and worker dependency

This example underscores that *human interpretive authority* must be intentionally protected to avoid over-dependence on automated logic.

9.2 Kenya – Cooperative Mediation as Ethical Infrastructure

Vignette

In rural Kenyan maize-growing cooperatives, AI-based pest advisory SMS systems have improved early detection of fall armyworm outbreaks. Yet farmers rarely act on the advisory messages individually. Instead, they *discuss the advisories collectively* before agreeing on a response. One farmer explained, "The phone tells me, but the group decides."

Analysis

Here, the cooperative acts as an *ethical intermediary*, converting system outputs into socially acceptable action. This reveals an insight:

- Ethical AI governance does not always occur at the *individual* level.
- In many regions, *decision-making is fundamentally communal.*

Thus, impact depends not only on algorithm design, but on whether the deployment *respects existing social decision structures*—a principle aligned with the institutional layer in *Figure 3*.

9.3 India – Water-Sharing Communities and Algorithmic Fairness

Vignette

In canal-irrigated regions of Maharashtra, decisions about when to release water are historically negotiated among farmers based on crop stage, household need, and reciprocity. When a pilot AI irrigation scheduler suggested "efficiency-optimal" allocation, farmers perceived the recommendation as *socially unfair* because it disrupted obligations to neighbors. The objection was not technological—it was *relational.*

Analysis

This case illustrates that *optimization without social context can destabilize communal trust.* Ethical automation requires acknowledgment that:

- Efficiency *is not the only value* in agricultural decision-making.
- Water allocation is not simply a technical problem—it is a *moral economy.*

Ethical impact here must be assessed using the continuous evaluation model illustrated in *Figure 4*, where community feedback is *a structural input*, not an afterthought.

9.4 Brazil – Agroecology, Biodiversity, and Stewardship Identity

Vignette

In agroforestry systems in Pará, farmers described their work not as "production" but as *care for the forest community.* When introduced to an AI forest-growth simulation tool, farmers used *it not to maximize yield*, but to experiment with *biodiversity regeneration scenarios.* One farmer said, "The computer helps me imagine futures I cannot yet see."

Analysis

Here, AI strengthens *ecological imagination,* reinforcing indigenous and agroecological stewardship identities. This suggests automation is most aligned with ethical goals when it *builds upon existing cultural commitments*, rather than replacing them (Table 8).

Table 8. Comparative ethical opportunities and risks of agricultural AI across contexts

Context	Ethical Opportunity	Ethical Risk
Brazil	Supporting ecological stewardship	Over-simplifying biodiversity into yield metrics
India	Supporting water governance dialogue	Undermining reciprocity norms
Kenya	Supporting cooperative decision practices	Creating dependency on external platforms
Canada	Supporting skill augmentation	Concentrating authority in system logic

PART II — ARTIFICIAL INTELLIGENCE AND AUTOMATION IN AGRICULTURAL DECISION-MAKING

10. Automation as a Reconfiguration of Agricultural Judgment

Automation in agriculture is frequently discussed in terms of efficiency gains: reduced input expenditure, minimized crop loss, optimized irrigation, and streamlined labor cycles. However, when decision processes are partially or fully automated, what changes is *not only the informational basis of decisions*, but the *locus of judgment itself*. Decision-making moves, in part, from embodied, experiential assessment toward *model-derived inference*.

For instance, in precision horticulture environments across *Ontario and British Columbia*, automated climate-control systems continuously adjust temperature and humidity conditions in greenhouses. While such systems reliably maintain optimal crop environments, growers describe a shift in *how they engage with the plants themselves*. Rather than detecting subtle leaf tension, stem color shifts, or soil scent changes, attention is often focused on interpreting dashboards and system alerts. The *site of agricultural interpretation shifts*—from plant to interface.

In contrast, in *smallholder rice and millet systems in Karnataka, Tamil Nadu, and parts of Nepal*, weather-advisory and disease prediction apps serve primarily as *interpretive supplements*. Farmers consult them, but often anchor final decisions in *collective deliberation*, seasonal memory, and observation of "sign indicators" such as ant trail density or morning soil warmth. Here, automation enters as *advice rather than authority*, maintaining the *situated and relational quality* of agricultural knowledge (Narayanan, 2022).

The ethical question, therefore, is not whether automation is present, but *how automation is positioned within the hierarchy of decision-making authority*. Systems can have the following modes of automation as detailed in Table 9.

11. Constraints, Risks, and the Uneven Geography of Automation

The impacts of automation differ significantly based on *region, resource access, infrastructure, and political economy*. Here, the earlier requested heatmap (now *Figure 5*) becomes relevant — as it visualizes how different agricultural regions possess *different capacities for integrating automation effectively*.

Figure 5. Comparative adoption pathways in agricultural AI

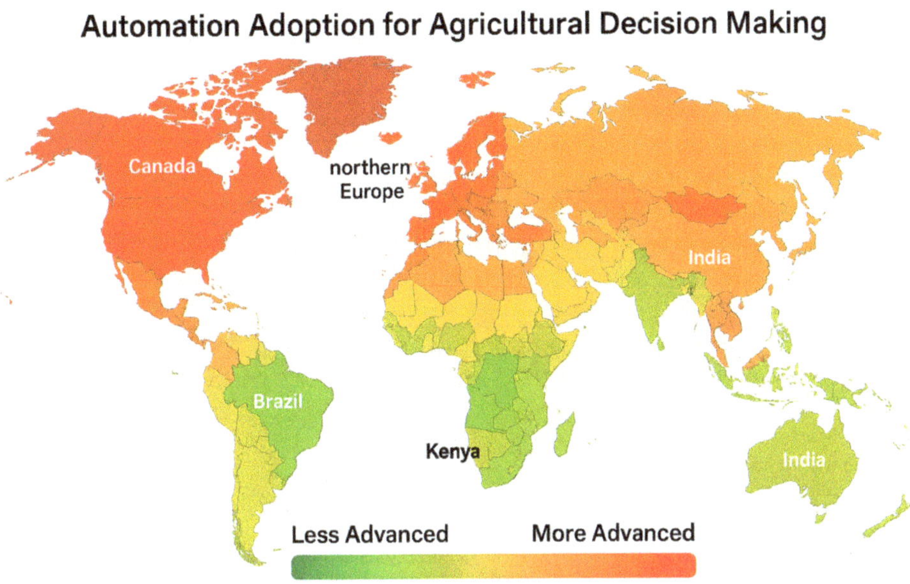

As illustrated in *Figure 5*, automation adoption is not a simple gradient from "less advanced" to "more advanced." Instead, it reflects historically produced differences in resource access, labor stability, ecological priorities, and decision centralization. High-input commercial agriculture in Canada and northern Europe is structured around mechanization and can integrate automation rapidly; smallholder agriculture in Kenya, India, and Brazil integrates automation selectively, balancing it against communal stewardship norms and fluctuating market risk.

Risks of uneven automation adoption heighten when:

- Platform recommendations cannot be challenged or audited,
- Private agritech firms control long-term storage of soil, rainfall, or yield data,
- Co-ops lack bargaining capacity or digital literacy training,
- Automation accelerates monocropping or chemical reliance,
- AI optimizes short-term yields while undermining *biodiversity and soil regeneration*.

In this way, automation becomes not only a *technical* intervention, but a *political one*, shaping how agricultural futures are imagined and distributed.

Case Examples: Canada, Kenya, India, Brazil

These cases identified in Table 10 demonstrates *different forms of automation governance.*

Table 10. Regional patterns in agricultural automation: outcomes and ethical tensions

Region	Automation Mode	Outcome	Ethical Tension
Canada (Controlled-environment horticulture)	Climate dashboards + fertigation automation	Increased production stability and seasonal predictability	Risk of grower deskilling if interpretation role is reduced
Kenya (Maize cooperatives)	Mobile-based pest advisory systems	Strengthened collective action against armyworm outbreaks	Data intermediaries gain negotiating power over seed + chemical supply
India (Canal irrigation governance, Maharashtra)	Water allocation optimization apps	Reduced water waste	Conflicts with reciprocity norms + local hydrological knowledge
Brazil (Agroforestry / Amazon reserves)	AI-based biomass + biodiversity monitoring	Supports forest protection and regenerative livelihoods	Must avoid converting multispecies stewardship into yield-maximization logic

Automation is sustainable only when it is integrated into existing social and ecological governance systems — not imposed as a replacement for them.

This reinforces the principle seen earlier in *Figure 1*, where *Human Oversight* is not optional—it is the *ethical backbone* of agricultural AI.

12. Automation, Climate Resilience, and Ecological Stewardship

As climate extremes intensify — including prolonged droughts, heat stress, and unpredictable weather — automation is increasingly framed as a *resilience strategy.* AI systems can support:

- Drought-adaptive irrigation scheduling
- Precision nutrient reduction to prevent soil degradation
- Heat stress mitigation for livestock
- Water table monitoring and aquifer protection

However, resilience is not achieved solely through *optimization*. It requires *diversity, redundancy, and adaptive flexibility* — characteristics found in agroecological and indigenous land stewardship systems (Altieri et al., 2021).

Ethical alignment occurs when automation strengthens ecological attentiveness:

- Monitoring tools that visualize *soil moisture gradients* encourage farmers to track hydrological balance.
- Biodiversity indexing systems (now used in parts of Brazil and Ecuador) track *pollinator presence* alongside yields.
- Regenerative AI decision platforms (emerging in parts of Kenya) recommend *intercropping strategies*, not just input purchases.

If automation shifts toward *regenerative intelligence*, rather than *extractive optimization*, its ethical value increases. Refer to Figure 6 below for further details.

Figure 6. Comparative heatmap of structural conditions shaping agricultural automation across regions

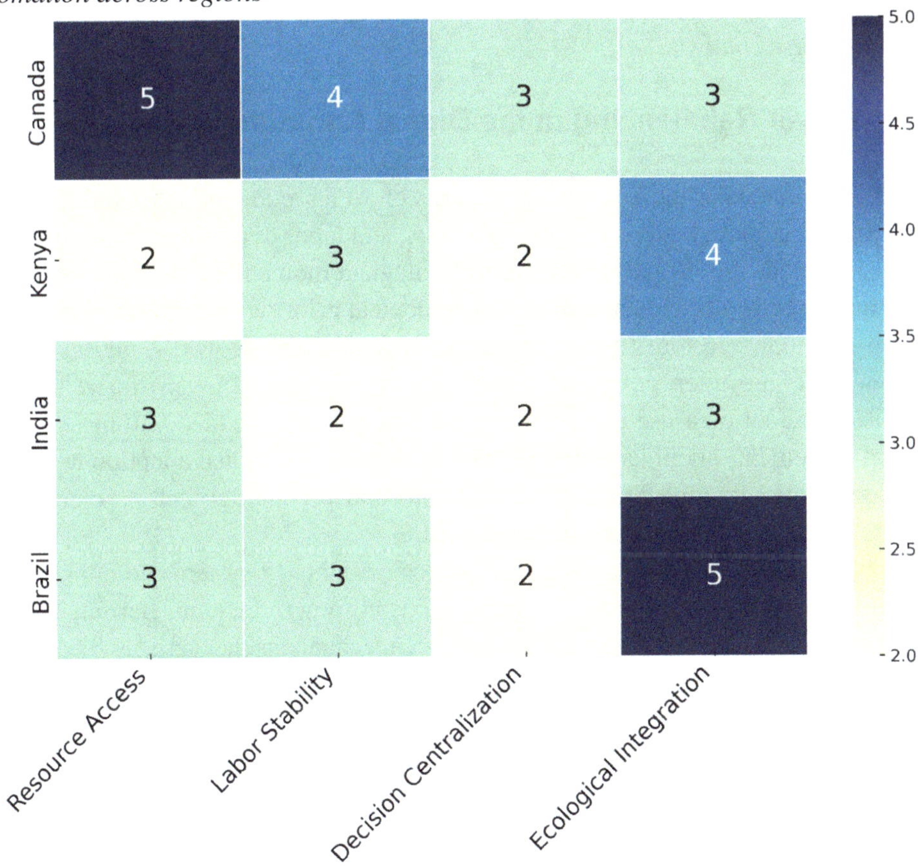

SECTION 13 – FUTURE ETHICAL PATHWAYS FOR AGRICULTURAL AI: TOWARD SHARED STEWARDSHIP

The trajectory of artificial intelligence in agriculture is neither purely technological nor purely social. Across the cases examined in this chapter—from controlled-environment agriculture in Canada to cooperative irrigation systems in Kenya—the ethical direction of AI has been shaped less by algorithmic sophistication than by the relationships and governance structures surrounding its use. What emerges is a set of ethical pathways oriented toward shared stewardship: arrangements in which efficiency, care, and accountability are held together rather than traded off against one another.

The pathways outlined below extend directly from the empirical and analytical foundations developed earlier in the chapter. They do not prescribe a single model for ethical agricultural AI, but instead highlight conditions under which automation can support human agency, ecological responsibility, and institutional accountability across diverse contexts.

13.1 Power Rebalancing in the Digital Agricultural Economy

As illustrated earlier in Figure 5, patterns of AI adoption are inseparable from governance culture, land tenure arrangements, and labor dynamics. While regions such as Canada and Brazil may share technological momentum, they diverge significantly in how decision authority and ecological priorities are distributed.

In Brazilian agroforestry and regenerative agriculture networks, digital monitoring tools are often governed collectively at cooperative or territorial levels. Decisions about data use and system deployment are embedded within existing social agreements about land stewardship. In contrast, Canadian adoption tends to be coordinated through producer associations and private agritech partnerships, where decision authority is more centralized.

This variation underscores a critical ethical insight: AI systems do not simply optimize outcomes; they encode and amplify existing power relations (Barrett, 2022). Ethical pathways therefore require governance mechanisms that actively rebalance power, including data sovereignty norms, collective bargaining over digital contracts, and community-level processes with the authority to accept, adapt, or refuse automation initiatives. Where such mechanisms are present, AI can reinforce local autonomy rather than erode it.

13.2 Strengthening Human–Machine Co-Decision Practices

Across the cases informing this chapter, ethical tensions most often surfaced at the boundary between recommendation and decision. When AI systems are framed as advisory, farmers and workers remain engaged as interpreters of their landscapes. When systems are framed as directive, disengagement and deskilling tend to follow.

As discussed earlier in Sections 4 and 6, interpretive authority in agricultural AI systems remains situated with human actors who contextualize, negotiate, and bear responsibility for outcomes. Ethical pathways therefore point toward active human–machine co-decision rather than passive oversight. This includes interfaces that communicate the reasoning behind recommendations, structured opportunities for contextual judgment, and seasonal forums—often at the cooperative level—for reviewing algorithmic performance.

Rather than positioning humans as monitors of automated systems, this approach recognizes interpretation as an ongoing responsibility central to ethical agricultural decision-making.

13.3 Climate Resilience as an Ethical Baseline

The acceleration of climate volatility has shifted the ethical baseline for agricultural AI. Across climate adaptation initiatives observed in Canada, Kenya, and parts of South America, technologies were adopted most sustainably when ecological indicators—such as soil health, water retention, and biodiversity—were treated as core performance criteria rather than secondary constraints.

Systems that increase short-term yields while undermining ecological resilience represent deferred risk rather than success. Ethical pathways therefore require multi-season evaluation horizons, ecological baselines embedded within datasets, and decision frameworks that weigh resilience alongside efficiency. This aligns with FAO guidance emphasizing that environmental stewardship must be evaluated in parallel with productivity outcomes (FAO, 2023).

13.4 Institutional Cultures That Normalize Ethical Reflection

Experience across agricultural policy and technology institutions suggests that ethical outcomes endure only when they become part of organizational culture. Formal policies establish boundaries, but everyday practices determine whether those boundaries are respected.

Institutions that support ethical AI adoption tend to invest in interdisciplinary training, normalize stakeholder consultation, and treat precaution as compatible with innovation rather than opposed to it (Leslie, 2020). While such cultural shifts are often incremental, they form the institutional infrastructure that allows ethical concerns to be identified and addressed before harm becomes entrenched.

In this sense, ethics functions less as an external audit and more as an internal habit of decision-making.

13.5 A Shared Governance Horizon

The final pathway points toward plural rather than uniform governance arrangements. No single ethical framework can account for the diversity of agricultural systems examined in this chapter. Instead, ethical agricultural AI depends on layered governance: national certification regimes, cooperative-level data negotiations, extension-based mediation, and transnational knowledge exchange through bodies such as FAO, CGIAR, and GPAI.

This shared governance horizon acknowledges that responsibility is distributed, perspectives are partial, and legitimacy is earned through participation rather than imposed through standardization. Ethical stewardship emerges through coordination across scales, rather than through centralized control.

Closing Note on Pathways

Taken together, these pathways do not prescribe a single future for agricultural AI. They offer direction rather than destination—grounded in empirical experience, attentive to power and culture, and oriented toward shared responsibility. In this sense, the future of ethical agricultural AI is less about technological foresight than about institutional and relational preparedness.

Conclusion

Ethical questions surrounding artificial intelligence and automation in agriculture cannot be resolved through technical performance alone. Across the cases examined in this chapter—spanning Canada, Kenya, India, and Brazil—it becomes clear that automation reshapes not only decision-making efficiency, but also authority, labor dignity, ecological values, and the circulation of knowledge. The Ethical Risk Lifecycle presented earlier (Figure 1) demonstrates that responsibility in agricultural AI is cumulative and ongoing, extending from data collection and model design through human interpretation, institutional governance, and long-term ecological outcomes. Ethical agricultural AI, therefore, is not a feature that can be added at deployment; it is a condition that must be sustained over time.

The policy implications of this insight are substantial. Governments must strengthen protections around data sovereignty and consent to prevent exploitative platform dependence. Agritech developers carry responsibility for designing systems that invite interpretation rather than suppress it, supporting explainability, override pathways, and participatory calibration. Cooperatives and producer organizations play a critical role in preserving collective bargaining power and contextual decision-making, particularly where individual farmers face asymmetrical market pressure. Academic and research institutions, in turn, must prioritize evaluative frameworks that include ecological health, biodiversity, and social resilience alongside productivity metrics. Without coordinated action across these actors, the efficiency gains of automation risk being achieved at the expense of autonomy, equity, and long-term sustainability.

Looking forward, the ethical horizon for agricultural AI is not defined by smarter systems alone, but by more careful ones. Agriculture has always been a practice grounded in observation, judgment, and shared responsibility toward land and community. If AI is to contribute meaningfully to its future, it must learn to support those traditions rather than overwrite them. The central question, then, is not simply what AI can do for agriculture, but what forms of agricultural life—and care—we are choosing to sustain.

DEDICATION

To my beloved father, Zafar Ahmad (Late)

whose quiet strength, unwavering integrity, and belief in the dignity of knowledge continue to guide every step of my professional endeavors. His values live in every line of this chapter, and in every effort I make to ensure that technology serves people, communities, and futures worth building.

REFERENCES

Altieri, M. A., Nicholls, C. I., & Montalba, R. (2021). Agroecology and the design of climate change–resilient farming systems. *Agronomy for Sustainable Development, 41*(1), 1–14. DOI: 10.1007/s13593-020-00666-4

Barrett, C. B. (2022). Overcoming global food security challenges through data-driven agriculture. *Nature Food, 3*(6), 409–414. DOI: 10.1038/s43016-022-00515-0 PMID: 37118052

Beerbaum, D., & Otto, B. (2023). Data governance in digital ecosystems: Concepts, challenges, and research directions. *Electronic Markets, 33*(1), 1–15. DOI: 10.1007/s12525-022-00563-9 PMID: 37035262

Birhane, A. (2021). Algorithmic injustice: A relational ethics approach. *Patterns (New York, N.Y.), 2*(2), 100205. DOI: 10.1016/j.patter.2021.100205 PMID: 33659914

Birhane, A., Prabhu, V., & Kahembwe, E. (2022). The values encoded in machine learning research. *Proceedings of the ACM Conference on Fairness, Accountability, and Transparency*, 173–184. https://doi.org/DOI: 10.1145/3531146.3533154

Carbonell, I. M. (2016). The ethics of big data in agriculture. *Internet Policy Review, 5*(1), 1–15. DOI: 10.14763/2016.1.405

Couldry, N., & Mejias, U. A. (2019). Data colonialism: Rethinking big data's relation to the contemporary subject. *Television & New Media, 20*(4), 336–349. DOI: 10.1177/1527476418796632

Food and Agriculture Organization of the United Nations (FAO). (2022). *Data governance in agriculture: Principles and case studies*. FAO.

Food and Agriculture Organization of the United Nations (FAO). (2023). *Responsible use of artificial intelligence in agriculture and food systems*. FAO., DOI: 10.4060/cc2453en

Garske, B., Bau, A., Ekardt, F., & Gawel, E. (2021). Digitalization and AI in sustainable agriculture: A review. *Sustainability*, *13*(16), 9033. DOI: 10.3390/su13169033

Harper, S., McLean, C., & Johnston, L. (2023). Human–AI collaboration in controlled-environment agriculture. *Agricultural Systems*, *209*, 103658. DOI: 10.1016/j.agsy.2023.103658

Hess, C., & Tadesse, G. (2022). Liability and accountability in automated agricultural decision systems. *AI and Ethics*, *2*(3), 389–402. DOI: 10.1007/s43681-021-00125-8

Junior, J. L., & Silva, P. R. (2022). Digital tools and ethical governance in Brazilian agroforestry systems. *Sustainability*, *14*(9), 5142. DOI: 10.3390/su14095142

Klerkx, L., Jakku, E., & Labarthe, P. (2019). A review of social science on digital agriculture. *NJAS Wageningen Journal of Life Sciences*, *90–91*, 100315. DOI: 10.1016/j.njas.2019.100315

Leslie, D. (2020). *Understanding artificial intelligence ethics and safety*. Alan Turing Institute.

Liang, F., & Fernandez, A. (2021). Responsibility gaps in automated decision-making. *Philosophy & Technology*, *34*(4), 957–978. DOI: 10.1007/s13347-020-00423-7

Mwangi, M., Otieno, D., & Kamau, P. (2023). Cooperative governance and digital advisory systems in Kenyan agriculture. *Journal of Rural Studies*, *98*, 45–57. DOI: 10.1016/j.jrurstud.2023.01.006

Narayanan, S. (2022). Data, power, and agrarian transformation in India. *The Journal of Peasant Studies*, *49*(6), 1271–1293. DOI: 10.1080/03066150.2021.1996358

Organisation for Economic Co-operation and Development (OECD). (2021). *AI policy observatory: Responsible AI practices*. OECD Publishing.

Organisation for Economic Co-operation and Development (OECD). (2022). *Artificial intelligence and agriculture: Opportunities and risks*. OECD Publishing., DOI: 10.1787/2baf5a32-

Patil, S., & Deshmukh, R. (2021). Irrigation decision-making and digital advisory tools in India. *Water Policy*, *23*(5), 1103–1119. DOI: 10.2166/wp.2021.084

Villalba, M., & Alarcón, D. (2022). Trust, technology, and participatory governance in agricultural innovation. *Technology in Society*, *68*, 101865. DOI: 10.1016/j.techsoc.2021.101865

Wright, S., & Chan, L. (2022). Participatory AI and data justice. *AI & Society*, *37*(4), 1521–1535. DOI: 10.1007/s00146-021-01278-8

Chapter 13
Addressing Global Challenges and Overcoming Barriers to the Widespread Adoption of Precision Agriculture Technologies

Farhat Yasmeen
http://orcid.org/0000-0003-1865-2296
University of Wah, Pakistan

Sajid Hussain
PMAS-Arid Agriculture University, Rawalpindi, Pakistan

Muhammad Iqbal
University of Chakwal, Pakistan

Maria Ehsan
PMAS-Arid Agriculture University, Rawalpindi, Pakistan

Naveed Iqbal Raja
PMAS-Arid Agriculture University, Rawalpindi, Pakistan

ABSTRACT

Precision Agriculture has emerged revolutionary approach to modern farming, integrating technologies such as Geographic Information Systems, Global Positioning

DOI: 10.4018/979-8-3373-7257-0.ch013

Systems, drones, artificial intelligence, Internet of Things, and big data analytics to achieve data-driven, site-specific crop management. This review explores how PA addresses critical global challenges, including food insecurity, resource depletion, and climate change, while promoting sustainable agricultural development. The review categorizes adoption barriers into economic, technological, knowledge-based, emphasizing disparities between developed and developing agricultural systems. Emerging innovations such as AI and machine learning for predictive analytics, blockchain for transparency, autonomous robotics, and digital twin ecosystems are reshaping agriculture toward sustainability and resilience. Precision agriculture thus represents not only a technological innovation but a strategic pathway toward a more productive, inclusive, and environmentally sustainable future for global farming systems.

1. INTRODUCTION

Precision Agriculture (PA) is a methodology that is very modern in terms of farming that utilizes different management systems, and therefore different management methods, across the entire agricultural continuum through technology-supported decision-making and PA individual site management (Al-Jaf et al., 2025). PA provides a data-driven decision model and management of diverse sites separately, thus allowing the farm manager to distribute products like water and chemical sprays or fertilizers according to the needs of the areas in a field (Samuel et al., 2025; Zainab & Astiko et al., 2025). Environmental, geographical, and climatic conditions, soil quality, and crop health are all considered by PA that efficiently processes raw data and uses it to derive strategic measures directed towards food security enhancement and ecosystem conservation (Banerjee & Bhat, 2025; Sousa et al., 2026). The PA system is strongly integrated with the very latest technologies, among others: remote sensing, drones, IoT, machine learning, and equipped mechanized farms with tractors (Aremora et al., 2025; Huang et al., 2025). The farming mechanization of the early 20th century represented the replacement of human labor with machines, thus increasing productivity within the industry; in this regard, PA offers even greater technological advancement. Optimizing resource use, product quality, and income earned by farmers and minimizing detrimental environmental effects extends further than yield improvement (Mehrotra et al. 2025).

The global population grows rapidly, increasing the strain on our agricultural systems to produce more food within limited land, water, and energy constraints. With the increasing demand, traditional farming practices are becoming increasingly outdated, and it is apparent that sustainable intensification is becoming a necessity. In this regard, precision agriculture is the most effective way to increase productivity and ensure the sustainable availability of food in the future (Lotsis et al. 2026; Zarboubi et al. 2025). Technologies of precision farming integrate the use of proximal and remote sensing along with IoT-based monitoring systems for real-time tracking of the condition of crops during different growth stages. Such systems can generate huge amounts of data, which, when adequately processed, allow for timely actions and reasonable decisions to be made (Wulandari et al. 2025). Adoption, however, continues to be uneven. Due to more significant financial resources and scale economies, large-scale farms are more prone to implementing precision agriculture than small-scale farmers. Indrabhai and Patel (2025) suggest that the inadequate resources, limited technical knowledge, and insufficient training of small-scale farmers are the primary barriers to precision agriculture. These barriers must be addressed to promote the adoption of smart and precision farming technologies.

2. GLOBAL CHALLENGES IN MODERN AGRICULTURE

The urgency of feeding the world's population by 2050 increased after the 2008 food price surge revealed the global food system's structural weaknesses. Though global agricultural production increases have alleviated some of the world's hunger, in the future, world income growth and global agricultural development will increase the demand for consumption of animal proteins, particularly in developing countries (Daszkiewicz, 2022; Jha et al., 2022). This dietary transition exerts even greater pressure on globally available land and water resources. Already, 38% of the world's land surface is being used for agriculture, of which 1.5 billion hectares are used for crop production and 3.4 billion hectares are allocated to pastureland (Wijerathna-Yapa & Pathirana, 2022). Agricultural land development estimates for 2050 predict the need for an additional 0.2 to 1 billion hectares of land for agricultural activities (Elechi et al., 2022). Much of the potential agricultural land is of lower quality and is located within forests and protected areas. The expansion of agricultural land threatens to increase global biodiversity loss and the unsustainability of ecosystems and the environment, which increases the urgency for land development for agricultural purposes to be curtailed (Hussain et al., 2025). Figure 1 illustrates the main challenges to modern agriculture. These include 1) increasing food demand resulting from population growth, 2) decreasing arable land due to soil degradation, and 3) increasing water scarcity due to climate change. 4) Greenhouse gas emissions led to environmental degradation.

To meet future food demands, it will be necessary to employ alternative strategies, such as stimulating yields from currently deficient and unproductive lands, improving food loss and waste, and changing food consumption patterns (Jha et al., 2021; Kay et al., 2022). In this case, dietary change is critical, as food systems based on plants produce fewer and less harmful environmental impacts than systems based on livestock, which are more land- and water-intensive (Jennings et al., 2023). Unfortunately, globally, there is an increasing consumption of highly processed foods, sugars, saturated fats, and livestock products. Furthermore, there is water scarcity, which is also a significant constraint for agricultural production. About 70% of freshwater withdrawn globally goes into agriculture, most of it into irrigation (Douzal et al., 2024). The combination of decreasing water supply and climate variability negatively impacts production by lowering the yields from both crops and pastures, particularly in rain-fed systems. In addition to this, agriculture is responsible for about 12% of the total greenhouse gas emissions, and simultaneously, soil is the largest and most critical carbon reservoir. This is the paradox of agriculture, which needs to be addressed: climate change mitigation versus impact (Haas et al., 2025; Mason-D'Croz & Herrero, 2025).

Figure 1. Global challenges related to agriculture

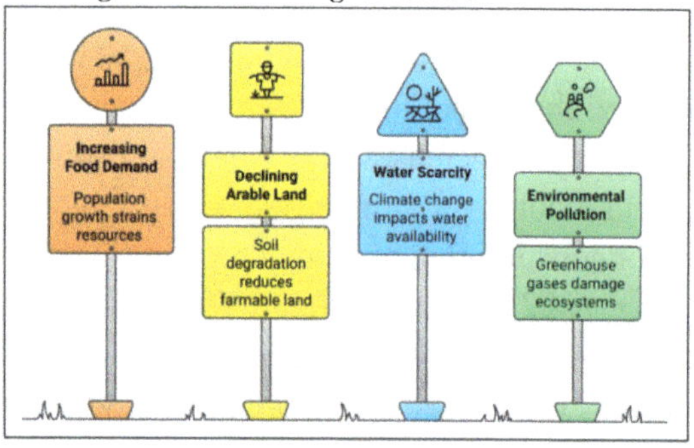

3. ROLE OF PRECISION AGRICULTURE IN TACKLING GLOBAL CHALLENGES

Due to the impact of climate change, rapid population growth, resource scarcity, and environmental degradation, global agriculture is under increasing stress. PA has advanced from a collection of field-level optimization tools to a highly developed, data-driven production system that can respond to these global issues (Padhiary et al., 2025; Sharma, 2025). Artificial intelligence, the IoT, cloud computing, and geospatial technologies have shifted agricultural information systems to adaptive and resilient farming systems (Chaurasia et al., 2026; David et al., 2024; Tatiya et al., 2025).

This transformation is best illustrated through modern nutrient management. After static and periodic soil sampling, modern nutrient mapping integrated geo-referenced soil data, IoT-enabled soil sensors, satellite information, and cloud analytics and has high resolution and dynamic capabilities (Bhamini et al., 2025; Liu et al., 2025). These approaches offer continuous monitoring of soil nitrogen, phosphorus, potassium, organic matter, and soil pH, and can do so at a resolution lower than a field. AI models use this data along with historical yields, weather, and crop growth to schedule fertilizer applications based on changing crop nutrient demands (Ahmad et al., 2025; Meena et al., 2025). Thus, the efficiency of fertilizer usage is enhanced, the amount of nutrients lost is decreased, and the issues linked to the environment are, therefore, partially solved, such as eutrophication and nitrate leaching (Goel & Vishnoi, 2025; Sahu, 2024).

The combination of AI, IoT, and cloud technologies has completely revolutionized the process of decision-making in farms. Sensors are constantly monitoring and measuring the soil moisture, temperature, and the overall microclimate and merging this data with the weather forecasts and remote sensing (Awate & Nagne, 2025; Surendran et al., 2024). This allows for decisions to be made on micro levels, such as in planting, irrigation, and input application. Through AI analytics, agribusinesses can be helped in the early diagnosis of crop stress, the outbreak of diseases and pests, and nutrient deficiencies, which in turn leads to timely interventions and avoids losses in yields. Real-time AI analytics combined with Variable Rate Technologies (VRTs) automatically alter the rates at which fertilizers, pesticides, and water are applied. Studies report that the adoption of site-specific PA practices increases crop productivity by 10-25% and decreases water, fertilizer, and energy expenditure by 20-35% (John & James, 2025; Kannan, 2025). With smart-time moisture control irrigation and analytics, PA facilitates climate adaptation, resource conservation, and sustainable intensification. This makes PA an ideal response to the growing global challenges of food security and environmental sustainability.

4. BARRIERS TO THE WIDESPREAD ADOPTION OF PRECISION AGRICULTURE TECHNOLOGIES

Although PA can improve productivity and sustainability, there has still not been any widespread global adoption of it. Digital skill scarcity, high investment cost, poor infrastructure, and inadequate policies all contribute to a lack of adoption (Nurmalitasari et al., 2025; Vinholis et al., 2025). These factors cause the most impact in developing and underdeveloped areas, where poor digital usage is caused by low-quality internet, inconsistent power, and lack of information (Lan & Ban, 2025; Shamsoddini et al., 2025). Table 1 summarizes the various dimensions of the barriers that constrain the adoption of this method, presenting the various unedited components of the European Union. Smallholders face the most significant impact of the previously named barriers, such as high economic costs, insufficient knowledge, poor infrastructure, unsteady markets, high social resistance, environmental issues, and a lack of trust, all of which limit the effective and widespread adoption of PA.

The adoption of PA tools has been impeded by high capital and operational costs. PA includes investments in new hardware and software, sensors, and continual maintenance, costing tens of thousands of dollars (Ahmed et al., 2025; Aremora et al., 2025). Small and medium-sized farmers cannot afford variable rate technology (VRA) and the Internet of Things (IoT) platforms. Weak subsidy programs, limited access to credit, and the absence of customized financing solutions stifle investment (Cheong et al., 2025; Nyathi & Mutale, 2025). While PA has potential for positive returns to farmers, the returns are highly contingent on crop type, farm size, and environmental factors, which resource-poor farmers are unable to mitigate.

There are also infrastructure obstacles. IoT systems rely on broadband, cellular, and electrical connectivity, which are all lacking in many rural communities. Operational incompatibilities between various systems and technologies (or "vendor lock-in") are some of the factors that contribute to farmers not being able to use the technology confidently. Such concerns are mentioned by various authors (Sharma & Garg, 2025). The insecurity caused by technology is most severe in the case of the agricultural sector, where no legal or regulatory frameworks exist and the law is uncertain about data ownership, privacy, and accountability. Farmers do not want to risk losing their data to non-accountability authorities, and this is why they are reluctant to surrender the data generated on their farms (Ahmad et al., 2025).

Knowledge and information deficiencies are further hurdles that keep the colonization rate low. Quite a few farmers view PA just as drones and similar tools, leaving aside the large and diverse possibilities, while many training systems neither focus on these new and developing digital tools nor do they take into consideration their rapid development (Manhas et al., 2026). Besides resistance, attachment to the old ways, and overall skepticism towards automation, the psychological factors have also contributed to the slow adoption process. Moreover, the adoption of PA technologies is usually restricted to larger-scale farming operations, which thus aggravates the inequalities and the digital divide in agriculture (Werulkar & Pande, 2025). The sphere of influence must have coordinated policy advocacy, equitable financing, sound infrastructure, and collaboration that fuses technological solutions with farmers' knowledge and the socio-cultural and vernacular context.

Table 1. Key challenges hindering the widespread adoption of precision agriculture technologies

Category of Barrier	Challenges	Impact on Adoption	Economic Impact	Social Implications	Technological Limitations	Policy Constraints	References
Economic	High initial investment costs	Limits accessibility for smallholder farmers	Reduce investment in technology	Limited financial viability	Inhibits full utilization	Lack of government incentives	(Kroupova et al., 2025; Lan & Ban, 2025; Munz & Schuele, 2022)
Knowledge	Lack of awareness and training	Reduces understanding of technology benefits	Potential loss in profitability	Slow adoption rates	Hinders technology integration	Insufficient extension services	(John et al., 2023; Vaishnavi, 2025)
Infrastructure	Limited access to necessary technology	Hinders efficient implementation and usage	Complicates decision-making	Limits real-time data analysis	Reduces operational efficiency	Poor support for digital tools	(Frona & Frona-Hadas, 2025; Hundal et al., 2023)
Market Conditions	Volatile market prices	Increases risk for investment in new technologies	Affects financial stability	Decreases agricultural output	Limits market access for precision	Unstable profit margins	(Masi et al., 2023; Forero et al., 2025)
Environmental Issues	Concerns over ecological impacts	Creates hesitance towards adopting precision practices	Affects sustainability initiatives	Influences public perception	Slows down innovative technology adoption	Regulatory barriers to implementation	(Ahmed et al., 2025; Bhattacharyya et al., 2020; Wijerathna-Yapa & Pathirana, 2022)
Social Resistance	Resistance to change in traditional practices	Limits technology acceptance and integration	Impacts social dynamics	Hinders collaboration	Affects collective action efforts	Cultural perceptions of technology	(Kendall et al., 2022; Manning, 2024)
Data Privacy	Concerns about data security	Affects trust in technology adoption	Reduces willingness to share data	Influences user engagement	Create barriers for data-driven decisions	Privacy regulations	(Atik, 2022; Jouanjean et al., 2020; Sullivan et al., 2024)

5. ECONOMIC BENEFITS AND CHALLENGES OF PRECISION AGRICULTURE

PA can revolutionize the farming industry by facilitating better operational efficiency, productivity, and environmental sustainability (Cogato & Schillaci, 2025; Jatav & Rajput, 2025). PA's economic impacts can be particularly intricate and uneven. PA can be economically beneficial to the farming industry. For instance, PA can enhance productivity (higher crop yields), optimize the use of farming inputs (resource use), and lower farming's environmental footprint (environmental damage). However, such economic advantages can be overshadowed due to the initial high investments in sophisticated farming gadgets. More to this point, Fayaz et al. (2025)

and Sinha et al. (2025) observe similar sentiments. In the global farming industry, Popp et al. (2025) most succinctly make the case that the economic gains and the costs of PA initiatives must be understood if the overall impacts of PA initiatives are to be understood.

One of the most important advantages of precision farming is the efficient management of input allocation using unmanned aerial vehicles (UAVs), machine learning models, and data-driven decision frameworks (Sharma & Gawade, 2025). These systems allow for the tailored application of fertilizer and pesticides; therefore, both cost and waste are improved (Zhao et al., 2025). For instance, in precision irrigation technology, water use is reduced by more than 40% (Gao et al., 2025). In addition, numerous types of research show cost reductions of 30% or more for fertilizers, pesticides, and other fuels, suggesting an input-to-cost ratio of 10-30% in conjunction with productivity improvements of 10-25% (Addagatla & Putta, 2025; Fernandes, 2025). Post-harvest losses are improved through monitoring and more quality control, and it improves the quality of the produce and the market value of the whole farm with more uniform and dependable farm income (Han et al., 2025). In addition, the demand for skilled labor in areas such as data analytics, agronomy, and mechatronics alters local employment structures (Miglani, 2025). Overall, Figure 2 shows the benefits of applying precision agriculture as a virtue of a net contributor to the economy, the environment, and the operational side of the farming business. PA increases the profitability of farms by diminishing the use of water and input and the cost of production while improving the quality of the produce. At the same time, it builds climate resilience, reduces environmental footprints, minimizes post-harvest losses, and fosters timely and improved data-driven decisions for sustainable farming systems.

Figure 2. Key benefits of precision agriculture

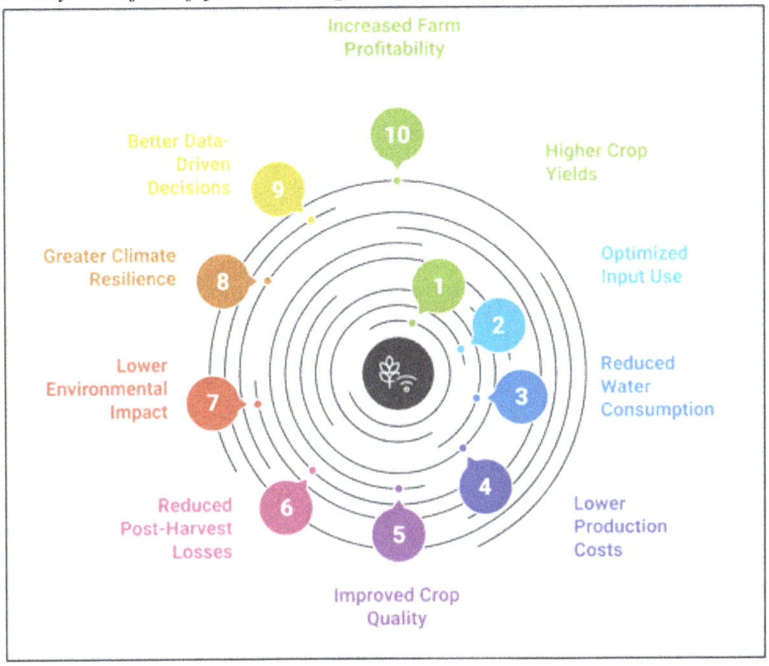

Even with the broad benefits that come with the use of new technology, the financial cost of adopting the technology remains the most important barrier when it comes to adopting new technology. The large amounts of capital needed to buy the new precision ag tools needed, like GPS, software, drones, and sensors, are often not available for smallholder farmers (Karunathilake et al., 2023; Munz & Schuele, 2022). An investment is often not worth the financial risk due to the long payback periods, and when no financial returns can be seen, it decreases the likelihood of an investment (McFadden et al., 2023). In less developed areas, most farms are small, and the limited available credit and no government subsidies make the situation worse (Guebsi et al., 2024; Nyoni et al., 2023). Many farmers are unable to adopt new technology due to the high costs, and the tools needed for precision agriculture, like drones, can cost over $15,000 (Dukat et al., 2025; Hundal et al., 2023). The outcomes of precision agriculture are also greatly determined by the size of the farm and the location of the farm. Larger visual obstacles often present themselves to larger farms that obtain competitive advantage through scale economies, whereas smaller producers bear more acute unit costs and skills constraints (Chinonyerem et al. 2025; Kendall et al. 2022). The more 'sceptical' school of thought will likely dominate until some combination of financial incentives, supportive policies, and capacity-building initiatives start to address the 'economic viability of precision agriculture' conundrum and break the adoption 'conundrum' (Masi et al. 2023; Padhiary et al. 2025). Figure 3 summarizes some of the barriers to the adoption of precision agriculture that include high initial costs, poor access to credit, underdeveloped digital infrastructure, low digital skills, and high costs of upkeep. The operational environment of precision agriculture is marked by poor interoperability, data privacy issues, and high uncertainty of returns on investment. Other obstacles include inadequate fitness for smallholders and the reluctance to change.

Figure 3. Key challenges in the adoption of precision agriculture

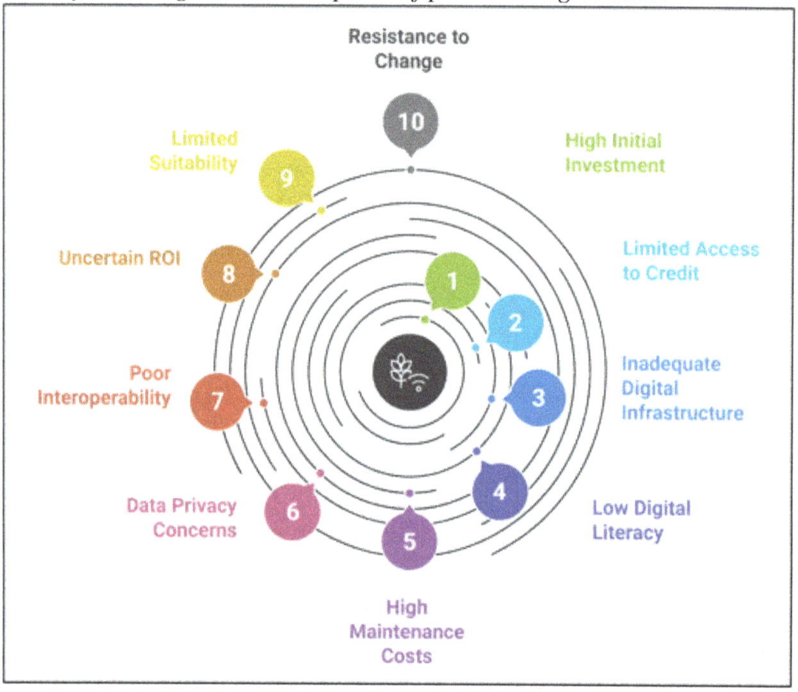

6. REGIONAL PERSPECTIVES ON ADOPTION BARRIERS

The extent to which PA technologies are adopted differs across regions due to varying infrastructure, farm size and structure, digital and financial literacy, and policies (Dibbern et al., 2025; Petrovic & Csambalik, 2025). Developed and developing regions show that barriers to adoption differ and are related to the structural differences between commercial-scale farming systems vis-à-vis smallholder-dominated agricultural systems (Aranguri et al., 2025; John et al., 2023). These regional differences are important when developing PA initiatives to be just and effective (Kroupova et al., 2025).

North America and Europe show significant adoption levels due to positive conditions for PA implementation: more considerable average farm sizes, advanced mechanization, and reliable broadband and electricity, alongside supportive institutions (Frona & Frona-Hadas, 2025; Ronzhin et al., 2025). Commercial farms tend to invest in technology, such as GPS-enabled machinery, yield monitors, and variable rate technologies. Automation and data-driven input optimization are spurred by high labor costs, rigid environmental policies, and increasing automation (Varzaru, 2025). However, developed regions have patchy adoption. Smaller and family-operated farms often bear barriers like those present in developing economies, such as high initial investment costs, unpredictable return levels, and inadequate capability to manage sophisticated digital systems (Baharin et al., 2024). Data ownership, governance, and interoperability issues continue to show that farm size is a more critical factor than national wealth (Kroupova et al., 2025).

Over a significant part of Asia, Africa, and Latin America, digital infrastructure is meager, and extension services are largely under-resourced. In these regions, fragmented landholdings and limited mechanization dominate agriculture (Matchaya et al., 2025). Smallholders dominate agriculture, and their production systems are poorly aligned with PA technologies (Nelson, 2025). As a result of unreliable electricity, digital illiteracy, and restricted broadband, IoT and cloud systems are limited (Raj & Prahadeeswaran, 2025). Because of these issues, the adoption of PA is often limited to pilot projects (Aliyeva et al., 2025). China partially demonstrates that low-cost, context-specific solutions can partially overcome these challenges. Selective adoption despite land fragmentation has been enabled by investments in digital platforms, rural infrastructure, and farmer training. Latin America exhibits a mixed pattern, with persistent smallholder exclusion due to land tenure, financial access, and inequality, but advanced adoption on large commercial farms (Dibbern et al., 2025).

7. BOUNDARIES OF TECHNOLOGY INTEGRATION

Although PA has notable positive effects on productivity, operational efficiency, and sustainability, its adoption in farming systems is limited and often constrained by structural, technical, and human factors. These boundaries emerge in part from the intricacies associated with individual technologies and the challenges of multiple digital technologies in response to varying farming systems, farming resources, and the skills of farmers (Atik, 2022; Sullivan et al., 2024). Figure 3 outlines the major constraints of PA adoption, such as poor interoperability, vendor lock-in, system complexity, skills gaps, digital infrastructure gaps, financial constraints, geographical inequities of service, and the overall weakness of agricultural extension and training services.

Interoperability is the most significant technical obstacle, as PA systems are dependent upon GPS, sensors, drones, and other advanced technology systems that are designed proprietary (Cordeiro and Ferreira 2025). These proprietary systems increase operational costs and cause redundant tasks. Vendor lock-in is another obstacle that prevents farmers from utilizing systems and features of differing manufacturers (Hackfort, 2023; Kulkarni and Bedekar, 2023). Even though open-source systems and standardized data protocols may make system interoperability possible, ineffective governance, poor industry coordination, and little incentive will continue to restrict the existing open-source systems and standardized protocols (Atik and Martens, 2021; Shafik, 2024).

The uptake of PA is also limited by human capacity constraints. Many farmers do not have the requisite skills to interpret data and troubleshoot in the digital realm, as many advanced technologies demand. This is especially true for farmers in developing countries (Sun et al., 2025). Initiatives such as digital extension programs and field demonstrations can bridge these gaps (Nafulah, 2020). Furthermore, the rural areas most in need of PA are the same areas where reliable electricity, internet, and mobile network availability are lacking (Emmanuel, 2024; Jouanjean et al., 2020). Although new peer-learning and mobile advisory models are emerging, implementation costs, insufficiently funded extension services, and the remaining underdeveloped gaps of the digital ecosystem and mobile services still limit scalable and inclusive adoption (Amraouy et al., 2025; Brewster et al., 2024).

Figure 4. Technology integration boundaries in precision agriculture

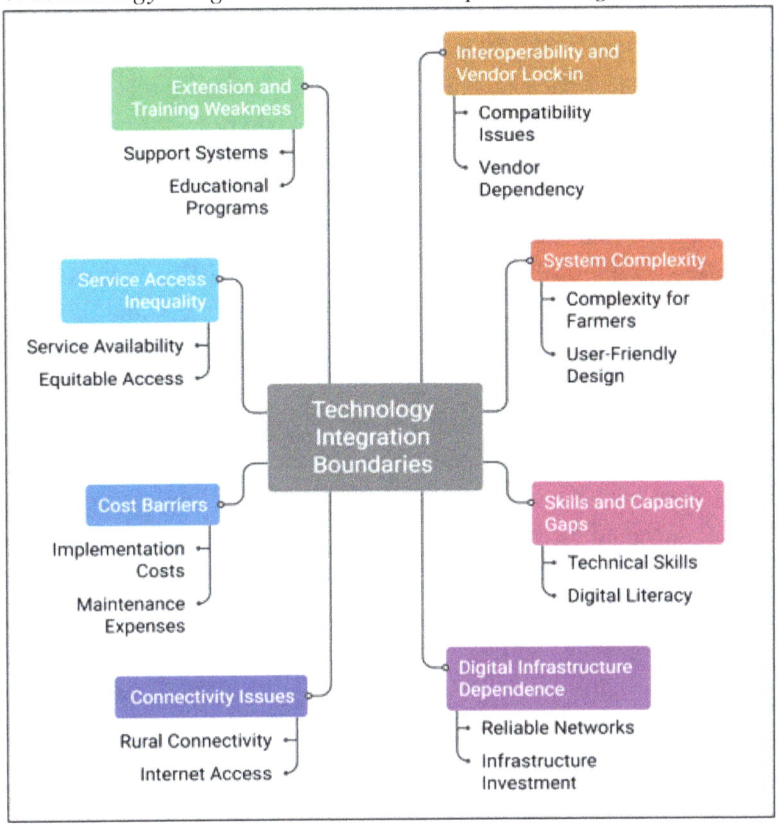

8. POLICY FRAMEWORKS AND GOVERNMENT SUPPORT FOR PA ADOPTION

The government has an important role in making PA technologies easier to take up and effectively be used. This role is played through financial incentives, support from institutions, and suitable regulations (Mohammadi et al., 2025; Tomashuk et al., 2025). It is the well-designed policies that, in conjunction with the financial and logistical support, can facilitate the entry of farmers, especially smallholder ones, into the soil that is rich in data through eco-friendly farming (Avila-Sanchez, 2025; Vaishnavi, 2025). The financial support in the form of subsidies, tax incentives, and low-interest rural credit schemes has effectively lowered the financial risks of investing in PA and has thus encouraged farmers to adopt new technologies (Frona & Frona-Hadas, 2025; Matchaya et al., 2025). It has been statistically proven that adoption of new technology is generally the case among farmers when policy support is available; thus, the primary constraint is not the resistance but the risk exposure (Nelson, 2025; Ronzhin et al., 2025).

One major issue in policy has been digital infrastructure that is lacking in the countryside. Reliable and stable broadband and electricity are the basic requirements for IoT systems, cloud analytics, and remote sensing, examples of PA applications (Dhir & Mishra, 2025; Ishore et al., 2026). The government can use targeted public-private partnerships and connectivity initiatives for the rural areas to overcome these gaps (Oklikah et al., 2025; Raj & Prahadeeswaran, 2025; Varzaru, 2025). The development of integrated digital agricultural frameworks in some advanced economies has been a big step, while there are still many developing countries lacking domestic strategies that would link infrastructure, finance, training, and institutional support (Degila et al., 2023; Mouedden et al., 2025).

The governance of data influences the outcome of adoption on a larger scale. Over time, farmers have become more open to expressing their concerns about data ownership, privacy, and the risk of technology vendors misusing the data (Matchaya et al., 2025; Nelson, 2025). Trust can be built through a legal framework that guarantees the ownership of data and establishes clear data use policies. Hence, it is necessary to have policies that guarantee equity and encourage the adoption of PA solutions that are not only cost-effective but also tailor-made to the specific context to achieve the facilitation of the inclusive and sustainable uptake across different agricultural systems (Avila-Sanchez, 2025; Mohammadi et al., 2025).

9. STRATEGIES TO OVERCOME ADOPTION BARRIERS

International agriculture contends with climate change, growing populations, and diminishing resources (Forero et al., 2025; Raj & Prahadeeswaran, 2025). The main challenges to PA adoption include lack of capability, poor structure, and insufficient finance (John et al., 2023; Manning, 2024). These constraints highlight the need for cross-border strategies to promote the adoption of PA that is more inclusive, low-cost, and scalable (Ghirardelli & Tarolli, 2025; Wang et al., 2025). In the context of Table 2, the IoT-driven climate-smart practices, drones, smart irrigation, and AI decision support are proven to be optimal from economic, environmental, and social perspectives for both crops and livestock systems. In addition, balanced the adoption of PA. The adoption of PA is significantly inhibited by the high initial and running costs (Ronzhin et al., 2025). The combined use of credit, subsidized loans, and public-private partnerships to lower the financial risk proves to be useful (Anjanappa, 2025; Olabanji & Chitakira, 2025). The accessibility of PA is increased by modular, service-oriented, and affordable PA (Matchaya et al., 2025; Nelson,

2025). The risk of vendor lock-in is less with low-cost IoT devices and open-source systems, as well as with shared services. This also led to the need for reliable Internet and power. Improvement of rural infrastructure is essential (Bhattacharyya et al., 2020; Ogunleke & Idowu, 2025). Activities such as training, peer learning, and digital literacy empower farmers (Chen, 2025; Jamtarat & Thongcha, 2025). The management of data rights, ethics, and sustainability fosters an inclusive and trusted adoption (Fusco et al., 2020).

Table 2. Strategic mechanisms and emerging innovations for inclusive and sustainable precision agriculture

Strategic Area	Key Actions / Innovations	Expected Outcomes	References
Sensor Technology in Dairy	Automated estrus & body condition sensors	Higher NPV and improved herd productivity	(Munz & Schuele, 2022; Vinholis et al., 2025)
IoT-Based Crop Management	Real-time soil and crop monitoring	Optimized irrigation and higher yields	(Hundal et al., 2023; Goel & Vishnoi, 2025; Mouedden et al., 2025)
AI-Driven Smart Farming	AI-based decision support systems	Efficient farm management decisions	(Raj & Prahadeeswaran, 2025; Wu & Zhong, 2025)
Precision Irrigation	Sensor-guided smart irrigation	Water savings and improved crop performance	(Awate & Nagne, 2025; Gao et al., 2025; Nurmalitasari et al., 2025)
Data Analytics for Yields	Big-data-based yield prediction models	Better forecasting and resource allocation	(David et al., 2024; Sharma & Gawade, 2025)
Drone-Based Monitoring	Aerial crop surveillance using drones	Early stress detection and timely action	(Guebsi et al., 2024; Nyoni et al., 2023; Han et al., 2025)
Climate-Smart Agriculture	Precision tools for climate adaptation	Resilience and sustainable production	(Pengpeng et al., 2025; Surendran et al., 2024)
Resource-Efficiency Framework	System-wide PA integration	Lower input costs and environmental impact	(Ahmed et al., 2025; Bhattacharyya et al., 2020; Wang et al., 2025)
Precision Livestock Systems	Smart feeding and health monitoring	Improved livestock health and output	(Lan & Ban, 2025; Padhiary et al., 2025)
Community Engagement	Training and local support programs	Higher adoption among smallholders	(Obaideen & Hassoun, 2025; Vinholis et al., 2025)
Sensor Technology in Dairy	Automated estrus & body condition sensors	Higher NPV and improved herd productivity	(Aranguri et al., 2025; Vaishnavi, 2025)

10. DATA PRIVACY AND ETHICS IN PRECISION AGRICULTURE

The growth of PA and its technologies not only benefits farmers and agricultural businesses but also creates significant ethical issues related to data privacy, ownership, and governance (Sharma et al., 2025; Uddin et al., 2024). Among the main producers of agricultural technology are the PA systems that can generate very detailed and sensitive data relative to farming operations, creating fears about misuse, monitoring, and unequal distribution of advantages (Gorain et al., 2025; Tzachor et al., 2022). Decision-making regarding data is often taken out of the farmers' hands, especially in cases involving proprietary platforms and non-transparent contracts that favor the tech companies (Onunga, 2025; Zangana et al., 2025). Therefore, to make sure the right thing is done, it will be necessary to have the governance frameworks that will set the rules, such as conferring and recognizing the data ownership rights clearly, requiring informed consent, and creating the portable data provisions (Garwe et al., 2025). All these steps in the right direction will not be enough unless there is very tight security around the data, as the interconnected devices have a comparatively higher chance of facing breaches and being tampered with (Ali et al., 2024; El-Sagheer & Hamd, 2025). The government should make it a rule for the security-by-design standards and breach reporting to be followed (Kumar, 2024). Another problem posed by using AI in PA is that of the bias and the lack of transparency, particularly when the data from smallholder farmers is likely to be underrepresented (Adewusi et al., 2024; Dara et al., 2022). The very nature of AI systems must be that they are both transparent and inclusive while at the same time allowing human oversight (Devasia et al., 2024). On the other hand, ethical governance must include considerations of the environment and labor that will lead to circular economy approaches and reskilling programs (Alawode et al., 2024). Responsible innovation in PA will, however, depend on building trust, equity, and strong policies that will not only prioritize farmer autonomy but also sustainability (Bogucki, 2025).

11. EMERGING TRENDS AND FUTURE DIRECTIONS

The agriculture sector is moving rapidly towards a fully digital ecosystem that smart farming utilizes, and technology is largely facilitating it. The integration of AI, IoT, blockchain, and other advanced technologies has resulted in analyzing and predicting the agricultural sector through a system that is not only capable of producing more but also sustainable outputs (Obaideen & Hassoun, 2025; Sharma et al., 2024; Wu & Zhong, 2025). Artificial intelligence is taking a leading role in technology since it is the main processor of the real-time data coming from IoT devices, drones, and satellites, which in turn greatly improves the decisions over irrigation, pest and nutrient management, etc. (Pengpeng et al., 2025; Veeramachaneni, 2025).

On the contrary, technologies like edge computing and federated learning grant local data processing capability, resulting in increased efficiency, privacy, and accessibility, particularly in rural areas (Akhund, 2025; Chelliah et al., 2024). Moreover, the blockchain acts as a watchdog of the entire agriculture technology ecosystem, injecting traceability and secure data exchange through the food value chain (Alanhdi & Toka, 2024; Tyagi, 2025). It advocates for use cases like tracking agricultural products from the farm to the table (farm-to-fork), verification of subsidies, and proofs of sustainable practices. However, the degree of such solutions' success will be tied to the development of lightweight blockchain protocols and seamless integration with IoT devices (Hemdan & Sayed, 2025).

Infused with AI, robotics and autonomous machines are revolutionizing fieldwork and doing it accurately the whole seeding and harvesting process, for instance (Okpala & Nwamekwe, 2025; Zia & Haleem, 2025). On top of that, digital twin technology is a great help in decision-making by providing simulations of the climate, resources, and economic variables in real time (Bhambri & Khang, 2025; Rojek et al., 2025). All these future trends taken together are precursors to a future of precision agriculture that will not only be resilient and inclusive but also will have long-term ecological and economic viability.

Conclusion

In this chapter, PA is highlighted as a major change in contemporary agriculture that could serve as an efficient solution to the problems of food scarcity, changes in climate, depletion of resources, and pollution of the environment. Through the implementation of site-specific, data-driven management, PA not only increases output but also lowers the demand for land, water, and the use of chemicals. On the contrary, the chapter highlights the different levels of PA benefits that individuals enjoy. Among other reasons that still limit the technology's use mainly to small farmers in the less developed regions are the extremely high investment costs, the absence of digital infrastructure, the lack of required skills, the concerns over data privacy, and the uncoordinated policies. These obstacles suggest that PA is still not only a technological intervention but also a socio-economic and governance challenge. The transition of this system to a sustainable and inclusive manner of functioning would ultimately depend on the presence of adequate policy support, the provision of affordable financing schemes, the establishment of the rural digital infrastructure, and the development of efficient farmer training and extension programs. Besides these, the scheme could also move from being precision-oriented to a resilient, inclusive, and regenerative agricultural system that will eventually help achieve global sustainability through the combination of ethical data governance and collaboration in innovation among governments, industry, researchers, and farming communities.

REFERENCES

Addagatla, D., & Putta, M. (2025). AI-driven harvest: Connecting farmers and consumers for a sustainable agricultural economy. *Home*, *1*(1), 13–21.

Adewusi, A. O., Asuzu, O. F., Olorunsogo, T., Iwuanyanwu, C., Adaga, E., & Daraojimba, D. O. (2024). AI in precision agriculture: A review of technologies for sustainable farming practices. *World Journal of Advanced Research and Reviews*, *21*(1), 2276–2285.

Ahmad, S., Wassay, M., & Hussain, M. A. (2025). *Cloud computing: Empowering the next generation of agricultural research*. Trends in Applied Sciences.

Ahmed, M., Raza, M. Y., Malik, N. A., & Malik, A. (2025). Climate-resilient agriculture (CRA): Pathway to sustainable development. In *Climate resilient and sustainable agriculture: Volume 1: Adaptation and mitigation strategies* (pp. 185-212). Springer Nature Switzerland.

Akhund, M. (2025). Validating federated learning performance in practice: An agricultural edge hardware testbed analysis (Doctoral dissertation, The University of Waikato).

Al-Jaf, I. H., Ibrahim, M. S., Hammadi, A. N., Al-Dulaimy, A. F., & EAlfahdawi, H. (2025). A review of smart agriculture and its impact on revolutionizing agricultural production.

Alanhdi, A., & Toka, L. (2024). A survey on integrating edge computing with AI and blockchain in maritime domain, aerial systems, IoT, and industry 4.0. *IEEE Access : Practical Innovations, Open Solutions*, *12*, 28684–28709.

Alawode, A., Blessing, A. O., & Chiamaka, O. T. (2024). Integrating IoT and AI in sustainable agriculture to mitigate environmental risk and financial misuse.

Ali, G., Mijwil, M. M., Buruga, B. A., Abotaleb, M., & Adamopoulos, I. (2024). A survey on artificial intelligence in cybersecurity for smart agriculture: State-of-the-art, cyber threats, artificial intelligence applications, and ethical concerns. *Mesopotamian Journal of Computer Science*, *2024*, 53–103.

Aliyeva, M., Huseynova, L., & Agabekova, G. (2025). The role of investment in the development of smart agricultural technologies in megacities.

Amraouy, O., Benbrahim, M., & Kabbaj, M. N. (2025). A blockchain-and self-sovereign identity-based collaborative framework for secure and verifiable cross-organizational data sharing in smart irrigation. *Smart Agricultural Technology*, •••, 101654.

Anjanappa, J. (2025). From barriers to solutions: Collaborative financing pathways for climate-resilient agri-SMEs in India. Available at *SSRN* 5377754.

Aranguri, M., Mera, H., Noblecilla, W., & Lucini, C. (2025). Digital literacy and technology adoption in agriculture: A systematic review of factors and strategies. *AgriEngineering*, *7*(9), 296.

Aremora, S., Sarfo, J. B., Abayomi Taiwo Fashina, K. M., & Balogun, M. O. A. (2025). Blockchain and AI integration for agricultural value chains in African markets.

Atik, C. (2022). Towards comprehensive European agricultural data governance: Moving beyond the "data ownership" debate. *IIC-International Review of Intellectual Property and Competition Law*, *53*(5), 701–742.

Atik, C., & Martens, B. (2021). Competition problems and governance of non-personal agricultural machine data: Comparing voluntary initiatives in the US and EU. *J. Intell. Prop. Info. Tech. & Elec. Com. L.*, *12*, 370.

Avila-Sanchez, H. (2025). Towards strengthening public policies for the digital transition in Mexican agriculture. In *Digital transition in agriculture and public policies* (p. 22640). Insights from Latin America.

Awate, P. L., & Nagne, A. D. (2025). Satellite imagery and GIS applications in precision agriculture. In *AI and data analytics in precision agriculture for sustainable development* (pp. 185–206). Springer Nature Switzerland.

Baharin, A. T., Ishak, N. A., Redzuan, N. A. L., Yusof, S. M., Sahadun, N. A., Hati, D. M., & Jamaluddin, S. P. S. (2024). Exploring the level of digital literacy and the adoption of precision farming technologies among smallholder paddy farmers in Kedah.

Banerjee, P., & Bhat, A. (2025). Cultivating justice: The economic-legal-technological nexus in advancing sustainable agriculture in the 21st century. *Journal of Humanities. Arts and Social Science*, *9*(12), 2222–2235.

Bhambri, P., & Khang, A. (2025). Edge computing for enhancing efficiency and sustainability in green transportation systems. In *Driving green transportation system through artificial intelligence and automation: Approaches, technologies and applications* (pp. 43–65). Springer Nature Switzerland.

Bhamini, K., Kumar, A., Magrey, A. H., & Kulsoom, S. (2025). The role of GIS and GPS in precision fertilization: Optimizing nutrient management for sustainable cropping. *International Journal of Research in Agronomy*, *8*, 260–267.

Bhattacharyya, P., Pathak, H., & Pal, S. (2020). Barriers to adaptation of climate-smart agriculture. In *Climate smart agriculture: Concepts, challenges, and opportunities* (pp. 155–167). Springer Singapore.

Bogucki, A. (2025). Applied trustworthy data and artificial intelligence governance in precision agriculture. In *Engineering and value change* (pp. 101–124). Springer Nature Switzerland.

Brewster, C., Kalatzis, N., Nouwt, B., Kruiger, H., & Verhoosel, J. (2024). Data sharing in agricultural supply chains: Using semantics to enable sustainable food systems. *Semantic Web*, *15*(4), 1207–1237.

Chaurasia, K., Yasmin, G., & Singh, D. (2026). Overview of geospatial technology and machine learning in agriculture. In *Agricultural Insights from Space* (pp. 1–27). Academic Press.

Chelliah, P. R., Rahmani, A. M., Colby, R., Nagasubramanian, G., & Ranganath, S. (Eds.). (2024). *Model optimization methods for efficient and edge AI: Federated learning architectures, frameworks and applications*. John Wiley & Sons.

Chen, X. (2025). The role of modern agricultural technologies in improving agricultural productivity and land use efficiency. *Frontiers in Plant Science*, *16*, 1675657.

Cheong, J. Q., Muttaqin, E. I., Wiranatakusuma, D. B., & Jusoh, M. H. (2025). Technology adoption and food security in ASEAN among women. In *International Conference on Mathematical and Statistical Physics, Computational Science, Education, and Communication (ICMSCE 2025)* (Vol. 13941, pp. 334-343). SPIE.

Chinonyerem, C. A., Ajibade, S. B., Rebecca, O. T., & Ahaiwe, E. C. (2025). The economic implications of precision agriculture adoption on farm profitability and resource use efficiency. *International Journal of Innovation Research and Advanced Studies*.

Cogato, A., & Schillaci, C. (2025). "From data to fork" data-driven science inspired by policy initiatives for the promotion of smart agricultural technologies. *Smart Agricultural Technology*, •••, 100928.

Cordeiro, M., & Ferreira, J. C. (2025). Beyond traceability: Decentralised identity and digital twins for verifiable product identity in agri-food supply chains. *Applied Sciences (Basel, Switzerland)*, *15*(11), 6062.

Dara, R., Fard, S. M. H., & Kaur, J. (2022). Recommendations for ethical and responsible use of artificial intelligence in digital agriculture. *Frontiers in Artificial Intelligence*, *5*, 884192.

Daszkiewicz, T. (2022). Food production in the context of global developmental challenges. *Agriculture*, *12*(6), 832.

David, A., Rao, P. S., Srivastava, D., Barthwal, A., Thomas, A., Thomas, T., & Thomas, C. A. (2024). Big data, AI, and geoinformatics for sustainable agriculture: An introduction. In *Geoinformatics* (pp. 131–161). Apple Academic Press.

Degila, J., Sodedji, F. A. K., Avakoudjo, H. G. G., Tahi, S. P. G., Houetohossou, S. C. A., Honfoga, A. C., & Assogbadjo, A. E. (2023). Digital agriculture policies and strategies for innovations in the agri-food systems—Cases of five West African countries. *Sustainability*, *15*(12), 9192.

Devasia, J. V., Deepanramkumar, P., Helensharmila, A., & Yenduri, G. (2024). Ethical considerations and social implications. In *Sustainable Farming through Machine Learning* (pp. 267–279). CRC Press.

Dhir, S. B., & Mishra, G. (2025). Digital agriculture mission: Unlocking opportunities and overcoming challenges for farmers using smartphones. *Journal of Information, Communication and Ethics in Society*.

Dibbern, T., Romani, L., & Massruhá, S. (2025). Drivers and barriers to digital agriculture adoption: A mixed-methods analysis of challenges and opportunities in Latin America. *Sustainability*, *17*(8), 3676.

Douzal, C., Mosnier, A., Chermain, C., Lynch, J., Harrison, P., Jones, S., ... & Smith, A. (2024). Transforming food and land systems to achieve the SDGs.

Dukat, D., Wahyuni, I., & Prihtiyantoro, W. (2025). The role of precision agriculture technologies in enhancing sustainable farming practices among smallholder farmers. *The Journal of Academic Science*, *2*(6), 1666–1677.

El-Sagheer, A. M., & Hamd, M. M. (2025). Ethical and regulatory considerations of machine learning in modern agriculture. In *Machine Learning for Plant Biology* (pp. 317-340).

Elechi, J. O. G., Nwiyi, I. U., & Adamu, C. S. (2022). Global food system transformation for resilience. In *Food Systems Resilience*. IntechOpen.

Emmanuel, M. (2024). Regulatory frameworks for cloud computing in agriculture in West and East Africa.

Fayaz, S., Kanth, R. H., Bhat, T. A., Dar, E. A., Shah, Z. A., Bashir, M., & Salem, A. (2025). Estimation of yield and nitrogen use efficiencies in hybrid maize varieties through site-specific nitrogen management based on leaf color chart (LCC). *Scientific Reports*, *15*(1), 4686.

Fernandes, S. (2025). The role of artificial intelligence in accelerating sustainable technological innovation. *Clareus Scientific Science and Engineering*, *2*, 32–40.

Forero, Á., Cruz, J. C., & Munoz, C. (2025). Adoption of agricultural innovations within the 'farm to fork' strategy: A realistic review of barriers, paradoxes, and avenues for change. *Sustainability*, *17*(21), 9493.

Frona, D., & Frona-Hadas, D. (2025). Digital transformation in rural Hungary: Infrastructure, skills, and uptake across regions. *Land (Basel)*, *14*(10), 2034.

Fusco, G., Melgiovanni, M., Porrini, D., & Ricciardo, T. M. (2020). How to improve the diffusion of climate-smart agriculture: What the literature tells us. *Sustainability*, *12*(12), 5168.

Gao, X., Qin, G., Yin, F., Liu, J., Lin, H., Wang, Z., & Wang, Z. (2025). Multi-objective optimization of cropping structures and irrigation strategies for sustainable agricultural development in arid regions. *Agricultural Water Management*, *321*, 109909.

Garwe, E. C., Arpaci, S., Moss, A. J., Bailey, K., Suchith, A., Oyewole, O. B., & Thondhlana, J. (2025). Ethical crossroads in agricultural data: Perspectives on trust, equity and sustainability. *Outlook on Agriculture*, *54*(4), 323–331.

Ghirardelli, A., & Tarolli, P. (2025). Bridging the gap between science, policy, stakeholders: Delineating a virtuous framework for agriculture under climate change scenarios. *iScience*, *28*(9).

Goel, R. K., & Vishnoi, S. (2025). Space-age evolution—Remote sensing and IoT for productive and sustainable agricultural landscape. *Sustainable Futures : An Applied Journal of Technology, Environment and Society*, *10*, 101280.

Gorain, S., Dutta, S., & Gawdiya, S. (2025). Ethical harvest: Ensuring fairness and privacy. In *AI-Driven IoT Solutions in Agriculture: New Insights* (pp. 169-193). Springer Nature Switzerland.

Guebsi, R., Mami, S., & Chokmani, K. (2024). Drones in precision agriculture: A comprehensive review of applications, technologies, and challenges. *Drones (Basel)*, *8*(11), 686.

Haas, M., Frankema, E., & Giller, K. E. (2025). Growing towards a food secure Africa in 2050: Reflections and pathways. *Pathways to African Food Security*, 267.

Hackfort, S. (2023). Unlocking sustainability? The power of corporate lock-ins and how they shape digital agriculture in Germany. *Journal of Rural Studies*, *101*, 103065.

Han, R., Liang, X., Shu, L., Jing, X., & Yang, F. (2025). TeaWeeding-Action: A vision-based dataset for weeding behavior recognition in tea plantations. *Frontiers in Plant Science*, *16*, 1722007.

Hemdan, E. E. D., & Sayed, A. (2025). Smart and secure healthcare with digital twins: A deep dive into blockchain, federated learning, and future innovations. *Algorithms*, *18*(7), 401.

Huang, I. Y., Beacham, A. M., Vickers, L. H., Mu, H., Tremma, O., Maritan, E., & Monaghan, J. M. (2025). Drivers, barriers, and grower perspectives of innovation adoption in the UK controlled environment agriculture sector. *Smart Agricultural Technology*, •••, 101748.

Hundal, G. S., Laux, C. M., Buckmaster, D., Sutton, M. J., & Langemeier, M. (2023). Exploring barriers to the adoption of internet of things-based precision agriculture practices. *Agriculture*, *13*(1), 163.

Hussain, M., Abdullah, M., Ashraf, M. N., Shaheen, A., Farooqi, Z. U. R., Abbas, A., & Tariq, Y. (2025). World food hunger in 2050 and nano solutions: Probabilities and prospects based on plant-based food. In *Nanomaterials for Enhanced Plant-Based Food Production* (pp. 11–21). Academic Press.

Indrabhai, M. N. A., & Patel, N. S. (2025). Agriculture and growth economy in new India. *Commerce Management & Economics, 55*.

Ishore, B., Bhargav, A. M., Patel, S. K., Kumar, V., Sinha, J., & Kumar, S. (2026). Advancing smart agriculture: How IoT and 5G technologies enhance precision farming and sustainability. *International Journal of Communication Systems*, *39*(1), e70310.

Jamtarat, S., & Thongcha, C. (2025). Agricultural sector transformation through a systematic literature review (SLR) approach: Innovations, challenges and future opportunities. [AIJoES]. *Anshara International Journal of Education and Science*, *2*(1), 1–7.

Jatav, H. S., & Rajput, V. D. (2025). Sustainable nutrient management under climate change. *Frontiers in Agronomy*, *7*, 1625151.

Jennings, R., Henderson, A. D., Phelps, A., Janda, K. M., & van den Berg, A. E. (2023). Five US dietary patterns and their relationship to land use, water use, and greenhouse gas emissions: Implications for future food security. *Nutrients, 15*(1), 215.

Jha, C. K., Singh, V., Stevanović, M., Dietrich, J. P., Mosnier, A., Weindl, I., & Lotze-Campen, H. (2022). The role of food and land use systems in achieving India's sustainability targets. *Environmental Research Letters, 17*(7), 074022.

Jha, C. K., Singh, V., Stevanovic, M., Dietrich, P., Saxena, S., Mosnier, A., ... & Schmidt-Traub, G. (2021). Pathways to sustainable land-use and food systems in India by 2050.

John, D., Hussin, N., Shahibi, M. S., Ahmad, M., Hashim, H., & Ametefe, D. S. (2023). A systematic review on the factors governing precision agriculture adoption among small-scale farmers. *Outlook on Agriculture, 52*(4), 469–485.

John, L. B., & James, G. J. (2025). Geospatial big data analytics for precision agriculture: Enhancing productivity and sustainability. *Journal of Scientific Reports, 10*(1), 116–129.

Jouanjean, M. A., Casalini, F., Wiseman, L., & Gray, E. (2020). Issues around data governance in the digital transformation of agriculture: The farmers' perspective.

Kannan, S. (2025). *Transforming agriculture for the digital age: Integrating artificial intelligence, cloud computing, and big data solutions for sustainable and smart farming systems*. Deep Science Publishing.

Karunathilake, E. M. B. M., Le, A. T., Heo, S., Chung, Y. S., & Mansoor, S. (2023). The path to smart farming: Innovations and opportunities in precision agriculture. *Agriculture, 13*(8), 1593.

Kay, M., Bunning, S., Burke, J., Boerger, V., Bojic, D., Bosc, P. M., ... & Ziadat, F. (2022). The state of the world's land and water resources for food and agriculture 2021. *Systems at breaking point.*

Kendall, H., Clark, B., Li, W., Jin, S., Jones, G. D., Chen, J., & Frewer, L. J. (2022). Precision agriculture technology adoption: A qualitative study of small-scale commercial "family farms" located in the North China Plain. *Precision Agriculture, 23*(1), 319–351.

Kroupova, Z. Z., Aulova, R., Rumankova, L., Bajan, B., Cechura, L., Simek, P., & Jarolimek, J. (2025). Drivers and barriers to precision agriculture technology and digitalisation adoption: Meta-analysis of decision choice models. *Precision Agriculture, 26*(1), 17.

Kulkarni, A., & Bedekar, M. (2023). Standardization in cloud computing: Unlocking the potential of a fragmented industry. In *International Conference on Advancements in Smart Computing and Information Security* (pp. 3-12). Springer Nature Switzerland.

Kumar, D. (2024). Ethical considerations of AI and IoT in farming. In *Agriculture 4.0* (pp. 319–333). CRC Press.

Lan, J., & Ban, Q. (2025). The farm-level economic and environmental benefits of precision agriculture technology adoption: A meta-analysis of global evidence. *Sustainability*, *17*(24), 11223.

Liu, J. J., Wu, H., & Riaz, I. (2025). Advanced technologies for smart fertilizer management in agriculture: A review. *IEEE Access : Practical Innovations, Open Solutions*.

Lotsis, S., Georgousis, I., & Papakostas, G. A. (2026). *AI, precision agriculture and tourism for sustainable regional development: The case of the Aegean Islands and Crete*. Sustainability.

Manhas, S., Sharma, T., Sharma, B., Khajuria, S., Thakur, N., Thakur, A., . . . Sharma, V. (2026). Climate-resilient agriculture: Strategies for nutritious food for all. In *One Planet, One Health, One Future* (pp. 357-386). Academic Press.

Manning, L. (2024). Innovating in an uncertain world: Understanding the social, technical and systemic barriers to farmers adopting new technologies. *Challenges*, *15*(2), 32.

Masi, M., Pasquale, J., Vecchio, Y., & Capitanio, F. (2023). Precision farming: Barriers of variable rate technology adoption in Italy. *Land (Basel)*, *12*(5), 1084.

Mason-D'Croz, D., & Herrero, M. (2025). Advances and future needs for modelling sustainable and just food systems transformations. *The Lancet. Planetary Health*.

Matchaya, G. C., Ebrahim, G. Y., Aheeyar, M., Ajayi, O. C., & Ajilore, O. (2025). Advancing digital agriculture for climate resilience and economic transformation in Malawi.

McFadden, J., Njuki, E., & Griffin, T. (2023). Precision agriculture in the digital era: Recent adoption on US farms.

Meena, N. K., Kumar, A., Yadav, V., Panigrahi, S., Jaganathan, T., Shekhar, G., & Narang, E. H. (2025). Emerging trends in precision horticulture integrating digital technologies for smart crop management. *International Journal of Advanced Biochemistry Research*, •••, 361–368.

Mehrotra, N., Phartiyal, M., Guha, S., Rangrass, S., & Mehrotra, N. (2025). AgDx—The Agri Digitalization Framework: A tool for transformation. *Journal of the Indian Institute of Science*, •••, 1–29.

Miglani, V. (2025). Use of AI in agricultural supply chain: A case of Gujarat warehouse owners and Bhusari cold storage. *Indian Journal of Agricultural Marketing, 39*(1spl), 178-179.

Mohammadi, A. R., Petrovic, E. K., & Perkins, N. (2025). Mapping forestry practices against sustainable development goal targets. *Frontiers in Sustainability*, *6*, 1534449.

Mouedden, M. B., Mabrouki, J., Azdem, D., & El Hajjaji, S. (2025). The development of intelligent agriculture, the contribution of the Internet of Things (IoT) to modernization and its implications for precision farming: A review. *Euro-Mediterranean Journal for Environmental Integration*, •••, 1–13.

Munz, J., & Schuele, H. (2022). Influencing the success of precision farming technology adoption—A model-based investigation of economic success factors in small-scale agriculture. *Agriculture*, *12*(11), 1773.

Nafulah, M. (2020). Digitalisation of agriculture: How to align and engage partners to move the industry into precision farming. Case study, maize farming in Belgium.

Nelson, A. (2025). The impact of digital technologies in the agro-allied sector: A systematic review of smart farming barriers in small and medium-scale enterprises.

Nurmalitasari, N., Nurchim, N., & Lestari, R. D. (2025). Artificial intelligence-driven solar smart irrigation for sustainable agriculture: Trends, challenges, and SDG implications–A systematic review. *Smart Agricultural Technology*, •••, 101665.

Nyathi, L. D., & Mutale, J. (2025). Leveraging smart technologies for sustainable economic development: Evidence from urban and rural communities in Bulawayo Province, Zimbabwe. *African Journal of Commercial Studies*, *6*(6), 21–34.

Obaideen, K., & Hassoun, A. (2025). Industry 4.0 for sustainable agriculture and food industry. In *Food Industry 4.0 and Food Sustainability* (pp. 25-48). Springer Nature Switzerland.

Ogunleke, A., & Idowu, F. O. (2025). Climate-smart agriculture and youth empowerment. Available at *SSRN* 5172794.

Okpala, C. C., & Nwamekwe, C. O. (2025). Artificial intelligence-augmented edge computing: Architectures, challenges, and future directions. *International Journal of Engineering Inventions*, *14*(9), 18–27.

Olabanji, M. F., & Chitakira, M. (2025). The adoption and scaling of climate-smart agriculture innovation by smallholder farmers in South Africa: A review of institutional mechanisms, policy frameworks and market dynamics. *WORLD (Oakland, Calif.)*, *6*(2), 51.

Onunga, J. (2025). Responsible innovation in AI and data governance in Kenyan agriculture: A review article. *Regulation*, *9*(6), 8–12.

Padhiary, M., Hoque, A., Prasad, G., Kumar, K., & Sahu, B. (2025). Precision agriculture and AI-driven resource optimization for sustainable land and resource management. In *Smart Water Technology for Sustainable Management in Modern Cities* (pp. 197–232). IGI Global Scientific Publishing.

Pengpeng, Y., Teng, F., Zhu, W., Shen, C., Chen, Z., & Song, J. (2025). Cloud–edge–device collaborative computing in smart agriculture: Architectures, applications, and future perspectives. *Frontiers in Plant Science*, *16*, 1668545.

Petrovic, B., & Csambalik, L. (2025). Enhancing precision agriculture for climate change mitigation in Visegrad countries: Factors shaping adaptation. *Land (Basel)*, *14*(2), 399.

Popp, L., Zhumasheva, S., Ibrishev, N., Islamov, E., & Nurgalieva, R. (2025). Engineering solutions for sustainable development in the agro-industrial sector. In *International Conference Innovation in Engineering* (pp. 246-265). Springer Nature Switzerland.

Raj, M., & Prahadeeswaran, M. (2025). Revolutionizing agriculture: A review of smart farming technologies for a sustainable future. *Discover Applied Sciences*, *7*(9), 937.

Rojek, I., Prokopowicz, P., Piechowiak, M., Kotlarz, P., Náprstková, N., & Mikołajewski, D. (2025). The impact of data analytics based on Internet of Things, edge computing, and artificial intelligence on energy efficiency in smart environments. *Applied Sciences (Basel, Switzerland)*, *16*(1), 225.

Ronzhin, A., Figurek, A., Surovtsev, V., & Dibirova, K. (2025). Digital transformation and precision farming as catalysts of rural development. *Land (Basel)*, *14*(7), 1464.

Sahu, B. (2024). Artificial intelligence and automation in smart agriculture: A comprehensive review of precision farming, all-terrain vehicles, IoT innovations, and environmental impact mitigation. *Int. J. Sci. Res. IJSR*, *13*(11), 656–665.

Samuel, M. K., Manickam, T., Omar, M. S., Zamir, M., Rashid, A., Sahil, N. M., ... & Farouk, N. (2025). From pedons to policy: The strategic role of soil mapping in Peninsular Malaysia's agricultural transformation.

Shafik, W. (2024). Barriers to implementing computational intelligence-based agriculture system. In *Computational Intelligence in Internet of Agricultural Things* (pp. 193–219). Springer Nature Switzerland.

Shamsoddini, A., Jelodar, S. M., & Rasouli, S. H. (2025). *Modeling opportunities and challenges for entrepreneurship development in the marketplaces of Sari County*. Village and Space Sustainable Development.

Sharma, A., Rani, S., & Sharma, A. (2025). Sustainable intelligence: Ethical issues in the evolution of intelligent systems. *Mathematical Computing and Sustainability: Predictive Modeling and Ramifications of Intelligent Systems, 57.*

Sharma, L., & Garg, P. K. (2025). *Digital twins and simulation technology: Concepts and applications.*

Sharma, M., Tomar, A., & Hazra, A. (2024). Edge computing for industry 5.0: Fundamentals, applications, and research challenges. *IEEE Internet of Things Journal, 11*(11), 19070–19093.

Sharma, R. D., & Gawade, S. (2025). Advancements in precision agriculture: A literature review of machine learning applications for crop monitoring and yield prediction. *Environmental Claims Journal*, •••, 1–45.

Sharma, S. K. (2025). Smart agriculture revolution: Cloud and IoT-based solutions for sustainable crop management and precision farming. In *AI-based advanced optimization techniques for edge computing* (pp. 253-288).

Sinha, A. K., Vaid, A., Gupta, V., Sharma, N., Gupta, S., Sharma, V. K., & Shalini, K. (2025). Productivity and economics of maize cultivation after wheat crop by different sowing methods under rain-fed conditions in Samba district of J&K. *Journal of Community Mobilization and Sustainable Development, 20*(2), 642–646.

Sousa, M., Alves, A., Antunes, R., Aguiar, M. L., Gaspar, P. D., & Pereira, N. (2026). Interoperable IoT/WSN sensing station with edge AI-enabled multi-sensor integration for precision agriculture. *Agriculture.*

Sullivan, C. S., Gemtou, M., Anastasiou, E., & Fountas, S. (2024). Building trust: A systematic review of the drivers and barriers of agricultural data sharing. *Smart Agricultural Technology, 8*, 100477.

Sun, B., Yu, J., Khattak, S. I., Tariq, S., & Zahid, M. (2025). Digital innovation, business model transformations, and agricultural SMEs: A PRISMA-based review of challenges and prospects. *Systems, 13*(8), 673.

Surendran, U., Nagakumar, K. C. V., & Samuel, M. P. (2024). Remote sensing in precision agriculture. In *Digital agriculture: A solution for sustainable food and nutritional security* (pp. 201–223). Springer International Publishing.

Tomashuk, I., Koliadenko, S., & Burdìak, M. (2025). The impact of digital innovations on the development of agricultural business. *Baltic Journal of Economic Studies*, *11*(1), 361–375.

Tyagi, A. K. (2025). Blockchain and digital twin applications in smart agriculture. In Tiwari, S. (Ed.), *CRC Press*.

Tzachor, A., Devare, M., King, B., & Avin, S., & Ó hÉigeartaigh, S. (2022). Responsible artificial intelligence in agriculture requires systemic understanding of risks and externalities. *Nature Machine Intelligence*, *4*(2), 104–109.

Uddin, M., Chowdhury, A., & Kabir, M. A. (2024). Legal and ethical aspects of deploying artificial intelligence in climate-smart agriculture. *AI & Society*, *39*(1), 221–234.

Vaishnavi, P. (2025). Strengthening climate-smart agriculture through digital extension: A systematic review of technologies, adoption, and policy implications. *International Journal of Advanced Biochemistry Research*, •••, 785–791.

Varzaru, A. A. (2025). Digital revolution in agriculture: Using predictive models to enhance agricultural performance through digital technology. *Agriculture*, *15*(3), 258.

Veeramachaneni, V. (2025). Edge computing: Architecture, applications, and future challenges in a decentralized era. *Recent Trends in Computer Graphics and Multimedia Technology*, *7*(1), 8–23.

Vinholis, M. D. M. B., Souza Filho, H. M., Carrer, M. J., Eusebio, G. D. S., & David, L. C. D. S. (2025). Twin transition in beef livestock farming: Adoption of digital technologies in Brazilian feedlots. *Journal of Agribusiness in Developing and Emerging Economies*, •••, 1–19.

Wang, Z., Dai, Y., Yang, L., & Yu, Z. (2025). Barriers to climate-smart agriculture adoption in Northeast China's Black Soil Region: Insights from a multidimensional framework. *Agriculture*, *15*(21), 2236.

Werulkar, J., & Pande, M. (2025). The adoption and efficacy of precision agriculture tools and technology in Amravati division: Exploring the insights of farmers and agtech industry. *Lex Localis*, *23*(S6), 3949–3985.

Wijerathna-Yapa, A., & Pathirana, R. (2022). Sustainable agro-food systems for addressing climate change and food security. *Agriculture*, *12*(10), 1554.

Wu, P., & Zhong, Y. (2025). Artificial intelligence in sustainable agriculture: Towards a socio-technical roadmap. *Smart Agricultural Technology*, •••, 101578.

Wulandari, D., Fajarianto, O., Prayitno, P. H., Kusuma, F. I., Gunawan, A., Syafruddin, A. B., & Basuki, A. (2025). Precision agriculture ecosystem innovation through IoT technology: A study of learning factory development for farmer empowerment in Ngajum Village. *Jurnal Syntax Imperatif: Jurnal Ilmu Sosial dan Pendidikan, 6*(5), 1299-1323.

Zainab, Z., & Astiko, W. (2025). Assessment and analysis of sustainable agroecosystem management through the implementation of an integrated farming system (IFS) in Lempenge Hamlet, Rempek Village. *Jurnal Biologi Tropis, 25*(4b), 559–567.

Zangana, H. M., Luckyardi, S., Mustafa, F. M., & Li, S. (2025). Ethical and secure AI applications in agricultural research: Challenges and opportunities. In *Ensuring secure and ethical STM research in the AI era* (pp. 43-76).

Zhao, X., Zhu, A. L., Liu, X., Li, H., Tao, H., Guo, X., & Liu, J. (2025). Current status, challenges, and opportunities for sustainable crop production in Xinjiang. *iScience*.

Zia, A., & Haleem, M. (2025). Bridging research gaps in industry 5.0: Synergizing federated learning, collaborative robotics, and autonomous systems for enhanced operational efficiency and sustainability. *IEEE Access : Practical Innovations, Open Solutions*.

Chapter 14
Embracing AI and IoT to Ensure a Sustainable Agricultural Future for Global Food Security

Randa Hassan Elsalahi
http://orcid.org/0000-0003-0033-3633
National Center for Research, Sudan

Hanan I. Mudawi
http://orcid.org/0000-0002-5155-7082
Environment and Natural Resources and Desertification Research Institute, National Center for Research, Sudan

Magdoline M. Ahmed
Institute of Land Use, Engineering, and Precision Farming Technology, Hungary

Rania Alrasheed
http://orcid.org/0000-0001-9267-0172
McGill University, Canada

Donald L. Smith
McGill University, Canada

ABSTRACT

Artificial Intelligence (AI) and the Internet of Things (IoT) are transforming modern agriculture by enabling precision farming through advanced sensors, real-time data analytics, and satellite imagery. These technologies allow precise monitoring and management of soil health, nutrients, water, crop diseases, and yields, supporting

DOI: 10.4018/979-8-3373-7257-0.ch014

site-specific interventions that optimize resources, increase productivity, and promote sustainability. Integrating AI analytics with IoT sensors provides farmers with detailed, real-time field insights beyond traditional methods. Controlled environments like greenhouses and vertical farms further enhance efficiency by regulating climate and nutrients, reducing food waste, and improving market access. Adoption challenges include high costs, technical complexity, data privacy, and sociocultural barriers, especially for smallholders in developing areas. Overcoming these through policy, education, and infrastructure is vital to unlock AI and IoT's full potential in sustainable food production and global food security amid growing pressures.

INTRODUCTION

Artificial Intelligence AI and the Internet of Things (IoT) are revolutionizing agriculture by enabling smarter and more precise farming practices that have the potential to drastically improve productivity, sustainability, and resource efficiency (AlZubi & Galyna, 2023). Smart farming has emerged as a vital element in promoting sustainable agricultural practices by leveraging technology to enhance efficiency and environmental stewardship (Wu et al., 2010). Conventional crop cultivation requires substantial time, labor, and financial resources (Roux et al., 2019). Given the escalating global food demand and challenges such as climate change, resource scarcity, and environmental degradation, AI and IoT integration offers innovative solutions for optimizing agricultural management (Dhal & Kar, 2024).

Building on these benefits, precision agriculture technologies play a crucial role in enhancing farm productivity by enabling more efficient and effective farming practices. Technologies like Global Positioning System (GPS)-guided equipment, remote sensing, Geographic Information Systems (GIS) and data analytics allow farmers to monitor field variability, assess soil and nutrient differences, and apply inputs more accurately (Kasenzu, 2024). The combination of IoT and AI technologies empowers precision agriculture by enabling real-time monitoring and precise application of water, fertilizers, and pesticides. Smart sensors and AI-driven analytics assist farmers in delivering inputs only where and when needed, reducing costs, preventing waste, and minimizing environmental impact for more sustainable farming (Miller et al., 2025; Muninarayanappa & Ranjan, 2024; Ragazou et al., 2022; Sajid et al., 2023).

Beyond outdoor farming, AI and IoT enhance controlled environment agriculture, such as greenhouses and vertical farms, where precise climate control and resource delivery maximize food quality and output. Despite these advances, barriers remain, including cost, digital literacy, infrastructure gaps, data privacy concerns, environmental impacts, and acceptance by smallholder farmers, especially in developing regions (Assimakopoulos et al., 2025). AI and IoT technologies have become transformative forces in agriculture, enabling smarter, more precise, and sustainable farming practices worldwide, as shown in Figure 1.

This chapter delves into the core elements of precision agriculture driven by AI and IoT, examining their applications in soil health monitoring, nutrient management, water conservation, pest and disease detection, yield forecasting, and controlled-environment farming. Additionally, it explores countries and companies that have embraced precision agriculture technologies. It also explores the technological, socioeconomic, environmental, and policy challenges that must be addressed to harness these innovations for global food security and sustainable agriculture in the 21st century.

Figure 1. Top 20 Countries leading research in the field of AI and IoT in precision agriculture (2013–2025): Evidence from Web of Science Data

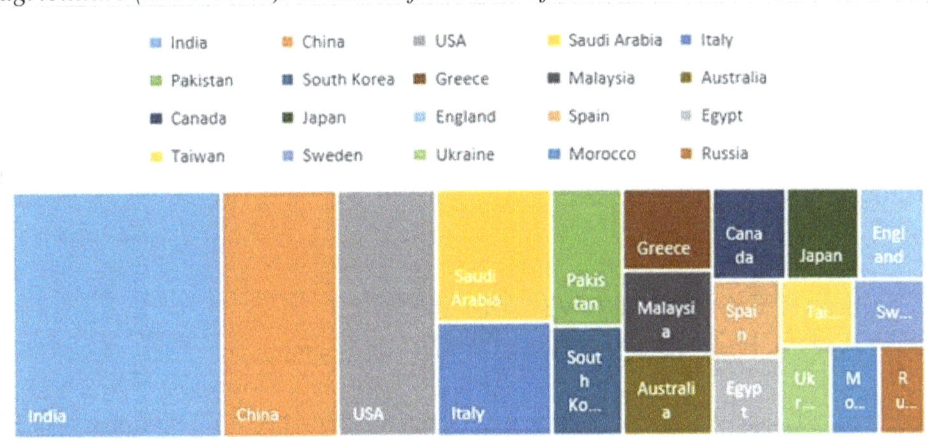

1. IMPORTANCE OF PRECISION AGRICULTURE FOR SUSTAINABILITY AND FOOD SECURITY

As the global population is expected to reach 10 billion by 2050, coupled with the growing scarcity of natural resources and environmental deterioration, adopting advanced and efficient food production systems and techniques is crucial to satisfy the exponential demand for food (Kirkaya, 2020). We need to employ sustainable production systems that simultaneously provide ample food supply to humanity and maintain natural resources and environments. The efficiency of these systems could be improved through resource optimization, achieved via precision agriculture that analyzes soil and crop data from sensors, GIS, and GPS. Precision agriculture has evolved into smart farming through integration with AI, IoT, and machine learning (ML). Such technologies as machine learning, data analytics, robotics, and IoT enhance sustainability by delivering real-time insights, automating labor-intensive tasks, and enabling better predictive and decision-making capabilities. In addition to sensors, drones, and satellite imagery, which enable farmers to track soil conditions, moisture, and crop development, yielding essential data that supports strategic agricultural decision-making. Moreover, real-time technologies also allow continuous tracking of key agricultural inputs such as seeds, fertilizers and pesticides, ensuring adequate supplies, preventing shortages, and improves overall production efficiency (Lysaght et al., 2019). Furthermore, AI algorithms combined with GIS are employed to streamline agricultural logistics and transportation planning, cutting operational expenses and lowering environmental footprints. For a proper sustainable agricultural system, it is essential to monitor the whole supply chain (SC) by implementing blockchain and similar technologies to enhance traceability, guarantee product integrity and adherence to food safety regulations throughout the entire SC from farm to table (Chen et al., 2023). Accordingly, it is crystal clear that IoT-supported precision agriculture has markedly improved both agricultural productivity and sustainability, especially within large-scale commercial agricultural practices (Dhanaraju et al., 2022; Miller et al., 2025).

2. THE ROLE OF AI AND IOT IN REVOLUTIONIZING AGRICULTURE

Practically speaking, agriculture is currently witnessing a revolutionary shift from traditional agricultural practices to embracing advanced technologies, like the IoT, AI, big data analysis, cloud computing and remote sensing, etc., thus the next agricultural phase is designated as Agriculture 4.0 (Ferrández-Pastor et al., 2016).

AI plays a crucial role in evaluating the ongoing soil conditions, fostering environmentally responsible soil management where the integration of advanced soil sensors with AI algorithms is paramount. Through data analysis, AI delivers insights into fertility, nutrient composition, and microbial activity as well, empowering farmers to adopt precise and efficient soil management strategies (Kaur et al., 2024). Developing AI models is crucial to handle diverse crops, including rare ones, ensuring all agricultural sectors benefit from the technology (Hussein, Jabbar, Mohammed, & Jasim, 2024). In addition, AI, through advancements in predictive analytics, can enhance resilience and reduce losses by enabling early warnings for climate risks such as droughts, floods, and pest outbreaks (Kanuru et al., 2021; Puspaningrum et al., 2022; Tanaka et al., 2024). Smart agriculture systems employ

IoT devices tailored to various farming tasks, where blockchain can represent a solution for issues of compatibility among equipment. Incorporating blockchain enables decentralized management and reduces dependence on centralized IoT control frameworks (Awan et al., 2020). Thus, the integration of AI and IoT in agriculture provides vast opportunities for future innovation by implementing further cutting-edge technologies like blockchain, advanced robotics, and gene editing, which can drive transformative changes in farming practices (Hussein, Jabbar, Mohammed, & Al-Jawahry, 2024). In the future, advancements in AI, 5G, blockchain, and climate-adaptive technologies are poised to amplify the influence of IoT in agriculture. Blockchain technology in precision agriculture enables seamless integration of digital innovations by eliminating organizational and institutional obstacles (Camaréna, 2020; Sharma & Shivandu, 2024).

A further advancement is the wireless sensor networks (WSNs), with the help of GPS- digital images and signal processing technologies, continuously supply updated information on crop growth and soil features. In the near future, IoT systems will be developed to the fifth-generation (5G) mobile communication technologies, enabling farmers to access real-time data seamlessly, regardless of location or time (Dhanaraju et al., 2022). Moreover, it is anticipated that 4.1 million data points will be established daily from farms by 2050 (Villa-Henriksen et al., 2020).

In the same context, Low Power Wide Area (LPWA) technology is set to advance smart agriculture due to its wide coverage, low power use, and cost efficiency. Cellular operators offer new services that, combined with stable IoT, can boost farmer income. Meanwhile, drones with advanced sensors and 3D cameras enhance land-management capabilities.(Dhanaraju et al., 2022).

3. MERGING AI AND IOT FOR SUSTAINABLE AGRICULTURE

AI and IoT are expected to markedly affect agriculture and food production by enhancing productivity, sustainability, and resilience to universal agricultural challenges (Ahmed & Shakoor, 2025). IoT-enabled smart farming offers real-time data harnessing, automation and precision agriculture techniques. This is achieved through interactive agriculture that facilitates seamless data access for users across various devices, including desktop computers and mobile phones (Chandhini, 2016). Such technologies give hope to eventual fine-tuning of the whole agricultural value chain - from sowing through monitoring up to processing, transportation, and marketing.

One significant aspect of agricultural sustainability is that IoT, AI, and renewable energy systems are essential advanced technologies that facilitate zero agricultural waste, incorporating circular systems that minimise waste and recycle it into biogas and organic fertilisers (Holm-Nielsen et al., 2009). (Holm-Nielsen et al., 2009).

Additionally, aquaponics is a synergistic system that integrates aquaculture and hydroponics as effective, waste-free farming methods. (Goddek et al., 2015).

4. KEY APPLICATIONS AND TECHNOLOGIES IN PRECISION AGRICULTURE

4.1. Soil Health Monitoring and Fertilization Management

The strategy of precision soil health and fertilization relies on accurate soil and nutrient monitoring, yet traditional assessment methods remain labor-intensive, inefficient, and limited in spatial coverage.

IoT and AI technologies like sensors, satellite images, and data analytics are pivotal in monitoring soil health and nutrients and in determining fertilizer quantity needed for plants (Nath, 2023). Sensors are key IoT tools for assessing soil health, measuring factors such as pH, conductivity, organic matter, erosion, compaction, nutrients, and moisture. Their data are processed by AI platforms to detect spatial and temporal patterns, providing a dynamic view of soil conditions across farms (Adak et al., 2023; Alawode et al., 2024). A hardware system combining Raspberry Pi, USB-connected pH, temperature, moisture, and NPK sensors, plus a Node MCU, produced accurate readings within acceptable error margins. This supports evidence-based decisions for irrigation, fertilization, and crop selection.(Gurdita et al., 2016).

A US Agricultural Research Service study in Iowa, Illinois, and Nebraska farms showed that precision nitrogen management using soil sensors and satellite imagery increased corn yields by 22% and reduced fertilizer use by 15%, improving efficiency and lowering environmental impact (Chen, 2025). IBM company, developed AgroPad, an AI-based paper device that instantly analyzes soil and water. Farmers can scan color-coded results from a small sample with a smartphone app to obtain precise chemical readings (Qayoom & Manzoor, 2022). Leveraging artificial intelligence, AI tools analyze metagenomic and transcriptomic data to reveal how soil microbes support nutrient cycling, disease control, and plant growth. By identifying key microbial taxa and functions, AI enhances microbiome-based strategies—like targeted inoculation and conservation tillage — leading to healthier soils, stronger crops, and reduced use of chemical inputs. (Zhao et al., 2023). In Nigeria, Folorunso et al., (2025) introduced GeaGrow, a smartphone app by the SmartSoil Team that recommends fertilizers, suggests crops like yams, maize, cassava, and rice, and predicts soil properties, with 99.96% accuracy, using the iSDAsoil API and ANN systems.

Agricultural machinery, such as tractors, is increasingly equipped with sensors and actuators to enable automation. GPS-guided tractors equipped with sensors and actuators can now perform tasks like planting and fertilizing with a high degree of precision and significantly reduce fuel consumption by optimizing routes and minimizing unnecessary field passes (Ahmed & Shakoor, 2025).

4.2. Water Management

According to the UNCCD, by 2030, desertification will affect 168 countries, with approximately 50% of the global population residing in regions of severe water scarcity (Dutta et al., 2025). Traditional irrigation practices, often inefficient and environmentally detrimental, leading to water wastage, energy inefficiency, nutrient leaching, and sometimes even reduced crop yields (Daraz et al., 2025; Kaur et al., 2025).

In precision agriculture, water management leverages data from diverse sources, including soil-moisture sensors, drones, weather forecasts, satellite imagery, GPS-based controllers, wireless communication, mobile applications, and cloud platforms (Daraz et al., 2025; Kaur et al., 2025).

An advanced IoT tool for water management is the agro-solar photovoltaic fusion system, which integrates soil-moisture and atmospheric sensors to monitor moisture, assess evapotranspiration, and use weather data for crop water needs (Daraz et al., 2025). The Crop Water Stress Index (CWSI) uses real-time climate sensors and satellite data to predict water needs and guide field-level management based on slope and soil variability (Dhanaraju et al., 2022).

A ThingSpeak-integrated irrigation system with capacitive soil-moisture sensors and solar-powered real-time data increased sweetcorn yield by 12.8%, water use efficiency by 24.5%, and productivity by 12.8% under semi-arid conditions, outperforming evapotranspiration-based methods (Sharma et al., 2025). IoT-enabled, energy-efficient irrigation, such as smart water wells, reduced energy use by 25% through real-time sensor data and automated control of pumping and water distribution (Ahmed & Shakoor, 2025). This level of precision allows standard use of constricted water resources, advocates sustainable farming practices and reduces the environmental impact of agriculture.

4.3. Disease Detection

Crop diseases and pests cause 20–40% annual yield losses (Nations), 2022), threatening sustainability and food security. Excessive reliance on pesticides, which harms the environment, coupled with the inaccuracy and expertise-dependence of manual detection, has prompted the adoption of AI- and IoT-based models, enabling precise, sustainable pest management by reducing chemical use and providing real-time data on crop health and pest activity (Morchid et al., 2023).

IoT-based devices - such as robots, wireless sensors, and drones enhance pest control through real-time monitoring, modeling, and disease forecasting, outperforming traditional methods (Dhanaraju et al., 2022). AI-powered image recognition systems have become instrumental in identifying and diagnosing plant diseases with high accuracy (Assimakopoulos et al., 2024b). This approach relies on detection and image processing, using remote-sensing imagery and field sensors to track plant health and pest incidence throughout the crop cycle. IoT-enabled automated traps capture, count, and identify insect species, uploading data to the cloud for analysis. Advances in robotics further enable agricultural robots equipped with multispectral imaging and precision sprayers to detect and manage pests with increased accuracy (Dhanaraju et al., 2022).

Convolutional neural networks (CNNs) have shown outstanding capability in image-based analysis, making them highly effective for the early detection of pests and plant diseases. By identifying symptoms at an early stage, CNNs help prevent infestations and infections from spreading across large agricultural areas (Nyakuri et al., 2025). GoogLeNet, AlexNet, VGG, ResNet, and Inception are among the most widely recognized CNN architectures. These models learn to visually identify and locate objects by analyzing large sets of sample images, in a way that loosely mimics how the human brain processes visual information (Thi Thoa et al., 2024). Earlier, (Kawasaki et al., 2015) introduced a disease detection system based on CNN, using leaf images to detect melon yellow spot virus and zucchini yellow mosaic virus in cucumber. The system achieved 94.9% accuracy in disease detection.

The advanced YOLOv5 model reduces computational overhead and captures long-range spatial dependencies, achieving 86.5% accuracy in detecting powdery mildew. By integrating the novel Involution Bottleneck architecture, complex datasets were effectively handled and the accuracy of disease outbreak prediction was improved (Assimakopoulos et al., 2024a).

Early recognition of weeds is fundamental for successful weed control. Although agrochemicals remain an effective option for growers, their use raises concerns about environmental pollution. To minimize these impacts, target-based spraying or mechanical weeding methods are applied. Over the years, many autonomous mechanical weeding systems have evolved to improve efficiency and crop safety. Åstrand and Baerveldt (2002) developed a mechanical weeder with dual vision units to navigate crop rows and identify crops by color and shape, lifting the tool automatically to avoid damage. Similarly, Griepentrog et al. (2006) introduced a rotating ring with rigid tines that moves independently to protect crops during weeding. More recently, Blue River Technology, integrated with John Deere, developed the AI-driven "See and Spray" robot, which distinguishes crops from weeds with 95%

accuracy, applies herbicide selectively, and optimizes nutrient application while protecting soil microbiomes (Chen, 2025).

4.4. Weather and Yield Monitoring

Timely climatic data are vital for predictive analysis in precision agriculture. Automatic weather stations (AWSs) equipped with multiple sensors collect and transmit information on key meteorological parameters—such as precipitation, crop yield, wind speed, air temperature, humidity, solar radiation, and rainfall—allowing yield monitors and weather systems to assess environmental factors affecting crop productivity (Fuentes-Peñailillo et al., 2024). These sensors can form measurement networks for specific applications and be integrated with microcontrollers to create modular, portable monitoring devices (Rogulski & Badyda, 2020).

Autonomous harvesters, robotic weeders, and sensor-equipped drones offer real-time field monitoring while minimising environmental impact. Machine learning supports gene selection, seed recommendations, and soil and rainfall predictions, while GPS-integrated yield-monitoring systems and mobile applications provide real-time maps of grain flow, moisture, and crop quality for informed decisions (Quy et al., 2022; Naorem et al., 2019).

These smart tools delve further into producing physical and physiological details. Image-based monitoring using RGB and satellite imagery helps estimate fruit size, color, and maturation to determine optimal harvest timing (Wang et al., 2017). Moreover, crop forecasting applies yield and environmental data to predictive models that enhance production planning and market management.

Currently, smart techniques and tools are gaining increasing attention and popularity as vital backup for precision agriculture, as shown in the following cases and applications by several reputed agricultural companies (Table 1).

Several types of yield monitors are employed, such as mass flow sensors used to predict corn yields and measures the harvest flow through grain elevator, which are sensitive to variations in flow rate and ground slope, and optical sensors, which are remote sensing devices that use light and interact with plants to gather data and provide insights into crop health and potential yield (Fuentes-Peñailillo et al., 2024). Soybean farmers achieved an 18% increase in yield by adopting the auto-steer planters, which enhanced seed spacing uniformity and depth control, highlighting the productivity benefits of precision farming tools (Chen, 2025).

Harvesting technologies are increasingly focusing on interactive agricultural robots (agrobots) for precise crop tasks, inculding fruit picking, weeding, and herbicide micro-dosing. Equipped with vision systems, GPS, and robotic arms, both task-specific (Harvest CROO, GUSS) and multi-purpose platforms (Farmdroid,

Husky) reflect the shift toward automation and versatile precision agriculture (Vrochidou et al., 2021).

Rogulski and Badyda (2020) reported that XAG, a leading Chinese smart agriculture company founded in 2007, developed an integrated solution combining unmanned aerial system (UAS) robots, automated steering, and management software to leverage AI in agriculture. XAG established over 2,300 Real-Time Kinematic base stations, linking China's BeiDou satellite navigation system across 35,000 villages. Its AI cloud platform processes large-scale data to optimize crop management, including seedling population analysis, anomaly detection, and yield estimation. AI algorithms assess plant density, detect diseases and pests via NDVI remote sensing, and estimate yield from aerial imagery of panicle counts. XAG's multifunctional drones apply AI-generated prescription maps for precise spraying, seeding, and fertilization with centimeter-level accuracy. In addition, UAS and solar-powered XIoT™ systems collect high-definition maps and real-time soil and weather data, transmitted to smart agriculture platforms. Farmers can remotely access georeferenced field information via apps, reducing manual scouting and improving efficiency, productivity, and resource use (Rogulski and Badyda (2020).

4.5 Smart-Controlled Environments in Agriculture

4.5.1 Greenhouses

Indoor crops are less affected by external conditions, but the most challenging aspect in greenhouses is careful monitoring of environmental constraints. Multiple measurements are required to predict various parameters for effective control of the local climate. In an IoT-based greenhouse, sensors, robots and automated climate control systems are used to monitor many parts of greenhouses and measure the internal parameters, such as humidity, temperature, pressure, mist, CO_2 level, UV light intensity, pH, EC values, nutrients and amount of pesticides (Dhanaraju et al., 2022).

A greenhouse wireless sensor network (WSN) is a distributed system of sensors and actuators connected wirelessly in a mesh topology, with nodes communicating via multi-hop links to a central gateway (Lamprinos et al., 2015). It collects data on environmental factors such as temperature, humidity, light, CO_2, soil moisture and plant health indicators inculding leaf temperature, chlorophyll, and nutrients. Soil moisture guides irrigation, while temperature and humidity readings support climate control through heating, ventilation, and cooling systems, ensuring optimal plant growth (Assimakopoulos et al., 2024b). In 2024, KUBO Greenhouse Projects announced the completion of a 24-acre greenhouse facility for WINDSET FARMS in Canada—the largest greenhouse of its kind in Delta, BC, Canada. The facility incorporates multiple sustainability-oriented technologies, including renewable hydroelectric power and biomass heating systems. With these innovations, Windset Farms is positioned to supply high-quality tomatoes and cucumbers year-round to markets across Western Canada, the United States, and Asia (Mordor Intelligence, 2025a).

4.5.2 Vertical Farming (VF)

VF can be categorized into three types, namely hydroponics, aeroponics and aquaponics (Rathor et al., 2024). These VFs patterns allow year-round, consistent production, independent of external weather conditions, ensuring a reliable supply of high-quality products. Vertical farming systems are high-tech glasshouses that provide precise control over temperature, light, and CO_2 concentration, enabling high productivity per unit (Hanafi et al., 2025).

A hydroponics system, is used to cultivate plants without soil, using a nutrient solution to support healthy growth and higher yields. The liquid-based hydroponics market is valued at USD 5.95 billion in 2025 and projected to reach USD 9.03 billion by 2030, with a CAGR of 20.7% (Malabadi et al., 2024). Growth is driven by urbanization, sustainability efforts, and advances in controlled-environment agriculture.

Crops grown hydroponically, such as tomatoes, lettuce and leafy greens, peppers, cucumbers, are geographically distributed in North America, Europe, Asia-Pacific, South America, the Middle East, and Africa (Mordor Intelligence, 2025b). Chinese cabbage is grown in a vertical hydroponic system powered by Arduino, Wi-Fi, UNO, ESP8266, and WeMos D1. Sensors monitor temperature, humidity, water and nutrient levels, and pH, with data integrated via the Blynk IoT platform for real-time viewing on a smartphone app (Hanafi et al., 2025). AeroFarms in New Jersey grows leafy greens and herbs via aeroponic VFs, supplying nutrients through misted sensors, yielding 390 times more per square foot and using 95% less water than conventional farms. Plenty in San Francisco cultivates leafy greens, berries, and vine vegetables using climate control systems, automated nutrient delivery, and LED lighting to optimize growth Choorappulakkal et al., 2024).

Aquaponics VF combines fish and plant in one system. Fish tank water nourishes the plants, gets cleaned, and is then returned to the fish production facility. The ANFIS model developed by Debroy and Seban (2022) demonstrated high accuracy in predicting tomato fruit production in aquaponic systems. It employed a feed-forward back propagation network trained on input variables including recirculating water temperature, dissolved oxygen, nitrate concentration, and pH, with tomato biomass as the output parameter.

Figure 2. Indoor and outdoor implementation of AI and IoT in precision agriculture

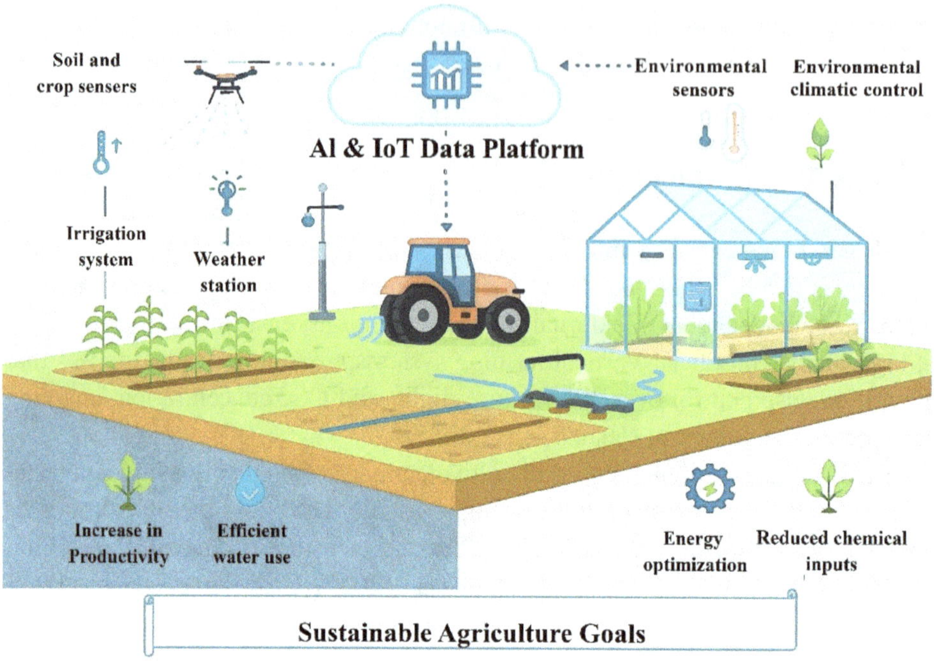

4.6 AI and IoT Advancing Organic Agriculture

To reveal microbial interactions in soil and their roles in plant nutrition and health, AI is increasingly used to uncover functional capabilities of soil microbial communities through metagenomic and transcriptomic data analysis. Interestingly, AI-powered soil microbiome assessments demonstrated their potential to support the development of agronomic strategies and sustainable agriculture (Gupta & Kumar Pal, 2025).

Furthermore, advanced high-throughput phenotyping techniques help breeders generate new high-yielding crop varieties adapted to local environmental conditions. Large-scale datasets obtained through visible and near-infrared hyperspectral imaging enhance the capabilities of ML techniques such as least squares regression (LSR), enabling highly accurate predictions of biochemical and physicochemical properties that outperform traditional vegetation indices (Gonzalez-de-Soto et al., 2016). In addition, other biotechnological advancements, such as CRISPR (Clustered Regularly Interspaced Short Palindromic Repeats) and various gene-editing techniques, are being utilized to cultivate crops with enhanced traits (Bhalerao et al., 2025). Moreover, integrating AI-driven practices, such as precision agriculture and smart pest control, improved biodiversity on organic farms (Kaur et al., 2024).

To conclude, AI is anticipated to significantly enhance the scalability, efficiency, and profitability of organic agriculture. Emerging developments may include autonomous farming systems driven by AI, such as blockchain-enabled traceability in organic SCs, and advanced bio-inspired AI models that improve soil health and pest resistance (Bharathi et al., 2024).

4.7 AI Technologies and Supply Chain in Precision Agriculture

AI technology is adding an extraordinary dimension to the whole SC in agriculture inculding soil health analysis, cultivation, fertilizer and pesticide application, harvest, packaging, marketing, storage, etc. Hence, saving time, money, effort and resources by leveraging precision through current data analysis and prediction, using historical data from various sources via cutting edge technology combinations (Assimakopoulos et al., 2024a).

It is worth mentioning that AI-driven solutions in agricultural logistics are crucial for continuous tracking of environmental parameters using IoT sensors, throughout transport and storage, consequently, helping preserve product integrity. Hence, AI algorithms enhance route planning and traffic patterns, thereby shortening delivery duration and lowering operational expenses (Rajabzadeh & Fatorachian, 2023). AI tools play a significant role in mitigating the risk of overproduction. Furthermore, AI-powered machines equippted with sensors, cameras, and GPS systems execute autonomous field operations, thereby reducing labor costs (Shiba & Miwa, 2022).

Regarding yield, AI helps producers cope with market demand, predict future market trends and optimize pricing strategies accordingly, maintaining balanced demand and supply trends (Assimakopoulos et al., 2024a). Moreover, agribusinesses and policymakers can leverage insights generated through AI integration to enhance strategic planning and support informed decision-making.

AI-intelligent energy management systems significantly lower energy usage and operational expenses by dynamically adjusting and optimizing heating, lighting, and ventilation based on real-time data (Assimakopoulos et al., 2024).

5. ADOPTION OF PRECISION AGRICULTURE

The adoption patterns of precision agriculture technologies commercialized in the early 21st century have remained uneven. Certain technologies, such as GNSS sprayer boom control and seeder row shutoff systems, have achieved widespread uptake, while others have experienced slower diffusion (McFadden et al., 2024;

Sanjay, 2024). Variable Rate Technology (VRT) is projected to grow at a compound CAGR of 13% through 2030, driven by rising fertilizer and chemical costs that encourage targeted application. Drone-based remote sensing, supported by affordable multispectral payloads, can reduce chemical usage by up to 67% when prescription maps guide sprayers (Mordor Intelligence, 2025c). While drones are widely used for imagery and data collection, their application for spraying remains limited due to regulatory constraints. Adoption is growing rapidly in countries with flexible regulations, such as China and Brazil, whereas strict rules in the EU and UK restrict spraying to specific scenarios. In the United States, use of drone-based input applications is increasing as regulations become more permissive (McFadden et al., 2024; Mordor Intelligence, 2025c). Embracing precision farming technologies is significantly influenced by farm size, farmers' age, household income, and long farming experience (Maloku, 2020). Table 1 illustrates the technologies adopted by their components, applications, and regions.

Table 1. Recent adoption of AI and IoT tools according to their components, applications, technology, and regions

Category	Technology	Details
By Components	Hardware Software Service	Automation and Control systems. Sensors. Antenna and Actuators Wed and cloud-based System integration and consulting Maintenance and support Managed service Assisted professional services
By Application	Yield monitoring. Field mapping. Croup scouting. Weather tracking and forecasting Inventory Management Farm labour management Others	
By Technology	Guidance Systems Remote sensing Variable-Rate Technology (map or sensor-based VRT. Drones and UAVS Robotics and Autonomous equipment Edge and cloud Analytic platforms. Others	GPS/GNSS/GISR) Handheld or ground based. satellite or aerial sensing. Variable-Rate Fertilizers, Seeding and Pesticides (Map and sensors)
By region	Northern America Europe Asia Pacific LAMEA	USA, Canada UK, Germany, France, rest of region China, India, Japan, rest of region Latin America, Middle East, Africa

Source:(Allied Market, 2021) and Mordor Intelligence, 2025c.

Over the past decade, developed countries, including the United States, Germany, the United Kingdom, and France have made substantial progress in precision farming by integrating information technologies and developing techniques tailored to regional conditions and farm structures, effectively addressing challenges in precision crop management (Maloku, 2020). In the Southern United States, 73.5% of farms employed precision farming technologies, with 1,331 of 1,811 surveyed participants using at least one such technology (Zhou et al., 2017). In the Asia–Pacific region, China, Japan, India, and Australia are among the leading adopters of precision agriculture technologies (Mordor Intelligence, 2025c). The highest adoption rates are reported in Australia, Canada, and Brazil, while countries such as Argentina, India, Egypt, South Africa, Malawi, and Ethiopia demonstrating moderate but growing levels of adoption (Kendall et al., 2021). However, according to Mordor Intelligence (2025c) (Figure 3), precision farming markets are growing in Africa, Asia and North American.

Figure 3. Precision farming, source: Mordor Intelligence, 2025c

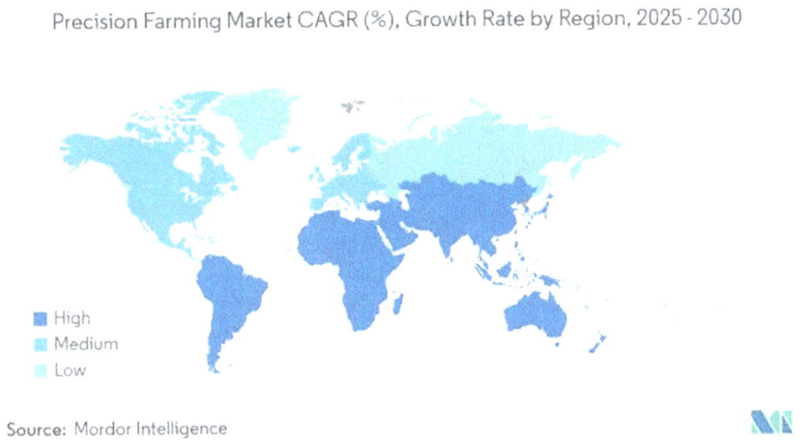

Artificial intelligence has become increasingly in demand throughout the agricultural sector. Recently, pioneer companies have developed long-term strategies to maintain their competitive advantage. These strategies are reflected in sustained research efforts harnessing the full potential of this powerful technology, thereby ensuring future organizational growth and innovation (Table 2).

Table 2. Companies at the forefront of smart and sustainable farming technologies

Company Name	Country/ Founded	Technology Type (AI / IoT)	Key applications	Reference
Arable Labs	USA 2001	Veris 3100 Veris 3150 MSP3	Soil electrical conductivity (ECa) mapping system. Measures pH, organic matter and soil electrical conductivity using dual wavelength optical sensor.	(Tayler, 2012)
Bayer	USA)2006)	The Climate Corporation (FieldView)	Integrates meteorological data with crop models, and AI to provide farmers with insights on planting, nutrient management, pest control, field monitoring, yield forecasting, and climate analysis.	(Mamoria & Shakya, 2023) https:// climatefieldview.com
XAG	China 2007	Drones with Multispectral/ thermal imaging AI analytics, targeted spraying, - Unmanned aerial system robots, -Solar-powered XIoT™ systems, -NDVI-based remote sensing.	Crop health monitoring, pest and disease detection, precision spraying, irrigation management, detection of soil and weather data	(Rogulski & Badyda, 2020)

continued on following page

Table 2. Continued

Company Name	Country/ Founded	Technology Type (AI / IoT)	Key applications	Reference
Deere & Company	USA (2010)	-AutoTrac™ Guidance: -Variable-Rate Application -Predictive Maintenance -Integration with IoT Platforms	Automated steering tractors and systems ensure overlap-free field coverage. Real-time adjustments to seeding, fertilization, and pesticide application Machine learning models analyze equipment data to forecast maintenance needs, reducing unexpected downtime. Data from Blue River's systems seamlessly integrates with the Operations Center, giving farmers even more granular control over field operations.	(Ahmed & Shakoor, 2025) https://www.datanext.ai/case-study/john-deere-iot-in-agriculture/
Blue River Technology (John Deer)	USA 2011	-AI-Driven Weed Control (See and Spray)	The robot uses computer vision and AI to distinguish crops from weeds, sprays herbicide with 95% accuracy and protects soil microbiomes and customizes nutrient application maps.	(Chen, 2025)

continued on following page

Table 2. Continued

Company Name	Country/ Founded	Technology Type (AI / IoT)	Key applications	Reference
Microsoft and the International Crop Research Institute for the Semi-Arid Tropics (ICRISAT) and Microsoft India.	India 2016	-Sowing App (AI-powered models)	Sowing date decisions based on weather conditions, soil moisture levels, and insights from historical information on crops, resulting in increased crop yields by 30%.	(Dasarath, 2016; Lakhiar et al., 2024)
AgNext Technologies	India 2016	QuaLIX (AI-based spectral and picture analytics)	-Real-time quality analysis of agricultural commodities such as grains, spices, tea, milk, and oilseeds. -Grain chemical and physical composition.	(Qayoom & Manzoor, 2022)
Intello Labs	India 2016	Intello Deep Intello Sort Intello Track	-Detection of Brix, pH, total soluble solids, dry matter, moisture content, and pesticide residues. -Sorting farm produce by colour, size, and visible defects, enabling multi -commodity sorting on a single, compact, and low-cost line. The AI platform depends on the farmer capturing images through the mobile app that provide fair grading based on color, size, and visual defects.	(Sengupta et al., 2021)

continued on following page

Table 2. Continued

Company Name	Country/ Founded	Technology Type (AI / IoT)	Key applications	Reference
OneSoil	Switzerla 2017	OneSoil AG (Mobile apps (Android/iO)	-Analyze satellite images, data from ground equipment, and drone footage use NDVI to assess plant health, track crop growth stages, identify optimal spraying times, and map routes to problem areas. -Support variable-rate decisions. -Provide weather forecasts, and display efficiency graphs and vegetation zones.	(Premachandra & Kumara, 2021). https://onesoil.ai/en
IBM	Brazil 2018	AgroPad	AI-based paper device provides instant soil and water quality analysis	(Qayoom & Manzoor, 2022) https://www.farmautomationtoday.com/news/data-analytics/ibms-instant-ai-soil-analysis-the-agropad.html
BASF	Geermay 2020	xarvio® FIELD MANAGER	A platform that combines satellites, drones, weather stations, and field sensors to collect real-time data, providing precise agronomic recommendations for each field or zone.	(Wecht et al., 2021)

6. CHALLENGES AND FUTURE INSIGHTS FOR AI AND IOT INTEGRATION IN PRECISION AGRICULTURE:

Despite the obvious benefits of integrating AI and IoT technologies in precision agriculture, there are several multifaceted challenges encompassing sociocultural, technological, economic, environmental and regulatory aspects involved within the process.

6.1. Sociocultural and Economics

Since the unveiling of AI technology, it has been an efficient tool for boosting precision agriculture. It is crucial to note that the targeted audience comprises farmers' from diverse sociocultural backgrounds, with varying educational levels, financial and farm scales worldwide. Hence, to boost the adoption of AI, IoT and ML technologies, a complex multitude of aspects should be addressed:

a- Willingness to change: many conventional farmers feel a strong attachment to inherited practices and exhibit little positive attitude toward change (Adewusi et al., 2024). Trust and reliability issues dominate this group of farmers towards IoT-based systems, coupled with a generational divide in which young farmers are much open and willing to adopt new technologies (Joshi, 2024).
b-Low level of technological adoption among a large portion of farmers, especially in developing countries (Adewusi et al., 2024). In India, low literacy levels, besides language barriers, hamper the understanding and interpretation of data from smart farming devices (Joshi, 2024).
c- High cost of precision farming techniques is a determining factor for a wider range of adoption, especially by local farmers in developing countries (Alazzai et al., 2024). Besides the limited access to credit and subsidies or government incentives (Joshi, 2024).
d- Risk of labor displacement associated with automation of cultural practices. This shift could alter rural economies that rely primarily on agricultural employment, resulting in labor displacement (Eastwood et al., 2019).

6.2. Technological

For adequate adoption, a technology should fulfil certain criteria concerning the nature, characteristics and needs of such groups. In this sense, the questions of technology affordability, ease of application and deployment, security and reliability arise.

a- The complexity of AI and RS systems is a considerable challenge various agricultural stakeholders (Fassnacht et al., 2024). Moreover, system integration complexity between diverse technologies like IoT, Big Data, and Edge Computing into a coherent framework (Vermesan & Friess, 2013), is crucial for seamless interconnectivity, data transfer efficiency, and real-time processing (Ahmed & Shakoor, 2025).

c- Network connectivity and access to advanced hardware may hamper the broad adoption of AI-driven precision agriculture, especially in areas with low levels of technology infrastructure (Adewusi et al., 2024). This remarkably hampers real-time data collection, seamless integration of IoT technology and consequently adoption in vast rural agricultural areas in developing countries with poor consistent internet connectivity, electricity and IoT infrastructure (Joshi, 2024).

d- Continuous processing of large-volume data flows from various sources needs extensive cloud-based infrastructure and complex data systems management. Centralized data control by large tech companies raises ethical issues concerning data ownership and farmer independence in decision-making (Ahmed & Shakoor, 2025). In addition to cybersecurity and the probable liability to hacking by private agritech companies (Joshi, 2024).

h- Scalability and adaptability of AI technologies under various agricultural conditions is a challenging area where most studies are carried out in laboratories or greenhouses (Bauer & Aschenbruck, 2018; Mohamed et al., 2021). This could also be aggravated by the heterogeneity of devices implemented in the system design, each requiring different set of operational and maintenance needs (Morchid et al., 2024).

6.3. Policy and Regulation Concerns

The adoption level of AI-driven precision agriculture could improve significantly through supportive policies and regulations, besides the interdisciplinary collaboration among different sectors. In this regard, several challenges are discussed below:

a- Absence of collaboration between the different stakeholders involving in the process such as AI researchers, agronomists, and farmers, is one of the fundamental challenges (Eastwood et al., 2019).

b- Lack of economic models that manage setup and operational expenses including, cost of IoT devices, recurring subscriptions, data processing, maintenance, and energy (Adewusi et al., 2024).

Unsatisfactory data privacy legislation and agricultural data ownership are topics of growing concern since AI systems efficiency calls for handling extensive agricultural information (Gupta et al., 2020). In addition to modest awareness of policy and organizational bodies towards ethical and effective use of AI and IoT in agriculture, with regard to different regional regulations and cultural practices (Hussein, Jabbar, Mohammed, & Al-Jawahry, 2024).

f- Lack of strong policy structure coupled with poorly-funded educational programs impede widespread adoption of AI in agriculture and the expected contribution to global food security (Aijaz et al., 2025).

6.4. Environmental Concerns

Sustainability of precision farming methods is very important to keep the balance between increasing yield and returns on one side and environmental conservation on the other (Alazzai et al., 2024).

a- The great intrinsic variability of soil and climatic aspects across the globe renders the universal spread of a uniform AI-based technology model in agriculture perhaps the most challenging aspect (Galaz et al., 2021; Rejeb et al., 2022; Talaviya et al., 2020).

b- Unsatisfactory environmental impact assessment studies of the long-term continuous use of smart technologies on soil health, crop diversity, ecosystems and biodiversity are crucial (Hussein, Jabbar, Mohammed, & Al-Jawahry, 2024). Added to the potential negative effects on insect biodiversity through long-term exposure to wireless communication devices (Sharma & Shivandu, 2024). Furthermore, malfunctioning sensors or IoT devices may result in the inadvertent overapplication of agricultural inputs, leading to soil deterioration and contamination (Saha et al., 2025).

Moreover, it is important to keep in mind that IoT and Edge Computing devices, though essential (Kalyani & Collier, 2021) for smart agriculture, heighten the energy footprint of agriculture (Ahmed & Shakoor, 2025).

6.5. Logistics and Supply Chain Concerns

AI is capable of optimizing logistics and agricultural SCs, which comprise various processes such as preproduction, production, storage, processing, transportation, distribution, retail, up to the final product and the end consumers. The SC system also involves multiple stakeholders, such as farmers, producers, processors, certification agencies, traders, government, retailers, distributors, and final consumers.

Sophisticated agriculture SCs, owing to the nature of perishability and great supply-demand instability of the products and high consumer awareness of produce nutritional and safety properties (Camaréna, 2020; Sharma et al., 2021). This leads to a scarcity of streamline logistical infrastructure, such as cold storage, resulting in food waste and diminished efficacy of IoT use in SC management (Joshi, 2024).

Substandard transparency in food SCs supports middlemen to accumulate smart farming returns (Joshi, 2024). Coupled with the commodity price manipulation by major corporations leveraging AI-powered market predictions, this favors large industrial farming operators at the expense of smallholder farmers. (Miller et al., 2025). This leads to complications in food traceability, verification and transparency within the market and food supply (Kumar et al., 2022).

7. REMEDIES

As previously discussed, several challenges affect the adoption of AI-based technologies in agriculture, including initial costs, data security, and the requirement for technological know-how. Potential labor displacement can be mitigated by a comprehensive understanding of the socio-economic impacts of AI and IoT on agricultural employment, especially in regions where farming is the primary source of livelihood.

To address digital literacy, satisfactorily funded educational programs are crucial to enhance widespread adoption of AI in agriculture and contribute to global food security (Aijaz et al., 2025). Clear guidelines and mechanisms are needed to manage agricultural data sharing and ownership, which affect collaboration (Adewusi et al., 2024). Governments should support these actions with robust data privacy laws to ensure farmers retain complete authority over their agricultural information, guaranteeing fair and ethical use of AI-generated insights (Kumar et al., 2022; Kumar & Kumar, 2024).

The development of economic models to manage IoT setup and operational expenses, including cost of devices, recurring subscriptions, data processing, maintenance, and energy will offset regional economic variation facing IoT integration (Adewusi et al., 2024).

Regulatory frameworks must be established to counter commodity price manipulations and monopoly by large companies to promote equitable AI-driven agricultural markets and prohibit anti-competitive uses of AI that harm local producers (Akkaş & Sokullu, 2017; Eshbayev et al., 2024).

To improve SC transparency, secure fair pricing, and reduce inefficiencies, logistics can be streamlined via the complementary use of blockchain and AI (Chen et al., 2023). Furthermore, incorporating blockchain will reduce dependence on centralized IoT control frameworks and enhance compatibility among different AI-based devices (Awan et al., 2020). A link between communication technologies and heterogeneous modules is useful to avoid falsified results stemming from network complexity and heterogeneity of equipment in most agricultural models (Morchid et al., 2024).

In addition to the integration of ML in agriculture, SC to assist farmers and relevant organizations with valuable observations on agricultural processes, resulting in increased productivity through data-driven platform decisions (Sharma & Shivandu, 2024)

A worldwide interdisciplinary collaboration is needed to facilitate the universal spread of a uniform AI-based technology model in agriculture to address the great intrinsic variability of soils and climatic conditions (Galaz et al., 2021; Rejeb et al., 2022; Talaviya et al., 2020). Hence, joined international efforts, collaboration and partnerships between public and private sectors will eventually expedite the adoption of AI technologies and establish a universal farming landscape to feed the growing world population and adapt to climate change as well (Mmbando, 2025).

8. FUTURE RESEARCH AND DEVELOPMENT

There are some crucial areas concerning the implementation of AI and IoT technologies in different segments of the agricultural value chain where research and development are needed to foster widespread adoption of technology and effectively shape the future of precision agriculture.

Various research institutions across the globe are dedicated to the improvement of AI and IoT technologies and counter the arising challenges to enhance their integration into smart agriculture, such as:

1- Designing a seamless technology interface is crucial to satisfy the needs of farmers of different backgrounds of technical know-how through understanding how AI and IoT can be integrated into existing agricultural infrastructures while maintaining the integrity of traditional practices (Alazzai et al., 2024).
2- The development of AI models specially to deal with the intrinsic variation of agricultural environments (Alazzai et al., 2024).
3- Advanced AI models are highly needed to deal with the wide crop diversity, especially rarely grown crops, facilitating technology reach by all sectors (Hussein, Jabbar, Mohammed, & Al-Jawahry, 2024).
4- Innovation of low-cost AI and IoT solutions, such as affordable sensors and accessible data analytics tools, supports small-scale farmers (Hussein, Jabbar, Mohammed, & Al-Jawahry, 2024).
5- Innovation of more energy-efficient sensors will remarkably improve carbon footprint monitoring (Ahmed & Shakoor, 2025).
6- Boosting drones efficacy in large-scale agricultural processes by expanding their battery viability and autonomous proficiencies (Alazzai et al., 2024).

7- Improvement of accuracy and reliability of precision agriculture systems through enhancement of IoT efficiency (Ahmed & Shakoor, 2025).
8- Nanotechnology has the potential to revolutionize precision by monitoring soil health and nutrient delivery precision (Ahmed & Shakoor, 2025).

CONCLUSIONS

The integration of AI and IoT technologies into precision agriculture presents unprecedented opportunities to enhance soil health monitoring, nutrient management, water-use efficiency, disease detection, and overall crop productivity. These innovations enable dynamic, data-driven decision-making that supports sustainable farming practices, conserves natural resources, and mitigates environmental impacts. However, significant challenges remain, including technological complexity, high costs, sociocultural barriers, infrastructure limitations, and regulatory concerns. Addressing these obstacles through interdisciplinary research, policy development, and inclusive education is critical to ensuring broad adoption, especially among smallholder farmers in developing regions. As advancements continue in sensor technology, automation, and data analytics, the future of agriculture promises greater resilience, efficiency and sustainability, contributing significantly to global food security and environmental management. Collaborative efforts across the public and private sectors will be essential to realize the full potential of these transformative technologies for a more productive and equitable agricultural landscape.

REFERENCES

Adak, S., Mandal, N., Maity, P. P., Mukhopadhyay, A., & Gunturi, A. (2023). Digital Soil Health: Unveiling the Potential of Soil Health Monitoring. *Food and Scientific Reports*, *4*(9), 96–103.

Adewusi, A. O., Asuzu, O. F., Olorunsogo, T., Iwuanyanwu, C., Adaga, E., & Daraojimba, D. O. (2024). AI in precision agriculture: A review of technologies for sustainable farming practices. *World Journal of Advanced Research and Reviews*, *21*(1), 2276–2285. DOI: 10.30574/wjarr.2024.21.1.0314

Ahmed, N., & Shakoor, N. (2025). Advancing Agriculture through IoT, Big Data, and AI: A Review of Smart Technologies Enabling Sustainability. *Smart Agricultural Technology, 100848*. Advance online publication. DOI: 10.1016/j.atech.2025.100848

Aijaz, N., Lan, H., Raza, T., Yaqub, M., Iqbal, R., & Pathan, M. S. (2025). Artificial intelligence in agriculture: Advancing crop productivity and sustainability. *Journal of Agriculture and Food Research, 101762*. Advance online publication. DOI: 10.1016/j.jafr.2025.101762

Alawode, A., Blessing, A., & Thankgod Chiamaka, O. (2024). Integrating IoT and AI in Sustainable Agriculture to Mitigate Environmental Risk and Financial Misuse. *International Journal of Research Publication and Reviews*, *5*, 2810–2828. DOI: 10.55248/gengpi.6.0625.2187

Alazzai, W. K., Abood, B. S. Z., Al-Jawahry, H. M., & Obaid, M. K. (2024). Precision farming: the power of AI and IoT technologies. E3S Web of Conferences.

Allied Market. R. (2021). Precision Agriculture Market size, competitive landscape, and trend analysis report; Global opportunity analysis and industry forecast, 2021–2030. https://www.alliedmarketresearch.com/precision-agriculture-market

AlZubi, A. A., & Galyna, K. (2023). Artificial intelligence and internet of things for sustainable farming and smart agriculture. *IEEE Access : Practical Innovations, Open Solutions*, *11*, 78686–78692. DOI: 10.1109/ACCESS.2023.3298215

Assimakopoulos, F., Vassilakis, C., Margaris, D., Kotis, K., & Spiliotopoulos, D. (2024a). Artificial Intelligence Tools for the Agriculture Value Chain: Status and Prospects. *Electronics (Basel)*, *13*(22), 4362. DOI: 10.3390/electronics13224362

Assimakopoulos, F., Vassilakis, C., Margaris, D., Kotis, K., & Spiliotopoulos, D. (2024b). The Implementation of “Smart” Technologies in the Agricultural Sector: A Review. *Information (Basel)*, *15*, 466. DOI: 10.3390/info15080466

Assimakopoulos, F., Vassilakis, C., Margaris, D., Kotis, K., & Spiliotopoulos, D. (2025). AI and related technologies in the fields of smart agriculture: A review. *Information (Basel), 16*(2), 100. DOI: 10.3390/info16020100

Åstrand, B., & Baerveldt, A.-J. (2002). An agricultural mobile robot with vision-based perception for mechanical weed control. *Autonomous Robots, 13*(1), 21–35.

Awan, S. H., Ahmed, S., Nawaz, A., Sulaiman, S., Zaman, K., Ali, M. Y., & Imran, S. (2020). BlockChain with IoT, an emergent routing scheme for smart agriculture. *International Journal of Advanced Computer Science and Applications, 11*(4), 420–429. DOI: 10.14569/IJACSA.2020.0110457

Bauer, J., & Aschenbruck, N. (2018). *Design and implementation of an agricultural monitoring system for smart farming. 2018 IoT Vertical and Topical Summit on Agriculture - Tuscany.* IOT Tuscany.

Bhalerao, S., Okoń, S., Rai, M., Jadhav, P., & Prokisch, J. (2025). Recent Advances in CRISPR/Cas9 Technology for Engineering Disease Resistance in Plants. *Acta Agrobotanica, 78*, 1–31. DOI: 10.5586/aa/201389

Bharathi, P., Karthickraja, A., & Saravanane, P. (2024). Advanced precision agriculture techniques for effective weed management cotton. In Arunkumar, R., Vignesh, K., Dhivya, C., & Bala Murugan, S. (Eds.), *Recent Trends in Agriculture and Allied Sciences* (pp. 162–171). JPS Scientific Publications.

Camaréna, S. (2020). Artificial intelligence in the design of the transitions to sustainable food systems. *Journal of Cleaner Production, 271*, 122574. DOI: 10.1016/j.jclepro.2020.122574

Chandhini, K. (2016). A literature study on agricultural production system using IoT as inclusive technology. *International Journal of Innovative Technology and Research, 4*, 2727–2731.

Chen, H. Y., Sharma, K., Sharma, C., & Sharma, S. (2023). Integrating explainable artificial intelligence and blockchain to smart agriculture: Research prospects for decision making and improved security. *Smart Agricultural Technology, 6*, 100350. DOI: 10.1016/j.atech.2023.100350

Chen, X. (2025). The role of modern agricultural technologies in improving agricultural productivity and land use efficiency. *Frontiers in Plant Science, 16*, 1675657.

Daraz, U., Bojnec, Š., & Khan, Y. (2025). Energy-Efficient Smart Irrigation Technologies: A Pathway to Water and Energy Sustainability in Agriculture. *Agriculture, 15*, 554. DOI: 10.3390/agriculture15050554

Dasarath, B. R. (2016). Microsoft, ICRISAT develop new sowing app for farmers using AI and Azure cloud. https://www.business-standard.com/article/companies/microsoft-icrisat-develop-new-sowing-app-for-farmers-using-al-and-azure-cloud-116060900752_1.html

Debroy, P., & Seban, L. (2022). A tomato fruit biomass prediction model for aquaponics system using machine learning algorithms. *IFAC-PapersOnLine*. Advance online publication. DOI: 10.1016/j.ifacol.2022.04.116

Dhal, S. B., & Kar, D. (2024). Transforming agricultural productivity with AI-driven forecasting: Innovations in food security and supply chain optimization. *Forecasting*, *6*(4), 925–951. DOI: 10.3390/forecast6040046

Dhanaraju, M., Chenniappan, P., Ramalingam, K., Pazhanivelan, S., & Kaliaperumal, R. (2022). Smart Farming: Internet of Things (IoT)-Based Sustainable Agriculture. *Agriculture*, *12*(10), 1745. DOI: 10.3390/agriculture12101745

Dutta, M., Gupta, D., Tharewal, S., Goyal, D., Sandhu, J. K., Kaur, M., AlZubi, A. A., & Alanazi, J. M. (2025). Internet of Things-Based Smart Precision Farming in Soilless Agriculture: Opportunities and Challenges for Global Food Security. *IEEE Access : Practical Innovations, Open Solutions*, •••, 34238–34268. DOI: 10.1109/ACCESS.2025.3540317

Eastwood, C., Klerkx, L., Ayre, M., & Dela Rue, B. (2019). Managing socio-ethical challenges in the development of smart farming: From a fragmented to a comprehensive approach for responsible research and innovation. *Journal of Agricultural & Environmental Ethics*, *32*(5), 741–768.

Fassnacht, F. E., White, J. C., Wulder, M. A., & Næsset, E. (2024). Remote sensing in forestry: Current challenges, considerations and directions. Forestry. *International Journal of Forestry Research*, *97*(1), 11–37. DOI: 10.1093/forestry/cpad024

Ferrández-Pastor, F. J., García-Chamizo, J. M., Nieto-Hidalgo, M., Mora-Pascual, J., & Mora-Martínez, J. (2016). Developing ubiquitous sensor network platform using internet of things: Application in precision agriculture. *Sensors (Basel)*, *16*(7), 1141. DOI: 10.3390/s16071141

Folorunso, A. T., & Otonne, A. (2025). The Impact of Financial Indicators and Stock Market Volatility on Stock Returns in Nigeria: Evidence from Panel Analysis. *African Journal of Economic Review*, *13*(2), 104–121.

Fuentes-Peñailillo, F., Gutter, K., Vega, R., & Silva, G. C. (2024). Transformative Technologies in Digital Agriculture: Leveraging Internet of Things, Remote Sensing, and Artificial Intelligence for Smart Crop Management. *Journal of Sensor and Actuator Networks*, *13*(4), 39. DOI: 10.3390/jsan13040039

Galaz, V., Centeno, M. A., Callahan, P. W., Causevic, A., Patterson, T., Brass, I., & Levy, K. (2021). Artificial intelligence, systemic risks, and sustainability. *Technology in Society*, *67*, 101741. DOI: 10.1016/j.techsoc.2021.101741

Goddek, S., Delaide, B., Mankasingh, U., Ragnarsdottir, K. V., Jijakli, H., & Thorarinsdottir, R. (2015). Challenges of sustainable and commercial aquaponics. *Sustainability*, *7*(4), 4199–4224.

Gonzalez-de-Soto, M., Emmi, L., Benavides, C., Garcia, I., & Gonzalez-de-Santos, P. (2016). Reducing air pollution with hybrid-powered robotic tractors for precision agriculture. *Biosystems Engineering*, *143*, 79–94. DOI: 10.1016/j.biosystemseng.2016.01.008

Griepentrog, H. W., Nørremark, M., & Nielsen, J. (2006, 2006/09/03). Autonomous intra-row rotor weeding based on GPS. *CIGR World Congress: Agricultural Engineering for a Better World*, Bonn, Germany.

Gupta, G., & Kumar Pal, S. (2025). Applications of AI in precision agriculture. *Discover Agriculture*, *3*(1), 61. DOI: 10.1007/s44279-025-00220-9

Gupta, M., Abdelsalam, M., Khorsandroo, S., & Mittal, S. (2020). Security and privacy in smart farming: Challenges and opportunities. *IEEE Access : Practical Innovations, Open Solutions*, *8*, 34564–34584. DOI: 10.1109/ACCESS.2020.2979358

Gurdita, A., Vovko, H., & Ungrin, M. (2016). A Simple and Low-Cost Monitoring System to Investigate Environmental Conditions in a Biological Research Laboratory. *PLoS One*, *11*(1), e0147140. DOI: 10.1371/journal.pone.0147140

Hanafi, A. M., Hussien, S. A., Elnahal, D. H., Ahmed, S. E. H., Salem, M. A., Zainhum, A. R., Elsayed, A. A., Ibrahim, M. A., & Sattar, Y. S. A. (2025). Revolutionizing Agriculture with IoT, Mobile Apps, and Computer Vision in Automated Hydroponic Greenhouses. International Journal of Engineering and Applied Sciences - October 6 University, 2(1), 1-16. DOI: 10.21608/ijeasou.346170.1029

Holm-Nielsen, J. B., Al Seadi, T., & Oleskowicz-Popiel, P. (2009). The future of anaerobic digestion and biogas utilization. *Bioresource Technology*, *100*(22), 5478–5484.

Hussein, A. H. A., Jabbar, K. A., Mohammed, A., & Al-Jawahry, H. M. (2024). AI and IoT in farming: A sustainable approach. E3S Web of Conferences.

Hussein, A. H. A., Jabbar, K. A., Mohammed, A., & Jasim, L. (2024). Harvesting the future: AI and IoT in agriculture. E3S Web of Conferences.

Joshi, J. (2024). Sustainable farming with IoT: A case study of smart agriculture practices in rural India. *International Journal for Multidisciplinary Research*, *6*(1), 1–11. https://www.ijfmr.com/papers/2024/1/38502.pdf

Kalyani, Y., & Collier, R. (2021). A systematic survey on the role of cloud, fog, and edge computing combination in smart agriculture. *Sensors (Basel)*, *21*(17), 5922. DOI: 10.3390/s21175922

Kanuru, L., Tyagi, A. K., Fernandez, T. F., Sreenath, N., & Mishra, S. (2021). Prediction of pesticides and fertilizers using machine learning and Internet of Things. *2021 International Conference on Computer Communication and Informatics (ICCCI)*.

Kasenzu, B. (2024). Role of precision agriculture technologies in enhancing farm productivity in Kenya. *American Journal of Agriculture*, *6*(2), 1–12. DOI: 10.47672/aja.2118

Kaur, N. R., Nehra, N. D., & Bhushanwar, N. K. (2025). Enhancing sustainability, climate resilience, and resource efficiency with IoT-Based Precision Agriculture. *International Journal of Research and Review in Applied Science, Humanities, and Technology*, *2*(2), 172–179. DOI: 10.71143/7db36796

Kaur, R., Sharma, D. I., & Saini, C. (2024). Smart farming: Empowering organic agriculture with AI. *International Journal for Research in Applied Science and Engineering Technology*, *12*(3), 805–809. DOI: 10.22214/ijraset.2024.58931

Kawasaki, Y., Uga, H., Kagiwada, S., & Iyatomi, H. (2015). Basic Study of Automated Diagnosis of Viral Plant Diseases Using Convolutional Neural Networks. In Bebis, G., Boyle, R., Parvin, B., Koracin, D., Pavlidis, I., Feris, R., McGraw, T., Elendt, M., Kopper, R., & Ragan, E. (Eds.), *Advances in Visual Computing* (Vol. 9475, pp. 638–645). Springer International Publishing., https://www.ict.polimi.it/network/proxy/

Kendall, H., Clark, B., Li, W., Jin, S., Jones, G. D., Chen, J., Taylor, J., Li, Z., & Frewer, L. J. (2021). Precision agriculture technology adoption: A qualitative study of small-scale commercial "family farms" located in the North China Plain. *Precision Agriculture*, *23*(1), 319–351. DOI: 10.1007/s11119-021-09839-2

Kirkaya, A. (2020). SMART FARMING-PRECISION AGRICULTURE TECHNOLOGIES AND PRACTICES. Journal of Scientific Perspectives. DOI: 10.26900/jsp.4.010

Kumar, A., Singh, S., Yadav, M., Bhuj, B., Dhar, S., Pruthi, N., Kumar, R., Bajpai, V., Rizwan, M., Jyoti, K., Thapa, R., Kumar, V., Kumar, H., Mishra, B., Kumar, V., Rajput, A., Singh, A., & Kumar, R. (2022). Artificial Intelligence, Internet of Things (IoT) and Smart Agriculture for Sustainable Farming: A Review. *Annals of Plant Sciences*, *11*(11), 5512–5564. https://annalsofplantsciences.com/index.php/aps/article/view/1005

Lakhiar, I. A., Yan, H., Zhang, C., Wang, G., & Hao, B. (2024). A review of precision irrigation water-saving technology under changing climate for enhancing water use efficiency, crop yield, and environmental footprints. *Agriculture*, *13*(2), 245–267. DOI: 10.3390/agriculture14071141

Lysaght, T., Lim, H. Y., Xafis, V., & Ngiam, K. Y. (2019). AI-assisted decision-making in healthcare: The application of an ethics framework for big data in health and research. *Asian Bioethics Review*, *11*(3), 299–314. DOI: 10.1007/s41649-019-00096-0

Maloku, D. (2020). Adoption of precision farming technologies: USA and EU situation. *SEA–Practical Application of Science*, *8*(22), 7–14.

Mamoria, P. R. K., & Shakya, R. (2023). The Role of Artificial Intelligence in Addressing Climate Change. In Vempaty, L. N., Raja, V., Mukherjee, R., Shriya, K., Maddela Prasanthi, M., & Laxmi, K. V. (Eds.), *Fusion of Knowledge: Multidisciplinary Perspectives in Research* (Vol. 4, pp. 31–36). Infinity Publication., DOI: 10.25215/9392917953.06

McFadden, J., Erickson, B., Lowenberg-DeBoer, J., & Milics, G. (2024, 2024/07/21-24). Global Adoption of Precision Agriculture: An Update on Trends and Emerging Technologies. *16th International Conference on Precision Agriculture*, Monticello, IL.

Miller, T., Mikiciuk, G., Durlik, I., Mikiciuk, M., Łobodzińska, A., & Śnieg, M. (2025). The IoT and AI in Agriculture: The Time Is Now—A Systematic Review of Smart Sensing Technologies. *Sensors (Basel)*, *25*(12), 3583. DOI: 10.3390/s25123583

Mmbando, G. S. (2025). Harnessing artificial intelligence and remote sensing in climate-smart agriculture: The current strategies needed for enhancing global food security. *Cogent Food & Agriculture*, *11*(1), 2454354. DOI: 10.1080/23311932.2025.2454354

Mohamed, E. S., Belal, A. A., Abd-Elmabod, S. K., El-Shirbeny, M. A., Gad, A., & Zahran, M. B. (2021). Smart farming for improving agricultural management. *The Egyptian Journal of Remote Sensing and Space Sciences*, *24*(3), 971–981.

Morchid, A., El Alami, R., Al Raezah, A. A., & Sabbar, Y. (2023). Applications of Internet of Things (IoT) and sensors technology to increase food security and agricultural sustainability: Benefits and challenges. *Ain Shams Engineering Journal, 15*(3), 102509. DOI: 10.1016/j.asej.2023.102509

Morchid, A., El Alami, R., Raezah, A. A., & Sabbar, Y. (2024). Applications of internet of things (IoT) and sensors technology to increase food security and agricultural Sustainability: Benefits and challenges. *Ain Shams Engineering Journal, 15*(3), 102509. DOI: 10.1016/j.asej.2023.102509

Mordor Intelligence, I. (2025a). Hydroponics - Market Share Analysis, Industry Trends & Statistics, Growth Forecasts (2025–2030). https://www.marketresearch.com/Mordor-Intelligence-LLP-v4018/Hydroponics-Share-Trends-Statistics-Growth-42554019/

Mordor Intelligence, I. (2025b). Precision Farming Market Size and Share Analysis - Growth Trends and Forecast (2025–2030). https://www.mordorintelligence.com/industry-reports/global-precision-farming-market-industry

Mordor Intelligence, I. (2025c). Smart Greenhouse Market Size & Share Analysis - Growth Trends & Forecasts (2024–2029). https://www.marketresearch.com/Mordor-Intelligence-LLP-v4018/Smart-Greenhouse-Size-Share-Growth-36614817/

Muninarayanappa, V., & Ranjan, R. (2024). Agriculture data analysis using parallel k-nearest neighbour classification algorithm. *Int J Reconfigurable & Embedded Syst ISSN, 2089*(4864), 4864. DOI: 10.11591/ijres.v13.i2

Naorem, L. D., Muthaiyan, M., & Venkatesan, A. (2019). Identification of dysregulated miRNAs in triple negative breast cancer: A meta-analysis approach. *Journal of Cellular Physiology, 234*(7), 11768–11779.

Nath, D. (2023). Smart Farming: Automation and Robotics in Agriculture. In. Nations), F. F. a. A. O. o. t. U. (2022). Understanding the context: pests and pesticide management. Food and Agriculture Organization of the United Nations. https://www.fao.org/pest-and-pesticide-management/about/understanding-the-context/en/?utm_source=chatgpt.com

Nyakuri, J., Nkundineza, P. C., Omar Gatera, O., Nkurikiyeyezu, K. K., & Mwitende, G. (2025). AI and IoT-powered edge device optimized for crop pest and disease detection. *Scientific Reports, 15*(1). Advance online publication. DOI: 10.1038/s41598-025-06452-5

Premachandra, J., & Kumara, P. (2021). Mobile Applications for Precision Agriculture Practices: A Review. 14th International Research Conference, Security, Stability and National Development in the New Normal, Ratmalana, Sri Lanka.

Puspaningrum, A., Sumarudin, A., & Putra, W. P. (2022). Irrigation prediction using machine learning in precision agriculture. 2022 5th International Conference of Computer and Informatics Engineering (IC2IE).

Qayoom, K., & Manzoor, S. (2022). Digital Agriculture: An Emerging Approach Towards Sustainability in Agriculture and Rural Development. Journal of Community Mobilization and Sustainable Development, 1(Seminar Special Issue), 150-158.

Quy, V. K., Hau, N. V., Anh, D. V., Quy, N. M., Ban, N. T., Lanza, S., & Muzirafuti, A. (2022). IoT-enabled smart agriculture: Architecture, applications, and challenges. *Applied Sciences (Basel, Switzerland)*, *12*(7), 3396.

Ragazou, K., Garefalakis, A., Zafeiriou, E., & Passas, I. (2022). Agriculture 5.0: A new strategic management mode for a cut cost and an energy efficient agriculture sector. *Energies*, *15*(9), 3113. DOI: 10.3390/en15093113

Rajabzadeh, M., & Fatorachian, H. (2023). Modelling factors influencing IoT adoption: With a focus on agricultural logistics operations. *Smart Cities*, *6*(6), 3266–3296.

Rathor, A. S., Choudhury, S., Sharma, A., Nautiyal, P., & Shah, G. (2024). Empowering vertical farming through IoT and AI-Driven technologies: A comprehensive review. *Heliyon*, *10*(15), e34998. DOI: 10.1016/j.heliyon.2024.e34998

Rejeb, A., Rejeb, K., Zailani, S., Keogh, J. G., & Appolloni, A. (2022). Examining the interplay between artificial intelligence and the agri-food industry. *Artificial Intelligence in Agriculture*, *6*, 111–128.

Rogulski, M., & Badyda, A. (2020). Investigation of Low-Cost and Optical Particulate Matter Sensors for Ambient Monitoring. *Atmosphere*, *11*, 1040. DOI: 10.3390/atmos11101040

Roux, J., Escriba, C., Fourniols, J.-Y., & Soto-Romero, G. (2019). A new bi-frequency soil smart sensing moisture and salinity for connected sustainable agriculture. *Journal of Sensor Technology*, *9*(3), 35–43.

S, E., Sandiya, H., Dharshini, D., Nanthini, D., Karthickraja, A., A, R., S, B., & Pasoubady, S. (2025). Revolutionizing Organic Farming With Artificial Intelligence: A Sustainable Future. In (pp. 113-121).

Saha, S., Ghosh, B., Barman, B., & Satpathy, B. (2025). Predictions for the Future of Digital Agriculture and Potential Developments. In (pp. 423-451).

Sajid, J., Hayawi, K., Malik, A. W., Anwar, Z., & Trabelsi, Z. (2023). A fog computing framework for intrusion detection of energy-based attacks on UAV-assisted smart farming. *Applied Sciences (Basel, Switzerland)*, *13*(6), 3857. DOI: 10.3390/app13063857

Sanjay, K. (2024). Sensor Based Technology Development: Advancements In Precision Agriculture. *International Journal of Advances in Engineering & Scientific Research*, *11*(1), 50–64. DOI: 10.53882/IJAESR.2023.1001004

Sengupta, S., Ghosh, A., Mallick, T. K., Chanda, C. K., Saha, H., Bose, I., Jana, J., & Sengupta, S. (2021). Model based generation prediction of SPV power plant due to weather stressed soiling. *Energies*, *14*(17), 5305.

Sharma, A., Jain, A., Gupta, P., & Chowdary, V. (2021). Machine learning applications for precision agriculture: A comprehensive review. *IEEE Access : Practical Innovations, Open Solutions*, *9*, 4843–4873. DOI: 10.1109/ACCESS.2020.3048415

Sharma, K., & Shivandu, S. K. (2024). Integrating artificial intelligence and Internet of Things (IoT) for enhanced crop monitoring and management in precision agriculture. *Sensors International*, *5*, 100292. DOI: 10.1016/j.sintl.2024.100292

Shiba, M., & Miwa, H. (2022). Path Control Algorithm for Weeding AI Robot. In *International Conference on Emerging Internetworking, Data & Web Technologies* (pp. 375-385). Springer International Publishing.

Talaviya, T., Shah, D., Patel, N., Yagnik, H., & Shah, M. (2020). Implementation of artificial intelligence in agriculture for optimisation of irrigation and application of pesticides and herbicides. *Artificial Intelligence in Agriculture*, *4*, 58–73.

Tanaka, T. S., Heuvelink, G. B., Mieno, T., & Bullock, D. S. (2024). Can machine learning models provide accurate fertilizer recommendations? *Precision Agriculture*, *25*(4), 1839–1856. DOI: 10.1007/s11119-024-10136-x

Tayler, M. (2012). New Technologies in Arable Farming. https://ruralleaders.co.nz/wp-content/uploads/2025/09/2012-Michael-Tayler.pdf

Thi Thoa, M., Nguyen, T.-D., Dang, H.-K., Nguyen, D.-T., & Nguyen, X.-T. (2024). Intelligent agricultural robotic detection system for greenhouse tomato leaf diseases using soft computing techniques and deep learning. *Scientific Reports*, *14*, 23887. DOI: 10.1038/s41598-024-75285-5

Vermesan, O., & Friess, P. (2013). *Internet of things: converging technologies for smart environments and integrated ecosystems*. River publishers.

Villa-Henriksen, A., Edwards, G. T., Pesonen, L. A., Green, O., & Sørensen, C. A. G. (2020). Internet of Things in arable farming: Implementation, applications, challenges and potential. *Biosystems Engineering*, *191*, 60–84.

Vrochidou, E., Tziridis, K., Nikolaou, A., Kalampokas, T., Papakostas, G. A., Pachidis, T. P., Mamalis, S., Koundouras, S., & Kaburlasos, V. G. (2021). An Autonomous Grape-Harvester Robot: Integrated System Architecture. *Electronics (Basel)*, *10*(9), 1056. DOI: 10.3390/electronics10091056

Wang, Z., Walsh, K. B., & Verma, B. (2017). On-tree mango fruit size estimation using RGB-D images. *Sensors (Basel)*, *17*, 2738. DOI: 10.3390/s17122738

Wecht, C., Nachtmann, M., & Koppenhagen, F. (2021). BASF: Precision Farming with Lark Bread Initiative. In Gassmann, O., & Ferrandina, F. (Eds.), *Connected Business: Create Value in a Networked Economy* (pp. 315–325). Springer., DOI: 10.1007/978-3-030-76897-3

Wu, J., Ping, L., Ge, X., Wang, Y., & Fu, J. (2010). Cloud storage as the infrastructure of cloud computing. 2010 International conference on intelligent computing and cognitive informatics, (ICICCI), Kuala Lumpur, Malaysia.

Zhao, L., Walkowiak, S., & Fernando, W. G. D. (2023). Artificial intelligence: A promising tool in exploring the phytomicrobiome in managing disease and promoting plant health. *Plants*. Advance online publication. DOI: 10.3390/plants12091852

Zhou, X., English, B. C., Larson, J. A., Lambert, D. M., Roberts, R. K., Boyer, C. N., & Martin, S. W. (2017). Precision Farming Adoption Trends in the Southern U.S. *Journal of Cotton Science*, *21*(2), 143–155.

Chapter 15
A Vision for the Future: The Next Wave of Innovation in Precision Agriculture

Afnan Khan Shinwari
http://orcid.org/0009-0008-7448-2942
University of Houston, USA

Salma Hameed
University of Jhang, Pakistan

Md. Shoeab Akhter
http://orcid.org/0009-0004-4582-2937
International University of Business, Agriculture, and Technology, Dhaka, Bangladesh

Faiz Ul Hassan
http://orcid.org/0009-0001-0758-1368
Universiti Kebangsaan Malaysia, Malaysia

Ayesha Sani
http://orcid.org/0000-0001-9718-8131
University of South Africa, South Africa

ABSTRACT

Precision agriculture is undergoing a fundamental paradigm shift, driven by the fusion of AI, autonomous robotics, advanced sensing, digital twins, and biotechnology. This chapter outlines the creation of integrated, data-driven ecosystems to enhance productivity, resource efficiency, and climate resilience. It details enabling technologies—from hyperspectral imaging and biodegradable sensors to AI platforms and nano-enabled inputs—and analyzes the architecture of interoperable

DOI: 10.4018/979-8-3373-7257-0.ch015

farm systems. Socio-economic dimensions, equitable access, and environmental sustainability are examined alongside scalable implementation pathways for all farm types. Critical research gaps in AI generalizability, robotic coordination, and adoption strategies are identified. The trajectory points toward adaptive, fully connected, and self-optimizing farming systems that harmonize food security with ecological stewardship.

1. INTRODUCTION

Precision agriculture has transformed modern day agronomy and brought discipline away from an experiential framework and toward a science driven and technology enabled paradigm. Over the last few decades, the sector has evolved from primitive GPS-enabled equipment and satellite imagery to a complex digital environment that combines artificial intelligence (AI), remote and proximal sensing, automation, robotics and cloud-based analytics. This evolution means a move away from homogenous field management practices towards highly granular, site-specific decision making that optimizes the utilization of inputs of inputs, improves the efficiency of use of resources and strengthens environmental sustainability. Presently the use of real-time flow of data generated from unmanned aerial vehicles, intelligent sensors, autonomous tractors, and soil/plant/atmosphere models has become more common among agronomists to produce actionable data from it (Lee 2022). Nevertheless, despite the large progress observed in productivity and sustainability, precision agriculture is now located at the edge of an even more disruptive technological age that promise to fundamentally restructure the global food systems.

The new wave of innovation which will emerge is not simply incremental, it is imperative. The agricultural sector faces challenges it cannot afford to overcome to then be tackled using traditional methodologies or the technology currently available (United Nations, 2023). Climate change is further increasing the frequency and intensity of droughts, floods and heatwaves and changes the ecology of pests and diseases resulting in a destabilization of yield consistency across the world (IPCC, 2022). Concurrently, the issue of soil nutrient depletion, decreased levels of organic carbon and pervasiveness of the contamination of agricultural lands by agrochemicals, microplastics and heavy metals pose a threat to long-term fertility (FAO, 2021). Global population growth is expected to bring food demand to rates that are impossible to support, requiring as much as a 70 per cent increase in production needs by 2050 (World Bank, 2023). Simultaneously, water scarcity affects almost every region, while the environmental cost of conventional agriculture presents itself in ever more ways, through biodiversity loss, greenhouse gas emissions and land degradation (European Environment Agency, 2022). These growing challenges highlight limitations of current tools in precision agriculture which may be effectively addressing local optimization but, due to their complexity, interconnectedness, and persistence of global stressors, ending up being poorly equipped to tackle them (Kumar et al., 2023).

Consequently, the future path of precision agriculture requires change of paradigm towards autonomous, climate-resilient and multi-scalable systems deeply coupled with earth observation, genomics and cutting-edge computational frameworks. The next ten years will see the coming together of nanotechnology, synthetic biology, swarm robotics, quantum sensing, edge AI and predictive digital twins - technologies that have the ability to transmute farms into intelligent and adaptive landscapes. This chapter outlines a visionary roadmap of these future innovations that will be based on evidence as well as questioning the transformative potential of these innovations and how they will affect sustainable, productive and resilient agricultural systems. In addition, the discussion is directed towards the important governance structures, socio-economic implications and policy frameworks needed to support a global shift towards future-ready precision farming, (OECD, 2023).

The purpose of this chapter is to outline the next generation of revolutionary innovations leading to revisions in precision agriculture over the next several decades. The objectives include analytical evaluation of the emerging technologies, evaluation of their potential contribution to food security and sustainability in the global food resource, and the proposal of a strategic roadmap to enable integration of the technologies into on-farm agricultural systems. Ultimately the aim is to ensure people have a forward-looking understanding of the changing agricultural landscape including researchers, policy makers and practitioners.

2. TRANSFORMATIVE TECHNOLOGIES SHAPING THE NEXT ERA

The next stage in precision agriculture is expected to be characterized by technologies enabled by precision that have never been so precise, responsive, and sustainable in a wide variety of agro-ecosystems (Todmal, 2023). As global agriculture increasingly moves towards automation, climatic resilience, along with systems-after-stone analysis, the readiness of an unspeakable sensing modalities, A-I-IOT ecology, straightforward data twins, and bio-nanotech cores, alongside biological regularities are rearranged. Collectively, these innovations are changing the way of doing things on farms-they are changing farm operations from reactive production environments to intelligent, predictive, self-optimizing environments (Miller et al., 2025).

2.1. Advanced Sensing and Data Systems

Advanced sensing platforms form the backbone of the next generation agriculture domains called precision agriculture by providing high resolution and continuous and multi-layered data, which is organizations cannot do without while making informed decisions (Mani et al., 2025). New sensor architectures from hyperspectral sensors, thermal sensors and flexible and biodegradable probes that go into soils provide a major jump in the ability to monitor plant and soil health.

Hyperspectral sensors provide the possibility to see subtle physiological changes in crop in advance of visible modifications initial and therefore performing early stress diagnosis and accurate intervention. Thermal sensors give you canopy temperature profile which is important to know the transpiration, irrigation requirements, and response to heat stress (Yan et al., 2025). Bio-degradable sensors are a solution to the growing issue of electronic waste because they allow monitoring temporary data without any long term impact in the environment. In-situ at depth probes which include electrochemical components, optical and microfluidic components provide data on the continuous measurement of nutrient dynamics, pH and moisture and microbial activity which are crucial for fine-tuning irrigation, fertilization and root zone management (Eldeeb et al., 2023).

The large amount of data produced by these sensors requires next-generation smart data pipelines that can process, manage and interpret information efficiently. The combination of 5G connectivity, edge computing, and cloud analytics makes it possible to have real-time data flows from the sensors used in the field to the decision support platforms. Spatio-temporal monitoring systems combine satellite imagery, drone-based observations and ground sensors to build dynamic multilayer maps of the conditions of the crops (Rehman et al., 2025). Machine learning embedded within these pipelines facilitate automated process of detecting anomalies, pattern recognition and predictive modelling activities thus reducing the amount of manual labor in data interpretation activity to a large extent.

Even so, there are a number of barriers to widespread adoption. High manufacturing and maintenance expenses of sensors is a limitation to their deployment, especially in low resource farming systems. Calibration issues are still a problem and critical, particularly in mixed fields where environmental factors change quickly. Sensor longevity is often discontinued by moisture, extreme temperature and mechanical stress. Moreover, to integrate diverse sensors into unified platforms, standardization of data formats and interoperability is needed (problems that have not been addressed in many regions) (Kim et al., 2025; Nalakurthi et al., 2024). Despite all these challenges, progress in low-cost materials, Artificial intelligence based self-calibration and modular system design are expected to speed up the adoption process for sensors across the world.

2.2. AI, IoT, and Autonomous Systems

Artificial intelligence (AI), internet of things (IoT) and autonomous systems are at the technological heart of the future smart farm. AI-driven decision support platforms are revolutionizing the management of crops with the ability to map using predictive modelling and anticipate crop stresses before they happen(Choudhary et al., 2025). Machine-learning algorithms look through real-time data from sensors, long-term climate trends, soil health patterns, and historical yield records and provide recommendations of what to do such as when to irrigate or which nutrients to apply or when to spray for pests and when to harvest the crop. Deep learning based systems help in defining crop diseases, insect infestations as well as nutrient deficiencies with high accuracy based on the image based diagnostics (Roy et al., 2025). These AI tools are becoming critical in order to achieve high levels of productivity with a minimum wastage of the inputs.

The coupling of AI and IoT networks is what creates a digitally interconnected farm ecosystem, where devices can communicate with each other and share information and automate operations. IoT enabled machinery - smart sprayers, automated irrigation systems, and climate controlled greenhouses - responds to the input from the sensors without any interaction with humans (Mansoor et al., 2025). These systems provide precise application of input and optimize water - nutrient efficiency in changing environmental conditions, as shown in *Figure 1.*

Despite making good progress, serious challenges prevent large scale deployment. Terrain variation such as slopes, uneven terrain and dense cropping systems make autonomous platforms navigation difficult. Weather conditions like heavy rain, fog and strong winds deduct from the accuracy of sensors and interrupt the work of drones. Regulatory barriers are still high especially for autonomous drones and robotic vehicles with strict regulations on flying heights, boundaries for a location, and finally the handling of vehicles by an operator. AI transparency is another issue - farmers need to have models that are interpretable so that trust can be established and accountability ensured in decision-making (Nagaraja et al., 2024). Overcoming these issues requires strong policies and sound machine adaptability and farmer training.

Figure 1. Schematics on use of AI, robotics and IoT in agriculture monitoring

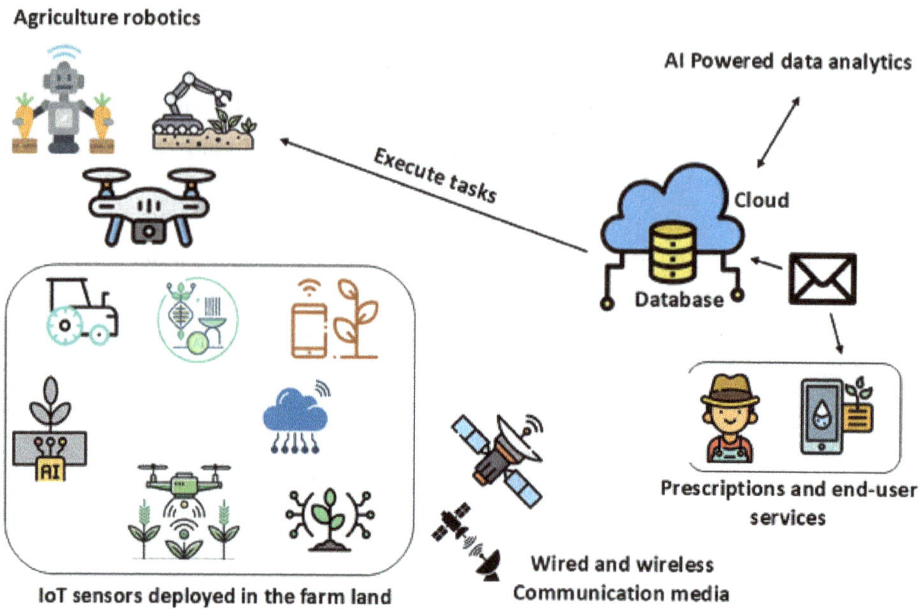

2.3. Digital Twins, Nano-Enabled Tools, and Biological Innovations

The agricultural sector is seeing the emergence of revolutionary innovations that merge the integration of digital intelligence, nano-scale engineering and biological lace into integrated production frameworks (Zhang et al., 2025). Digital twins or virtual copies of entire fields, farms or production systems are being at the center of future precision agriculture. These are dynamic models that continuously update using actual sensors in the world to simulate crop growth, water movement and nutrient cycle, pest emergence and climate interaction. Digital twins allow scenario testing, given farmers test the effects of different irrigation strategies, fertilizer regimes or harvesting plans, without having to physically test them. They also improve optimization of resources through yield response prediction under different climate and management scenarios (Awais et al., 2025). As the computational power grows, digital twins will become powerful decision-making engines that will be able to predict long term soil health, carbon sequestration and economic profitability.

Nanotechnology is shaping the next generation nano enabled tools for agricultural efficiency. Nanofertilizers allow for improved nutrient use efficiency through control over the timing of release, delivery and leaching losses. Nano pesticides offer enhanced penetration, the stability of the pesticide, and lower environmental toxicity. Nano biosensors - tiny devices incorporated with nanoparticles - may detect pathogens, poisons and lack of nutrients at ultra-low concentrations, which makes early intervention possible (Bist et al., 2025). These technologies allow to minimize input wastage, minimize exposure of chemicals, and promote sustainable intensification. But worries about nanoparticles accumulation and ecological interactions as well as the subject regulatory manufacturer's governance makes careful measure needs (Mustapha et al., 2025).

Biological innovations, especially those concerning microbes and the microbiota (microbiome), are being recognized as precision tools that can be used to improve soil fertility, plant resilience and nutrient cycling. Beneficial microbes, such as nitrogen fixers, phosphate solubilizes, plant growth promoting rhizobacteria and biocontrol agents can be adapted to a specific crop-soil environment (Yusuf et al., 2025). Advances in synthetic biology have now allowed for the engineering of microbes with improved metabolic capabilities, stress tolerance and biosensing capabilities. Bio - based precision tools such as microbial consortia, which are intended to function under specific climatic or soil conditions, are a sustainable alternative to synthetic agrochemicals. These innovations are allies of the regenerative agriculture and contribute to better sequestering of carbon in soil and to a decrease in the dependence on external inputs (McCarty & Ledesma-Amaro, 2019).

Table 1. Key technologies, applications, and future directions in next-generation precision agriculture

Technology	Primary Applications	Enabling Tools	Limitations	Future Directions	References
Advanced Sensors (Hyperspectral, Thermal, Biodegradable)	Crop monitoring, stress detection, soil health analysis	IoT, 5G, Edge computing	High cost, calibration, lifespan, integration	Low-cost sensors, self-calibration, multi-parameter sensing	(Alahmad et al., 2023)
AI & Decision-Support Systems	Predictive modeling, irrigation/fertilizer optimization, pest/disease management	Machine learning, cloud analytics, spatio-temporal modeling	Model transparency, adaptation to diverse crops	Generalizable AI, real-time adaptive decision-making	(Cate, 2022)
Robotics & Autonomous Systems	Weeding, seeding, harvesting, drone–UGV coordination	LiDAR, RTK-GPS, multispectral vision, swarm algorithms	Terrain, weather, regulation, energy efficiency	Multi-agent coordination, soil-interactive robots, self-repairing machinery	(Yin et al., 2021)

continued on following page

Table 1. Continued

Technology	Primary Applications	Enabling Tools	Limitations	Future Directions	References
Digital Twins	Farm simulation, scenario testing, yield forecasting	Cloud platforms, sensor fusion, AI models	Data standardization, computation, real-time updating	Fully integrated predictive ecosystems, real-time decision loops	(Peladarinos et al., 2023)
Nano-enabled Tools	Nano-fertilizers, nano-pesticides, smart delivery of agrochemicals	Nanomaterials, controlled release systems	Environmental safety, regulation, long-term effects	Sustainable nano-inputs, eco-friendly formulations, precision delivery	(Kekeli et al., 2025)
Biological Innovations	Microbial inoculants, engineered microbiomes	Synthetic biology, biosensors	Environmental adaptation, field performance variability	Soil–plant–microbe integration, bio-based precision tools	(Sen et al., 2025)
Connectivity & Data Ecosystems	Real-time monitoring, farm orchestration	5G/LoRaWAN, cloud-edge hybrid, APIs	Limited rural coverage, data privacy/security	Scalable networks, data marketplaces, interoperable standards	
Socioeconomic Models	Equitable access, cooperative platforms, financing	RaaS/SaaS, co-ops, training programs	Adoption gap, digital literacy, affordability	Inclusive technology access, capacity-building, participatory research	(Sobolev et al., 2021)

3. INTEGRATED PRECISION AGRICULTURE ECOSYSTEMS

As precision agriculture moves into its next step of transformation, the focus is moving from the individual technologies to integrated ecosystems which can harmonies sensors, automation, analytics and biological systems into an integrated operational framework (Mgendi, 2024). The farms of the future will operate as networks of interconnected systems where devices speak to each other without any problems, information is flowing seamlessly through different platforms, and decision-making is autonomous at multiple levels. This integrated model is critical in ensuring climate resilience, efficiency and sustainability both at the field and regional levels (Khan et al., 2024). The next sections examine the architecture and operational coordination and economic structures which will characterize the next generation of precision agriculture ecosystems.

3.1. System Architecture and Interoperability

Next-generation precision agriculture demands a-connection of the worlds A unified digital architecture where sensors, robots, AI programmers and programs, digital twins and farming equipment communicate over one, interconnected platform, AI Driven Decision Support System, Digital Twins and Field Equipment (Taha et al., 2025). In today's farming systems, most of the agricultural devices are working in isolation from each other using different communication standards, their own software, and non-compatible data formats. This fragmentation hugely reduces the ability to incorporate data from multiple sources and hinders the promotion of automated and self-optimizing farms. Consequently, the future is open, modular and scalable system architectures to unify different technologies (Atik & Martens, 2021).

A major element in such an architecture is that of standardizing data formats and communications protocols (ISO farm data standards, MQTT, LoRaWAN, OPC-ua, etc.). These standards enable the data connections of various components of hardware and software - as well as soil sensors, drones, tractors, digital twins and climate models - to exchange information with each other in a harmonious way (Roccatello et al., 2025). Interoperability should not touch only connection: systems should be able to read and use data from different sources, in real-time. This requires API-based integration frameworks to allow for flexible data exchange between equipment manufacturers, software developers and farm management systems (López-Morales et al., 2020).

Unified platforms will most likely run through modular layers of services, each with a specific set of operations - data ingestion, storage, analytics, visualization and action execution. This type of architecture makes it easier to integrate new sensors, robots, or software tools on the farm with plug-and-play capabilities so farmers can incorporate them without changing entirely the existing equipment partnerships on the farm(Bimonte et al., 2025). Furthermore, interoperability assures that innovations (whether ideas conceived by global manufactures, local entrepreneurs or research institutions) can feed into a farm's digital ecosystem. Overcoming interoperability issues is consequently crucial if the full potential of intelligent agriculture is to be realized, redundancy is avoided, and systems can be automated on a system-wide basis (Roussaki et al., 2023).

3.2. Farm-Level Orchestration and Decision Making

Integrated precision agriculture ecosystems have a high dependence on multi-layer decision making architectures that enable operations to be coordinated across the cloud, edge and on-farm servers. Each of these layers play a critical role in making sure that the autonomous systems can process at high speed, being more resilient and efficient systems (Yu et al., 2025). Cloud platforms such as storage of massive amounts of data, complex artificial intelligence (AI) analytics, integration with other databases such as weather models, satellite imagery, and market intelligence. They support long term Forecasting, Digital Twins, and System-wise optimization . However, the cloud-based operations cannot be the only answer to the problem and tasks that need to react to events in real time, such as autonomous navigation, quick anomaly detection, or robotic swarming (Yang et al., 2024).

For time sensitive operations, edge computing becomes important. Edge devices - local servers, gateways or embedded processors - quickly crunch numbers Ember's Sewell, the most important thing to keep in mind is that edge computing only can take place when there are devices capable of crunching numbers. This leads to immediate reaction on the dynamic field situations, e.g. adapting the irrigation rate, directing robots around obstacles or carrying out specific pest control activities (Gong et al., 2025). Edge systems decrease network latency and preserve operational continuity even in case of connectivity breaks. The third layer of orchestration is the on-farm servers and control units which serve as the command centres for robots, drones and IoT networks. These systems provide the mechanism of ensuring that autonomous tractor, UAV, swarm and irrigation device systems work in coordination and not in isolation (Kalyani & Collier, 2021). For example, drones scouting the field can trigger variable rate sprayers; robots that can detect weeds can notify other robots to do spot treatment; soil moisture data can automatically adjust irrigation valves and fertigation pumps, as shown in **Figure 2**. As digital twins are being made more and more sophisticated, these models will be continuously synchronized with real-time data to simulate results and to inform operational strategies (Łągiewska & Panek-Chwastyk, 2025).

Effective orchestration also involves hierarchical decision logic (higher level models for setting objective and constraints such as optimize for yield, minimize water) support lower level system operation (things where there is an element of autonomy, being controlled by localized systems such as storage and retrieval) (Escribà-Gelonch et al., 2024). This distributed coordination is similar to biological ecosystems where many agents all work independently but make a contribution to collective optimization. As a result, with robotics, sensing technologies and AI working together without any problems, farms will evolve into fully autonomous and self-regulating farms, which can adapt to stress, predict threats and optimize productivity with minimal human intervention (Nkwocha et al., 2025).

Figure 2. Conceptual framework of a Digital Twin-driven farm orchestration system. The diagram illustrates the closed-loop integration between the physical farm and its digital counterpart, highlighting the transition from raw sensor data to multi-layered decision support and automated action

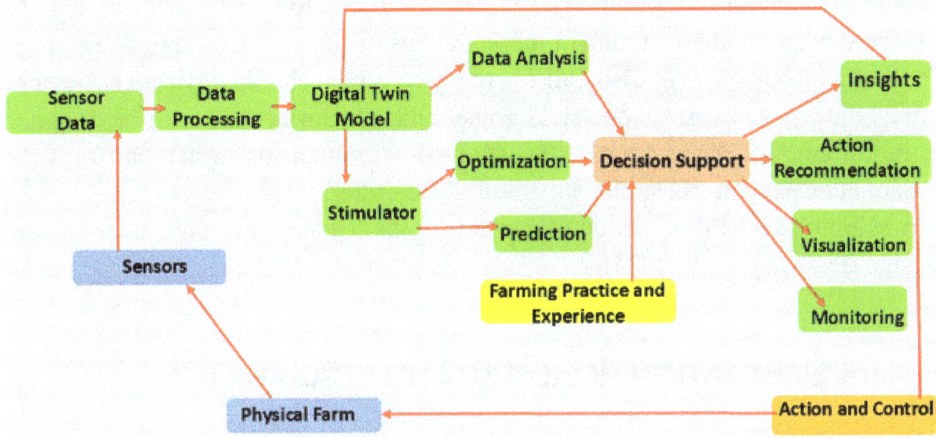

3.3. Economic Models and Technology Access

For integrated ecosystems of precision agriculture to have a global impact, innovative economic models need to be in place to guarantee the accessibility of most people. The high capital costs linked to highly-technology robots, sensing systems and analytics in a cloud often limit adoption in smallholders and farmers in developing economies. To overcome these barriers, there is a growing need to move then to service- and cooperative-based and economic models (Abraham et al., 2022; Mizik, 2023).

Another important concept is the development of data marketplaces, where farmers are able to monetize their field data or exchange information with researchers, cooperatives and agribusinesses (Raj & Prahadeeswaran, 2025). By establishing economic value based on data, there are incentives to farmers to use the IoT systems and be part of digital ecosystems. Data cooperatives could even strengthen collective bargaining power in terms of empowering whole groups of farmers to negotiate for better service rates, access to technologies and analytics (Cesco et al., 2023).

In many areas we can expect cooperative sharing systems - like those for machinery - to play an important role in democratizing access to high-tech equipment. Autonomous drones, sensors, and robotic systems can be controlled by co-operatives and rented by farmers according to the seasonal needs (Lytridis et al., 2021). Public-private partnerships, government subsidies and development programs will be essential in the implementation of these systems in context with limited resources.

Being affordable will also call for the design of technologies better suited to smallholders - sensors, robots (field grade) and AI interfaces need to be designed quite differently. Without such a broad approach, precision agriculture runs the risk of increasing the technological divide. Therefore, future innovation must make affordability, scalability and equitable access the leading attributes of ecosystem design (Padhiary et al., 2025).

4. IMPLEMENTATION ROADMAP AND SCALABILITY

The shift towards integrated and intelligent ecosystems of precision agriculture requires more than technological innovation, but careful and balanced elaboration of an implementation plan, which is carefully structured, scalable and compatible with the heterogeneity of farming systems, socio- economical contexts and environmental constraints (Hurrah et al., 2023). The next wave of innovation must be engineered across from smallholder landscape to be large commercial enterprise to mixed farming regions seamlessly. A strategic roadmap for adoption, key infrastructure and capacity building requirements and risk management approaches to ensure longer term resilience and operational continuity are all described encompassed in this section.

4.1. Adoption Pathways Across Farm Types

The scaling up of precision agriculture technologies requires adaptation of its pathways of adoption to the realities of the smallholder farms, medium-sized producers and large industries. A strategy that works everywhere in the world cannot allow for the diversity of agricultural systems; therefore, the adoption must be context specific, flexible and phased (Kroupová et al., 2024).

Smallholders, especially in Asia, Africa and Latin America, face financial, infrastructural and educational obstacles to mitigate technological implementation. For these farmers, the next iteration of precision agriculture technologies must focus at low cost and with short complexity where the impact is high. Mobile based services of precision - such as Artificial Intelligence (AI), driven advisory applications, climate smart advice and drone on demand services - form a pragmatic point of attack (Vijayakumar et al., 2025).

Mixed and transitional farms need a hybrid model, where digital technologies are introduced in phased stages starting with easily accessible technologies such as mobile platforms, satellite-based monitoring systems and low-cost sensors, and moving from there towards advanced robotics and automation technologies (Hussain et al., 2023). Flexible adoption pathways should allow farmers in transitional economies to step-by-step increase the size of their systems, avoiding overstretching of financial or operational capacities.

4.2 Infrastructure and Capacity Requirements

The prospective ecosystem of precision agriculture depends on efficacious infrastructures and competent labor and institutional endorsement for its successful operation. Connectivity at the foundation Network Infrastructure connecting rural regions to 5G networks, IoT communications using LoRaWAN and data transmission via satellite forms a sine qua non for the unimpeded flow of data across farms of any size. In areas of poor connectivity, entities providing hybrid networks to include offline edge computing and intermittent cloud synchronization have opportunities to provide resiliency (Mesfin et al., 2024).

Sensor networks are the backbone of data farming. Establishment of reliable networks is not only given by the deployment of various types of sensors, i.e. soil probes, canopy sensors, weather devices and nano biosensors but also by the setting up of local calibration centers or support units for their maintenance (Rahu et al., 2022). Making sure sensors are reliable and last a while is critical to making the right decisions. A major bottleneck in the global adoption is the lack of skilled labor who are able to manage and interpret the technological systems on which precision agriculture is built. Hence, capacity-building programs should be upscaled at the national and regional scales. Training effort should be focused on farmers, technicians, agronomists, data analysts and extension workers. Digital literacy programs, remote training platforms and field-based workshops are imperative to breed an actual workforce in place who can use complicated systems (Rajpurohit et al., 2023).

They also facilitate public-private partnerships, which increases the pace of technology transfer and design, and regional customization. In many countries institutional support - including enabling policies, support for financial subsidies and regulatory frameworks - is essential to enable innovation, while ensuring that such innovation is safe and ethically used. Investment in digital infrastructure and extension services, as well as research, are particularly important for creating a balance between technologies in demand and the ability to afford them.

4.2. Managing Risks and Uncertainties

The adoption of integrated ecosystems in precision agriculture brings new categories of risk which require early mitigation efforts if long-term resilience and sustainability are to be ensured. Technological failures, like failures of equipment, sensors, or communication can hamper normal agricultural operations. Accordingly, redundancy strategies, such as the use of dual sensors and redundant communication networks and distributed computing architectures, are crucial to reduce such system vulnerability (Anthony, 2023)

Cybersecurity threats are an important and increasing challenge. With ever-evolving digital technologies being used, farms are potential targets for cyberattacks that may disrupt the operations of farms, manipulate data or compromise food supply chains. Implementing multilayer cybersecurity protocols, encryption of data transfer, secure authentication systems, and regularly implemented security auditors are very important. Farmers and technicians also need to be trained about data protection practices(Yakkala et al., 2025).

Data errors, caused by faulty sensors, poor calibration or faulty algorithms, can result in misguided decisions that can cause crops, soil or profitability damage. Integrating data validation layers, automated error detection algorithms and continuous retraining of the models can go a long way in mitigating such risks. Digital twins can also aid validation through cross comparisons of actual data from a field by comparing them with simulated outputs (Reppa et al., 2016).

A growing concern is on technological overdependency, where too much reliance on technology may affect the expertise and decision-making skills of the farmer. To mitigate this, human in the loop systems should continue to be an integral part of operational design mechanisms, so that farmers retain oversight and understanding of key processes(Gamage et al., 2024).

Resilience strategies also need to be able to cope with climatic and market uncertainties. By combining predictive modelling, alternative cropping strategies and adaptive management tools, farms can better cope with extreme weather and input and economic shocks as well as increase their resilience to change. Ultimately, developing a strong risk management infrastructure (though integration of technological controls, human expertise, and enabling institutional policy frameworks) is key to the success of ensuring long-term and sustainable operation of precision agriculture ecosystems (Singh et al., 2025).

5. ROBOTICS AND AUTONOMOUS SYSTEMS

Robotics and autonomous systems are becoming one of the key building blocks of next-generation precision agriculture and are helping to ensure a move toward labor-efficient, sustainable, high throughput farm operations. Modern farms are increasingly using robotic platforms that can do anything from monitoring crops to harvesting, weeding and checking soil. As these systems are evolving, they incorporate intelligent sensing capability, real-time analytics and artificial intelligence-based decision making to improve their operational efficiency and minimize the reliance on manual labor and chemical inputs, as represented pictorially in *Figure 3* (Srivastav & Das, 2025).

5.1. Types of Agricultural Robots

Weeding Robots: Weeding robots are autonomous robots that use a combination of machine vision, mechanical devices and precision sprayers to identify and destroy weeds with minimal amounts of herbicides. Such robots work in the in-row and between the rows and thus reduce chemical load and enhance the health of the crop and reduce the labor cost. They pose particular value in those organic farming systems where the use of chemical herbicides are restricted.

Autonomous Tractors - Self-driving tractors with geofencing, RTK-GPS navigation and obstacle detection technology allow continuous and accurate operations in the field including tillage, seeding and spraying. Their ability to work through the day and night, even with high levels of accuracy and less labor, makes them ideal for large scale commercial farms.

UAV-UGV Cooperative Systems: Unmanned Aerial Vehicles (UAVs) and ground robots (UGVs) are hybrid working systems. Drones are used to acquire real-time multispectral and thermal information, detect hotspots and direct UGVs towards spot irrigation, fertilizer placement and mechanical weeding. This multi-agent collaboration adds a lot of efficiency in operations and location precision (Fontani et al., 2025).

Figure 3. Operational architecture of agricultural robotic systems. The diagram illustrates the integration of core modules, including stability factors, navigation systems (GPS, LIDAR, Vision), and power sources, which collectively enable autonomous applications such as seeding, weeding, spraying, and harvesting

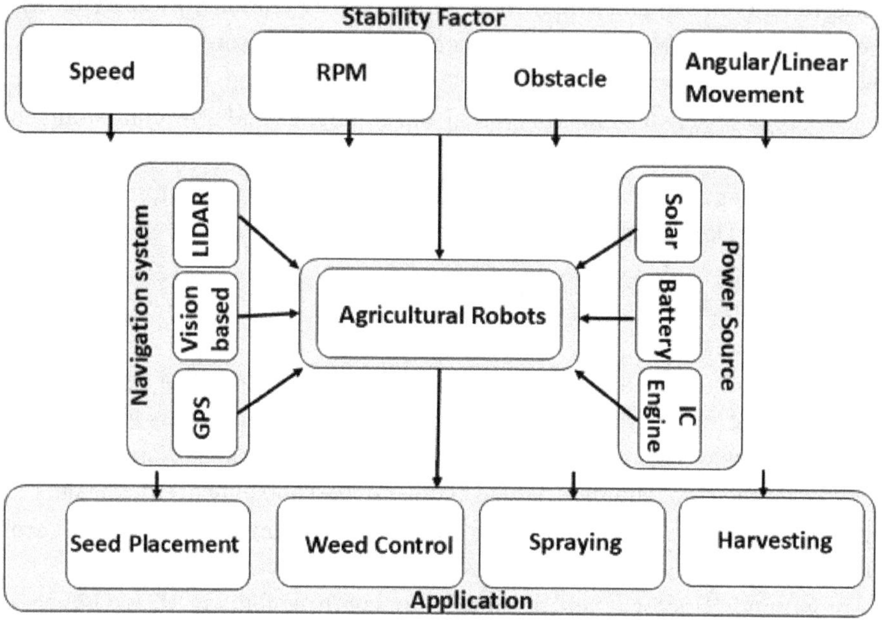

5.2. Key Enabling Technologies

The efficacy of autonomous systems depends on the set of advanced technologies. LIDAR allows the mapping of the operational environment in three dimensions and thus allows safe navigation in complex terrains. Real-Time Kinematic (RTK) positioning is a centimeter-accurate positioning required for precision applications such as planting, spraying, and navigation (Patkar & Mehendale, 2025). Multispectral and hyperspectral vision systems can access detailed information on crop health, leading to disease identification and the estimation of biomass. Collectively, all these innovations enable robots to act autonomously with considerable confidence under dynamic conditions of field environment

5.3 Deployment Models: RaaS, Co-ops, and Fleet Management

Widespread access to agricultural robotics is made possible few times by innovative ways robots are deployed. Robotics -as a service (RaaS): This enables farmers to hire robotic units on demand and thus lower the capital barriers. Cooperative models of ownership allow fleets of robots to be shared between groups of smallholders, which offers a way to optimize their usage and, hence, systematically. Fleet management platforms guarantee the coordination of scheduling and remote diagnostics and predictive maintenance as well as real-time monitoring of autonomous units over large areas of the farm (Maqbool, 2025).

6. POLICY, ETHICS AND SOCIOECONOMIC IMPLICATIONS

The swift growth of precision agriculture, driven by artificial intelligence - fueled technologies, autonomous robotics, nanotechnologies and digital twin integration, reflects the absolute need to resolve the policy, ethical and socioeconomic dimensions that are critical for sustainable and equitable deployment of these technologies (Padhiary et al., 2025). Advanced agricultural technologies hold great promise for achieving much more in terms of productivity, resource efficiency, and climate change adaptation, but without the necessary frameworks of governance, people-oriented strategies for adoption, and international cooperation, we will all risk the extreme impacts of exacerbated inequality and escalated environmental pressures, and undermine people's trust in the future of agricultural innovation. This section considers the issues of governance frameworks, labor and equity issues, environmental protection and the importance of international harmonization.

6.1. Governance and Regulation

The emergence of autonomous systems, nanotechnologies and artificial intelligence (AI) based decision support systems requires wide-ranging regulatory policies that simultaneously address safety and ethical issues (Batool et al., 2025). Guidelines for autonomous machinery (drones, self-tractors, swarms of robots) need to be defined and introduced as they need to ensure the safety of their operations, define liability regimes and ensure adherence to airspace and land use legislation. Nanotechnology based inputs like nano-fertilizers, nano-pesticides and nano biosensors call for biosafety rules that assess potential environmental and human health risks of population accumulation and particle mobility and ecological interactions of nanoparticles (Tang et al., 2024).

Data governance is another dimension that is extremely important. Precision agriculture produces huge data sets including data on productivity at the field level, soil not only in terms of fertility but also in respect to the genetics of the crops and practices followed by the farmer (Ruder & Wittman, 2025). Ensuring that this data is not accessed by anyone unauthorized, adhering to privacy laws, and safeguarding against cyberattacks is a must in order to maintain trust. Moreover, the use of Artificial Intelligence (AI)-driven decision models brings questions one closer to oversight of transparency, interpretability, and question of bias. Policy makers and developers are required to put standards for explainable AI in place with respect to agriculture in order to empower farmers, regulators, and other stakeholders to understand, validate, and trust algorithm recommendations. Effective governance also requires a Multi-Stakeholder Approach that provides for integration of national governments, regional regulatory bodies, private companies, research institutions and farmer associations. This integration represents the balancing of policies between innovation incentives and social responsibility and environmental protection on one side, and equal access to them, on the other (Maryono et al., 2024).

6.2. Equity, Labor, and Environmental Concerns

Precision agriculture technologies are offering great opportunities to increase efficiency, but they are also likely to increase social and economic inequalities if not done in an inclusive way. Advanced robotics, artificial intelligence improvement platforms, models of digital twins often need a higher capital investment which is responsible for smallholders and resource limited farmers (Hamrani et al., 2025). To avoid technological discrepancies policies should invest in affordable tools, ownership models through cooperation, service-based access (e.g. RaaS, SaaS) and community level digital training programs . The work situation in agriculture too is changing rapidly. Automation does decrease the need for people to work on the ground, yet at the same time, it increases the need for skilled operators and data analysts as well as robotic technicians and agronomists that can control the advanced systems. Upskilling programs and vocational training need to be part of national agriculture strategies so that disadvantaged displaced workers are not left without an alternative option and to make general participation in the new digital economy possible.

Environmental sustainability: Environmental sustainability needs to be at the center of developing precision agriculture. Although autonomous systems and nano-enabled tools will decrease the overuse of chemicals, energy, and water, if not executed correctly, they may pose new ecological threats (Goel et al., 2024). Policies should promote environment first frameworks that favor soil health, biodiversity, water preservation, and carbon storage, ensuring that technological adoption for agricultural production would be in the right direction, meaning contributing to regenerative and climate smart agriculture and not increasing the stress on the environment (Bharali et al., 2024).

6.3. International Harmonization and Public Acceptance

Precision agriculture is globally distributed by industry nature, as innovations, businesses, and environmental effects have no limits with countries (Rani, 2025). Coordinated international standards for autonomous farming devices, artificial intelligence (AI) models, nanomaterials and biosafety protocols are indispensable in promotion of technology transfer, avoiding technology barriers in the size of regulations, and ensuring competitive fairness (Mgendi, 2024). Organizations such as the Food and Agriculture Organization, commonly known as FAO, the International Organization for Standardization or ISO and regional trade blocs are playing critical roles to promote aspects of interoperability and standardization, certification and good practice frameworks (Ma et al., 2023).

Equally important is the engagement of the public and building trust. Farmers, consumers and environmental advocates need to be included in discussions about the adoption of technology. Transparency and open access reporting in field trials and in evaluating ethical considerations of artificial intelligence in decision-making, robotics, nanotechnology and microbial intervention might enable wider acceptance by society (Rostami & Salehi, 2024). Educational campaigns and participatory platforms that help stakeholders to get an understanding of benefits, risks, and long-term impacts, and minimize skepticism and resistance to change.

International harmonization also enables joint research and innovation, enabling nations to learn from each other and to adapt successful lessons and jointly develop technologies that are safe, sustainable and equitable. Such cooperation ensures that precision agriculture helps ensure national productivity and also their contribution towards ensuring global food security, climate resilience and social equity.

7. RESEARCH GAPS

Although great progress has been made in precision agriculture, there are still considerable research areas that remain to be filled in to realize the full potential of intelligent, autonomous, and sustainable agriculture systems. From a technical perspective, it would require huge efforts to develop robust and low-cost sensors that can be used in a wide range of different environmental conditions and extreme climatic conditions with minimal demands for calibration and maintenance. Equally critical is the creation of generalizable models for artificial intelligence (AI) and machine learning that can be adapted to heterogeneous soils and crop varieties and climate zones without being retrained to such great extent. Swarm robotics coordination is still an arena for research, with issues of multi- agent navigation, task allocation, real-time decision-making and energy efficient operation, these need to be addressed using innovative algorithms and communication protocols.

Long - term data fusion strategies incorporating multispectral imaging, soil probes, UAV and UGV observations, as well as environmental variables, are vital to develop actionable, high resisters of the data which could guide predictive farm management without data redundancy and inconsistency. From both an agronomic and ecological point of view, there is a real need for the understanding of interactions between advanced technologies and biological systems. Research is needed to understand the dynamics of soil-microbe-plant interactions under conditions of autonomous machinery, precision irrigation, and inputs built with nanomaterials, for several growing seasons. The long term environmental effects of nano agrochemicals including possible soil accretion, microbial changes, and ecosystem dynamics are poorly characterized, as are ecological effects of repeated robotic traffic on soil structure and compaction and biodiversity. Determining these dynamics is critical in order to ensure that interventions with technology increase, rather than compromise, agroecosystem resilience.

In parallel, large gaps in socioeconomic and capacity-building efforts still exist. Digital literacy capacity of farmers and agricultural technicians still varies unevenly, especially in the case of smallholders and emerging economies - restricting the adoption of solutions related to AI, robotics and IoT. Research strategies must therefore be inclusive, participatory and co-designed with local communities to be as practical, usable and socially accepted as possible. Access to innovation financing, service-based deployment models, and cooperative platforms is also a priority, so that there may be equitable access to the process of precision agriculture and not worsen the existing inequalities.

Finally, multidisciplinary studies that integrate technical, ecological, and socio-economic dimensions are necessary in order to provide comprehensive frameworks about sustainable adoption and to guide policies development and educational and training programs. Addressing these gaps in research in their totality will ensure that precision agriculture results not only in the maximization of productivity and resource efficiency, but will also keep environmental integrity, social equity and build resilient, future-ready food systems that will be able to meet the demands of a changing climate and a growing global population.

8. FUTURE DIRECTION

Anticipating the decade from 2030 onward, precision agriculture is about to go through a massive change. Emerging technologies are anticipated to replace agronomic operations, to reduce the environmental footprints and to redraw the adaptive responses to agronomic challenges. Real-time monitoring systems-embracing living monitors that will be capable of sensing the nutrition, pathogenic presence and microclimatic variables of the soil-will give farmers an unprecedented granularity of crop surveillance.

Autonomous crop systems, combining aerial drones, land-based robotic systems, and intelligent irrigation systems will all work together to provide unmatched maintenance of plant vitality, resources, and yields. Self-healing machinery that uses smart materials and intelligent robotic based actuators is expected to reduce operational downtime and related costs. Concurrently, soil-interaction robots will work with root systems, microbial consortia and sub-surface matrices to ensure targeted delivery of nutrients, water or biological control agents, to ensure a seamless integration of biological and mechanized intervention.

Adaptive artificial intelligence will support complete data acquisition, multi scale learning, predictive diagnostics, yield prognostication and resource optimization. Such systems are expected to create intelligent and resilient agroecosystems in which data analytics, automation and biological dynamics can operate together. Next-generation agriculture is likely to be self-optimizing and adaptive to changing climate conditions and thus increase sustainability and contribute to the reduction of global challenges such as increasing population and scarce resources. The integration of these technologies along with strong socio - economic governance, regulatory frameworks and focused training programmes will make agricultural practices proactive and therefore ensure efficient food production, soil health, ecosystem stewardship and equitable access to technological innovations. In the foreseeable future, arable landscapes will change into smart, adaptive ecosystems that are able to learn and regulate themselves autonomously. Such transformation is about to make food systems more resilient and keep pace with the expected increase in global population.

9. CONCLUSION

The next wave of precision agriculture marks a paradigm change of traditional, input intensive agricultural systems towards intelligent, autonomous, data summed, and high impact systems that solve significant challenges human population is experiencing as a result of climate change, resource depletion and global food security. Advanced sensing platforms, prestigious coexistent decision room, digital twins, robotics, and bio g nano tech ones complexly assist as of unprecedented monitoring, predictive modeling, and field endowment. Integrated precision agriculture ecosystems that interlacing cloud, edge and on-farm computational levels, to enable operations related to autonomous machinery, unmanned aerial vehicles, and irrigation network is part of the know-how to optimize productivity at the need of minimizing environmental impacts. The successful implementation of these technologies is not only a question of technical innovation, but also of powerful governance frameworks, models of inclusion and access and strategies for labor and environmental protection. Smallholders and emerging economies require tools that are low in cost, coop platforms, and digital literacy programs, while the commercial kind of farms will benefit from high throughput levels of automation, high densities of sensors, and even data forecasting. Forward-looking research that addresses crucial technical, ecological and socio- economical gaps concerning the services and usage of marine robots requires the successful development of sensors and algorithms for use in such environments, including substantial generalization of AI to open the field to fundamental interpretation of data for novel uses and methods, swarm algorithms, soil- microbe technology interaction studies and final nano agrochemical safety issues, and critical participatory adoption programs. Horizon technologies - such as living sensors, self-repairing machinery, the soil and robots and completely integrated adaptive AI platforms - offer a vision of a time when farms would be connected, resilient, predictive and sustainable. By balancing the technological, ecological, and social aspects, precision agriculture can become an autonomous and self-optimizing system to maximize, preserve natural resources, and form the basis for global food security. Ultimately, the next generation of farming envisioned is a fusion of innovation, sustainability, and justice to convert agricultural landscapes to become intelligent, climate-adaptive ecosystems to sustain the world's growing population and preserve the world.

REFERENCES

Abraham, M., Verteramo Chiu, L., Joshi, E., Ali Ilahi, M., & Pingali, P. (2022). Aggregation models and small farm commercialization – A scoping review of the global literature. *Food Policy*, *110*, 102299. DOI: 10.1016/j.foodpol.2022.102299

Alahmad, T., Neményi, M., & Nyéki, A. (2023). Applying IoT Sensors and Big Data to Improve Precision Crop Production: A Review. *Agronomy (Basel)*, *13*(10), 2603.

Anthony, R. (2023). *Risk and Uncertainty in Agricultural Decision Making*. In (pp. 85–93).

Atik, C., & Martens, B. (2021). Competition problems and governance of non-personal agricultural machine data: Comparing voluntary initiatives in the US and EU. *J. Intell. Prop. Info. Tech. & Elec. Com. L.*, *12*, 370.

Awais, M., Wang, X., Hussain, S., Aziz, F., & Mahmood, M. Q. (2025). Advancing Precision Agriculture Through Digital Twins and Smart Farming Technologies: A Review. *AgriEngineering*, *7*(5), 137.

Batool, A., Zowghi, D., & Bano, M. (2025). AI governance: A systematic literature review. *AI and Ethics*, *5*(3), 3265–3279. DOI: 10.1007/s43681-024-00653-w

Bharali, D., Bhuyan, S., Rout, S., Behera, H. S., & Gupta, S. (2024). Integrating policy frameworks with soil and plant conservation economics: Strategies for sustainable agricultural practices. Agriculture Archives: an International Journal.

Bimonte, S., Bellocchi, G., Pinet, F., Chalhoub, G., Sakr, M., & Skrzypczyński, P. (2025). Data engineering for sustainable agriculture: Developments, challenges, and case studies of a novel IoRT architecture. *Journal of Big Data*, *12*(1), 195. DOI: 10.1186/s40537-025-01253-z

Bist, D. R., Chapagaee, P., Kunwar, A., Pant, B. D., Khatri, L., & Mandal, A. (2025). Nanotechnology in Agriculture: A Review of Innovations in Crop Protection and Food Security. *Advances in Agriculture*, *2025*(1), 8892001. DOI: 10.1155/aia/8892001

Cate, M. (2022). Precision Agriculture and AI-Driven Decision Support Systems Improving Yield Forecasting and Resource Efficiency.

Cesco, S., Sambo, P., Borin, M., Basso, B., Orzes, G., & Mazzetto, F. (2023). Smart agriculture and digital twins: Applications and challenges in a vision of sustainability. *European Journal of Agronomy*, *146*, 126809. DOI: 10.1016/j.eja.2023.126809

Choudhary, V., Guha, P., Pau, G., & Mishra, S. (2025). An Overview of Smart Agriculture Using Internet of Things (IoT) and Web Services. *Environmental and Sustainability Indicators*, *26*, 100607. DOI: 10.1016/j.indic.2025.100607

Eldeeb, M. A., Dhamu, V. N., Paul, A., Muthukumar, S., & Prasad, S. (2023). Espial: Electrochemical Soil pH Sensor for In Situ Real-Time Monitoring. *Micromachines*, *14*(12), 2188.

Escribà-Gelonch, M., Liang, S., van Schalkwyk, P., Fisk, I., Long, N. V. D., & Hessel, V. (2024). Digital Twins in Agriculture: Orchestration and Applications. *Journal of Agricultural and Food Chemistry*, *72*(19), 10737–10752. DOI: 10.1021/acs.jafc.4c01934

Fontani, M., Luglio, S. M., Gagliardi, L., Peruzzi, A., Frasconi, C., Raffaelli, M., & Fontanelli, M. (2025). A Systematic Review of 59 Field Robots for Agricultural Tasks: Applications, Trends, and Future Directions. *Agronomy (Basel)*, *15*(9), 2185.

Gamage, A., Gangahagedara, R., Subasinghe, S., Gamage, J., Guruge, C., Senaratne, S., Randika, T., Rathnayake, C., Hameed, Z., Madhujith, T., & Merah, O. (2024). Advancing sustainability: The impact of emerging technologies in agriculture. *Current Plant Biology*, *40*, 100420. DOI: 10.1016/j.cpb.2024.100420

Goel, A., Masurkar, S., & Pathade, G. R. (2024). An Overview of Digital Transformation and Environmental Sustainability: Threats, Opportunities, and Solutions. *Sustainability*, *16*(24), 11079.

Gong, R., Zhang, H., Li, G., & He, J. (2025). Edge Computing-Enabled Smart Agriculture: Technical Architectures, Practical Evolution, and Bottleneck Breakthroughs. *Sensors (Basel)*, *25*(17), 5302.

Hamrani, A., Allouhi, A., Bouarab, F. Z., & Jayachandran, K. (2025). AI and Robotics in Agriculture: A Systematic and Quantitative Review of Research Trends (2015–2025). *Crops (Basel)*, *5*(5), 75.

Hurrah, N. N., Khan, E., & Parah, S. A. (2023). Smart Ecosystems for Sustainable Development: Opportunities, Challenges, and Solutions. In Parah, S. A., Hurrah, N. N., & Khan, E. (Eds.), *Intelligent Multimedia Signal Processing for Smart Ecosystems* (pp. 3–28). Springer International Publishing., DOI: 10.1007/978-3-031-34873-0_1

Hussain, N., Ali, Z., & Shafique, M. (2023). HYBRID APPROACHES TO SUSTAINABLE AGRICULTURE: BRIDGING TRADITIONAL FARMING KNOWLEDGE WITH MODERN BIOTECHNOLOGICAL INNOVATIONS. Scientific. *Research Reports (Montgomery)*, *1*(01), 1–21.

Kalyani, Y., & Collier, R. (2021). A Systematic Survey on the Role of Cloud, Fog, and Edge Computing Combination in Smart Agriculture. *Sensors (Basel), 21*(17), 5922.

Kekeli, M. A., Wang, Q., & Rui, Y. (2025). The Role of Nano-Fertilizers in Sustainable Agriculture: Boosting Crop Yields and Enhancing Quality. *Plants, 14*(4), 554.

Khan, S., Mazhar, T., Shahzad, T., Khan, M. A., Guizani, S., & Hamam, H. (2024). Future of sustainable farming: Exploring opportunities and overcoming barriers in drone-IoT integration. *Discover Sustainability, 5*(1), 470. DOI: 10.1007/s43621-024-00736-y

Kim, D., Zarei, M., Lee, S., Lee, H., Lee, G., & Lee, S. G. (2025). Wearable Standalone Sensing Systems for Smart Agriculture. *Advancement of Science, 12*(16), 2414748. DOI: 10.1002/advs.202414748

Kroupová, Z. Ž., Aulová, R., Rumánková, L., Bajan, B., Čechura, L., Šimek, P., & Jarolímek, J. (2024). Drivers and barriers to precision agriculture technology and digitalisation adoption: Meta-analysis of decision choice models. *Precision Agriculture, 26*(1), 17. DOI: 10.1007/s11119-024-10213-1

Łągiewska, M., & Panek-Chwastyk, E. (2025). Integrating Remote Sensing and Autonomous Robotics in Precision Agriculture: Current Applications and Workflow Challenges. *Agronomy (Basel), 15*(10), 2314.

López-Morales, J. A., Martínez, J. A., & Skarmeta, A. F. (2020). Digital Transformation of Agriculture through the Use of an Interoperable Platform. *Sensors (Basel), 20*(4), 1153.

Lytridis, C., Kaburlasos, V. G., Pachidis, T., Manios, M., Vrochidou, E., Kalampokas, T., & Chatzistamatis, S. (2021). An Overview of Cooperative Robotics in Agriculture. *Agronomy (Basel), 11*(9), 1818.

Ma, W., Marini, M., & Rahut, D. (2023). Farmers' Organizations and Sustainable Development: An Introduction. *Annals of Public and Cooperative Economics, 94*. Advance online publication. DOI: 10.1111/apce.12449

Mani, M., Baskar, D., Janakiraman, S., & Rajendran, S. (2025). MODERN AGRICULTURE SYSTEMS: ENHANCING PRECISION FARMING THROUGH ADVANCED AERIAL TECHNIQUES. *Agrociencia*, •••, 1–15. DOI: 10.47163/agrociencia.v59i8.3551

Mansoor, S., Iqbal, S., Popescu, S. M., Kim, S. L., Chung, Y. S., & Baek, J.-H. (2025). Integration of smart sensors and IOT in precision agriculture: trends, challenges and future prospectives [Review]. Frontiers in Plant Science, Volume 16 - 2025. DOI: 10.3389/fpls.2025.1587869

Maqbool, A. (2025). The Evolution of Agricultural Robotics: A Comprehensive Review and Future Challenges. *International Journal for Research in Applied Science and Engineering Technology.*

Maryono, M., Killoes, A. M., Adhikari, R., & Abdul Aziz, A. (2024). Agriculture development through multi-stakeholder partnerships in developing countries: A systematic literature review. *Agricultural Systems, 213*, 103792. DOI: 10.1016/j.agsy.2023.103792

McCarty, N. S., & Ledesma-Amaro, R. (2019). Synthetic Biology Tools to Engineer Microbial Communities for Biotechnology. *Trends in Biotechnology, 37*(2), 181–197. DOI: 10.1016/j.tibtech.2018.11.002

Mesfin, S., Kefale, H., & Gelaye, Y. (2024). Application of Precision Agriculture Technologies for Sustainable Crop Production and Environmental Sustainability: A Systematic Review. The Scientific World Journal, 2024, ID 2126734. DOI: 10.1155/2024/2126734

Mgendi, G. (2024). Unlocking the potential of precision agriculture for sustainable farming. *Discover Agriculture, 2*(1), 87. DOI: 10.1007/s44279-024-00078-3

Miller, T., Mikiciuk, G., Durlik, I., Mikiciuk, M., Łobodzińska, A., & Śnieg, M. (2025). The IoT and AI in Agriculture: The Time Is Now—A Systematic Review of Smart Sensing Technologies. *Sensors (Basel), 25*(12), 3583.

Mizik, T. (2023). How can precision farming work on a small scale? A systematic literature review. *Precision Agriculture, 24*(1), 384–406. DOI: 10.1007/s11119-022-09934-y

Mustapha, L. S., Obayomi, O. V., Lau, S. Y., & Obayomi, K. S. (2025). The role of nanotechnology in agricultural systems with emphasis on water management. *Inorganic Chemistry Communications, 182*, 115507. DOI: 10.1016/j.inoche.2025.115507

Nagaraja, G., Shoba, H., Ms, S., & Pn, K. (2024). The impact of robotics and drones on agricultural efficiency and productivity. *International Journal of Research in Agronomy, 7*(9), 1001–1009.

Nalakurthi, N. V. S. R., Abimbola, I., Ahmed, T., Anton, I., Riaz, K., Ibrahim, Q., Banerjee, A., Tiwari, A., & Gharbia, S. (2024). Challenges and Opportunities in Calibrating Low-Cost Environmental Sensors. *Sensors (Basel), 24*(11), 3650.

Nkwocha, C. L., Adewumi, A., Folorunsho, S. O., Eze, C., Jjagwe, P., Kemeshi, J., & Wang, N. (2025). A Comprehensive Review of Sensing, Control, and Networking in Agricultural Robots: From Perception to Coordination. *Robotics (Basel, Switzerland), 14*(11), 159.

Padhiary, M., Kumar, A., & Sethi, L. N. (2025). Emerging technologies for smart and sustainable precision agriculture. *Discover Robotics*, *1*(1), 6. DOI: 10.1007/s44430-025-00006-0

Patkar, V., & Mehendale, N. (2025). Autonomous ground vehicles: Technological advancements, implementation challenges, and future directions. *International Journal of Intelligent Robotics and Applications*, *9*(3), 1044–1081. DOI: 10.1007/s41315-024-00418-y

Peladarinos, N., Piromalis, D., Cheimaras, V., Tserepas, E., Munteanu, R. A., & Papageorgas, P. (2023). Enhancing Smart Agriculture by Implementing Digital Twins: A Comprehensive Review. *Sensors (Basel)*, *23*(16). Advance online publication. DOI: 10.3390/s23167128

Rahu, M., Karim, S., Shams, R., Hoshu, A. A., & Chandio, A. (2022). Wireless Sensor Networks-based Smart Agriculture: Sensing Technologies, Application and Future Directions. *Sukkur IBA Journal of Emerging Technologies*, *5*, 18–32. DOI: 10.30537/sjet.v5i2.1104

Raj, M., & Prahadeeswaran, M. (2025). Revolutionizing agriculture: A review of smart farming technologies for a sustainable future. *Discover Applied Sciences*, *7*(9), 937. DOI: 10.1007/s42452-025-07561-6

Rajpurohit, T., Singh, D., & Kumar, R. (2023). Capacity Building in Agriculture. In (pp. 139–152).

Rani, S. (2025). Precision Agriculture for Sustainable Development: Concepts and Applications. In Rani, S., Dutta, S., Rocha, Á., & Cengiz, K. (Eds.), *AI and Data Analytics in Precision Agriculture for Sustainable Development* (pp. 1–9). Springer Nature Switzerland., DOI: 10.1007/978-3-031-93087-4_1

Rehman, W. u., Koondhar, M. A., Afridi, S. K., Albasha, L., Smaili, I. H., Touti, E., Aoudia, M., Zahrouni, W., Mahariq, I., & Ahmed, M. M. R. (2025). The role of 5G network in revolutionizing agriculture for sustainable development: A comprehensive review. *Energy Nexus*, *17*, 100368. DOI: 10.1016/j.nexus.2025.100368

Reppa, V., Polycarpou, M., & Panayiotou, C. (2016). Sensor Fault Diagnosis. *Foundations and Trends in Systems and Control*, *3*, 1–248. DOI: 10.1561/2600000007

Roccatello, E., Pagano, A., Levorato, N., & Rumor, M. (2025). State of the Art in Internet of Things Standards and Protocols for Precision Agriculture with an Approach to Semantic Interoperability. *Network (Bristol, England)*, *5*(2), 14.

Rostami, K., & Salehi, L. (2024). Rural cooperatives social responsibility in promoting Sustainability-oriented Activities in the agricultural sector: Nexus of community, enterprise, and government. *Sustainable Futures : An Applied Journal of Technology, Environment and Society*, *7*, 100150. DOI: 10.1016/j.sftr.2023.100150

Roussaki, I., Doolin, K., Skarmeta, A., Routis, G., Lopez-Morales, J. A., Claffey, E., Mora, M., & Martinez, J. A. (2023). Building an interoperable space for smart agriculture. *Digital Communications and Networks*, *9*(1), 183–193. DOI: 10.1016/j.dcan.2022.02.004

Roy, P., Padhiary, M., Hoque, A., Sahu, B., Roy, D., & Kumar, K. (2025). Machine Learning for Precision Agriculture and Crop Yield Optimization: Techniques and Applications. In (pp. 189–233). DOI: 10.4018/979-8-3693-9208-9.ch007

Ruder, S.-L., & Wittman, H. (2025). Agricultural data governance from the ground up: Exploring data justice with agri-food movements. *Big Data & Society*, *12*(1), 20539517251330182. DOI: 10.1177/20539517251330182

Sen, A., Poddar, R., Sarkar, A., Saha, S., & Asif, S. M. (2025). Latest advancements in microbial inoculants for enhanced agricultural productivity. In Kumar, A., Singh, J., Queijeiro López, A. M., & Kharwar, R. N. (Eds.), *Microbial Inoculants* (pp. 181–201). Academic Press., DOI: 10.1016/B978-0-443-23560-3.00012-7

Singh, S., Khatana, K., Singh, Y., & Mishra, A. K. (2025). Enhancing Resilience and Sustainability in Farming Practices. In Mishra, A. K., Sharma, S., Mishra, A., & Roy, A. (Eds.), *Transition to Regenerative Agriculture: Principles and Indicators of Soil Health Management* (pp. 187–203). Springer Nature Singapore., DOI: 10.1007/978-981-96-1421-9_9

Sobolev, A. V., Pakhomov, V. M., Brilon, A. V., Konnova, O. A., & Minchenckova, I. V. (2021). Cooperative Platform in the Modern Economy. In Bogoviz, A. V., Suglobov, A. E., Maloletko, A. N., Kaurova, O. V., & Lobova, S. V. (Eds.), *Frontier Information Technology and Systems Research in Cooperative Economics* (pp. 25–31). Springer International Publishing., DOI: 10.1007/978-3-030-57831-2_3

Srivastav, A. K., & Das, P. (2025). Agricultural Robotics and Autonomous Systems. In Srivastav, D. A. K., & Das, D. P. (Eds.), *Biotechnology and IoT in Agriculture and Food Production: Green Innovation* (pp. 201–225). Apress., DOI: 10.1007/979-8-8688-1469-3_22

Taha, M. F., Mao, H., Zhang, Z., Elmasry, G., Awad, M. A., Abdalla, A., Mousa, S., Elwakeel, A. E., & Elsherbiny, O. (2025). Emerging Technologies for Precision Crop Management Towards Agriculture 5.0: A Comprehensive Overview. *Agriculture*, *15*(6), 582.

Tang, Y., Zhao, W., Zhu, G., Tan, Z., Huang, L., Zhang, P., Gao, L., & Rui, Y. (2024). Nano-Pesticides and Fertilizers: Solutions for Global Food Security. *Nanomaterials (Basel, Switzerland)*, *14*(1), 90.

Todmal, A. (2023). Precision Agriculture: Technology-Driven Approaches for Efficient Farming. In (pp. 46–53). DOI: 10.25215/9358791152.07

Vijayakumar, S., Murugaiyan, V., Ilakkiya, S., Kumar, V., Sundaram, R. M., & Kumar, R. M. (2025). Opportunities, challenges, and interventions for agriculture 4.0 adoption. *Discover Food*, *5*(1), 265. DOI: 10.1007/s44187-025-00576-3

Yakkala, V. S., Nusimala, K. V., Gayathri, B., Kanamarlapudi, S., Aravinth, S. S., Salau, A. O., & Srithar, S. (2025). Deep learning-based crop health enhancement through early disease prediction. *Cogent Food & Agriculture*, *11*(1), 2423244. DOI: 10.1080/23311932.2024.2423244

Yan, J., Wu, H., Diao, Z., Miao, Y., Zhang, B., & Zhao, C. (2025). Recent Developments and Applications of Crop Disease Detection, Prediction, and Early Warning: A review. *Engineering*. Advance online publication. DOI: 10.1016/j.eng.2025.10.032

Yang, D., Wu, J., & He, Y. (2024). Optimizing the Agricultural Internet of Things (IoT) with Edge Computing and Low-Altitude Platform Stations. *Sensors (Basel)*, *24*(21), 7094.

Yin, H., Cao, Y., Marelli, B., Zeng, X., Mason, A., & Cao, C. (2021). Smart Agriculture Systems: Soil Sensors and Plant Wearables for Smart and Precision Agriculture (Adv. Mater. 20/2021). *Advanced Materials*, *33*, 2170156. DOI: 10.1002/adma.202170156

Yu, P., Teng, F., Zhu, W., Shen, C., Chen, Z., & Song, J. (2025). Cloud–edge–device collaborative computing in smart agriculture: architectures, applications, and future perspectives [Review]. Frontiers in Plant Science, Volume 16 - 2025. DOI: 10.3389/fpls.2025.1668545

Yusuf, A., Li, M., Zhang, S. Y., Odedishemi-Ajibade, F., Luo, R. F., Wu, Y. X., Zhang, T. T., Yunusa Ugya, A., Zhang, Y., & Duan, S. (2025). Harnessing plant-microbe interactions: Strategies for enhancing resilience and nutrient acquisition for sustainable agriculture. *Front Plant Sci*, *16*, 1503730. DOI: 10.3389/fpls.2025.1503730

Zhang, R., Zhu, H., Chang, Q., & Mao, Q. (2025). A Comprehensive Review of Digital Twins Technology in Agriculture. *Agriculture*, *15*(9), 903.

Compilation of References

Abbas, A., Zhang, Z., Zheng, H., Alami, M. M., Alrefaei, A. F., Abbas, Q., Naqvi, S. A. H., Rao, M. J., Mosa, W. F. A., Abbas, Q., Hussain, A., Hassan, M. Z., & Zhou, L. (2023). Drones in Plant Disease Assessment, Efficient Monitoring, and Detection: A Way Forward to Smart Agriculture. *Agronomy (Basel)*, *13*(6), 1524.

Abbasi, R., Martinez, P., & Ahmad, R. (2022). The digitization of the agricultural industry – a systematic literature review on agriculture 4.0. *Smart Agricultural Technology*, *2*, 100042. DOI: 10.1016/J.ATECH.2022.100042

Abbas, S. Z., Wang, J.-Y., Wang, H., Wang, J.-X., Wang, Y.-T., & Yong, Y.-C. (2022). Recent advances in soil microbial fuel cells based self-powered biosensor. *Chemosphere*, *303*, 135036. DOI: 10.1016/j.chemosphere.2022.135036

Abbass, H., & Mostaghim, S. (2025). The road forward with swarm systems. *Philosophical Transactions. Series A, Mathematical, Physical, and Engineering Sciences*, *383*(2289), 20240135. DOI: 10.1098/rsta.2024.0135 PMID: 39880030

Abbes, A. B., Jarray, N., & Farah, I. R. (2024). Advances in remote sensing based soil moisture retrieval: Applications, techniques, scales and challenges for combining machine learning and physical models. *Artificial Intelligence Review*, *57*(9), 224. DOI: 10.1007/s10462-024-10734-1

Abdelmoneim, A. A., Kimaita, H. N., Al Kalaany, C. M., Derardja, B., Dragonetti, G., & Khadra, R. (2025). IoT Sensing for Advanced Irrigation Management: A Systematic Review of Trends, Challenges, and Future Prospects. *Sensors (Basel)*, *25*(7), 2291. https://www.mdpi.com/1424-8220/25/7/2291

Abdulhussain, S. H., Mahmmod, B. M., Alwhelat, A., Shehada, D., Shihab, Z. I., Mohammed, H. J., Abdulameer, T. H., Alsabah, M., Fadel, M. H., Ali, S. K., Abbood, G. H., Asker, Z. A., & Hussain, A. (2025). A Comprehensive Review of Sensor Technologies in IoT: Technical Aspects, Challenges, and Future Directions. *Computers*, *14*(8), 342.

Abi, R. (2025). Ethical and Explainable AI in Data Science for Transparent Decision-Making Across Critical Business Operations. *International Journal of Research Publication and Reviews*, *6*(6), 50–72. DOI: 10.55248/GENGPI.6.0625.2126

Abiri, R., Rizan, N., Balasundram, S. K., Shahbazi, A. B., & Abdul-Hamid, H. (2023). Application of digital technologies for ensuring agricultural productivity. *Heliyon*, *9*(12), e22601. DOI: 10.1016/J.HELIYON.2023.E22601 PMID: 38125472

Abraham, M., Verteramo Chiu, L., Joshi, E., Ali Ilahi, M., & Pingali, P. (2022). Aggregation models and small farm commercialization – A scoping review of the global literature. *Food Policy*, *110*, 102299. DOI: 10.1016/j.foodpol.2022.102299

Adak, S., Mandal, N., Maity, P. P., Mukhopadhyay, A., & Gunturi, A. (2023). Digital Soil Health: Unveiling the Potential of Soil Health Monitoring. *Food and Scientific Reports*, *4*(9), 96–103.

Adamo, T., Caivano, D., Colizzi, L., Dimauro, G., & Guerriero, E. (2025). Optimization of irrigation and fertigation in smart agriculture: An IoT-based microservices framework. *Smart Agricultural Technology*, *11*, 100885. DOI: 10.1016/J.ATECH.2025.100885

Addagatla, D., & Putta, M. (2025). AI-driven harvest: Connecting farmers and consumers for a sustainable agricultural economy. *Home*, *1*(1), 13–21.

Adewusi, A. O., Asuzu, O. F., Olorunsogo, T., Iwuanyanwu, C., Adaga, E., & Daraojimba, D. O. (2024). AI in precision agriculture: A review of technologies for sustainable farming practices. *World Journal of Advanced Research and Reviews*, *21*(1), 2276–2285.

Adhinata, F. D. Wahyono, & Sumiharto, R. (2024). A comprehensive survey on weed and crop classification using machine learning and deep learning. *Artificial Intelligence in Agriculture, 13*, 45-63. 10.1016/j.aiia.2024.06.005

Adkisson, M., Kimmel, J. C., Gupta, M., & Abdelsalam, M. (2021). *Autoencoder-based Anomaly Detection in Smart Farming Ecosystem.* https://arxiv.org/abs/2111.00099

Adli, H. K., Remli, M. A., Wan Salihin Wong, K. N. S., Ismail, N. A., González-Briones, A., Corchado, J. M., & Mohamad, M. S. (2023). Recent Advancements and Challenges of AIoT Application in Smart Agriculture: A Review. *Sensors (Basel)*, *23*(7), 3752. https://www.mdpi.com/1424-8220/23/7/3752

Agrawal, K., & Kumar, N. (2026). Novel agricultural techniques: an artificial intelligence and machine learning approach. *Dietary, Sensory and Gastronomic Applications*, 39–61. https://doi.org/DOI: 10.1016/B978-0-443-33345-3.00006-7

Ahmad, A., Liew, A. X. W., Venturini, F., Kalogeras, A., Candiani, A., Di Benedetto, G., Ajibola, S., Cartujo, P., Romero, P., Lykoudi, A., De Grandis, M. M., Xouris, C., Lo Bianco, R., Doddy, I., Elegbede, I., D'Urso Labate, G. F., García del Moral, L. F., & Martos, V. (2024). AI can empower agriculture for global food security: Challenges and prospects in developing nations. *Frontiers in Artificial Intelligence, 7*. Advance online publication. DOI: 10.3389/frai.2024.1328530

Ahmad, L., & Nabi, F. (2021). *Agriculture 5.0: Artificial Intelligence, IoT and Machine Learning*. CRC Press.

Ahmad, S., Wassay, M., & Hussain, M. A. (2025). *Cloud computing: Empowering the next generation of agricultural research.* Trends in Applied Sciences.

Ahmad, U., Nasirahmadi, A., Hensel, O., & Marino, S. (2022). Technology and Data Fusion Methods to Enhance Site-Specific Crop Monitoring. *Agronomy (Basel), 12*, 555. DOI: 10.3390/agronomy12030555

Ahmed, H. A. ; G., Zeng, M.-D. ;, Applications, Y., Mushtaq, M. A., Ghulam, H., Ahmed, M.-D., & Zeng, Y. (2024). Applications of Artificial Intelligence in Wheat Breeding for Sustainable Food Security. *Sustainability 2024, Vol. 16, 16*(13). /DOI: 10.3390/SU16135688

Ahmed, M., Raza, M. Y., Malik, N. A., & Malik, A. (2025). Climate-resilient agriculture (CRA): Pathway to sustainable development. In *Climate resilient and sustainable agriculture: Volume 1: Adaptation and mitigation strategies* (pp. 185-212). Springer Nature Switzerland.

Ahmed, A., Saleem, S. R., Tahir, M. N., Manzoor, S. H., Zaman, Q., Zhang, Z., & Ahmed, R. (2025). Design and development of industrial prototype of spot specific orchard sprayer using ultrasonic sensing system and advanced pressure control mechanism. *Computers and Electronics in Agriculture, 237*, 110690. 10.1016/j.compag.2025.110690

Ahmed, M. (2024). Genetic Mapping and Marker-Assisted Selection: Tools for Accelerating Crop Improvement. *Adv. Crop Sci. Technol., 12*(5), 1–2.

Ahmed, M. T., Ahmed, M. W., & Kamruzzaman, M. (2025). A systematic review of explainable artificial intelligence for spectroscopic agricultural quality assessment. *Computers and Electronics in Agriculture, 235*. Advance online publication. DOI: 10.1016/J.COMPAG.2025.110354

Ahmed, N., & Shakoor, N. (2025). Advancing agriculture through IoT, Big Data, and AI: A review of smart technologies enabling sustainability. *Smart Agricultural Technology, 10*, 100848. DOI: 10.1016/J.ATECH.2025.100848

Ahmed, N., & Shakoor, N. (2025). Advancing Agriculture through IoT, Big Data, and AI: A Review of Smart Technologies Enabling Sustainability. *Smart Agricultural Technology, 100848*. Advance online publication. DOI: 10.1016/j.atech.2025.100848

Ahvar, E., Ahvar, S., & Lee, G. M. (2024). Artificial Intelligence of Things: Architectures, Applications, and Challenges. In Ziegler, S., Radócz, R., Quesada Rodriguez, A., & Matheu Garcia, S. N. (Eds.), *Springer Handbook of Internet of Things* (pp. 443–462). Springer International Publishing.

Aijaz, N., Lan, H., Raza, T., Yaqub, M., Iqbal, R., & Pathan, M. S. (2025). Artificial intelligence in agriculture: Advancing crop productivity and sustainability. *Journal of Agriculture and Food Research, 20*, 101762. DOI: 10.1016/J.JAFR.2025.101762

Aishwarya, K. A., & Kumar, P. (2025). Sensors-Based Irrigation for Increasing Crop and Water Productivity. In A. Kumar, V. K. Singh, Y. Singh, S. K. Singh, & P. Kumar (Eds.), *Modern Technology for Sustainable Agriculture* (pp. 101-113). Cham: Springer Nature Switzerland.

Akbar, J. U. M., Kamarulzaman, S. F., Muzahid, A. J. M., & Rahman, Md. A., & Uddin, M. (2024). A Comprehensive Review on Deep Learning Assisted Computer Vision Techniques for Smart Greenhouse Agriculture. *IEEE Access : Practical Innovations, Open Solutions, 12*, 4485–4522. DOI: 10.1109/ACCESS.2024.3349418

Akhter, Md. S., Jahan, I., Tasnim, M., Miah, M. S., & Karim, A. J. M. S. (2025). AI-Driven Collaborative Robotics in Agriculture: Optimizing Soil and Crop Yield. In K. K. Reddy C., B. Saju, A. Hulus, T. Sibalija, & H. Nakouri (Eds.), *Advances in Computational Intelligence and Robotics* (pp. 57–88). IGI Global Scientific Publishing. https://doi.org/DOI: 10.4018/979-8-3373-4207-8.ch003

Akhund, M. (2025). Validating federated learning performance in practice: An agricultural edge hardware testbed analysis (Doctoral dissertation, The University of Waikato).

Akkem, Y., Biswas, S. K., & Varanasi, A. (2025). Role of Explainable AI in Crop Recommendation Technique of Smart Farming. *International Journal of Intelligent Systems and Applications, 17*(1), 31–52. DOI: 10.5815/IJISA.2025.01.03

Al khatib, A. M. G., & Alshaib, B. M. (2025). Smart Irrigation Systems: Optimizing Water Use with AI. *Transforming Agriculture through Artificial Intelligence for Sustainable Food Systems*, 73–93. /DOI: 10.1007/978-981-96-4795-8_5

Alahe, M. A., Chang, Y., Kemeshi, J., Won, K., Yang, X., & Wei, L. (2025). Real-Time agricultural image encryption algorithm using AES on edge computing devices. *Computers and Electronics in Agriculture*, *237*, 110594. DOI: 10.1016/j.compag.2025.110594

Alahmad, T., Neményi, M., & Nyéki, A. (2023). Applying IoT Sensors and Big Data to Improve Precision Crop Production: A Review. *Agronomy (Basel)*, *13*(10), 2603.

Alanhdi, A., & Toka, L. (2024). A survey on integrating edge computing with AI and blockchain in maritime domain, aerial systems, IoT, and industry 4.0. *IEEE Access : Practical Innovations, Open Solutions*, *12*, 28684–28709.

Al-Atawi, A. A., Alyahyan, S., Alatawi, M. N., Sadad, T., Manzoor, T., Farooq-I-Azam, M., & Khan, Z. H. (2023). Stress Monitoring Using Machine Learning, IoT and Wearable Sensors. *Sensors (Basel)*, *23*(21), 8875. DOI: 10.3390/S23218875

Alawode, A., Blessing, A. O., & Chiamaka, O. T. (2024). Integrating IoT and AI in sustainable agriculture to mitigate environmental risk and financial misuse.

Alawode, A., Blessing, A. O., & Chiamaka, O. T. (2025a). Integrating IoT and AI in Sustainable Agriculture to Mitigate Environmental Risk and Financial Misuse. *International Journal of Research Publication and Reviews*, *6*(6), 2810–2828. DOI: 10.55248/GENGPI.6.0625.2187

Alazzai, W. K., Abood, B. S. Z., Al-Jawahry, H. M., & Obaid, M. K. (2024). Precision farming: the power of AI and IoT technologies. E3S Web of Conferences.

Al-Bemani, A., Al-Mahruqi, A., Balaji, G., & Suresh, U. (2025). A Relative Analysis for Plant Disease Detection with AI-Driven Techniques: Optimizing Precision Agriculture. *Procedia Computer Science*, *258*, 839–851.

Albiero, D., Pontin Garcia, A., Kiyoshi Umezu, C., & Leme de Paulo, R. (2022). Swarm robots in mechanized agricultural operations: A review about challenges for research. *Computers and Electronics in Agriculture*, *193*, 106608. 10.1016/j.compag.2021.106608

Alfarisy, A. A., Chen, Q., & Guo, M. (2018). Deep learning based classification for Paddy pests & diseases recognition. *Proceedings of 2018 International Conference on Mathematics and Artificial Intelligence*, 21-25.

Ali, A., Chana, I., & Buyya, R. (2019). *A fungus detection system for greenhouses using wireless visual sensor networks and machine learning*

Ali, A., Hussain, T., Tantashutikun, N., Hussain, N., & Cocetta, G. (2023). Application of Smart Techniques, Internet of Things and Data Mining for Resource Use Efficient and Sustainable Crop Production. *Agriculture 2023, Vol. 13, 13*(2). DOI: 10.3390/AGRICULTURE13020397

Ali, G., Mijwil, M. M., Buruga, B. A., Abotaleb, M., & Adamopoulos, I. (2024). A survey on artificial intelligence in cybersecurity for smart agriculture: State-of-the-art, cyber threats, artificial intelligence applications, and ethical concerns. *Mesopotamian Journal of Computer Science, 2024*, 53–103.

Aliyeva, M., Huseynova, L., & Agabekova, G. (2025). The role of investment in the development of smart agricultural technologies in megacities.

Ali, Z., Muhammad, A., Lee, N., Waqar, M., & Lee, S. W. (2025). Artificial Intelligence for Sustainable Agriculture: A Comprehensive Review of AI-Driven Technologies in Crop Production. *Sustainability (Switzerland), 17*(5). Advance online publication. DOI: 10.3390/SU17052281/S1

Ali, Z., Muhammad, A., Lee, N., Waqar, M., & Lee, S. W. (2025). Artificial Intelligence for Sustainable Agriculture: A Comprehensive Review of AI-Driven Technologies in Crop Production. *Sustainability, 17*(5), 2281. https://www.mdpi.com/2071-1050/17/5/2281

Al-Jaf, I. H., Ibrahim, M. S., Hammadi, A. N., Al-Dulaimy, A. F., & EAlfahdawi, H. (2025). A review of smart agriculture and its impact on revolutionizing agricultural production.

Allied Market. R. (2021). Precision Agriculture Market size, competitive landscape, and trend analysis report; Global opportunity analysis and industry forecast, 2021–2030. https://www.alliedmarketresearch.com/precision-agriculture-market

Almalki, F. A., & Angelides, M. C. (2024). A synthesis of machine learning and internet of things in developing autonomous fleets of heterogeneous unmanned aerial vehicles for enhancing the regenerative farming cycle. *Computing, 106*(12), 4167–4192.

Al-Musawi, Z. K., Vona, V., & Kulmány, I. M. (2025). Utilizing Different Crop Rotation Systems for Agricultural and Environmental Sustainability: A Review. *Agronomy 2025, Vol. 15, 15*(8). DOI: 10.3390/AGRONOMY15081966

Alpyssov, A., Uzakkyzy, N., Talgatbek, A., Moldasheva, R., Bekmagambetova, G., Yessekeyeva, M., Kenzhaliev, D., Yerzhan, A., & Tolstoy, A. (2023). Assessment of plant disease detection by deep learning. *Eastern-European Journal of Enterprise Technologies, 1*(2 (121)), 41–48.

Alqurashi, H., Bouabdallah, F., & Khairullah, E. (2023). SCAP SigFox: A Scalable Communication Protocol for Low-Power Wide-Area IoT Networks. *Sensors (Basel), 23*(7). Advance online publication. DOI: 10.3390/s23073732

Alsabt, R., Alkhaldi, W., Adenle, Y. A., & Alshuwaikhat, H. M. (2024). Optimizing waste management strategies through artificial intelligence and machine learning - An economic and environmental impact study. *Cleaner Waste Systems, 8*, 100158. DOI: 10.1016/J.CLWAS.2024.100158

Alsadik, B., Ellsäßer, F. J., Awawdeh, M., Al-Rawabdeh, A., Almahasneh, L., Oude Elberink, S., Abuhamoor, D., & Al Asmar, Y. (2024). Remote Sensing Technologies Using UAVs for Pest and Disease Monitoring: A Review Centered on Date Palm Trees. *Remote Sensing, 16*(23), 4371.

Al-Shammary, A. A. G., Al-Shihmani, L. S. S., Fernández-Gálvez, J., & Caballero-Calvo, A. (2024). Optimizing sustainable agriculture: A comprehensive review of agronomic practices and their impacts on soil attributes. *Journal of Environmental Management, 364*, 121487. DOI: 10.1016/J.JENVMAN.2024.121487 PMID: 38889650

Altaf, A., Gull, S., Shah, A. Z., Faheem, M., Saeed, A., Khan, I., Al Amin Miah, A., Zhu, X., & Zhu, M. (2021). ADVANCED GENETIC STRATEGIES FOR IMPROVING RICE YIELD. *Journal of Global Innovations in Agricultural Sciences, 9*(4), 167–172. DOI: 10.22194/JGIAS/9.9520

Altammar, K. A. (2023). A review on nanoparticles: Characteristics, synthesis, applications, and challenges. *Frontiers in Microbiology, 14*, 1155622. DOI: 10.3389/FMICB.2023.1155622

Altieri, M. A., Nicholls, C. I., & Montalba, R. (2021). Agroecology and the design of climate change–resilient farming systems. *Agronomy for Sustainable Development, 41*(1), 1–14. DOI: 10.1007/s13593-020-00666-4

Álvarez-Herrera, J. G., Jaime-Guerrero, M., & Fernández-Pérez, C. J. (2025). Spatial Variability and Geostatistical Modeling of Soil Physical Properties Under Eucalyptus globulus Plantations. *Geomatics 2025, Vol. 5, 5*(3). DOI: 10.3390/GEOMATICS5030041

Alves, R. G., Maia, R. F., & Lima, F. (2023). Development of a Digital Twin for smart farming: Irrigation management system for water saving. *Journal of Cleaner Production, 388*, 135920. DOI: 10.1016/j.jclepro.2023.135920

Alzakari, S. A., Aldrees, A., Umer, M., Cascone, L., Innab, N., & Ashraf, I. (2024). Artificial intelligence-driven predictive framework for early detection of still birth. *SLAS Technology*, *29*(6). Advance online publication. DOI: 10.1016/J.SLAST.2024.100203

AlZubi, A. A., & Galyna, K. (2023). Artificial intelligence and internet of things for sustainable farming and smart agriculture. *IEEE Access : Practical Innovations, Open Solutions*, *11*, 78686–78692. DOI: 10.1109/ACCESS.2023.3298215

Amraouy, O., Benbrahim, M., & Kabbaj, M. N. (2025). A blockchain-and self-sovereign identity-based collaborative framework for secure and verifiable cross-organizational data sharing in smart irrigation. *Smart Agricultural Technology*, •••, 101654.

Anam, I., Arafat, N., Hafiz, M. S., Jim, J. R., Kabir, M. M., & Mridha, M. F. (2024). A systematic review of UAV and AI integration for targeted disease detection, weed management, and pest control in precision agriculture. *Smart Agricultural Technology*, *9*, 100647. DOI: 10.1016/J.ATECH.2024.100647

Anastasiou, E., Ntakos, G., Kanakari, E., Bitsika, S., Gemtou, M., Katsaragakis, M., Soudris, D., Volioti, C., Arvanitou, E. M., Folina, M. T., Maikantis, T., Kanidou, E. P., Fountouli, M., Ampatzoglou, A., Tsiogkas, N., Villa-Henriksen, A., Pedersen, S. M., Tamirat, T. W., Milella, A., & Fountas, S. (2025). Applications of robotics and extended reality in agriculture: A review. *Smart Agricultural Technology*, *12*, 101521. DOI: 10.1016/J.ATECH.2025.101521

Anjanappa, J. (2025). From barriers to solutions: Collaborative financing pathways for climate-resilient agri-SMEs in India. Available at *SSRN* 5377754.

Ansari, S. A., & Vidyarthi, V. K. (2025). Use of Internet of Things in water resources applications: Challenges and future directions: a critical review. *Discov. Internet Things*, *5*(1), 96.

Anthony, R. (2023). *Risk and Uncertainty in Agricultural Decision Making*. In (pp. 85–93).

Aranguri, M., Mera, H., Noblecilla, W., & Lucini, C. (2025). Digital literacy and technology adoption in agriculture: A systematic review of factors and strategies. *AgriEngineering*, *7*(9), 296.

Aremora, S., Sarfo, J. B., Abayomi Taiwo Fashina, K. M., & Balogun, M. O. A. (2025). Blockchain and AI integration for agricultural value chains in African markets.

Arief, M. A. A., Kim, H., Kurniawan, H., Nugroho, A. P., Kim, T., & Cho, B. K. (2023). Chlorophyll Fluorescence Imaging for Early Detection of Drought and Heat Stress in Strawberry Plants. *Plants 2023, Vol. 12, 12*(6). DOI: 10.3390/PLANTS12061387

Arikapudi, R., & Vougioukas, S. G. (2023). Robotic Tree-fruit harvesting with arrays of Cartesian Arms: A study of fruit pick cycle times. *Computers and Electronics in Agriculture, 211*, 108023. 10.1016/j.compag.2023.108023

Arshad, J., Rehman, A. U., & Abbas, H. (2022). Implementation of a LoRaWAN based smart agriculture decision support system for optimum crop yield. *Sustainability, 14*(2), 827.

Asih, I., Purba, H., & Sitorus, T. (2020). KEY PERFORMANCE INDICATORS: A SYSTEMATIC LITERATURE REVIEW. *8*, 142-155.

Aslan, M. F., Sabanci, K., Aslan, B., Aslan, M. F., Sabanci, K., & Aslan, B. (2024). Artificial Intelligence Techniques in Crop Yield Estimation Based on Sentinel-2 Data: A Comprehensive Survey. *Sustainability 2024, Vol. 16, 16*(18). DOI: 10.3390/SU16188277

Aslan, M. F., Durdu, A., Sabanci, K., Ropelewska, E., & Gültekin, S. S. (2022). A Comprehensive Survey of the Recent Studies with UAV for Precision Agriculture in Open Fields and Greenhouses. *Applied Sciences (Basel, Switzerland), 12*(3), 1047.

Assimakopoulos, F., Vassilakis, C., Margaris, D., Kotis, K., & Spiliotopoulos, D. (2024a). Artificial Intelligence Tools for the Agriculture Value Chain: Status and Prospects. *Electronics (Basel), 13*(22), 4362. DOI: 10.3390/electronics13224362

Assimakopoulos, F., Vassilakis, C., Margaris, D., Kotis, K., & Spiliotopoulos, D. (2024b). The Implementation of "Smart" Technologies in the Agricultural Sector: A Review. *Information (Basel), 15*, 466. DOI: 10.3390/info15080466

Assimakopoulos, F., Vassilakis, C., Margaris, D., Kotis, K., & Spiliotopoulos, D. (2025). AI and related technologies in the fields of smart agriculture: A review. *Information (Basel), 16*(2), 100. DOI: 10.3390/info16020100

Åstrand, B., & Baerveldt, A.-J. (2002). An agricultural mobile robot with vision-based perception for mechanical weed control. *Autonomous Robots, 13*(1), 21–35.

Ataei Kachouei, M., Kaushik, A., & Ali, Md. A. (2023). Internet of Things-Enabled Food and Plant Sensors to Empower Sustainability. *Advanced Intelligent Systems, 5*(12), 2300321. DOI: 10.1002/aisy.202300321

Ataei, P., Takhtravan, A., Gheibi, M., Chahkandi, B., Faramarz, M. G., Wacławek, S., Fathollahi-Fard, A. M., & Behzadian, K. (2024). An intelligent decision support system for groundwater supply management and electromechanical infrastructure controls. *Heliyon, 10*(3), e25036. DOI: 10.1016/J.HELIYON.2024.E25036 PMID: 38317976

Atapattu, A. J., Perera, L. K., Nuwarapaksha, T. D., Udumann, S. S., & Dissanayaka, N. S. (2024). Challenges in Achieving Artificial Intelligence in Agriculture. *Artificial Intelligence Techniques in Smart Agriculture*, 7–34. https://doi.org/DOI: 10.1007/978-981-97-5878-4_2

Atik, C. (2022). Towards comprehensive European agricultural data governance: Moving beyond the "data ownership" debate. *IIC-International Review of Intellectual Property and Competition Law*, *53*(5), 701–742.

Atik, C., & Martens, B. (2021). Competition problems and governance of non-personal agricultural machine data: Comparing voluntary initiatives in the US and EU. *J. Intell. Prop. Info. Tech. & Elec. Com. L.*, *12*, 370.

Atiku, S., Jeremiah, A., Semente, E., & Boateng, F. (2024). *Sustainable development in industry and society 5.0: Governance, management, and financial implications: Governance, management, and financial implications*. IGI Global.

Attaran, M., & Celik, B. G. (2023). Digital Twin: Benefits, use cases, challenges, and opportunities. *Decision Analytics Journal, 6*, 100165. 10.1016/j.dajour.2023.100165

Avila-Sanchez, H. (2025). Towards strengthening public policies for the digital transition in Mexican agriculture. In *Digital transition in agriculture and public policies* (p. 22640). Insights from Latin America.

Awais, M., Wang, X., Hussain, S., Aziz, F., & Mahmood, M. Q. (2025). Advancing Precision Agriculture Through Digital Twins and Smart Farming Technologies: A Review. *AgriEngineering, 7*(5), 137. https://www.mdpi.com/2624-7402/7/5/137

Awais, M., Wang, X., Hussain, S., Aziz, F., & Mahmood, M. Q. (2025). Advancing Precision Agriculture Through Digital Twins and Smart Farming Technologies: A Review. *AgriEngineering*, *7*(5), 137.

Awan, A. G., & Yaqoob, R. (2023). Economic value of introducing technology to improve productivity: An ARDL approach. *Innovation and Green Development, 2*(3), 100069. 10.1016/j.igd.2023.100069

Awan, S. H., Ahmed, S., Nawaz, A., Sulaiman, S., Zaman, K., Ali, M. Y., & Imran, S. (2020). BlockChain with IoT, an emergent routing scheme for smart agriculture. *International Journal of Advanced Computer Science and Applications*, *11*(4), 420–429. DOI: 10.14569/IJACSA.2020.0110457

Awate, P. L., & Nagne, A. D. (2025). Satellite imagery and GIS applications in precision agriculture. In *AI and data analytics in precision agriculture for sustainable development* (pp. 185–206). Springer Nature Switzerland.

Ayadi, R., Forouheshfar, Y., & Moghadas, O. (2025). Enhancing system resilience to climate change through artificial intelligence: A systematic literature review. *Frontiers in Climate*, *7*, 1585331. DOI: 10.3389/FCLIM.2025.1585331/XML

Ayorinde, A. (2025). Sustainable Farming Through AI-IoT Synergy: Mitigating Ecological Damage and Financial Resource Leakage. *International Journal of Research Publication and Reviews*, *6*, 7845–7862. DOI: 10.55248/gengpi.6.0625.2273

Azlan, Z. H. Z., Junaini, S. N., Khan, A. S., & Mustafa, W. A. (2025). Data-Driven Agriculture: Unveiling the Power of Internet of Things and Data Analytics For Transformative Farming Practices. *Journal of Sensors*, *2025*(1), 6205646. 10.1155/js/6205646

Azmi, H. N., Hajjaj, S. S. H., Gsangaya, K. R., Sultan, M. T. H., Mail, M. F., & Hua, L. S. (2021). Design and fabrication of an agricultural robot for crop seeding. *Materials Today: Proceedings*, *81*(2), 283–289. DOI: 10.1016/J.MATPR.2021.03.191

Babla, M., Katwal, U., Yong, M. T., Jahandari, S., Rahme, M., Chen, Z. H., & Tao, Z. (2022). Value-added products as soil conditioners for sustainable agriculture. *Resources, Conservation and Recycling*, *178*. Advance online publication. DOI: 10.1016/j.resconrec.2021.106079

Badidi, E. (2023). Edge AI for Early Detection of Chronic Diseases and the Spread of Infectious Diseases: Opportunities, Challenges, and Future Dircctions. *Future Internet*, *15*(11), 370. DOI: 10.3390/fi15110370

Baharin, A. T., Ishak, N. A., Redzuan, N. A. L., Yusof, S. M., Sahadun, N. A., Hati, D. M., & Jamaluddin, S. P. S. (2024). Exploring the level of digital literacy and the adoption of precision farming technologies among smallholder paddy farmers in Kedah.

Balkrishna, A., Pathak, R., Kumar, S., Arya, V., & Singh, S. K. (2023). A comprehensive analysis of the advances in Indian Digital Agricultural architecture. *Smart Agricultural Technology*, *5*, 100318.

Baluch, S. M., Wang, L., Faiz, M. A., Li, H., Chen, Y., Wang, L., & Li, M. (2025). Adaptation simulation and planning for crop yield under climate change: Integrating AquaCrop and DSSAT to project drought-induced yield risks in the Sanjiang Plain. *Agricultural Water Management, 319*, 109818. DOI: 10.1016/J.AGWAT.2025.109818

Banasik, A., Kanellopoulos, A., Claassen, G., Bloemhof-Ruwaard, J. M., & Van der Vorst, J. G. (2017). Closing loops in agricultural supply chains using multi-objective optimization: A case study of an industrial mushroom supply chain. *International Journal of Production Economics, 183*, 409–420.

Banerjee, P., & Bhat, A. (2025). Cultivating justice: The economic-legal-technological nexus in advancing sustainable agriculture in the 21st century. *Journal of Humanities. Arts and Social Science, 9*(12), 2222–2235.

Barbone, A., Bicocchi, N., Martinelli, M., Morandi, R., & Picone, M. (2026). On-device AI and digital twins: A synergistic approach to intelligent cyber-physical systems. *Future Generation Computer Systems, 175*, 108068. 10.1016/j.future.2025.108068

Barbosa Júnior, M. R., Santos, R. G. D., Sales, L. A., & Oliveira, L. P. (2024). Advancements in Agricultural Ground Robots for Specialty Crops: An Overview of Innovations, Challenges, and Prospects. *Plants, 13*(23). Advance online publication. DOI: 10.3390/plants13233372 PMID: 39683165

Barrett, C. B. (2022). Overcoming global food security challenges through data-driven agriculture. *Nature Food, 3*(6), 409–414. DOI: 10.1038/s43016-022-00515-0 PMID: 37118052

Batool, A., Zowghi, D., & Bano, M. (2025). AI governance: A systematic literature review. *AI and Ethics, 5*(3), 3265–3279. DOI: 10.1007/s43681-024-00653-w

Bauer, J., & Aschenbruck, N. (2018). *Design and implementation of an agricultural monitoring system for smart farming. 2018 IoT Vertical and Topical Summit on Agriculture - Tuscany*. IOT Tuscany.

Bayar, J., Ali, N., Cao, Z., Ren, Y., & Dong, Y. (2025). Artificial Intelligence of Things (AIoT) for Precision Agriculture: Applications in Smart Irrigation, Nutrient and Pest Management. *Smart Agricultural Technology, 101629*. Advance online publication. DOI: 10.1016/J.ATECH.2025.101629

Bazargani, K., & Deemyad, T. (2024). Automation's Impact on Agriculture: Opportunities, Challenges, and Economic Effects. *Robotics (Basel, Switzerland), 13*(2). Advance online publication. DOI: 10.3390/ROBOTICS13020033

Becerra-Encinales, J. F., Bernal-Hernandez, P., Beltrán-Giraldo, J. A., Cooman, A. P., Reyes, L. H., & Cruz, J. C. (2024). Agricultural extension for adopting technological practices in developing countries: A scoping review of barriers and dimensions. *Sustainability, 16*(9), 3555.

Bechar, A., & Vigneault, C. (2017). Agricultural robots for field operations. Part 2: Operations and systems. *Biosystems Engineering, 153*, 110–128. 10.1016/j.biosystemseng.2016.11.004

Beerbaum, D., & Otto, B. (2023). Data governance in digital ecosystems: Concepts, challenges, and research directions. *Electronic Markets, 33*(1), 1–15. DOI: 10.1007/s12525-022-00563-9 PMID: 37035262

Behera, S. (2025). *Drones and Aerial Imaging for Crop Health and Soil Analysis.* In (pp. 64–88).

Beisekenov, N., Banakinaou, W., Ajayi, A. D., Hasegawa, H., & Tadao, A. (2025). Remote sensing-based soil organic carbon monitoring using advanced machine learning techniques under conservation agriculture systems. *Smart Agricultural Technology, 11*, 101036.

Bekmurzaeva, R., Kalimullin, R., & Aguzarova, F. (2024). Application of drones and artificial intelligence to monitor and protect natural ecosystems. *BIO Web of Conferences, 140.* https://doi.org/DOI: 10.1051/bioconf/202414001008

Benchis, M. P., Shahzad, K., & Dan, S. (2025). Comparative analysis of blockchain adoption in the public and private sectors. A technology-organization-environment (TOE) framework approach. *Journal of Innovation & Knowledge, 10*(4), 100746. DOI: 10.1016/J.JIK.2025.100746

Beniwal, G., & Singhrova, A. (2022). A systematic literature review on IoT gateways. *Journal of King Saud University. Computer and Information Sciences, 34*(10, Part B), 9541–9563. 10.1016/j.jksuci.2021.11.007

Berger, K., Machwitz, M., Kycko, M., Kefauver, S. C., Van Wittenberghe, S., Gerhards, M., Verrelst, J., Atzberger, C., van der Tol, C., Damm, A., Rascher, U., Herrmann, I., Paz, V. S., Fahrner, S., Pieruschka, R., Prikaziuk, E., Buchaillot, M. L., Halabuk, A., Celesti, M., & Schlerf, M. (2022). Multi-sensor spectral synergies for crop stress detection and monitoring in the optical domain: A review. *Remote Sensing of Environment, 280*, 113198. DOI: 10.1016/J.RSE.2022.113198

Bessaad, N., Atsyo, S., Hanjie, D., Walkine, I., Dhillon, R., & He, L. (2026). Autonomous mowing in agriculture: Current status, needs, and future opportunities. *Smart Agricultural Technology, 13*, 101796. DOI: 10.1016/J.ATECH.2026.101796

Bettucci, F., Sozzi, M., Benetti, M., & Sartori, L. (2024). A data-driven approach to agricultural machinery working states analysis during ploughing operations. *Smart Agricultural Technology*, *8*, 100511. DOI: 10.1016/J.ATECH.2024.100511

Bhalerao, S., Okoń, S., Rai, M., Jadhav, P., & Prokisch, J. (2025). Recent Advances in CRISPR/Cas9 Technology for Engineering Disease Resistance in Plants. *Acta Agrobotanica*, *78*, 1–31. DOI: 10.5586/aa/201389

Bhambri, P., & Khang, A. (2025). Edge computing for enhancing efficiency and sustainability in green transportation systems. In *Driving green transportation system through artificial intelligence and automation: Approaches, technologies and applications* (pp. 43–65). Springer Nature Switzerland.

Bhamini, K., Kumar, A., Magrey, A. H., & Kulsoom, S. (2025). The role of GIS and GPS in precision fertilization: Optimizing nutrient management for sustainable cropping. *International Journal of Research in Agronomy*, *8*, 260–267.

Bhangar, N., & Kashem, A. (2023). IoT and AI for Next-Generation Farming: Opportunities, Challenges, and Outlook. 8.

Bharali, D., Bhuyan, S., Rout, S., Behera, H. S., & Gupta, S. (2024). Integrating policy frameworks with soil and plant conservation economics: Strategies for sustainable agricultural practices. Agriculture Archives: an International Journal.

Bharathi, P., Karthickraja, A., & Saravanane, P. (2024). Advanced precision agriculture techniques for effective weed management cotton. In Arunkumar, R., Vignesh, K., Dhivya, C., & Bala Murugan, S. (Eds.), *Recent Trends in Agriculture and Allied Sciences* (pp. 162–171). JPS Scientific Publications.

Bhargava, A., Sachdeva, A., Sharma, K., Alsharif, M. H., Uthansakul, P., & Uthansakul, M. (2024). Hyperspectral imaging and its applications: A review. *Heliyon*, *10*(12), e33208. 10.1016/j.heliyon.2024.e33208 PMID: 39021975

Bharti, A., Jain, U., & Chauhan, N. (2024). From lab to field: Nano-biosensors for real-time plant nutrient tracking. *Plant Nano Biology*, *9*, 100079. DOI: 10.1016/J.PLANA.2024.100079

Bhattacharyya, P., Pathak, H., & Pal, S. (2020). Barriers to adaptation of climate-smart agriculture. In *Climate smart agriculture: Concepts, challenges, and opportunities* (pp. 155–167). Springer Singapore.

Bhooshan, N., Raman, M. S., Gupta, S., Suyal, G., Singh, A., & Sharma, A. (2024). Revolutionizing agriculture: Role of agricultural mechanization and global trends in farming technology. *Current Science*, *126*(10), 1209–1216. DOI: 10.18520/CS/V126/I10/1209-1216

Bhujel, A., Basak, J., Khan, F., & Kim, H. (2020). Sensor Systems for Greenhouse Microclimate Monitoring and Control: A Review. *Journal of Biosystems Engineering, 45*. Advance online publication. DOI: 10.1007/s42853-020-00075-6

Bilal, M., Shah, A. A., Abbas, S., & Khan, M. A. (2025). High-Performance Deep Learning for Instant Pest and Disease Detection in Precision Agriculture. *Food Science & Nutrition, 13*(9), e70963. DOI: 10.1002/fsn3.70963 PMID: 40964151

Bimonte, S., Bellocchi, G., Pinet, F., Chalhoub, G., Sakr, M., & Skrzypczyński, P. (2025). Data engineering for sustainable agriculture: Developments, challenges, and case studies of a novel IoRT architecture. *Journal of Big Data, 12*(1), 195. DOI: 10.1186/s40537-025-01253-z

Binu Mol, T. V. (2024). A Study on Raspberry PI, Arduino and IoT with Its Application in E-Controlling UPS. *Journal of Electrical Systems, 20*(2), 2785–2791. DOI: 10.52783/JES.6266

Birhane, A. (2021). Algorithmic injustice: A relational ethics approach. *Patterns (New York, N.Y.), 2*(2), 100205. DOI: 10.1016/j.patter.2021.100205 PMID: 33659914

Birhane, A., Prabhu, V., & Kahembwe, E. (2022). The values encoded in machine learning research. *Proceedings of the ACM Conference on Fairness, Accountability, and Transparency*, 173–184. https://doi.org/DOI: 10.1145/3531146.3533154

Bissadu, K. D., Sonko, S., & Hossain, G. (2025). Society 5.0 enabled agriculture: Drivers, enabling technologies, architectures, opportunities, and challenges. *Information Processing in Agriculture, 12*(1), 112–124. DOI: 10.1016/J.INPA.2024.04.003

Bist, D. R., Chapagaee, P., Kunwar, A., Pant, B. D., Khatri, L., & Mandal, A. (2025). Nanotechnology in Agriculture: A Review of Innovations in Crop Protection and Food Security. *Advances in Agriculture, 2025*(1), 8892001. DOI: 10.1155/aia/8892001

Bleicher, F., Biermann, D., Drossel, W. G., Moehring, H. C., & Altintas, Y. (2023). Sensor and actuator integrated tooling systems. *CIRP Annals, 72*(2), 673–696. DOI: 10.1016/j.cirp.2023.05.009

Bogucki, A. (2025). Applied trustworthy data and artificial intelligence governance in precision agriculture. In *Engineering and value change* (pp. 101–124). Springer Nature Switzerland.

Botero-Valencia, J., García-Pineda, V., Valencia-Arias, A., Valencia, J., Reyes-Vera, E., Mejia-Herrera, M., & Hernández-García, R. (2025). Machine Learning in Sustainable Agriculture: Systematic Review and Research Perspectives. *Agriculture, 15*(4), 377. https://www.mdpi.com/2077-0472/15/4/377

Boyes, H., & Watson, T. (2022). Digital twins: An analysis framework and open issues. *Computers in Industry, 143*, 103763. DOI: 10.1016/j.compind.2022.103763

Brewster, C., Kalatzis, N., Nouwt, B., Kruiger, H., & Verhoosel, J. (2024). Data sharing in agricultural supply chains: Using semantics to enable sustainable food systems. *Semantic Web, 15*(4), 1207–1237.

Burns, T., Cosgrove, J., & Doyle, F. (2019). A Review of Interoperability Standards for Industry 4.0. *Procedia Manufacturing, 38*, 646–653. 10.1016/j.promfg.2020.01.083

Bwambale, E., Abagale, F. K., & Anornu, G. K. (2022). Smart irrigation monitoring and control strategies for improving water use efficiency in precision agriculture: A review. *Agricultural Water Management, 260*, 107324. DOI: 10.1016/j.agwat.2021.107324

Bwambale, E., Abagale, F. K., & Anornu, G. K. (2023). Data-Driven Modelling of Soil Moisture Dynamics for Smart Irrigation Scheduling. *Smart Agricultural Technology, 5*, 100251. DOI: 10.1016/j.atech.2023.100251

Byamukama, W., Ssemakula, E., Ssekandi, J., Shaman, A., Businge, P., Kalibwani, R., Atukunda, C., & Turyasingura, B. (2025). A Review of ICT Integration in Agricultural Extension Services. A Global Perspective. *East African Journal of Agriculture and Biotechnology, 8*(1), 292–312. DOI: 10.37284/EAJAB.8.1.849

Cabrera, V. E., Bewley, J., Breunig, M., Breunig, T., Cooley, W., De Vries, A., Fourdraine, R., Giordano, J. O., Gong, Y., Greenfield, R., Hu, H., Lenkaitis, A., Niu, M., Noronha, E. A. F., & Sullivan, M. (2025). Data Integration and Analytics in the Dairy Industry: Challenges and Pathways Forward. *Animals 2025, Vol. 15, 15*(3). https://doi.org/DOI: 10.3390/ANI15030329

Camaréna, S. (2020). Artificial intelligence in the design of the transitions to sustainable food systems. *Journal of Cleaner Production, 271*, 122574.

Cañadas-Aránega, F., Moreno, J. C., Blanco-Claraco, J. L., Giménez, A., Rodríguez, F., & Sánchez-Hermosilla, J. (2024). Autonomous collaborative mobile robot for greenhouses: Design, development, and validation tests. *Smart Agricultural Technology, 9*, 100606. 10.1016/j.atech.2024.100606

Cao, A., Guo, L., & Li, H. (2025). Understanding farmer cooperatives' intention to adopt digital technology: Mediating effect of perceived ease of use and moderating effects of internet usage and training. *International Journal of Agricultural Sustainability, 23*(1). Advance online publication. DOI: 10.1080/14735903.2025.2464523

Capurro, M. C., Ham, J. M., Kluitenberg, G. J., Comas, L., & Andales, A. A. (2024). A novel sap flow system to measure maize transpiration using a heat pulse method. *Agricultural Water Management*, *301*, 108963. DOI: 10.1016/J.AGWAT.2024.108963

Carbone, G., Gurtatta, A. S., & Malyshev, D. (2025). Development and implementation of a hybrid visual prediction algorithm for robotic smart tomato harvesting. *Engineering Applications of Artificial Intelligence*, *161*, 112261. DOI: 10.1016/J.ENGAPPAI.2025.112261

Carbonell, I. M. (2016). The ethics of big data in agriculture. *Internet Policy Review*, *5*(1), 1–15. DOI: 10.14763/2016.1.405

Cäsar, M., Pawelke, T., Steffan, J., & Terhorst, G. (2022). A survey on Bluetooth Low Energy security and privacy. *Computer Networks*, *205*, 108712. DOI: 10.1016/j.comnet.2021.108712

Cate, M. (2022). Precision Agriculture and AI-Driven Decision Support Systems Improving Yield Forecasting and Resource Efficiency.

Cesco, S., Sambo, P., Borin, M., Basso, B., Orzes, G., & Mazzetto, F. (2023). Smart agriculture and digital twins: Applications and challenges in a vision of sustainability. *European Journal of Agronomy*, *146*, 126809. DOI: 10.1016/j.eja.2023.126809

Chad, F. (2025). *Challenges of AI Adoption in Smallholder Farming: Infrastructure*. Cost, and Digital Literacy.

Chai, J., Zeng, H., Li, A., & Ngai, E. W. T. (2021). Deep learning in computer vision: A critical review of emerging techniques and application scenarios. *Machine Learning with Applications*, *6*, 100134. 10.1016/j.mlwa.2021.100134

Chakrabarty, S., Deb, C. K., Marwaha, S., Haque, M. A., Kamil, D., Bheemanahalli, R., & Shashank, P. R. (2026). Application of artificial intelligence in insect pest identification - A review. *Artificial Intelligence in Agriculture, 16*(1), 44-61. 10.1016/j.aiia.2025.06.005

Chakraborty, S., & Newton, A. C. (2011). Climate change, plant diseases and food security: An overview. *Plant Pathology*, *60*(1), 2–14. DOI: 10.1111/j.1365-3059.2010.02411.x

Chandel, N. S., Rajwade, Y. A., Dubey, K., Chandel, A. K., Subeesh, A., & Tiwari, M. K. (2022). Water Stress Identification of Winter Wheat Crop with State-of-the-Art AI Techniques and High-Resolution Thermal-RGB Imagery. *Plants*, *11*(23), 3344. DOI: 10.3390/plants11233344

Chandhini, K. (2016). A literature study on agricultural production system using IoT as inclusive technology. *International Journal of Innovative Technology and Research, 4*, 2727–2731.

Chang, X., He, H., Cheng, L., Yang, X., Li, S., Yu, M., Zhang, J., & Li, J. (2024). Combined Application of Chemical and Organic Fertilizers: Effects on Yield and Soil Nutrients in Spring Wheat under Drip Irrigation. *Agronomy 2024, Vol. 14, 14*(4). DOI: 10.3390/AGRONOMY14040655

Chaoub, A., Lall, B., Bhatia, V., Kliks, A., Mendes, L., Rabie, K., & Zorzi, M. (2021). 6G for Bridging the Digital Divide: Wireless Connectivity to Remote Areas. *IEEE Wireless Communications*. Advance online publication. DOI: 10.1109/MWC.001.2100137

Charalampopoulos, I., Droulia, F., Charalampopoulos, I., & Droulia, F. (2024). A Pathway towards Climate Services for the Agricultural Sector. *Climate 2024, Vol. 12, 12*(2). DOI: 10.3390/CLI12020018

Charles-Mary, C., & Agbo, A. (2025). AN IOT-BASED ENVIRONMENTAL MONITORING SYSTEM USING TEMPERATURE, HUMIDITY, AIR QUALITY, LIGHT INTENSITY, AND RAIN DETECTION SENSORS WITH ESP 32 MICROCONTROLLER. *International Journal of Engineering*.

Chatrabhuj, M. K., Mishra, U., & Rathnayake, U. (2025). Application of artificial intelligence in agri-tech, environmental and biodiversity conservation. *Array, 26*, 100412. 10.1016/j.array.2025.100412

Chattopadhyay, D., & Balachandra, Y., P, A., Pagire, K. S., Sahu, T. K., Panotra, N., Manisankar, G., Charitha, K., & Singh, B. veer. (2024). Precision Agriculture Technologies for Early Detection of Crop Pests and Diseases. *Uttar Pradesh Journal of Zoology, 45*(20), 328–342. DOI: 10.56557/UPJOZ/2024/V45I204588

Chaudhari, S. K., Patra, A., Dey, P., Bal, S. K., Gorantiwar, S., & Parsad, R. (2022). Sensor based Monitoring for Improving Agricultural Productivity and Sustainability-A Review. *Journal of the Indian Society of Soil Science, 70*(2), 121–141. DOI: 10.5958/0974-0228.2022.00013.5

Chaudhry, Z., & Rashid, H. (2011). ISOLATION AND CHARACTERIZATION OF RALSTONIA SOLANACEARUM FROM INFECTED TOMATO PLANTS OF SOAN SKESAR VALLEY OF PUNJAB. *Pakistan Journal of Botany, 43*(6), 2979–2985.

Chaurasia, K., Yasmin, G., & Singh, D. (2026). Overview of geospatial technology and machine learning in agriculture. In *Agricultural Insights from Space* (pp. 1–27). Academic Press.

Chavanne, X., & Frangi, J.-P. (2024). A Sensor to Monitor Soil Moisture, Salinity, and Temperature Profiles for Wireless Networks. *Journal of Sensor and Actuator Networks, 13*(3), 32. https://www.mdpi.com/2224-2708/13/3/32

Chelliah, P. R., Rahmani, A. M., Colby, R., Nagasubramanian, G., & Ranganath, S. (Eds.). (2024). *Model optimization methods for efficient and edge AI: Federated learning architectures, frameworks and applications*. John Wiley & Sons.

Cheng, H.-C., Shu, M.-H., & Huang, J.-C. (2024). Economic strategies for efficient use of natural resources: The impact of carbon taxation and fiscal policy. *Resources Policy*, *92*, 104927. 10.1016/j.resourpol.2024.104927

Chen, H. Y., Sharma, K., Sharma, C., & Sharma, S. (2023). Integrating explainable artificial intelligence and blockchain to smart agriculture: Research prospects for decision making and improved security. *Smart Agricultural Technology*, *6*, 100350. DOI: 10.1016/j.atech.2023.100350

Chen, J., Paw, U. K. T., North, M., & Franklin, J. F. (2024). The contributions of microclimatic information in advancing ecosystem science. *Agricultural and Forest Meteorology*, *355*, 110105. DOI: 10.1016/J.AGRFORMET.2024.110105

Chen, Q., Vaudour, E., Richer-de-Forges, A. C., & Arrouays, D. (2025). Spectral indices in remote sensing of soil: Definition, popularity, and issues. A critical overview. *Remote Sensing of Environment*, *329*, 114918. DOI: 10.1016/J.RSE.2025.114918

Chen, X. (2025). The role of modern agricultural technologies in improving agricultural productivity and land use efficiency. *Frontiers in Plant Science*, *16*, 1675657. DOI: 10.3389/FPLS.2025.1675657/XML PMID: 41036393

Chen, X., Zhang, H., & Wong, C. U. I. (2025). Dynamic Monitoring and Precision Fertilization Decision System for Agricultural Soil Nutrients Using UAV Remote Sensing and GIS. *Agriculture*, *15*(15), 1627. https://www.mdpi.com/2077-0472/15/15/1627

Chen, Z., Yin, J., Farhan, S. M., Liu, L., Zhang, D., Zhou, M., & Cheng, J. (2026). A comprehensive review of obstacle avoidance for autonomous agricultural machinery in multi-operational environment. *Artificial Intelligence in Agriculture*, *16*(1), 139–163. DOI: 10.1016/J.AIIA.2025.10.001

Cheong, J. Q., Muttaqin, E. I., Wiranatakusuma, D. B., & Jusoh, M. H. (2025). Technology adoption and food security in ASEAN among women. In *International Conference on Mathematical and Statistical Physics, Computational Science, Education, and Communication (ICMSCE 2025)* (Vol. 13941, pp. 334-343). SPIE.

Chinonyerem, C. A., Ajibade, S. B., Rebecca, O. T., & Ahaiwe, E. C. (2025). The economic implications of precision agriculture adoption on farm profitability and resource use efficiency. *International Journal of Innovation Research and Advanced Studies.*

Chirwa, M., & Boikanyo, D. (2022). The role of effective communication in successful strategy implementation. *Acta Commercii, 22*. Advance online publication. DOI: 10.4102/ac.v22i1.1020

Choudhary, A. (2024). Internet of Things: a comprehensive overview, architectures, applications, simulation tools, challenges and future directions. *Discover Internet of Things 2024 4:1, 4*(1), 31-. DOI: 10.1007/S43926-024-00084-3

Choudhary, A. (2024). Internet of Things: A comprehensive overview, architectures, applications, simulation tools, challenges and future directions. *Discover Internet of Things*, *4*(1), 31. DOI: 10.1007/s43926-024-00084-3

Choudhary, V., Guha, P., Pau, G., & Mishra, S. (2025). An overview of smart agriculture using internet of things (IoT) and web services. *Environmental and Sustainability Indicators*, *26*, 100607. DOI: 10.1016/J.INDIC.2025.100607

Choudhary, V., Guha, P., Pau, G., & Mishra, S. (2025). An Overview of Smart Agriculture Using Internet of Things (IoT) and Web Services. *Environmental and Sustainability Indicators*, *26*, 100607. DOI: 10.1016/j.indic.2025.100607

Choudhury, S., & Banerjee, B. (2024). Deep Learning: A Catalyst for Sustainable Agriculture Transformation. In M. S. Raval, S. Chaudhary, J. Adinarayana, & W. Guo (Eds.), *Harnessing Data Science for Sustainable Agriculture and Natural Resource Management* (pp. 173-199). Springer Nature Singapore. https://doi.org/ DOI: 10.1007/978-981-97-7762-4_8

Chouhan, S. S., Patel, R. K., Singh, U. P., & Tejani, G. G. (2025). Integrating drone in Agriculture: Addressing technology, challenges, solutions, and applications to drive economic growth. *Remote Sensing Applications: Society and Environment*, *38*, 101576. 10.1016/j.rsase.2025.101576

Chowdhury, K., Rahman, M. A., Mallik, S. K., Chowdhury, L. N., & Nova, N. A. (2025). Bridging the Global Digital Divide in Agriculture: The Role of AI in Equitable Technology Access. *International Journal of Sustainable Development and Planning*, *20*(6), 2469–2481. DOI: 10.18280/IJSDP.200616

Chowdhury, S., Sen, S., & Janardhanan, S. (2022). *Comparative Analysis and Calibration of Low Cost Resistive and Capacitive Soil Moisture Sensor*. https://arxiv.org/abs/2210.03019

Cogato, A., Brščić, M., Guo, H., Marinello, F., & Pezzuolo, A. (2021). Challenges and Tendencies of Automatic Milking Systems (AMS): A 20-Years Systematic Review of Literature and Patents. *Animals (Basel)*, *11*(2). Advance online publication. DOI: 10.3390/ani11020356 PMID: 33572673

Cogato, A., & Schillaci, C. (2025). "From data to fork" data-driven science inspired by policy initiatives for the promotion of smart agricultural technologies. *Smart Agricultural Technology*, •••, 100928.

Čolaković, A., Haskovic Dzubur, A., & Karahodža, B. (2021). Wireless communication technologies for the Internet of Things. *Science Engineering and Technology*, *1*, 1–14. DOI: 10.54327/set2021/v1.i1.3

Coleman, S., Kerr, D., & Zhang, Y. (2022). Image Sensing and Processing with Convolutional Neural Networks. *Sensors (Basel)*, *22*(10). Advance online publication. DOI: 10.3390/s22103612 PMID: 35632021

Colizzi, L., Dimauro, G., Guerriero, E., & Lomonte, N. (2025). Artificial intelligence and IoT for water saving in agriculture: A systematic review. *Smart Agricultural Technology*, *11*, 101008. DOI: 10.1016/J.ATECH.2025.101008

Cordeiro, M., & Ferreira, J. C. (2025). Beyond traceability: Decentralised identity and digital twins for verifiable product identity in agri-food supply chains. *Applied Sciences (Basel, Switzerland)*, *15*(11), 6062.

Couldry, N., & Mejias, U. A. (2019). Data colonialism: Rethinking big data's relation to the contemporary subject. *Television & New Media*, *20*(4), 336–349. DOI: 10.1177/1527476418796632

Cowan, N., Ferrier, L., Spears, B., Drewer, J., Reay, D., & Skiba, U. (2022). CEA Systems: the Means to Achieve Future Food Security and Environmental Sustainability? [Review]. *Frontiers in Sustainable Food Systems, Volume 6 - 2022*. https://doi.org/DOI: 10.3389/fsufs.2022.891256

Csikós, N., & Tóth, G. (2023). Concepts of agricultural marginal lands and their utilisation: A review. *Agricultural Systems*, *204*, 103560. 10.1016/j.agsy.2022.103560

Curry, G. N., Nake, S., Koczberski, G., Oswald, M., Rafflegeau, S., Lummani, J., Peter, E., & Nailina, R. (2021). Disruptive innovation in agriculture: Socio-cultural factors in technology adoption in the developing world. *Journal of Rural Studies*, *88*, 422–431. 10.1016/j.jrurstud.2021.07.022

D, D. C., AC, P. chandra, M, R., TS, B., MR, H., & B, S. (2026). *Artificial Intelligence enabled robotics and automation in modern agriculture*. /DOI: 10.21203/RS.3.RS-8601053/V1

da Silveira, F., Barbedo, J. G. A., Baysoy, M. K., & Caba, Z. T. (2025). Digital Technologies Transforming the Agri-Food Value Chain in the Era of Agriculture 4.0. In Hassoun, A. (Ed.), *Food Industry 4.0 and Food Sustainability* (pp. 125–145). Springer Nature Switzerland.

Dara, R., Fard, S. M. H., & Kaur, J. (2022). Recommendations for ethical and responsible use of artificial intelligence in digital agriculture. *Frontiers in Artificial Intelligence*, *5*, 884192.

Daraz, U., Bojnec, Š., & Khan, Y. (2025). Energy-Efficient Smart Irrigation Technologies: A Pathway to Water and Energy Sustainability in Agriculture. *Agriculture*, *15*, 554. DOI: 10.3390/agriculture15050554

Dargan, S., Kumar, M., Ayyagari, M. R., & Kumar, G. (2020). A Survey of Deep Learning and Its Applications: A New Paradigm to Machine Learning. *Archives of Computational Methods in Engineering*, *27*(4), 1071–1092. DOI: 10.1007/s11831-019-09344-w

Dasarath, B. R. (2016). Microsoft, ICRISAT develop new sowing app for farmers using AI and Azure cloud. https://www.business-standard.com/article/companies/microsoft-icrisat-develop-new-sowing-app-for-farmers-using-al-and-azure-cloud-116060900752_1.html

Das, S., Dutta, S., Roy Choudhury, M., Garai, S., Mukherjee, S., Sengupta, S., Jana, S., Dey, S., Dhar, A., Dutta, S., & Awasthi, A. (2025). Regenerating rural soil and ecosystems: A 15-year systematic review of emerging methods and technologies. *The Science of the Total Environment*, *990*, 179926. DOI: 10.1016/J.SCITOTENV.2025.179926

Daszkiewicz, T. (2022). Food production in the context of global developmental challenges. *Agriculture*, *12*(6), 832.

David, I., Shao, G., Gomes, C., Tilbury, D., & Zarkout, B. (2025, 2025//). Interoperability of Digital Twins: Challenges, Success Factors, and Future Research Directions. Leveraging Applications of Formal Methods, Verification and Validation. Application Areas, Cham.

David, A., Rao, P. S., Srivastava, D., Barthwal, A., Thomas, A., Thomas, T., & Thomas, C. A. (2024). Big data, AI, and geoinformatics for sustainable agriculture: An introduction. In *Geoinformatics* (pp. 131–161). Apple Academic Press.

Dayioğlu, M. A., & Türker, U. (2021). Digital transformation for sustainable future-agriculture 4.0: A review. *Tarim Bilimleri Dergisi, 27*(4), 373–399. DOI: 10.15832/ANKUTBD.986431

de Carvalho Silva, J., Rodrigues, J., Alberti, A., Šolić, P., & Aquino, A. (2017). *LoRaWAN - A Low Power WAN Protocol for Internet of Things: a Review and Opportunities.*

Debroy, P., & Seban, L. (2022). A tomato fruit biomass prediction model for aquaponics system using machine learning algorithms. *IFAC-PapersOnLine.* Advance online publication. DOI: 10.1016/j.ifacol.2022.04.116

Degila, J., Sodedji, F. A. K., Avakoudjo, H. G. G., Tahi, S. P. G., Houetohossou, S. C. A., Honfoga, A. C., & Assogbadjo, A. E. (2023). Digital agriculture policies and strategies for innovations in the agri-food systems—Cases of five West African countries. *Sustainability, 15*(12), 9192.

Delfani, P., Thuraga, V., Banerjee, B., & Chawade, A. (2024). Integrative approaches in modern agriculture: IoT, ML and AI for disease forecasting amidst climate change. *Precision Agriculture, 25*(5), 2589–2613.

Demissie, W. A., Sebastiani, L., & Rossetto, R. (2026). Integration of artificial intelligence and remote sensing for crop yield prediction and crop growth parameter estimation in Mediterranean agroecosystems: Methodologies, emerging technologies, research gaps, and future directions. *European Journal of Agronomy, 173*, 127894. DOI: 10.1016/J.EJA.2025.127894

Deng, R., Tao, M., Xing, H., Yang, X., Liu, C., Liao, K., & Qi, L. (2021). Automatic diagnosis of rice diseases using deep learning. *Frontiers in Plant Science, 12.* Advance online publication. DOI: 10.3389/fpls.2021.701038

De, S., & Moessner, K. (2023). Sensor Networks: Physical and Social Sensing in the IoT. *Sensors (Basel), 23*(3). Advance online publication. DOI: 10.3390/s23031451 PMID: 36772505

Desikan, J., Singh, S. K., Jayanthiladevi, A., Bhushan, S., Rishiwal, V., & Kumar, M. (2025). Hybrid Machine Learning-Based Fault-Tolerant Sensor Data Fusion and Anomaly Detection for Fire Risk Mitigation in IIoT Environment. *Sensors 2025, Vol. 25, 25*(7). DOI: 10.3390/S25072146

Devasia, J. V., Deepanramkumar, P., Helensharmila, A., & Yenduri, G. (2024). Ethical considerations and social implications. In *Sustainable Farming through Machine Learning* (pp. 267–279). CRC Press.

Dey, S., Widanagamage, N., Achar, S., Debangshi, U., Palla, S., Kim, J., & Jha, G. (2025). *Precision Agriculture Tools, Techniques, and Future Directions for Climate Resilience*. 89–115. DOI: 10.1007/978-981-96-7699-6_5

Dhal, S. B., & Kar, D. (2024). Transforming agricultural productivity with AI-driven forecasting: Innovations in food security and supply chain optimization. *Forecasting, 6*(4), 925–951. DOI: 10.3390/forecast6040046

Dhanaraju, M., Chenniappan, P., Ramalingam, K., Pazhanivelan, S., Kaliaperumal, R., Dhanaraju, M., Chenniappan, P., Ramalingam, K., Pazhanivelan, S., & Kaliaperumal, R. (2022). Smart Farming: Internet of Things (IoT)-Based Sustainable Agriculture. *Agriculture 2022, Vol. 12, 12*(10). DOI: 10.3390/AGRICULTURE12101745

Dhanaraju, M., Chenniappan, P., Ramalingam, K., Pazhanivelan, S., & Kaliaperumal, R. (2022). Smart Farming: Internet of Things (IoT)-Based Sustainable Agriculture. *Agriculture, 12*(10), 1745.

Dhir, S. B., & Mishra, G. (2025). Digital agriculture mission: Unlocking opportunities and overcoming challenges for farmers using smartphones. *Journal of Information, Communication and Ethics in Society*.

Dhirani, L. L., Mukhtiar, N., Chowdhry, B. S., & Newe, T. (2023). Ethical Dilemmas and Privacy Issues in Emerging Technologies: A Review. *Sensors (Basel), 23*(3). Advance online publication. DOI: 10.3390/s23031151 PMID: 36772190

Dian, R., Li, S., Sun, B., & Guo, A. (2021). Recent advances and new guidelines on hyperspectral and multispectral image fusion. *Information Fusion, 69*, 40–51. 10.1016/j.inffus.2020.11.001

Diao, Z., Chen, L., Yang, Y., Liu, Y., Yan, J., He, S., & Zhang, B. (2025). Localization technologies for smart agriculture and precision farming: A review. *Computers and Electronics in Agriculture, 236*, 110464. 10.1016/j.compag.2025.110464

Díaz, J., Quiñonez, Y., De-la-Hoz-Franco, E., Butt-Aziz, S., Mercado, T., Salcedo, D., Díaz, J., Quiñonez, Y., De-la-Hoz-Franco, E., Butt-Aziz, S., Mercado, T., & Salcedo, D. (2025). Information and Communication Technologies Used in Precision Agriculture: A Systematic Review. *AgriEngineering 2025, Vol. 7, 7*(6). DOI: 10.3390/AGRIENGINEERING7060167

Dibbern, T., Romani, L., Massruhá, S., Dibbern, T., Romani, L., & Massruhá, S. (2025). Drivers and Barriers to Digital Agriculture Adoption: A Mixed-Methods Analysis of Challenges and Opportunities in Latin American. *Sustainability 2025, Vol. 17, 17*(8). /DOI: 10.3390/SU17083676

Dibbern, T., Romani, L. A. S., & Massruhá, S. M. F. S. (2024). Main drivers and barriers to the adoption of Digital Agriculture technologies. *Smart Agricultural Technology*, *8*, 100459.

Dibbern, T., Romani, L., & Massruhá, S. (2025). Drivers and barriers to digital agriculture adoption: A mixed-methods analysis of challenges and opportunities in Latin America. *Sustainability*, *17*(8), 3676.

Ding, S., Liu, J., & Yue, M. (2021). The Use of ZigBee Wireless Communication Technology in Industrial Automation Control. *Wireless Communications and Mobile Computing*, *2021*(1), 8317862. 10.1155/2021/8317862

Diro, A., Kaisar, S., Saini, A., Fatima, S., Hiep, P. C., & Erba, F. (2025). Workplace security and privacy implications in the GenAI age: A survey. *Journal of Information Security and Applications*, *89*, 103960. DOI: 10.1016/J.JISA.2024.103960

Djidrov, M. Z. (2024). Challenges in Damage Detection for Real-time Monitoring and Predictive Maintenance Methodology. *SAR Journal - Science and Research*, 310–316. DOI: 10.18421/SAR74-04

Dolatabadian, A., Neik, T. X., Danilevicz, M. F., Upadhyaya, S. R., Batley, J., & Edwards, D. (2025). Image-based crop disease detection using machine learning. *Plant Pathology*, *74*(1), 18–38. DOI: 10.1111/ppa.14006

Domínguez-Bolaño, T., Campos, O., Barral, V., Escudero, C. J., & García-Naya, J. A. (2022). An overview of IoT architectures, technologies, and existing open-source projects. *Internet of Things : Engineering Cyber Physical Human Systems*, *20*, 100626. 10.1016/j.iot.2022.100626

Dong, J., & Xu, J. (2025). Driving agricultural strength through digital transformation. *Scientific Reports*. Advance online publication. DOI: 10.1038/s41598-025-29878-3

Dong, Y., Sloan, G., & Chappuies, J. (2024). Open-source time-lapse thermal imaging camera for canopy temperature monitoring. *Smart Agricultural Technology, 7*, 100430. DOI: 10.1016/J.ATECH.2024.100430

Douzal, C., Mosnier, A., Chermain, C., Lynch, J., Harrison, P., Jones, S., ... & Smith, A. (2024). Transforming food and land systems to achieve the SDGs.

Droukas, L., Doulgeri, Z., Tsakiridis, N. L., Triantafyllou, D., Kleitsiotis, I., Mariolis, I., Giakoumis, D., Tzovaras, D., Kateris, D., & Bochtis, D. (2023). A Survey of Robotic Harvesting Systems and Enabling Technologies. *Journal of Intelligent & Robotic Systems, 107*(2), 21. DOI: 10.1007/s10846-022-01793-z PMID: 36721646

Dsouza, A., Newman, L., Graham, T., & Fraser, E. D. G. (2023). Exploring the landscape of controlled environment agriculture research: A systematic scoping review of trends and topics. *Agricultural Systems, 209*, 103673. 10.1016/j.agsy.2023.103673

Duguma, A. L., & Bai, X. (2025). How the internet of things technology improves agricultural efficiency. *Artificial Intelligence Review, 58*(2). Advance online publication. DOI: 10.1007/S10462-024-11046-0

Dukat, D., Wahyuni, I., & Prihtiyantoro, W. (2025). The role of precision agriculture technologies in enhancing sustainable farming practices among smallholder farmers. *The Journal of Academic Science, 2*(6), 1666–1677.

Dury, J., Schaller, N., Garcia, F., Reynaud, A., & Bergez, J. E. (2011). Models to support cropping plan and crop rotation decisions. A review. *Agronomy for Sustainable Development, 32*(2), 567–580.

Dutta, M., Gupta, D., Tharewal, S., Goyal, D., Sandhu, J. K., Kaur, M., AlZubi, A. A., & Alanazi, J. M. (2025). Internet of Things-Based Smart Precision Farming in Soilless Agriculture: Opportunities and Challenges for Global Food Security. *IEEE Access : Practical Innovations, Open Solutions*, •••, 34238–34268. DOI: 10.1109/ACCESS.2025.3540317

Dutta, S. (2024). Lean Manufacturing and Process Optimization: Enhancing Efficiency in Modern Production. *International Journal of Scientific Research in Computer Science. Engineering and Information Technology, 10*, 265–273. DOI: 10.32628/CSEIT241051024

Du, Z., Yang, L., Zhang, D., Cui, T., He, X., Xiao, T., Xie, C., & Li, H. (2022). Corn variable-rate seeding decision based on gradient boosting decision tree model. *Computers and Electronics in Agriculture, 198*, 107025. 10.1016/j.compag.2022.107025

Earthdata (NASA). (2022, August 8). *The value of data in monitoring the health of crops.* https://www.earthdata.nasa.gov/dashboard/stories/sat-data-agriculture

Eastwood, C., Klerkx, L., Ayre, M., & Dela Rue, B. (2019). Managing socio-ethical challenges in the development of smart farming: From a fragmented to a comprehensive approach for responsible research and innovation. *Journal of Agricultural & Environmental Ethics, 32*(5), 741–768.

Edje, A. E., Abd Latiff, M. S., & Chan, W. H. (2023). IoT data analytic algorithms on edge-cloud infrastructure: A review. *Digital Communications and Networks, 9*(6), 1486–1515. DOI: 10.1016/j.dcan.2023.10.002

Ehrampoosh, A., Hettiarachchi, P., Koirala, A., Hassan, J., Islam, N., Ray, B., Nabi, M. N., Tolba, M., Mazid, A. M., Xu, C. Y., Ashwath, N., Dzitac, P., & Moore, S. (2025). Intelligent weed management using aerial image processing and precision herbicide spraying: An overview. *Crop Protection (Guildford, Surrey), 194*, 107206. DOI: 10.1016/J.CROPRO.2025.107206

El Alami, H., Sidna, J., Baina, A., & Najid, A. (2020). *Analysis and evaluation of communication Protocols for IoT Applications*.

El khediri, S., Benfradj, A., Thaljaoui, A., Moulahi, T., Ullah Khan, R., Alabdulatif, A., & Lorenz, P. (2024). Integration of artificial intelligence (AI) with sensor networks: Trends, challenges, and future directions. *Journal of King Saud University - Computer and Information Sciences, 36*(1), 101892. 10.1016/j.jksuci.2023.101892

Eldeeb, M. A., Dhamu, V. N., Paul, A., Muthukumar, S., & Prasad, S. (2023). Espial: Electrochemical Soil pH Sensor for In Situ Real-Time Monitoring. *Micromachines, 14*(12), 2188.

Elechi, J. O. G., Nwiyi, I. U., & Adamu, C. S. (2022). Global food system transformation for resilience. In *Food Systems Resilience*. IntechOpen.

Elhamali, S., Musbah, H., Zawi, L., Shuwehdi, A., Faris, H., & Mahdawe, A. (2025). Artificial intelligence meets laser technology: A review of recent advances. *Results in Surfaces and Interfaces, 19*, 100484. 10.1016/j.rsurfi.2025.100484

El-Sagheer, A. M., & Hamd, M. M. (2025). Ethical and regulatory considerations of machine learning in modern agriculture. In *Machine Learning for Plant Biology* (pp. 317-340).

Eltarabily, M. G., Amer, A., Farzamian, M., Bouksila, F., Elkiki, M., & Selim, T. (2024). Time-Lapse Electromagnetic Conductivity Imaging for Soil Salinity Monitoring in Salt-Affected Agricultural Regions. *Land 2024, Vol. 13, 13*(2). DOI: 10.3390/LAND13020225

Emmanuel, M. (2024). Regulatory frameworks for cloud computing in agriculture in West and East Africa.

Enescu, F. M., Dr guşin, S. A., Boştinaru, R. N., Marica, M. C., Nicolăescu, C. G., Frîncu, N. C., & Bizon, N. (2025). Energy market: A review of current solutions, trends, issues, and future directions based on the blockchain, internet of things and artificial intelligence. *Results in Engineering, 28*, 107736. DOI: 10.1016/J. RINENG.2025.107736

Ennaji, O., Vergütz, L., & El Allali, A. (2023). Machine learning in nutrient management: A review. *Artificial Intelligence in Agriculture, 9*, 1-11. 10.1016/j. aiia.2023.06.001

Ermias Benti, N., Mesfin, ·, Chaka, D., Addisu, ·, Semie, G., Warkineh, B., & Soromessa, · Teshome. (2024). Transforming agriculture with Machine Learning, Deep Learning, and IoT: perspectives from Ethiopia—challenges and opportunities. *Discover Agriculture 2024 2:1, 2*(1), 63-. /DOI: 10.1007/S44279-024-00066-7

Escribà-Gelonch, M., Liang, S., van Schalkwyk, P., Fisk, I., Long, N. V. D., & Hessel, V. (2024). Digital Twins in Agriculture: Orchestration and Applications. *Journal of Agricultural and Food Chemistry, 72*(19), 10737–10752. DOI: 10.1021/ acs.jafc.4c01934

Eyasin, M. S. H., Sobhani, M. E., Nasrin, S., Al Rafi, A. S., & Muzahidul Islam, A. K. M. (2025). CropSynergy: Harnessing IoT Solutions for Smart and Efficient Crop Management. *Crop Design*, 100127. 10.1016/j.cropd.2025.100127

Fan, X., Che, H., Wang, Y., Chen, D., & Hu, X. (2025). Effects of deep storage irrigation on summer maize yield formation and safety irrigation threshold for maximizing flood resources utilization. *Agricultural Water Management, 319*, 109805. DOI: 10.1016/J.AGWAT.2025.109805

Fao, , CWeber, , R. (n.d.). *Policy Analysis*. Advance online publication. DOI: 10.4060/cc9152en

Faqir, Y., Qayoom, A., Erasmus, E., Schutte-Smith, M., & Visser, H. G. (2024). A review of the application of advanced soil and plant sensors in the agriculture sector. *Computers and Electronics in Agriculture, 226*, 109385. DOI: 10.1016/J. COMPAG.2024.109385

Faqir, Y., Qayoom, A., Erasmus, E., Schutte-Smith, M., & Visser, H. G. (2024). A review on the application of advanced soil and plant sensors in the agriculture sector. *Computers and Electronics in Agriculture, 226*, 109385. 10.1016/j.compag.2024.109385

Farooq, M. S., Riaz, S., Abid, A., Umer, T., & Zikria, Y. B. (2020). Role of IoT Technology in Agriculture: A Systematic Literature Review. *Electronics (Basel), 9*(2), 319. https://www.mdpi.com/2079-9292/9/2/319

Fassnacht, F. E., White, J. C., Wulder, M. A., & Næsset, E. (2024). Remote sensing in forestry: Current challenges, considerations and directions. Forestry. *International Journal of Forestry Research, 97*(1), 11–37. DOI: 10.1093/forestry/cpad024

Fayaz, S., Kanth, R. H., Bhat, T. A., Dar, E. A., Shah, Z. A., Bashir, M., & Salem, A. (2025). Estimation of yield and nitrogen use efficiencies in hybrid maize varieties through site-specific nitrogen management based on leaf color chart (LCC). *Scientific Reports, 15*(1), 4686.

Fernandes, S. (2025). The role of artificial intelligence in accelerating sustainable technological innovation. *Clareus Scientific Science and Engineering, 2*, 32–40.

Ferrández-Pastor, F. J., García-Chamizo, J. M., Nieto-Hidalgo, M., & Mora-Martínez, J. (2018). Precision Agriculture Design Method Using a Distributed Computing Architecture on Internet of Things Context. *Sensors (Basel), 18*(6), 1731. PMID: 29843386

Ferrández-Pastor, F. J., García-Chamizo, J. M., Nieto-Hidalgo, M., Mora-Pascual, J., & Mora-Martínez, J. (2016). Developing ubiquitous sensor network platform using internet of things: Application in precision agriculture. *Sensors (Basel), 16*(7), 1141. DOI: 10.3390/s16071141

Ferrero, F., Rolando, G., Tabacco, E., & Borreani, G. (2024). Water Use Efficiency in Agriculture: From the Water Footprint Concept to Water Management Policies and Practices. In Minella, M., Bianco Prevot, A., & Maurino, V. (Eds.), *Water Reuse and Unconventional Water Resources: A Multidisciplinary Perspective* (pp. 483–510). Springer Nature Switzerland., DOI: 10.1007/978-3-031-67739-7_20

Fevgas, G., Lagkas, T., Papadopoulos, P., Sarigiannidis, P., & Argyriou, V. (2025). Integrating thermal infrared and RGB imaging for early detection of water stress in lettuces with comparative analysis of IoT sensors. *Smart Agricultural Technology, 11*, 100881. DOI: 10.1016/j.atech.2025.100881

Ficili, I., Giacobbe, M., Tricomi, G., & Puliafito, A. (2025). From Sensors to Data Intelligence: Leveraging IoT, Cloud, and Edge Computing with AI. *Sensors (Basel), 25*(6). Advance online publication. DOI: 10.3390/S25061763

Fikry, I., Gheith, M., & Eltawil, A. (2019). A new mathematical model for the deterministic crop rotation planning problem. *2019 IEEE International Conference on Industrial Engineering and Engineering Management (IEEM)*, 314-318.

Folorunso, A. T., & Otonne, A. (2025). The Impact of Financial Indicators and Stock Market Volatility on Stock Returns in Nigeria: Evidence from Panel Analysis. *African Journal of Economic Review, 13*(2), 104–121.

Fontani, M., Luglio, S. M., Gagliardi, L., Peruzzi, A., Frasconi, C., Raffaelli, M., & Fontanelli, M. (2025). A Systematic Review of 59 Field Robots for Agricultural Tasks: Applications, Trends, and Future Directions. *Agronomy (Basel), 15*(9), 2185.

Food and Agriculture Organization of the United Nations (FAO). (2022). *Data governance in agriculture: Principles and case studies*. FAO.

Food and Agriculture Organization of the United Nations (FAO). (2023). *Responsible use of artificial intelligence in agriculture and food systems*. FAO., DOI: 10.4060/cc2453en

Forero, Á., Cruz, J. C., & Munoz, C. (2025). Adoption of agricultural innovations within the 'farm to fork' strategy: A realistic review of barriers, paradoxes, and avenues for change. *Sustainability, 17*(21), 9493.

Fraser, E. D. G., & Campbell, M. (2019). Agriculture 5.0: Reconciling Production with Planetary Health. *One Earth, 1*(3), 278–280. DOI: 10.1016/j.oneear.2019.10.022

Freitas, J., & Silva, P. (2022). Sustainable Agricultural Systems for Fruit Orchards: The Influence of Plant Growth Promoting Bacteria on the Soil Biodiversity and Nutrient Management. *Sustainability, 14*(21), 13952.

Frona, D., & Frona-Hadas, D. (2025). Digital transformation in rural Hungary: Infrastructure, skills, and uptake across regions. *Land (Basel), 14*(10), 2034.

Fuentes-Peñailillo, F., Gutter, K., Vega, R., & Silva, G. C. (2024). Transformative Technologies in Digital Agriculture: Leveraging Internet of Things, Remote Sensing, and Artificial Intelligence for Smart Crop Management. *Journal of Sensor and Actuator Networks, 13*(4), 39. https://www.mdpi.com/2224-2708/13/4/39

Fuentes-Peñailillo, F., Ortega-Farías, S., Acevedo-Opazo, C., Rivera, M., & Araya-Alman, M. (2024). A Smart Crop Water Stress Index-Based IoT Solution for Precision Irrigation of Wine Grape. *Sensors (Basel), 24*(1), 25. DOI: 10.3390/s24010025

Fusco, G., Melgiovanni, M., Porrini, D., & Ricciardo, T. M. (2020). How to improve the diffusion of climate-smart agriculture: What the literature tells us. *Sustainability, 12*(12), 5168.

Galaz, V., Centeno, M. A., Callahan, P. W., Causevic, A., Patterson, T., Brass, I., & Levy, K. (2021). Artificial intelligence, systemic risks, and sustainability. *Technology in Society, 67*, 101741. DOI: 10.1016/j.techsoc.2021.101741

Gamage, A., Gangahagedara, R., Subasinghe, S., Gamage, J., Guruge, C., Senaratne, S., Randika, T., Rathnayake, C., Hameed, Z., Madhujith, T., & Merah, O. (2024). Advancing sustainability: The impact of emerging technologies in agriculture. *Current Plant Biology*, *40*, 100420. DOI: 10.1016/J.CPB.2024.100420

Gao, X., Liu, J., Lin, H., Javed, T., Yin, F., Chen, R., Wen, Y., Zhang, J., Yi, K., & Wang, Z. (2025). Using machine learning techniques to evaluate the impact of future climate change on wheat yields in Xinjiang, China. *Agricultural Water Management*, *317*, 109646. DOI: 10.1016/J.AGWAT.2025.109646

Gao, X., Qin, G., Yin, F., Liu, J., Lin, H., Wang, Z., & Wang, Z. (2025). Multi-objective optimization of cropping structures and irrigation strategies for sustainable agricultural development in arid regions. *Agricultural Water Management*, *321*, 109909.

García, L., Parra, L., Jimenez, J. M., Lloret, J., & Lorenz, P. (2020). IoT-Based Smart Irrigation Systems: An Overview on the Recent Trends on Sensors and IoT Systems for Irrigation in Precision Agriculture. *Sensors (Basel)*, *20*(4), 1042. https://www.mdpi.com/1424-8220/20/4/1042

García-Martín, J. P., Torralba, A., Hidalgo-Fort, E., Daza, D., & González-Carvajal, R. (2023). IoT solution for smart water distribution networks based on a low-power wireless network, combined at the device-level: A case study. *Internet of Things : Engineering Cyber Physical Human Systems*, *22*, 100746. DOI: 10.1016/j.iot.2023.100746

Gardezi, M., Joshi, B., Rizzo, D. M., Ryan, M., Prutzer, E., Brugler, S., & Dadkhah, A. (2024). Artificial intelligence in farming: Challenges and opportunities for building trust. *Agronomy Journal*, *116*(3), 1217–1228.

Gargaro, M., Murphy, R. J., & Harris, Z. M. (2023). Let-us investigate; A meta-analysis of influencing factors on lettuce crop yields within controlled-environment agriculture systems. *Plants*, *12*(14), 2623.

Garg, G., Gupta, S., Mishra, P., Vidyarthi, A., Singh, A., & Ali, A. (2023). CROP-CARE: An Intelligent Real-Time Sustainable IoT System for Crop Disease Detection Using Mobile Vision. *IEEE Internet of Things Journal*, *10*(4), 2840–2851. DOI: 10.1109/JIOT.2021.3109019

Garg, T., Kumar, U., Yadav, H., & Jain, A. (2025). Fostering Sustainable Growth via the Fusion of Artificial Intelligence and Computer Vision. In Whig, P., Silva, N., Elngar, A. A., Aneja, N., & Sharma, P. (Eds.), *Sustainable Development through Machine Learning, AI and IoT* (pp. 102–109). Springer Nature Switzerland., DOI: 10.1007/978-3-031-71729-1_10

Garske, B., Bau, A., Ekardt, F., & Gawel, E. (2021). Digitalization and AI in sustainable agriculture: A review. *Sustainability, 13*(16), 9033. DOI: 10.3390/su13169033

Garwe, E. C., Arpaci, S., Moss, A. J., Bailey, K., Suchith, A., Oyewole, O. B., & Thondhlana, J. (2025). Ethical crossroads in agricultural data: Perspectives on trust, equity and sustainability. *Outlook on Agriculture, 54*(4), 323–331.

Gavhane, K. P., Kiran, P. R., Pradhan, N. C., Mandal, S., Kumari, K., & Parray, R. A. (2026). Precision agriculture and unconventional food sources. *Health, Nutrition and Sustainability*, 627–648. /DOI: 10.1016/B978-0-443-32920-3.00024-0

Gawande, A., Sherekar, S., & Gawande, R. (2024). Early prediction of grape disease attack using a hybrid classifier in association with IoT sensors. *Heliyon, 10*(19), e38093. DOI: 10.1016/J.HELIYON.2024.E38093

GC, R. K., Hall, R. P., GC, R. K., & Hall, R. P. (2020). The Commercialization of Smallholder Farming—A Case Study from the Rural Western Middle Hills of Nepal. *Agriculture 2020, Vol. 10, 10*(5). DOI: 10.3390/AGRICULTURE10050143

Georgiadis, G. P., Elekidis, A. P., & Georgiadis, M. C. (2019). Optimization-Based Scheduling for the Process Industries: From Theory to Real-Life Industrial Applications. *Processes (Basel, Switzerland), 7*(7), 438. https://www.mdpi.com/2227-9717/7/7/438

Gerhards, M., Rock, G., Schlerf, M., & Udelhoven, T. (2016). Water stress detection in potato plants using leaf temperature, emissivity, and reflectance. *International Journal of Applied Earth Observation and Geoinformation, 53*, 27–39. 10.1016/j.jag.2016.08.004

Getahun, S., Kefale, H., & Gelaye, Y. (2024). Application of Precision Agriculture Technologies for Sustainable Crop Production and Environmental Sustainability: A Systematic Review. *TheScientificWorldJournal*, (1), 2126734. PMID: 39421732

Ghafar, A. S. A., Hajjaj, S. S. H., Gsangaya, K. R., Sultan, M. T. H., Mail, M. F., & Hua, L. S. (2023). Design and development of a robot for spraying fertilizers and pesticides for agriculture. *Materials Today: Proceedings, 81*, 242–248. 10.1016/j.matpr.2021.03.174

Ghag, N., Sonar, H., Jagtap, S., & Trollman, H. (2024). Unlocking AI's potential in the food supply chain: A novel approach to overcoming barriers. *Journal of Agriculture and Food Research, 18*, 101349. DOI: 10.1016/J.JAFR.2024.101349

Ghazal, S., Munir, A., & Qureshi, W. S. (2024). Computer vision in smart agriculture and precision farming: Techniques and applications. *Artificial Intelligence in Agriculture, 13*, 64-83. 10.1016/j.aiia.2024.06.004

Ghazal, S., Munir, A., & Qureshi, W. S. (2024). Computer vision in smart agriculture and precision farming: Techniques and applications. *Artificial Intelligence in Agriculture, 13*, 64–83. DOI: 10.1016/J.AIIA.2024.06.004

Ghirardelli, A., & Tarolli, P. (2025). Bridging the gap between science, policy, stakeholders: Delineating a virtuous framework for agriculture under climate change scenarios. *iScience, 28*(9).

Ghorbani, M., Eskandari-Damaneh, H., Cotton, M., Ghoochani, O. M., & Borji, M. (2021). Harnessing indigenous knowledge for climate change-resilient water management – lessons from an ethnographic case study in Iran. *Climate and Development, 13*(9), 766–779. DOI: 10.1080/17565529.2020.1841601

Ghraizi, D., Talj, R., & Francis, C. (2023). An Overview of Decision-Making in Autonomous Vehicles. *IFAC-PapersOnLine, 56*(2), 10971–10983. 10.1016/j.ifacol.2023.10.793

Gibson, R. B. (2013). Avoiding sustainability trade-offs in environmental assessment. *Impact Assessment and Project Appraisal, 31*(1), 2–12.

Gkrimpizi, T., Peristeras, V., & Magnisalis, I. (2023). Classification of Barriers to Digital Transformation in Higher Education Institutions: Systematic Literature Review. *Education Sciences 2023, Vol. 13, 13*(7). DOI: 10.3390/EDUCSCI13070746

Goddek, S., Delaide, B., Mankasingh, U., Ragnarsdottir, K. V., Jijakli, H., & Thorarinsdottir, R. (2015). Challenges of sustainable and commercial aquaponics. *Sustainability, 7*(4), 4199–4224.

Goel, A., Masurkar, S., & Pathade, G. R. (2024). An Overview of Digital Transformation and Environmental Sustainability: Threats, Opportunities, and Solutions. *Sustainability, 16*(24), 11079.

Goel, R. K., & Vishnoi, S. (2025). Space-age evolution—Remote sensing and IoT for productive and sustainable agricultural landscape. *Sustainable Futures : An Applied Journal of Technology, Environment and Society, 10*, 101280.

Gong, R., Zhang, H., Li, G., & He, J. (2025). Edge Computing-Enabled Smart Agriculture: Technical Architectures, Practical Evolution, and Bottleneck Breakthroughs. *Sensors 2025, Vol. 25, 25*(17). DOI: 10.3390/S25175302

Gong, L., Yan, J., Chen, Y., An, J., He, L., Zheng, L., & Zou, Z. (2022). An IoT-based intelligent irrigation system with data fusion and a self-powered wide-area network. *Journal of Industrial Information Integration, 29*, 100367. DOI: 10.1016/j.jii.2022.100367

Gong, R., Zhang, H., Li, G., & He, J. (2025). Edge Computing-Enabled Smart Agriculture: Technical Architectures, Practical Evolution, and Bottleneck Breakthroughs. *Sensors (Basel)*, *25*(17), 5302. https://www.mdpi.com/1424-8220/25/17/5302 PMID: 40942729

Gonzalez-de-Soto, M., Emmi, L., Benavides, C., Garcia, I., & Gonzalez-de-Santos, P. (2016). Reducing air pollution with hybrid-powered robotic tractors for precision agriculture. *Biosystems Engineering*, *143*, 79–94. DOI: 10.1016/j.biosystemseng.2016.01.008

González-Rodríguez, V. E., Izquierdo-Bueno, I., Cantoral, J. M., Carbú, M., & Garrido, C. (2024). Artificial Intelligence: A Promising Tool for Application in Phytopathology. *Horticulturae 2024, Vol. 10, 10*(3). /DOI: 10.3390/HORTICULTURAE10030197

Gookyi, D. A. N., Wulnye, F. A., Arthur, E. A. E., Ahiadormey, R. K., Agyemang, J. O., Agyekum, K. O. B. O., & Gyaang, R. (2024). TinyML for smart agriculture: Comparative analysis of TinyML platforms and practical deployment for maize leaf disease identification. *Smart Agricultural Technology*, *8*, 100490. DOI: 10.1016/J.ATECH.2024.100490

Gorain, S., Dutta, S., & Gawdiya, S. (2025). Ethical harvest: Ensuring fairness and privacy. In *AI-Driven IoT Solutions in Agriculture: New Insights* (pp. 169-193). Springer Nature Switzerland.

Gordon, W. J., & Rudin, R. S. (2022). Why APIs? Anticipated value, barriers, and opportunities for standards-based application programming interfaces in healthcare: Perspectives of US thought leaders. *JAMIA Open*, *5*(2), ooac023. PMID: 35474716

Griepentrog, H. W., Nørremark, M., & Nielsen, J. (2006, 2006/09/03). Autonomous intra-row rotor weeding based on GPS. *CIGR World Congress: Agricultural Engineering for a Better World*, Bonn, Germany.

Grillo, H., Alemany, M., Ortiz, A., & Fuertes-Miquel, V. (2017). Mathematical modelling of the order-promising process for fruit supply chains considering the perishability and subtypes of products. *Applied Mathematical Modelling*, *49*, 255–278.

Grinin, L., Grinin, A., & Korotayev, A. (2024). *Global Technological Transformations Since the Stone Age: Theory and History*. 57–106. DOI: 10.1007/978-3-031-56764-3_2

Groumpos, P. P. (2021). A Critical Historical and Scientific Overview of all Industrial Revolutions. *IFAC-PapersOnLine*, *54*(13), 464–471. 10.1016/j.ifacol.2021.10.492 PMID: 38620687

Guaragnella, G. (2024). *RGB and Thermal Camera Integration for Advanced Perception of an Agricultural Robot* [Laurea, Politecnico di Torino]. https://webthesis.biblio.polito.it/31892/

Guebsi, R., Mami, S., & Chokmani, K. (2024). Drones in Precision Agriculture: A Comprehensive Review of Applications, Technologies, and Challenges. *Drones 2024, Vol. 8, 8*(11). DOI: 10.3390/DRONES8110686

Guebsi, R., Mami, S., & Chokmani, K. (2024). Drones in precision agriculture: A comprehensive review of applications, technologies, and challenges. *Drones (Basel), 8*(11), 686.

Guiné, R. P. F. (2024). The Challenges and Strategies of Food Security under Global Change. *Foods, 13*(13). Advance online publication. DOI: 10.3390/foods13132083

Gültekin, R. (2023). Future Agricultural Irrigation Technologies and Water Management. In (pp. 181-198).

Gund, R., Badgujar, C. M., Samiappan, S., & Jagadamma, S. (2025). Application of Digital Twin Technology in Smart Agriculture: A Bibliometric Review. *Agriculture 2025, Vol. 15, 15*(17). DOI: 10.3390/AGRICULTURE15171799

Gupta, G., & Kumar Pal, S. (2025). Applications of AI in precision agriculture. *Discov. Agric., 3*(1), 61.

Gupta, G., & Kumar Pal, S. (2025). Applications of AI in precision agriculture. *Discover Agriculture, 3*(1), 61. DOI: 10.1007/s44279-025-00220-9

Gupta, M., Abdelsalam, M., Khorsandroo, S., & Mittal, S. (2020). Security and privacy in smart farming: Challenges and opportunities. *IEEE Access : Practical Innovations, Open Solutions, 8*, 34564–34584. DOI: 10.1109/ACCESS.2020.2979358

Gurdita, A., Vovko, H., & Ungrin, M. (2016). A Simple and Low-Cost Monitoring System to Investigate Environmental Conditions in a Biological Research Laboratory. *PLoS One, 11*(1), e0147140. DOI: 10.1371/journal.pone.0147140

Haas, M., Frankema, E., & Giller, K. E. (2025). Growing towards a food secure Africa in 2050: Reflections and pathways. *Pathways to African Food Security*, 267.

Habib, M., Singh, S., Jan, S., Jan, K., & Bashir, K. (2025). The future of the future foods: Understandings from the past towards SDG-2. *NPJ Science of Food, 9*(1), 138. DOI: 10.1038/S41538-025-00484-X PMID: 40651949

Hackfort, S. (2023). Unlocking sustainability? The power of corporate lock-ins and how they shape digital agriculture in Germany. *Journal of Rural Studies, 101*, 103065.

Hafeez, A., Husain, M. A., Singh, S. P., Chauhan, A., Khan, M. T., Kumar, N., & Soni, S. K. (2023). Implementation of drone technology for farm monitoring & pesticide spraying: A review. *Information Processing in Agriculture, 10*(2), 192–203. DOI: 10.1016/j.inpa.2022.02.002

Hakami, R. A., Naser, R. S. M., El-Bakkali, M., Othman, M. D. M., Yahya, M. S., Raweh, S., Mohammed, A., & Belghyti, D. (2024). Groundwater quality deterioration evaluation for irrigation using several indices and geographic information systems: A case study. *Desalination and Water Treatment, 320.* Advance online publication. DOI: 10.1016/J.DWT.2024.100645

Hall, E. L., Kruger, R. P., Dwyer, S. J., Hall, D. L., Mclaren, R. W., & Lodwick, G. S. (1971). A Survey of Preprocessing and Feature Extraction Techniques for Radiographic Images. *IEEE Transactions on Computers, C–20*(9), 1032–1044. DOI: 10.1109/T-C.1971.223399

Hamadani, H., Rashid, S. M., Parrah, J. D., Khan, A. A., Dar, K. A., Ganie, A. A., Gazal, A., Dar, R. A., & Ali, A. (2021). *Traditional farming practices and its consequences. Microbiota and Biofertilizers* (Vol. 2). Ecofriendly Tools for Reclamation of Degraded Soil Environs., DOI: 10.1007/978-3-030-61010-4_6

Hamdan, M. F., Noor, S. N. M., Abd-Aziz, N., Pua, T. L., & Tan, B. C. (2022). Green Revolution to Gene Revolution: Technological Advances in Agriculture to Feed the World. *Plants, 11*(10), 1297. DOI: 10.3390/PLANTS11101297 PMID: 35631721

Hamrani, A., Allouhi, A., Bouarab, F. Z., & Jayachandran, K. (2025). AI and Robotics in Agriculture: A Systematic and Quantitative Review of Research Trends (2015–2025). *Crops (Basel), 5*(5), 75.

Hanafi, A. M., Hussien, S. A., Elnahal, D. H., Ahmed, S. E. H., Salem, M. A., Zainhum, A. R., Elsayed, A. A., Ibrahim, M. A., & Sattar, Y. S. A. (2025). Revolutionizing Agriculture with IoT, Mobile Apps, and Computer Vision in Automated Hydroponic Greenhouses. International Journal of Engineering and Applied Sciences - October 6 University, 2(1), 1-16. DOI: 10.21608/ijeasou.346170.1029

Han, R., Liang, X., Shu, L., Jing, X., & Yang, F. (2025). TeaWeeding-Action: A vision-based dataset for weeding behavior recognition in tea plantations. *Frontiers in Plant Science, 16*, 1722007.

Haq, S. I. U., Wang, G., Khan, S. N., Song, C., Ma, C., Zhang, X., & Lan, Y. (2025). Multi-dimensional optical remote sensing in agriculture: Spectral, angular, and spatial scaling for crop stress monitoring. *Smart Agricultural Technology, 12*, 101583. 10.1016/j.atech.2025.101583

Haque, A., Islam, N., Samrat, N. H., Dey, S., & Ray, B. (2021). Smart farming through responsible leadership in Bangladesh: Possibilities, opportunities, and beyond. *Sustainability, 13*(8), 4511.

Harahap, V., Kamp, L. M., & Ubacht, J. (2024). A decision support scheme for solving the mobile coverage gap in rural areas in developing countries – Demonstrated with a case in Indonesia. *Technological Forecasting and Social Change, 207*, 123600. DOI: 10.1016/j.techfore.2024.123600

Hardin, R. G., Barnes, E. M., Delhom, C. D., Wanjura, J. D., & Ward, J. K. (2022). Internet of things: Cotton harvesting and processing. *Computers and Electronics in Agriculture, 202*, 107294. DOI: 10.1016/j.compag.2022.107294

Hargrove, W. L., Heyman, J. M., Mayer, A., Mirchi, A., Granados-Olivas, A., Ganjegunte, G., Gutzler, D., Pennington, D. D., Ward, F. A., Chavira, L. G., Sheng, Z., Kumar, S., Villanueva-Rosales, N., & Walker, W. S. (2023). The future of water in a desert river basin facing climate change and competing demands: A holistic approach to water sustainability in arid and semi-arid regions. *Journal of Hydrology. Regional Studies, 46*, 101336. 10.1016/j.ejrh.2023.101336

Harper, S., McLean, C., & Johnston, L. (2023). Human–AI collaboration in controlled-environment agriculture. *Agricultural Systems, 209*, 103658. DOI: 10.1016/j.agsy.2023.103658

Hasan, H. R., Musamih, A., Salah, K., Jayaraman, R., Omar, M., Arshad, J., & Boscovic, D. (2024). Smart agriculture assurance: IoT and blockchain for trusted, sustainable produce. *Computers and Electronics in Agriculture, 224*, 109184. DOI: 10.1016/J.COMPAG.2024.109184

Hashemi, S.-Z., Darzi-Naftchali, A., Karandish, F., Ritzema, H., & Solaimani, K. (2024). Enhancing agricultural sustainability with water and crop management strategies in modern irrigation and drainage networks. *Agricultural Water Management, 305*, 109110. DOI: 10.1016/j.agwat.2024.109110

Held, L. K., Vermeylen, L., Dignath, D., Notebaert, W., Krebs, R. M., & Braem, S. (2024). Reinforcement learning of adaptive control strategies. *Communications Psychology, 2*(1), 8. DOI: 10.1038/s44271-024-00055-y PMID: 39242891

Hemdan, E. E. D., & Sayed, A. (2025). Smart and secure healthcare with digital twins: A deep dive into blockchain, federated learning, and future innovations. *Algorithms, 18*(7), 401.

Hernández, H. A., Mondragón, I. F., González, S. R., & Pedraza, L. F. (2025). Reconfigurable agricultural robotics: Control strategies, communication, and applications. *Computers and Electronics in Agriculture*, *234*, 110161. 10.1016/j.compag.2025.110161

Hess, C., & Tadesse, G. (2022). Liability and accountability in automated agricultural decision systems. *AI and Ethics*, *2*(3), 389–402. DOI: 10.1007/s43681-021-00125-8

He, T., Li, M., & Jin, D. (2025). Deep learning-based time series prediction for precision field crop protection. *Frontiers in Plant Science*, *16*, 1575796. DOI: 10.3389/FPLS.2025.1575796/BIBTEX

Heydari, M. (2024). Cultivating sustainable global food supply chains: A multifaceted approach to mitigating food loss and waste for climate resilience. *Journal of Cleaner Production*, *442*, 141037. 10.1016/j.jclepro.2024.141037

Hider, U., & Zafer, B. (2024). Emerging Threats in Cybersecurity: A Comprehensive Analysis of Modern Attack Vectors.

Hnatiuc, M., Ghita, S., Alpetri, D., Ranca, A., Artem, V., Dina, I., Cosma, M., & Abed Mohammed, M. (2023). Intelligent grapevine disease detection using IoT sensor network. *Bioengineering (Basel, Switzerland)*, *10*(9), 1021. DOI: 10.3390/bioengineering10091021

Holm-Nielsen, J. B., Al Seadi, T., & Oleskowicz-Popiel, P. (2009). The future of anaerobic digestion and biogas utilization. *Bioresource Technology*, *100*(22), 5478–5484.

Hosny, K. M., El-Hady, W. M., & Samy, F. M. (2025). Technologies, Protocols, and applications of Internet of Things in greenhouse Farming: A survey of recent advances. *Information Processing in Agriculture*, *12*(1), 91–111. DOI: 10.1016/J.INPA.2024.04.002

Huang, W., & Wang, X. (2024). The Impact of Technological Innovations on Agricultural Productivity and Environmental Sustainability in China. *Sustainability 2024, Vol. 16*, *16*(19). DOI: 10.3390/SU16198480

Huang, G., Du, S., & Wang, D. (2023). GNSS techniques for real-time monitoring of landslides: A review. *Satellite Navigation*, *4*(1), 5.

Huang, I. Y., Beacham, A. M., Vickers, L. H., Mu, H., Tremma, O., Maritan, E., & Monaghan, J. M. (2025). Drivers, barriers, and grower perspectives of innovation adoption in the UK controlled environment agriculture sector. *Smart Agricultural Technology*, •••, 101748.

Huang, S., Wang, B., Li, X., Zheng, P., Mourtzis, D., & Wang, L. (2022). Industry 5.0 and society 5.0-Comparison, complementation and Co-evolution. *Journal of Manufacturing Systems*, *64*, 424–428.

Huda, S. S., Akhtar, A., Ahmed, E., Samiul Hoq, K. M., & Islam, M. N. (2026). Artificial intelligence in agriculture across south Asia: Technology adoption, improvements, and sustainability outcomes. *Sustainable Futures : An Applied Journal of Technology, Environment and Society*, *11*, 101620. DOI: 10.1016/J.SFTR.2025.101620

Hundal, G. S., Laux, C. M., Buckmaster, D., Sutton, M. J., & Langemeier, M. (2023). Exploring barriers to the adoption of internet of things-based precision agriculture practices. *Agriculture*, *13*(1), 163.

Hurrah, N. N., Khan, E., & Parah, S. A. (2023). Smart Ecosystems for Sustainable Development: Opportunities, Challenges, and Solutions. In Parah, S. A., Hurrah, N. N., & Khan, E. (Eds.), *Intelligent Multimedia Signal Processing for Smart Ecosystems* (pp. 3–28). Springer International Publishing., DOI: 10.1007/978-3-031-34873-0_1

Husein, M., Putro, H. P., Rajagukguk, J. R., & Putranto, K. E. (2024). The Role of Artificial Intelligence in Improving the Efficiency of the Company's Supply Chain. *International Journal of Engineering. Science and Information Technology*, *4*(4), 156–172. DOI: 10.52088/IJESTY.V4I4.596

Hussain, M., Abdullah, M., Ashraf, M. N., Shaheen, A., Farooqi, Z. U. R., Abbas, A., & Tariq, Y. (2025). World food hunger in 2050 and nano solutions: Probabilities and prospects based on plant-based food. In *Nanomaterials for Enhanced Plant-Based Food Production* (pp. 11–21). Academic Press.

Hussain, N., Ali, Z., & Shafique, M. (2023). HYBRID APPROACHES TO SUSTAINABLE AGRICULTURE: BRIDGING TRADITIONAL FARMING KNOWLEDGE WITH MODERN BIOTECHNOLOGICAL INNOVATIONS. Scientific. *Research Reports (Montgomery)*, *1*(01), 1–21.

Hussein, A. H. A., Jabbar, K. A., Mohammed, A., & Al-Jawahry, H. M. (2024). AI and IoT in farming: A sustainable approach. E3S Web of Conferences.

Hussein, A. H. A., Jabbar, K. A., Mohammed, A., & Jasim, L. (2024). Harvesting the future: AI and IoT in agriculture. E3S Web of Conferences.

Hussein, J. B., Workneh, T. S., Kassim, A., Ntsowe, K., Melesse, S. F., & El-Mesery, H. S. (2025). A review on the impact of big data analytics in transforming agricultural practices, food processing, and preservation strategies. *Applied Food Research*, *5*(2), 101234. DOI: 10.1016/J.AFRES.2025.101234

Huynh, H. X., Tran, L. N., & Duong-Trung, N. (2023). Smart greenhouse construction and irrigation control system for optimal Brassica Juncea development. *PLoS One, 18*(10), e0292971. DOI: 10.1371/journal.pone.0292971

Idoje, G., Dagiuklas, T., & Iqbal, M. (2021). Survey for smart farming technologies: Challenges and issues. *Computers & Electrical Engineering, 92*, 107104. DOI: 10.1016/j.compeleceng.2021.107104

Imran, & Li, J. (2025). UAV Design for Agricultural Applications. In Imran & J. Li (Eds.), *UAV Aerodynamics and Crop Interaction: Revolutionizing Modern Agriculture with Drone* (pp. 169–202). Springer Nature Singapore. DOI: 10.1007/978-981-96-8402-1_6

Indolia, S., Goswami, A. K., Mishra, S. P., & Asopa, P. (2018). Conceptual Understanding of Convolutional Neural Network- A Deep Learning Approach. *Procedia Computer Science, 132*, 679–688. 10.1016/j.procs.2018.05.069

Indrabhai, M. N. A., & Patel, N. S. (2025). Agriculture and growth economy in new India. *Commerce Management & Economics, 55.*

Ingrao, C., Strippoli, R., Lagioia, G., & Huisingh, D. (2023). Water scarcity in agriculture: An overview of causes, impacts and approaches for reducing the risks. *Heliyon, 9*(8), e18507. DOI: 10.1016/j.heliyon.2023.e18507

Inthasuth, T., Kaewjumras, Y., Somwong, S., & Boonsong, W. (2025). Comparative analysis of ZigBee, LoRa, and NB-IoT in a smart building: Advantages, limitations, and integration possibilities. [IJRES]. *International Journal of Reconfigurable and Embedded Systems, 14*, 165–175. DOI: 10.11591/ijres.v14.i1.pp165-175

Iqbal, N., Shahzad, M. U., Sherif, E. S. M., Tariq, M. U., Rashid, J., Le, T. V., & Ghani, A. (2024). Analysis of Wheat-Yield Prediction Using Machine Learning Models under Climate Change Scenarios. *Sustainability 2024, Vol. 16, 16*(16). https://doi.org/DOI: 10.3390/SU16166976

Iqbal, N., Mumtaz, R., Shafi, U., & Zaidi, S. M. H. (2021). Gray level co-occurrence matrix (GLCM) texture based crop classification using low altitude remote sensing platforms. *PeerJ. Computer Science, 7*, e536. DOI: 10.7717/peerj-cs.536

Ishore, B., Bhargav, A. M., Patel, S. K., Kumar, V., Sinha, J., & Kumar, S. (2026). Advancing smart agriculture: How IoT and 5G technologies enhance precision farming and sustainability. *International Journal of Communication Systems, 39*(1), e70310.

Islam, M. T., Sepanloo, K., Woo, S., Woo, S. H., & Son, Y. J. (2025). A Review of the Industry 4.0 to 5.0 Transition: Exploring the Intersection, Challenges, and Opportunities of Technology and Human–Machine Collaboration. *Machines 2025, Vol. 13, 13*(4). DOI: 10.3390/MACHINES13040267

Islam, M. R., Oliullah, K., Kabir, M. M., Alom, M., & Mridha, M. F. (2023). Machine learning enabled IoT system for soil nutrients monitoring and crop recommendation. *Journal of Agriculture and Food Research, 14*, 100880. DOI: 10.1016/J.JAFR.2023.100880

Islam, Md. R., Abdullah, H. M., Rahman, M. F., Islam, M., Tuhin, A. K., Ashiquzzaman, M., Islam, K. S., & Geisseler, D. (2025). Mitigation of Water-Deficit Stress in Soybean by Seaweed Extract: The Integrated Approaches of UAV-Based Remote Sensing and a Field Trial. *Drones (Basel), 9*(7), 487. DOI: 10.3390/drones9070487

Islam, S. F., Uddin, M. S., & Bansal, J. (2022). *Harvesting Robots for Smart Agriculture.* In (pp. 1–13). DOI: 10.1007/978-981-16-9991-7_1

Izadi, D., Abawajy, J. H., Ghanavati, S., & Herawan, T. (2015). A Data Fusion Method in Wireless Sensor Networks. *Sensors (Basel), 15*(2), 2964–2979. DOI: 10.3390/s150202964

Jabed, M. A., & Azmi Murad, M. A. (2024). Crop yield prediction in agriculture: A comprehensive review of machine learning and deep learning approaches, with insights for future research and sustainability. *Heliyon, 10*(24), e40836. DOI: 10.1016/J.HELIYON.2024.E40836

Jafar, A., Bibi, N., Naqvi, R. A., Sadeghi-Niaraki, A., & Jeong, D. (2024). Revolutionizing agriculture with artificial intelligence: Plant disease detection methods, applications, and their limitations. *Frontiers in Plant Science, 15*. Advance online publication. DOI: 10.3389/fpls.2024.1356260

Jafarbiglu, H., & Pourreza, A. (2022). A comprehensive review of remote sensing platforms, sensors, and applications in nut crops. *Computers and Electronics in Agriculture, 197*, 106844. DOI: 10.1016/J.COMPAG.2022.106844

Jain, M., Fishman, R., Mondal, P., Galford, G. L., Bhattarai, N., Naeem, S., & Lall, U., Balwinder-Singh, & DeFries, R. S. (2021). Groundwater depletion will reduce cropping intensity in India. *Science Advances, 7*(9). Advance online publication. DOI: 10.1126/SCIADV.ABD2849 PMID: 33627418

Jain, R. (2025). Artificial Intelligence in Precision Agriculture: Advanced Systems for Crop Management and Farm Optimization. *International Journal of Scientific Research in Computer Science. Engineering and Information Technology, 11*(1), 1849–1857. DOI: 10.32628/CSEIT251112183

Jamtarat, S., & Thongcha, C. (2025). Agricultural sector transformation through a systematic literature review (SLR) approach: Innovations, challenges and future opportunities. [AIJoES]. *Anshara International Journal of Education and Science, 2*(1), 1–7.

Jan, B., Mohapatra, S., Samal, M., Choudhary, B., & Basist, P. (2025). Role of artificial intelligence in food and agriculture industries: An updated review. *Nutrire, 50*(2), 96. DOI: 10.1186/s41110-025-00392-w

Jatav, H. S., & Rajput, V. D. (2025). Sustainable nutrient management under climate change. *Frontiers in Agronomy, 7*, 1625151.

Javaid, M., Haleem, A., Singh, R. P., & Suman, R. (2021). Substantial capabilities of robotics in enhancing industry 4.0 implementation. *Cognitive Robotics, 1*, 58-75. 10.1016/j.cogr.2021.06.001

Javaid, M., Haleem, A., Khan, I. H., & Suman, R. (2023). Understanding the potential applications of Artificial Intelligence in Agriculture Sector. *Advanced Agrochem, 2*(1), 15–30. DOI: 10.1016/J.AAC.2022.10.001

Javaid, M., Haleem, A., Singh, R. P., & Suman, R. (2022). Enhancing smart farming through the applications of Agriculture 4.0 technologies. *International Journal of Intelligent Networks, 3*, 150–164. DOI: 10.1016/J.IJIN.2022.09.004

Javed, K., Smagghe, G., Wang, Q., Javed, H., & Wang, Y. (2025). Artificial intelligence in crop protection: Revolutionizing agriculture for a sustainable future. *Information Processing in Agriculture*. Advance online publication. DOI: 10.1016/J.INPA.2025.12.003

Javidan, S. M., Banakar, A., Rahnama, K., Vakilian, K. A., & Ampatzidis, Y. (2024). Feature engineering to identify plant diseases using image processing and artificial intelligence: A comprehensive review. *Smart Agricultural Technology, 8*, 100480.

Jenkins, M., & Block, D. E. (2024). A Review of Methods for Data-Driven Irrigation in Modern Agricultural Systems. *Agronomy (Basel), 14*(7), 1355.

Jennings, R., Henderson, A. D., Phelps, A., Janda, K. M., & van den Berg, A. E. (2023). Five US dietary patterns and their relationship to land use, water use, and greenhouse gas emissions: Implications for future food security. *Nutrients, 15*(1), 215.

Jha, C. K., Singh, V., Stevanovic, M., Dietrich, P., Saxena, S., Mosnier, A., ... & Schmidt-Traub, G. (2021). Pathways to sustainable land-use and food systems in India by 2050.

Jha, C. K., Singh, V., Stevanović, M., Dietrich, J. P., Mosnier, A., Weindl, I., & Lotze-Campen, H. (2022). The role of food and land use systems in achieving India's sustainability targets. *Environmental Research Letters*, *17*(7), 074022.

Jia, H., Chen, W., Su, Z., Sun, Y., Qian, Z., & Huang, L. (2025). AI-Driven Cooperative Control for Autonomous Tractors and Implements: A Comprehensive Review. *AgriEngineering*, *7*(11). Advance online publication. DOI: 10.3390/AGRIENGINEERING7110394

Jiang, L., Xu, B., Husnain, N., Wang, Q., Jiang, L., Xu, B., Husnain, N., & Wang, Q. (2025). Overview of Agricultural Machinery Automation Technology for Sustainable Agriculture. *Agronomy 2025, Vol. 15, 15*(6). https://doi.org/DOI: 10.3390/AGRONOMY15061471

Jiang, M., Huang, Z., Qiu, L., Huang, W., & Yen, G. G. (2018). Transfer Learning-Based Dynamic Multiobjective Optimization Algorithms. *IEEE Transactions on Evolutionary Computation*, *22*(4), 501–514. DOI: 10.1109/TEVC.2017.2771451

Jin, S., Meng, X., Dardanelli, G., & Zhu, Y. (2024). Multi-Global Navigation Satellite System for Earth Observation: Recent Developments and New Progress. *Remote Sensing*, *16*(24), 4800.

Jin, W., Wang, N., Zhang, L., Tian, X., Shi, B., & Zhao, B. (2025). A Review of AI-Driven Automation Technologies: Latest Taxonomies, Existing Challenges, and Future Prospects. *Computers, Materials & Continua*, *84*(3), 3961–4018. 10.32604/cmc.2025.067857

John, D. A., & Babu, G. R. (2021). Lessons From the Aftermaths of Green Revolution on Food System and Health. *Frontiers in Sustainable Food Systems*, *5*, 644559. DOI: 10.3389/FSUFS.2021.644559 PMID: 34212131

John, D., Hussin, N., Shahibi, M. S., Ahmad, M., Hashim, H., & Ametefe, D. S. (2023). A systematic review on the factors governing precision agriculture adoption among small-scale farmers. *Outlook on Agriculture*, *52*(4), 469–485.

John, L. B., & James, G. J. (2025). Geospatial big data analytics for precision agriculture: Enhancing productivity and sustainability. *Journal of Scientific Reports*, *10*(1), 116–129.

Jonkman, J., Barbosa-Póvoa, A. P., & Bloemhof, J. M. (2019). Integrating harvesting decisions in the design of agro-food supply chains. *European Journal of Operational Research*, *276*(1), 247–258.

Jørgensen, L. N., Matzen, N., Hansen, J. G., Semaskiene, R., Korbas, M., Danielewicz, J., Glazek, M., Maumene, C., Rodemann, B., Weigand, S., Hess, M., Blake, J., Clark, B., Kildea, S., Batailles, C., Ban, R., Havis, N., & Treikale, O. (2018). Four azoles' profile in the control of Septoria, yellow rust and brown rust in wheat across Europe. *Crop Protection (Guildford, Surrey)*, *105*, 16–27. DOI: 10.1016/j.cropro.2017.10.018

Joshi, J. (2024). Sustainable farming with IoT: A case study of smart agriculture practices in rural India. *International Journal for Multidisciplinary Research*, *6*(1), 1–11. https://www.ijfmr.com/papers/2024/1/38502.pdf

Jouanjean, M. A., Casalini, F., Wiseman, L., & Gray, E. (2020). Issues around data governance in the digital transformation of agriculture: The farmers' perspective.

Judith, J., Tamilselvi, R., Beham, M. P., Lakshmi, S. S. P., Panthakkan, A., Mansoori, S. A., & Ahmad, H. A. 2025. Remote Sensing Based Crop Health Classification Using NDVI and Fully Connected Neural Networks. arXiv preprint arXiv:2504.10522.

Junior, J. L., & Silva, P. R. (2022). Digital tools and ethical governance in Brazilian agroforestry systems. *Sustainability*, *14*(9), 5142. DOI: 10.3390/su14095142

Kalpana, P., Smitha, L., Madhavi, D., Shaik, A., Kalpana, G., & Kodati, S. (2024). A Smart Irrigation System Using the IoT and Advanced Machine Learning Model: A Systematic Literature Review. *International Journal of Computational and Experimental Science and Engineering*, *10*. Advance online publication. DOI: 10.22399/ijcesen.526

Kalyani, Y., & Collier, R. (2021). A Systematic Survey on the Role of Cloud, Fog, and Edge Computing Combination in Smart Agriculture. *Sensors (Basel)*, *21*(17), 5922.

Kalyani, Y., & Collier, R. (2021). A systematic survey on the role of cloud, fog, and edge computing combination in smart agriculture. *Sensors (Basel)*, *21*(17), 5922. DOI: 10.3390/s21175922

Kalyani, Y., Vorster, L., Whetton, R., & Collier, R. (2024). Application Scenarios of Digital Twins for Smart Crop Farming through Cloud–Fog–Edge Infrastructure. *Future Internet*, *16*(3). Advance online publication. DOI: 10.3390/FI16030100

Kamal, M., & Bablu, T. (2023). Mobile Applications Empowering Smallholder Farmers. *An Analysis of the Impact on Agricultural Development.*, *8*, 36–52.

Kanaga Priya, P., Reethika, A., Vijaykumar, G., & James Deva Koresh, H. (2025). Robotics-assisted precision and sustainable irrigation, harvesting, and fertilizing processes. *Hyperautomation in Precision Agriculture: Advancements and Opportunities for Sustainable Farming*, 217–226. /DOI: 10.1016/B978-0-443-24139-0.00018-7

Kannan, S. (2025). *Transforming agriculture for the digital age: Integrating artificial intelligence, cloud computing, and big data solutions for sustainable and smart farming systems*. Deep Science Publishing.

Kanuru, L., Tyagi, A. K., Fernandez, T. F., Sreenath, N., & Mishra, S. (2021). Prediction of pesticides and fertilizers using machine learning and Internet of Things. *2021 International Conference on Computer Communication and Informatics (ICCCI)*.

Karami, A., & Karami, M. (2025). Edge computing in big data: Challenges and benefits. *International Journal of Data Science and Analytics*, *20*(7), 6183–6226. DOI: 10.1007/s41060-025-00855-3

Karanisa, T., Achour, Y., Ouammi, A., & Sayadi, S. (2022). Smart greenhouses as the path towards precision agriculture in the food-energy and water nexus: Case study of Qatar. *Environment Systems & Decisions*, *42*(4), 521–546.

Kariyanna, B., & Sowjanya, M. (2024). Unravelling the use of artificial intelligence in management of insect pests. *Smart Agricultural Technology*, *8*, 100517. DOI: 10.1016/J.ATECH.2024.100517

Karmakar, P., Teng, S. W., Murshed, M., Pang, S., Li, Y., & Lin, H. (2024). Crop monitoring by multimodal remote sensing: A review. *Remote Sensing Applications: Society and Environment*, *33*, 101093. DOI: 10.1016/J.RSASE.2023.101093

Karunathilake, E. M. B. M., Le, A. T., Heo, S., Chung, Y. S., & Mansoor, S. (2023). The path to smart farming: Innovations and opportunities in precision agriculture. *Agriculture*, *13*(8), 1593.

Kasenzu, B. (2024). Role of precision agriculture technologies in enhancing farm productivity in Kenya. *American Journal of Agriculture*, *6*(2), 1–12. DOI: 10.47672/aja.2118

Katie, B. (2024). Internet of Things (IoT) for Environmental Monitoring. *International Journal of Computing and Engineering*, *6*, 29–42. DOI: 10.47941/ijce.2139

Kaur, B., Mansi, Dimri, S., Singh, J., Mishra, S., Chauhan, N., Kukreti, T., Sharma, B., Singh, S. P., Arora, S., Uniyal, D., Agrawal, Y., Akhtar, S., Rather, M. A., Naik, B., Kumar, V., Gupta, A. K., Rustagi, S., & Preet, M. S. (2023). Insights into the harvesting tools and equipment's for horticultural crops: From then to now. *Journal of Agriculture and Food Research*, *14*, 100814. DOI: 10.1016/J.JAFR.2023.100814

Kaur, N. R., Nehra, N. D., & Bhushanwar, N. K. (2025). Enhancing sustainability, climate resilience, and resource efficiency with IoT-Based Precision Agriculture. *International Journal of Research and Review in Applied Science, Humanities, and Technology*, *2*(2), 172–179. DOI: 10.71143/7db36796

Kaur, R., Sharma, D. I., & Saini, C. (2024). Smart farming: Empowering organic agriculture with AI. *International Journal for Research in Applied Science and Engineering Technology*, *12*(3), 805–809. DOI: 10.22214/ijraset.2024.58931

Kawasaki, Y., Uga, H., Kagiwada, S., & Iyatomi, H. (2015). Basic Study of Automated Diagnosis of Viral Plant Diseases Using Convolutional Neural Networks. In Bebis, G., Boyle, R., Parvin, B., Koracin, D., Pavlidis, I., Feris, R., McGraw, T., Elendt, M., Kopper, R., & Ragan, E. (Eds.), *Advances in Visual Computing* (Vol. 9475, pp. 638–645). Springer International Publishing., https://www.ict.polimi.it/network/proxy/

Kay, M., Bunning, S., Burke, J., Boerger, V., Bojic, D., Bosc, P. M., ... & Ziadat, F. (2022). The state of the world's land and water resources for food and agriculture 2021. *Systems at breaking point.*

Kekeli, M. A., Wang, Q., & Rui, Y. (2025). The Role of Nano-Fertilizers in Sustainable Agriculture: Boosting Crop Yields and Enhancing Quality. *Plants*, *14*(4), 554.

Kendall, H., Clark, B., Li, W., Jin, S., Jones, G. D., Chen, J., & Frewer, L. J. (2022). Precision agriculture technology adoption: A qualitative study of small-scale commercial "family farms" located in the North China Plain. *Precision Agriculture*, *23*(1), 319–351.

Khadatkar, A., Sujit, P. B., Agrawal, R., Vishwanath, K., Sawant, C. P., Magar, A. P., & Chaudhary, V. P. (2025). WeeRo: Design, development and application of a remotely controlled robotic weeder for mechanical weeding in row crops for sustainable crop production. *Results in Engineering*, *26*, 105202. 10.1016/j.rineng.2025.105202

Khalifeh, A., Al-Qaisi, A., Aldahdouh, K., Al-Sit, W. T., Olaimat, A. A. L., Alouneh, S., & Darabkh, K. A. (2025). Low power wide area network (LPWAN) protocols: Enablers for future wireless networks. *Results in Engineering*, *27*, 105866. 10.1016/j.rineng.2025.105866

Khanna, M. (2021). Digital transformation of the agricultural sector: Pathways, drivers and policy implications. *Applied Economic Perspectives and Policy*, *43*(4), 1221–1242.

Khan, R. A. (2020). Response of Wheat (Triticum aestivum L.) to Zinc Sulphate and Copper Sulphate under salt stress. *Pure and Applied Biology*, *9*(4). Advance online publication. DOI: 10.19045/bspab.2020.90281

Khan, S., Mazhar, T., Shahzad, T., Khan, M. A., Guizani, S., & Hamam, H. (2024). Future of sustainable farming: Exploring opportunities and overcoming barriers in drone-IoT integration. *Discover Sustainability*, *5*(1), 470. DOI: 10.1007/s43621-024-00736-y

Kim, D., Zarei, M., Lee, S., Lee, H., Lee, G., & Lee, S. G. (2025). Wearable Stand-alone Sensing Systems for Smart Agriculture. *Advancement of Science*, *12*(16), 2414748. DOI: 10.1002/advs.202414748

Kiran, D. R. (2019). Internet of Things. In Kiran, D. R. (Ed.), *Production Planning and Control* (pp. 495–513). Butterworth-Heinemann.

Kirkaya, A. (2020). SMART FARMING-PRECISION AGRICULTURE TECHNOLOGIES AND PRACTICES. Journal of Scientific Perspectives. DOI: 10.26900/jsp.4.010

Kirkham, M. B. (2023c). Field capacity, wilting point, available water, and the nonlimiting water range. In (pp. 169-189).

Kirkham, M. B. (2023a). Tensiometers. In Kirkham, M. B. (Ed.), *Principles of Soil and Plant Water Relations* (3rd ed., pp. 65–81). Academic Press.

Klerkx, L., Jakku, E., & Labarthe, P. (2019). A review of social science on digital agriculture. *NJAS Wageningen Journal of Life Sciences*, *90–91*, 100315. DOI: 10.1016/j.njas.2019.100315

Kohut, N.. (2024). Detecting plant diseases using machine learning models. *Sustainability*, *17*(1), 132.

Kohzuma, K., & Miyamoto, K.-i. (2024). Analysis of plant physiological responses based on leaf color changes through the development and application of a wireless plant sensor. *Sensing and Bio-Sensing Research*, *46*, 100688. DOI: 10.1016/j.sbsr.2024.100688

Krishna, V., Akhila, S., & Rakesh, K. (2024). Technology Adoption Barriers and Economic Solutions. In (pp. 212-227).

Krishnamurthi, R., Kumar, A., Gopinathan, D., Nayyar, A., & Qureshi, B. (2020). An Overview of IoT Sensor Data Processing, Fusion, and Analysis Techniques. *Sensors (Basel)*, *20*(21), 6076. https://www.mdpi.com/1424-8220/20/21/6076

Krisnawijaya, N., Tekinerdogan, B., Catal, C., & Tol, P. P. J. (2022). Data analytics platforms for agricultural systems: A systematic literature review. *Computers and Electronics in Agriculture, 195*, 106813. DOI: 10.1016/j.compag.2022.106813

Kritzinger, W., Karner, M., Traar, G., Henjes, J., & Sihn, W. (2018). Digital Twin in manufacturing: A categorical literature review and classification. *IFAC-PapersOnLine, 51*(11), 1016–1022. 10.1016/j.ifacol.2018.08.474

Kroupova, Z. Z., Aulova, R., Rumankova, L., Bajan, B., Cechura, L., Simek, P., & Jarolimek, J. (2025). Drivers and barriers to precision agriculture technology and digitalisation adoption: Meta-analysis of decision choice models. *Precision Agriculture, 26*(1), 17.

Kufel, J., Bargieł-Łączek, K., Kocot, S., Koźlik, M., Bartnikowska, W., Janik, M., Czogalik, Ł., Dudek, P., Magiera, M., Lis, A., Paszkiewicz, I., Nawrat, Z., Cebula, M., & Gruszczyńska, K. (2023). What Is Machine Learning, Artificial Neural Networks and Deep Learning?-Examples of Practical Applications in Medicine. *Diagnostics (Basel), 13*(15). Advance online publication. DOI: 10.3390/diagnostics13152582 PMID: 37568945

Kulanthaivelu, R. K., Iyyanar, S., & Ramakrishnan, S. (2022). Climate change and agricultural losses in India. *American Journal of Economics and Sociology, 81*(2), 339–358.

Kulkarni, A., & Bedekar, M. (2023). Standardization in cloud computing: Unlocking the potential of a fragmented industry. In *International Conference on Advancements in Smart Computing and Information Security* (pp. 3-12). Springer Nature Switzerland.

Kull, K., Asad, B., Khan, M. A., Naseer, M. U., Kallaste, A., & Vaimann, T. (2025). Faults, Failures, Reliability, and Predictive Maintenance of Grid-Connected Solar Systems: A Comprehensive Review. *Applied Sciences 2025, Vol. 15, 15*(21). DOI: 10.3390/APP152111461

Kumar, A., Tiwari, R. G., & Trivedi, N. K. (2023). Smart farming: Design and implementation of an IoT-based automated irrigation system for precision agriculture. *2023 3rd International Conference on Innovative Sustainable Computational Technologies (CISCT)*, 1-6.

Kumar, A., Singh, S. R., Yadav, M. C., Yadav, V. K., Channi, A. S., Chouria, A., Kumar, N., Dall, H., & Nazir, T. (2025). The Role of Variable Rate Technology (VRT) in Modern Agriculture: A Review. *Journal of Advances in Biology & Biotechnology, 28*(6), 20–39. DOI: 10.9734/JABB/2025/V28I62370

Kumar, A., Singh, S., Yadav, M., Bhuj, B., Dhar, S., Pruthi, N., Kumar, R., Bajpai, V., Rizwan, M., Jyoti, K., Thapa, R., Kumar, V., Kumar, H., Mishra, B., Kumar, V., Rajput, A., Singh, A., & Kumar, R. (2022). Artificial Intelligence, Internet of Things (IoT) and Smart Agriculture for Sustainable Farming: A Review. *Annals of Plant Sciences, 11*(11), 5512–5564. https://annalsofplantsciences.com/index.php/aps/article/view/1005

Kumar, D. (2024). Ethical considerations of AI and IoT in farming. In *Agriculture 4.0* (pp. 319–333). CRC Press.

Kumari, K., Mirzakhani Nafchi, A., Mirzaee, S., & Abdalla, A. (2025). AI-Driven Future Farming: Achieving Climate-Smart and Sustainable Agriculture. *AgriEngineering 2025, Vol. 7, 7*(3). /DOI: 10.3390/AGRIENGINEERING7030089

Kumari, S., Ali, N., Dagati, M., & Dong, Y. (2025). IoT-Enabled Soil Moisture and Conductivity Monitoring Under Controlled and Field Fertigation Systems. *AgriEngineering 2025, Vol. 7, 7*(7). DOI: 10.3390/AGRIENGINEERING7070207

Kumari, S., & Arowolo, M. O. (2024). Internet of things (iot): Concepts, protocols, and applications. In *Emerging Technologies and Security in Cloud Computing* (pp. 19–52). IGI Global Scientific Publishing.

Kumari, S., Raturi, S., Kulshrestha, S., Chauhan, K., Dhingra, S., András, K., Thu, K., Khargotra, R., & Singh, T. (2023). A comprehensive review on various techniques used for synthesizing nanoparticles. *Journal of Materials Research and Technology, 27*, 1739–1763. DOI: 10.1016/J.JMRT.2023.09.291

Kumar, M., Malik, V., Tomar, B., Tomar, S., Kumari, M., & Gupta, S. (2023). *Agricultural Mechanization Development.* In (pp. 1-11).

Kumar, V., Sharma, K. V., Kedam, N., Patel, A., Kate, T. R., & Rathnayake, U. (2024). A comprehensive review on smart and sustainable agriculture using IoT technologies. *Smart Agricultural Technology, 8*, 100487. 10.1016/j.atech.2024.100487

Kumar, V., Sharma, K. V., Kedam, N., Patel, A., Kate, T. R., & Rathnayake, U. (2024a). A comprehensive review of smart and sustainable agriculture using IoT technologies. *Smart Agricultural Technology, 8*, 100487. DOI: 10.1016/J.ATECH.2024.100487

Kure, S. (2022). Sensors, Actuators and Microcontrollers. In (pp. 9).

Kusumastuti, R. D., Donk, D. P., & Teunter, R. (2016). Crop-related harvesting and processing planning: A review. *International Journal of Production Economics, 174*, 76–92.

Łągiewska, M., & Panek-Chwastyk, E. (2025). Integrating Remote Sensing and Autonomous Robotics in Precision Agriculture: Current Applications and Workflow Challenges. *Agronomy (Basel)*, *15*(10), 2314.

Laha, S. R., Pattanayak, B. K., Pattnaik, S., Mishra, D., Kumar Nayak, D. S., & Dash, B. B. (2023). An IOT-Based Soil Moisture Management System for Precision Agriculture: Real-Time Monitoring and Automated Irrigation Control. *Proceedings of the 4th International Conference on Smart Electronics and Communication, ICOSEC 2023*, 451–455. DOI: 10.1109/ICOSEC58147.2023.10276266

Lakhiar, I. A., Yan, H., Zhang, C., Wang, G., He, B., Hao, B., Han, Y., Wang, B., Bao, R., Syed, T. N., Chauhdary, J. N., Rakibuzzaman, Md., Lakhiar, I. A., Yan, H., Zhang, C., Wang, G., He, B., Hao, B., Han, Y., … Rakibuzzaman, Md. (2024). A Review of Precision Irrigation Water-Saving Technology under Changing Climate for Enhancing Water Use Efficiency, Crop Yield, and Environmental Footprints. *Agriculture 2024, Vol. 14, 14*(7). DOI: 10.3390/AGRICULTURE14071141

Lakhiar, I. A., Yan, H., Zhang, C., Wang, G., & Hao, B. (2024). A review of precision irrigation water-saving technology under changing climate for enhancing water use efficiency, crop yield, and environmental footprints. *Agriculture*, *13*(2), 245–267. DOI: 10.3390/agriculture14071141

Lalhriatpuii, Ruchi, & Wasson, V. (2025). *Comprehensive Exploration of IoT Communication Protocol: CoAP, MQTT, HTTP, LoRaWAN and AMQP.* Paper presented at the Machine Learning Algorithms, Cham.

Lan, J., & Ban, Q. (2025). The Farm-Level Economic and Environmental Benefits of Precision Agriculture Technology Adoption: A Meta-Analysis of Global Evidence. *Sustainability 2025, Vol. 17, 17*(24). /DOI: 10.3390/SU172411223

Lan, J., & Ban, Q. (2025). The farm-level economic and environmental benefits of precision agriculture technology adoption: A meta-analysis of global evidence. *Sustainability*, *17*(24), 11223.

Latif, G., Abdelhamid, S. E., Mallouhy, R. E., Alghazo, J., & Kazimi, Z. A. (2022). Deep learning utilization in agriculture: Detection of rice plant diseases using an improved CNN model. *Plants*, *11*(17), 2230. DOI: 10.3390/plants11172230

Lausch, A., Bumberger, J., Jung, A., Pause, M., Selsam, P., Zhou, T., & Herzog, F. (2025). Monitoring Agricultural Land Use Intensity with Remote Sensing and Traits. *Agriculture*, *15*(21), 2233. https://www.mdpi.com/2077-0472/15/21/2233

Lee, S. H., Kim, M. S., Kim, J. G., & Kim, S. O. (2020). Use of Soil Enzymes as Indicators for Contaminated Soil Monitoring and Sustainable Management. *Sustainability 2020, Vol. 12, 12*(19). DOI: 10.3390/SU12198209

Leng, J., Sha, W., Wang, B., Zheng, P., Zhuang, C., Liu, Q., Wuest, T., Mourtzis, D., & Wang, L. (2022). Industry 5.0: Prospect and retrospect. *Journal of Manufacturing Systems*, *65*, 279–295.

Leslie, D. (2020). *Understanding artificial intelligence ethics and safety*. Alan Turing Institute.

Liang, F., & Fernandez, A. (2021). Responsibility gaps in automated decision-making. *Philosophy & Technology*, *34*(4), 957–978. DOI: 10.1007/s13347-020-00423-7

Liben, F., Abera, W., Chernet, M. T., Ebrahim, M., Tilaye, A., Erkossa, T., Degefie, D. T., Mponela, P., Kihara, J., & Tamene, L. (2024). Site-specific fertilizer recommendation using data-driven machine learning enhanced wheat productivity and resource use efficiency. *Field Crops Research*, *313*, 109413. DOI: 10.1016/J.FCR.2024.109413

Lidder, P., Cattaneo, A., & Chaya, M. (2025). Innovation and technology for achieving resilient and inclusive rural transformation. *Global Food Security*, *44*, 100827. DOI: 10.1016/J.GFS.2025.100827

Li, J., Rodriguez, D., Zhang, D., & Ma, K. (2015). Crop rotation model for contract farming with constraints on similar profits. *Computers and Electronics in Agriculture*, *119*, 12–18.

Li, L., Zhang, S., & Wang, B. (2021). Plant disease detection and classification by deep learning-A review. *IEEE Access : Practical Innovations, Open Solutions*, *9*, 56683–56698.

Li, M. (2025). Beyond Conventional Drones: A Review of Unconventional Rotary-Wing UAV Design. *Drones (Basel)*, *9*(5), 323.

Lindemann, B., Müller, T., Vietz, H., Jazdi, N., & Weyrich, M. (2021). A survey on long short-term memory networks for time series prediction. *Procedia CIRP*, *99*, 650–655. DOI: 10.1016/j.procir.2021.03.088

Liopa-Tsakalidi, A., Thomopoulos, V., Barouchas, P., Boursianis, A. D., & Goudos, S. K. (2024). A LoRaWAN-based IoT platform for smart irrigation in olive groves. *Smart Agricultural Technology*, *9*, 100673. DOI: 10.1016/j.atech.2024.100673

Liu, J. J., Wu, H., & Riaz, I. (2025). Advanced technologies for smart fertilizer management in agriculture: A review. *IEEE Access : Practical Innovations, Open Solutions*.

Liu, J., Fang, Y., Wang, G., Liu, B., & Wang, R. (2023). The aging of farmers and its challenges for labor-intensive agriculture in China: A perspective on farmland transfer plans for farmers' retirement. *Journal of Rural Studies, 100*, 103013. 10.1016/j.jrurstud.2023.103013

Liu, Q., Yu, R., Suo, H., Cai, Y., Chen, L., & Jiang, H. (2025). Autonomous Driving in Agricultural Machinery: Advancing the Frontier of Precision Agriculture. *Actuators, 14*(9), 464.

Liu, R., Luan, T. H., Qu, Y., Xiang, Y., Gao, L., & Zhao, D. (2025). Internet of Digital Twin: Framework, Applications and Enabling Technologies. *IEEE Communications Surveys and Tutorials*.

Liu, X., Zhang, Q., Min, W., Geng, G., & Jiang, S. (2025). Solutions and challenges in AI-based pest and disease recognition. *Computers and Electronics in Agriculture, 238*, 110775. 10.1016/j.compag.2025.110775

Liu, X., Zhao, Z., & Rezaeipanah, A. (2025). Intelligent and automatic irrigation system based on internet of things using fuzzy control technology. *Scientific Reports, 15*(1), 14577. DOI: 10.1038/s41598-025-98137-2

Longo, L., Brcic, M., Cabitza, F., Choi, J., Confalonieri, R., Ser, J. D., Guidotti, R., Hayashi, Y., Herrera, F., Holzinger, A., Jiang, R., Khosravi, H., Lecue, F., Malgieri, G., Páez, A., Samek, W., Schneider, J., Speith, T., & Stumpf, S. (2024). Explainable Artificial Intelligence (XAI) 2.0: A manifesto of open challenges and interdisciplinary research directions. *Information Fusion, 106*, 102301. 10.1016/j.inffus.2024.102301

López-Morales, J. A., Martínez, J. A., & Skarmeta, A. F. (2020). Digital Transformation of Agriculture through the Use of an Interoperable Platform. *Sensors (Basel), 20*(4), 1153.

Lotsis, S., Georgousis, I., & Papakostas, G. A. (2026). *AI, precision agriculture and tourism for sustainable regional development: The case of the Aegean Islands and Crete*. Sustainability.

Louzada, R. O., Bergier, I., Bolfe, É. L., & Barbedo, J. G. A. (2025). Integrating GIS and remote sensing for soil attributes mapping in degraded pastures of the Brazilian Cerrado. *Soil Advances, 3*, 100044. DOI: 10.1016/J.SOILAD.2025.100044

Lowenberg-DeBoer, J., & Erickson, B. (2019). Setting the Record Straight on Precision Agriculture Adoption. *Agronomy Journal, 111*(4), 1552–1569.

Lu, Y., Hegyi, A., Maria Salomons, A., & Wang, H. (2025). Reference RL: Reinforcement learning with reference mechanism and its application in traffic signal control. *Information Sciences*, *689*, 121485. 10.1016/j.ins.2024.121485

Lysaght, T., Lim, H. Y., Xafis, V., & Ngiam, K. Y. (2019). AI-assisted decision-making in healthcare: The application of an ethics framework for big data in health and research. *Asian Bioethics Review*, *11*(3), 299–314. DOI: 10.1007/s41649-019-00096-0

Lytridis, C., Kaburlasos, V. G., Pachidis, T., Manios, M., Vrochidou, E., Kalampokas, T., & Chatzistamatis, S. (2021). An Overview of Cooperative Robotics in Agriculture. *Agronomy (Basel)*, *11*(9), 1818.

M. B. Akanbi, K. J. Adedotun, A. K. Raji, & I. K. Banjoko. (2025). Optimizing Water Resource Management in Agriculture Using AI-Powered Solar Irrigation Systems. *Journal of Science Innovation and Technology Research*. https://doi.org/ DOI: 10.70382/AJSITR.V7I9.011

Mabhaudhi, T., Dirwai, T. L., Taguta, C., Sikka, A., & Lautze, J. (2023). Mapping Decision Support Tools (DSTs) on agricultural water productivity: A global systematic scoping review. *Agricultural Water Management*, *290*, 108590. DOI: 10.1016/J.AGWAT.2023.108590

Majdalawieh, M., Martins, C., Radi, M., Alaraj, M., & Khan, S. (2025). Precision agriculture in the age of AI: A systematic review of machine learning methods for crop disease detection. *Smart Agricultural Technology*, *12*, 101491. DOI: 10.1016/J.ATECH.2025.101491

Majigi, M. U., Idris, I., Abdulhamid, S. M., & Ikuesan, R. A. (2025). Big data transfer service architecture for cloud data centers: problems, methods, applications, and future trends. *Discover Computing*, *28*(1), 163-. DOI: 10.1007/S10791-025-09682 3/TABLES/9

Majumdar, P., Mitra, S., Bhattacharya, D., & Bhushan, B. (2025). Enhancing sustainable 5G powered agriculture 4.0: Summary of low power connectivity, internet of UAV things, AI solutions and research trends. *Multimedia Tools and Applications*, *84*(17), 17389–17433. DOI: 10.1007/s11042-024-19728-1

Majumder, S., Kasirao, G., & Himavarsha, P., Himanshi, Chaudhary, S., Sekopo, K., . . . Verma, J. (2023). Assessing Low-cost Capacitive Soil Moisture Sensors: Accurate, Affordable, and IoT-ready Solutions for Soil Moisture Monitoring. *International Journal of Environment and Climate Change*, *13*, 2233–2242. DOI: 10.9734/ijecc/2023/v13i113386

Malapane, O. L., Musakwa, W., & Chanza, N. (2024). Indigenous agricultural practices employed by the Vhavenda community in the Musina local municipality to promote sustainable environmental management. *Heliyon, 10*(13), e33713. DOI: 10.1016/J.HELIYON.2024.E33713 PMID: 39040358

Malik, P., Singh Tomar, A., Trivedi, A., Paliwal, A., & Goyal, A. (2025)... *Optimizing Fertilizer Application Using Machine Learning for Precision Agriculture., 13*, 3.

Maloku, D. (2020). Adoption of precision farming technologies: USA and EU situation. *SEA–Practical Application of Science, 8*(22), 7–14.

Mamabolo, E., Mashala, M. J., Mugari, E., Mogale, T. E., Mathebula, N., Mabitsela, K., & Ayisi, K. K. (2025). Application of precision agriculture technologies for crop protection and soil health. *Smart Agricultural Technology, 12*, 101270. 10.1016/j.atech.2025.101270

Mamoria, P. R. K., & Shakya, R. (2023). The Role of Artificial Intelligence in Addressing Climate Change. In Vempaty, L. N., Raja, V., Mukherjee, R., Shriya, K., Maddela Prasanthi, M., & Laxmi, K. V. (Eds.), *Fusion of Knowledge: Multidisciplinary Perspectives in Research* (Vol. 4, pp. 31–36). Infinity Publication., DOI: 10.25215/9392917953.06

Mana, A. A., Allouhi, A., Hamrani, A., Rahman, S., el Jamaoui, I., & Jayachandran, K. (2024). Sustainable AI-based production agriculture: Exploring AI applications and implications in agricultural practices. *Smart Agricultural Technology, 7*, 100416. DOI: 10.1016/J.ATECH.2024.100416

Mandal, S., Yadav, A., Panme, F. A., Devi, K. M., & Kumar, S. M. S. (2024). Adaption of smart applications in agriculture to enhance production. *Smart Agricultural Technology, 7*, 100431. 10.1016/j.atech.2024.100431

Manhas, S., Sharma, T., Sharma, B., Khajuria, S., Thakur, N., Thakur, A., . . . Sharma, V. (2026). Climate-resilient agriculture: Strategies for nutritious food for all. In *One Planet, One Health, One Future* (pp. 357-386). Academic Press.

Mani, M., Baskar, D., Janakiraman, S., & Rajendran, S. (2025). MODERN AGRICULTURE SYSTEMS: ENHANCING PRECISION FARMING THROUGH ADVANCED AERIAL TECHNIQUES. *Agrociencia*, •••, 1–15. DOI: 10.47163/agrociencia.v59i8.3551

Manning, L. (2024). Innovating in an uncertain world: Understanding the social, technical and systemic barriers to farmers adopting new technologies. *Challenges, 15*(2), 32.

Manono, B. O., Mwami, B., Mutavi, S., & Nzilu, F. (2026). Precision Farming with Smart Sensors: Current State, Challenges and Future Outlook. *Sensors (Basel), 26*(3), 882. DOI: 10.3390/S26030882

Mansoor, S., Iqbal, S., Popescu, S. M., Kim, S. L., Chung, Y. S., & Baek, J.-H. (2025). Integration of smart sensors and IOT in precision agriculture: trends, challenges and future prospectives [Review]. *Frontiers in Plant Science, Volume 16 - 2025.* https://doi.org/DOI: 10.3389/fpls.2025.1587869

Manwatkar, A. B. KeshavraoDeshmukh, R., Junnarkar, A. A., JagdishKamble, S., Kachare, P. K., &PrashantLatkar, S. (2024). Securing industry 4.0-Based precision agriculture. *2024 IEEE 4th International Conference on ICT in Business Industry & Government (ICTBIG)*, 1-7.

Maqbool, A. (2025). The Evolution of Agricultural Robotics: A Comprehensive Review and Future Challenges. *International Journal for Research in Applied Science and Engineering Technology.*

Marinello, F., Zou, X., Liu, Z., Zhu, X., Zhang, W., Qian, Y., Li, Y., Karunathilake, E. M. B. M., Le, A. T., Heo, S., Chung, Y. S., & Mansoor, S. (2023). The Path to Smart Farming: Innovations and Opportunities in Precision Agriculture. *Agriculture 2023, Vol. 13, 13*(8). DOI: 10.3390/AGRICULTURE13081593

Maró, Z. M., Nagy, J., Molnár, E. M., & Mizik, T. (2025). Challenges and potential solutions to employment issues in the agri-food sector of developed countries - A systematic literature review. *Sustainable Futures : An Applied Journal of Technology, Environment and Society, 10*, 100895. 10.1016/j.sftr.2025.100895

Maryono, M., Killoes, A. M., Adhikari, R., & Abdul Aziz, A. (2024). Agriculture development through multi-stakeholder partnerships in developing countries: A systematic literature review. *Agricultural Systems, 213*, 103792. DOI: 10.1016/j.agsy.2023.103792

Masi, M., Pasquale, J., Vecchio, Y., & Capitanio, F. (2023). Precision farming: Barriers of variable rate technology adoption in Italy. *Land (Basel), 12*(5), 1084.

Mason-D'Croz, D., & Herrero, M. (2025). Advances and future needs for modelling sustainable and just food systems transformations. *The Lancet. Planetary Health.*

Masset, E., Kapoor Malhotra, S., Gupta, N., Bhandari, R., White, H., MacDonald, H., Puskur, R., Singaraju, N., & Sharma Waddington, H. (2023). PROTOCOL: The impact of agricultural mechanisation on women's economic empowerment: A mixed-methods systematic review. *Campbell Systematic Reviews, 19*(3), e1334. DOI: 10.1002/CL2.1334 PMID: 37361554

Massoud, M., Kineber, A., Elshaboury, N., Abunada, Z., Arashpour, M., Alatroush, M., Mostafa, S., & Alhusban, M. (2026). Identifying and assessing the internet of things implementation barriers for sustainable building projects: SEM-ANN approach. *KSCE Journal of Civil Engineering*, *30*(1), 100310. DOI: 10.1016/J.KSCEJ.2025.100310

Mastoi, M. S., Zhuang, S., Munir, H. M., Haris, M., Hassan, M., Usman, M., & Ro, J.-S. (2022). An in-depth analysis of electric vehicle charging station infrastructure, policy implications, and future trends. *Energy Reports*, *8*, 11504–11529. DOI: 10.1016/j.egyr.2022.09.011

Mastretta-Yanes, A., Tobin, D., Bellon, M. R., von Wettberg, E., Cibrián-Jaramillo, A., Wegier, A., Monroy-Sais, A. S., Gálvez-Reyes, N., Ruiz-Arocho, J., & Chen, Y. H. (2024). Human management of ongoing evolutionary processes in agroecosystems. *Plants, People, Planet*, *6*(6), 1190–1206. DOI: 10.1002/PPP3.10521

Masud Cheema, J., Aqib, M., Elbeltagi, A., Rashid Saleem, S., Hussain, S., Wang, Z., Xiao, S., Wang, J., Parab, A., & Patel, S. (2025). Reinforcement Learning-Based Agricultural Fertilization and Irrigation Considering N2O Emissions and Uncertain Climate Variability. *AgriEngineering 2025, Vol. 7, 7*(8). /DOI: 10.3390/AGRIENGINEERING7080252

Matchaya, G. C., Ebrahim, G. Y., Aheeyar, M., Ajayi, O. C., & Ajilore, O. (2025). Advancing digital agriculture for climate resilience and economic transformation in Malawi.

Matese, A., Czarnecki, J. M. P., Samiappan, S., & Moorhead, R. (2024). Are unmanned aerial vehicle-based hyperspectral imaging and machine learning advancing crop science? *Trends in Plant Science*, *29*(2), 196–209. PMID: 37802693

Maurya, D., Maurya, S. K., Kumar, M., Chaubey, C., Gupta, D., Patel, K., Mehta, A., & Yadav, R. (2024). A Review on Precision Agriculture: An Evolution and Prospect for the Future. *International Journal of Plant and Soil Science*, *36*, 363–374. DOI: 10.9734/IJPSS/2024/v36i54534

Ma, W., Marini, M., & Rahut, D. (2023). Farmers' Organizations and Sustainable Development: An Introduction. *Annals of Public and Cooperative Economics*, *94*. Advance online publication. DOI: 10.1111/apce.12449

Ma, W., Zhou, X., Boansi, D., Horlu, G. S. A., & Owusu, V. (2024). Adoption and intensity of agricultural mechanization and their impact on non-farm employment of rural women. *World Development*, *173*, 106434. DOI: 10.1016/J.WORLDDEV.2023.106434

Mba, P. C., Njoku, J. N., & Uyeh, D. D. (2025). Enhancing resilience in specialty crop production in a changing climate through smart systems adoption. *Smart Agricultural Technology*, *11*, 100897. DOI: 10.1016/J.ATECH.2025.100897

McCarty, N. S., & Ledesma-Amaro, R. (2019). Synthetic Biology Tools to Engineer Microbial Communities for Biotechnology. *Trends in Biotechnology*, *37*(2), 181–197. DOI: 10.1016/j.tibtech.2018.11.002

McFadden, J., Njuki, E., & Griffin, T. (2023). Precision agriculture in the digital era: Recent adoption on US farms.

McFadden, J., Erickson, B., Lowenberg-DeBoer, J., & Milics, G. (2024, 2024/07/21-24). Global Adoption of Precision Agriculture: An Update on Trends and Emerging Technologies. *16th International Conference on Precision Agriculture*, Monticello, IL.

Meena, N. K., Kumar, A., Yadav, V., Panigrahi, S., Jaganathan, T., Shekhar, G., & Narang, E. H. (2025). Emerging trends in precision horticulture integrating digital technologies for smart crop management. *International Journal of Advanced Biochemistry Research*, •••, 361–368.

Mehrotra, N., Phartiyal, M., Guha, S., Rangrass, S., & Mehrotra, N. (2025). AgDx—The Agri Digitalization Framework: A tool for transformation. *Journal of the Indian Institute of Science*, •••, 1–29.

Mekki, K., Bajic, E., Chaxel, F., & Meyer, F. (2019). A comparative study of LPWAN technologies for large-scale IoT deployment. *ICT Express*, *5*(1), 1–7. 10.1016/j.icte.2017.12.005

Melesse, T. Y. (2025). Digital twin-based applications in crop monitoring. *Heliyon*, *11*(2), e42137. 10.1016/j.heliyon.2025.e42137 PMID: 39911426

Melhim, S. H., & Isaifan, R. J. (2025). The Energy-Economy Nexus of Advanced Air Pollution Control Technologies: Pathways to Sustainable Development. *Energies*, *18*(9), 2378. DOI: 10.3390/en18092378

Mendes, J., Pinho, T. M., Neves Dos Santos, F., Sousa, J., Peres, E., Cunha, J., Cunha, M., & Morais, R. (2020). Smartphone Applications Targeting Precision Agriculture Practices—A Systematic Review. *Agronomy (Basel)*, *10*, 855. DOI: 10.3390/agronomy10060855

Mesfin, S., Kefale, H., & Gelaye, Y. (2024). Application of Precision Agriculture Technologies for Sustainable Crop Production and Environmental Sustainability: A Systematic Review. The Scientific World Journal, 2024, ID 2126734. DOI: 10.1155/2024/2126734

Meshram, A. T., Vanalkar, A. V., Kalambe, K. B., & Badar, A. M. (2022). Pesticide spraying robot for precision agriculture: A categorical literature review and future trends. *Journal of Field Robotics*, *39*(2), 153–171.

Meshram, A., Kadam, R., Shinde, G., & Khalge, M. (2024). *Sensor Based Agriculture and Horticulture Crop Production(Application of Sensors).* DOI: 10.20944/PREPRINTS202401.0544.V1

Mešić, A., Jurić, M., Donsì, F., Maslov Bandić, L., & Jurić, S. (2024). Advancing climate resilience: technological innovations in plant-based, alternative and sustainable food production systems. *Discover Sustainability 2024 5:1, 5*(1), 423-. DOI: 10.1007/S43621-024-00581-Z

Mgendi, G. (2024). Unlocking the potential of precision agriculture for sustainable farming. *Discover Agriculture*, *2*(1), 87. DOI: 10.1007/s44279-024-00078-3

Mienye, I. D., Swart, T. G., & Obaido, G. (2024). Recurrent Neural Networks: A Comprehensive Review of Architectures, Variants, and Applications. *Information (Basel)*, *15*(9), 517. https://www.mdpi.com/2078-2489/15/9/517

Miglani, V. (2025). Use of AI in agricultural supply chain: A case of Gujarat warehouse owners and Bhusari cold storage. *Indian Journal of Agricultural Marketing*, *39*(1spl), 178-179.

Miller, T., Mikiciuk, G., Durlik, I., Mikiciuk, M., Łobodzińska, A., Śnieg, M., Miller, T., Mikiciuk, G., Durlik, I., Mikiciuk, M., Łobodzińska, A., & Śnieg, M. (2025a). The IoT and AI in Agriculture: The Time Is Now—A Systematic Review of Smart Sensing Technologies. *Sensors 2025, Vol. 25, 25*(12). DOI: 10.3390/S25123583

Miller, T., Mikiciuk, G., Durlik, I., Mikiciuk, M., Łobodzińska, A., & Śnieg, M. (2025). The IoT and AI in Agriculture: The Time Is Now—A Systematic Review of Smart Sensing Technologies. *Sensors (Basel)*, *25*, 3583. DOI: 10.3390/s25123583 PMID: 40573470

Milrad, S. (2018). 11 - Satellite Imagery. In Milrad, S. (Ed.), *Synoptic Analysis and Forecasting* (pp. 145–161). Elsevier., 10.1016/B978-0-12-809247-7.00011-9

Mirzakhaninafchi, H., Singh, M., Dixit, A. K., Prakash, A., Sharda, S., Kaur, J., Nafchi, A. M., Mirzakhaninafchi, H., Singh, M., Dixit, A. K., Prakash, A., Sharda, S., Kaur, J., & Nafchi, A. M. (2022). Performance Assessment of a Sensor-Based Variable-Rate Real-Time Fertilizer Applicator for Rice Crop. *Sustainability 2022, Vol. 14, 14*(18). DOI: 10.3390/SU141811209

Mishra, H. (2025). Artificial Intelligence Technologies in Agriculture. *AgriTech Revolution: Next-Gen Solutions and Challenges in Modern Agriculture*, 371–396. /DOI: 10.1007/978-981-95-1268-3_19

Mishra, H., & Kayusi, F. (2025). Artificial intelligence and soil conservation: An overview. *Journal of Scientific Agriculture*, 229–249. https://doi.org/DOI: 10.25081/JSA.2025.V9.9661

Mishra, H., Gautam, S., Dwivedi, S., Raza, M. Y., Ali, H., & Ikram, S. (2025). *Evaluation of Climate Resilient Agriculture Practices in Subcontinent*. 225–278. DOI: 10.1007/978-3-032-00190-0_11

Mishra, P., Singh, P. P., Singh, S. K., & Verma, H. (2019). Sustainable agriculture and benefits of organic farming to special emphasis on PGPR. *Role of Plant Growth Promoting Microorganisms in Sustainable Agriculture and Nanotechnology*, 75–87. DOI: 10.1016/B978-0-12-817004-5.00005-1

Mishra, H., & Mishra, D. (2023). Artificial Intelligence and Machine Learning in Agriculture: Transforming Farming Systems (pp. 1–16). https://doi.org/DOI: 10.5281/zenodo.15157536

Mizik, T. (2023). How can precision farming work on a small scale? A systematic literature review. *Precision Agriculture*, *24*(1), 384–406. DOI: 10.1007/s11119-022-09934-y

Mmbando, G. S. (2025). Harnessing artificial intelligence and remote sensing in climate-smart agriculture: The current strategies needed for enhancing global food security. *Cogent Food & Agriculture*, *11*(1), 2454354. DOI: 10.1080/23311932.2025.2454354

Mohamed, E. S., Belal, A. A., Abd-Elmabod, S. K., El-Shirbeny, M. A., Gad, A., & Zahran, M. B. (2021). Smart farming for improving agricultural management. *The Egyptian Journal of Remote Sensing and Space Sciences*, *24*(3), 971–981.

Mohammadi, A. R., Petrovic, E. K., & Perkins, N. (2025). Mapping forestry practices against sustainable development goal targets. *Frontiers in Sustainability*, *6*, 1534449.

Mohsan, S. A. H., Othman, N. Q. H., Li, Y., Alsharif, M. H., & Khan, M. A. (2023). Unmanned aerial vehicles (UAVs): Practical aspects, applications, open challenges, security issues, and future trends. *Intelligent Service Robotics*, *16*(1), 109–137. DOI: 10.1007/s11370-022-00452-4 PMID: 36687780

Mohyuddin, G., Khan, M. A., Haseeb, A., Mahpara, S., Waseem, M., & Saleh, A. M. (2024). Evaluation of Machine Learning Approaches for Precision Farming in Smart Agriculture System: A Comprehensive Review. *IEEE Access : Practical Innovations, Open Solutions*, *12*, 60155–60184. DOI: 10.1109/ACCESS.2024.3390581

Monamodi, P., Ndoro, J. T., Matiwane, M. Ben, Monamodi, P., Ndoro, J. T., & Matiwane, M. Ben. (2025). The Impact of Innovative Irrigation System Use on Crop Yield Among Smallholder Farmers in Mbombela Local Municipality, South Africa. *Agriculture 2025, Vol. 15*, *15*(16). DOI: 10.3390/AGRICULTURE15161755

Mondal, P., & Basu, M. (2009). Adoption of precision agriculture technologies in India and in some developing countries: Scope, present status and strategies. *Progress in Natural Science*, *19*, 659–666.

Monika, T, Thangadurai, K, Sivaguru, R, & Abdulkalamazad, G, & Lakshmi kanth R. (2025). AI-Driven Pest and Disease Detection in Smart Farming Systems. [IRJAEM]. *International Research Journal on Advanced Engineering and Management*, *3*(06), 2315–2319. DOI: 10.47392/IRJAEM.2025.0365

Montoya-Cavero, L.-E., Díaz de León Torres, R., Gómez-Espinosa, A., & Escobedo Cabello, J. A. (2022). Vision systems for harvesting robots: Produce detection and localization. *Computers and Electronics in Agriculture*, *192*, 106562. 10.1016/j.compag.2021.106562

Morchid, A., El Alami, R., Al Raezah, A. A., & Sabbar, Y. (2023). Applications of Internet of Things (IoT) and sensors technology to increase food security and agricultural sustainability: Benefits and challenges. *Ain Shams Engineering Journal*, *15*(3), 102509. DOI: 10.1016/j.asej.2023.102509

Morchid, A., Et-taibi, B., Oughannou, Z., Alami, R. E., Qjidaa, H., Jamil, M. O., Boufounas, E.-M., & Abid, M. R. (2025). IoT-enabled smart agriculture for improving water management: A smart irrigation control using embedded systems and Server-Sent Events. *Scientific African*, *27*, e02527.

Morchid, A., Jebabra, R., Khalid, H. M., El Alami, R., Qjidaa, H., & Ouazzani Jamil, M. (2024). An IoT-based smart irrigation management system to enhance agricultural water security using embedded systems, telemetry data, and cloud computing. *Results in Engineering*, *23*, 102829. DOI: 10.1016/J.RINENG.2024.102829

Morchid, A., Jebabra, R., Khalid, H. M., El Alami, R., Qjidaa, H., & Ouazzani Jamil, M. (2024). IoT-based smart irrigation management system to enhance agricultural water security using embedded systems, telemetry data, and cloud computing. *Results in Engineering*, *23*, 102829. DOI: 10.1016/j.rineng.2024.102829

Morchid, A., Marhoun, M., El Alami, R., & Boukili, B. (2024). Intelligent detection for sustainable agriculture: A review of IoT-based embedded systems, cloud platforms, DL, and ML for plant disease detection. *Multimedia Tools and Applications*, *83*(28), 70961–71000. DOI: 10.1007/s11042-024-18392-9

Mordor Intelligence, I. (2025a). Hydroponics - Market Share Analysis, Industry Trends & Statistics, Growth Forecasts (2025–2030). https://www.marketresearch.com/Mordor-Intelligence-LLP-v4018/Hydroponics-Share-Trends-Statistics-Growth-42554019/

Mordor Intelligence, I. (2025b). Precision Farming Market Size and Share Analysis - Growth Trends and Forecast (2025–2030). https://www.mordorintelligence.com/industry-reports/global-precision-farming-market-industry

Mordor Intelligence, I. (2025c). Smart Greenhouse Market Size & Share Analysis - Growth Trends & Forecasts (2024–2029). https://www.marketresearch.com/Mordor-Intelligence-LLP-v4018/Smart-Greenhouse-Size-Share-Growth-36614817/

Mouedden, M. B., Mabrouki, J., Azdem, D., & El Hajjaji, S. (2025). The development of intelligent agriculture, the contribution of the Internet of Things (IoT) to modernization and its implications for precision farming: A review. *Euro-Mediterranean Journal for Environmental Integration*, •••, 1–13.

Mrisho, L. M., Mbilinyi, N. A., Ndalahwa, M., Ramcharan, A. M., Kehs, A. K., McCloskey, P. C., Murithi, H., Hughes, D. P., & Legg, J. P. (2020). Accuracy of a smartphone-based object detection model, PlantVillage Nuru, in identifying the foliar symptoms of the viral diseases of Cassava–CMD and CBSD. *Frontiers in Plant Science*, *11*. Advance online publication. DOI: 10.3389/fpls.2020.590889

Mubarok, M. U., Sari, M. I., Wibowo, Y. G., & Mathew, R. (2024). Comparative Study of Artificial Intelligence (AI) Utilization in Digital Marketing Strategies Between Developed and Developing Countries: A Systematic Literature Review. *Ilomata International Journal of Management*, *6*(1), 156–173. DOI: 10.61194/IJJM.V6I1.1534

Muhammed, D., Ahvar, E., Ahvar, S., Trocan, M., Montpetit, M. J., & Ehsani, R. (2024). Artificial Intelligence of Things (AIoT) for smart agriculture: A review of architectures, technologies, and solutions. *Journal of Network and Computer Applications*, *228*, 103905. DOI: 10.1016/J.JNCA.2024.103905

Muhammed, D., Ahvar, E., Ahvar, S., Trocan, M., Montpetit, M.-J., & Ehsani, R. (2024). Artificial Intelligence of Things (AIoT) for smart agriculture: A review of architectures, technologies and solutions. *Journal of Network and Computer Applications*, *228*, 103905. DOI: 10.1016/j.jnca.2024.103905

Mukhtar, S., Arbabi, A., & Viegas, J. (2025). Advances in Spectral Imaging: A Review of Techniques and Technologies. *IEEE Access : Practical Innovations, Open Solutions, 13*, 35848–35902. DOI: 10.1109/ACCESS.2025.3544476

Muninarayanappa, V., & Ranjan, R. (2024). Agriculture data analysis using parallel k-nearest neighbour classification algorithm. *Int J Reconfigurable & Embedded Syst ISSN, 2089*(4864), 4864. DOI: 10.11591/ijres.v13.i2

Munz, J., & Schuele, H. (2022). Influencing the success of precision farming technology adoption—A model-based investigation of economic success factors in small-scale agriculture. *Agriculture, 12*(11), 1773.

Mushtaq, M. A., Ateeq, M., Ikram, M., Alam, S. M., Kaleem, M. M., Ashraf, M. A., Asim, M., Almutairi, K. F., Seleiman, M. F., & Shireen, F. (2025). Securing fruit trees future: AI-driven early warning and predictive systems for abiotic stress in changing climate. *Plant Stress, 17*, 100953. DOI: 10.1016/J.STRESS.2025.100953

Mustapha, L. S., Obayomi, O. V., Lau, S. Y., & Obayomi, K. S. (2025). The role of nanotechnology in agricultural systems with emphasis on water management. *Inorganic Chemistry Communications, 182*, 115507. DOI: 10.1016/j.inoche.2025.115507

Mustaza, S. M., Mat Pauzi, N. A., Zainal, N., Mohd Zaman, M. H., & Mohamed Moubark, A. (2025). Artificial Intelligence in Precision Agriculture: A Review. *Jurnal Kejuruteraan, 37*(2), 1025–1047. DOI: 10.17576/JKUKM-2025-37(2)-38

Mu, W., Kleter, G. A., Bouzembrak, Y., Dupouy, E., Frewer, L. J., Radwan Al Natour, F. N., & Marvin, H. J. P. (2024). Making food systems more resilient to food safety risks by including artificial intelligence, big data, and internet of things into food safety early warning and emerging risk identification tools. *Comprehensive Reviews in Food Science and Food Safety, 23*(1), e13296. DOI: 10.1111/1541-4337.13296

Mwangi, M., Otieno, D., & Kamau, P. (2023). Cooperative governance and digital advisory systems in Kenyan agriculture. *Journal of Rural Studies, 98*, 45–57. DOI: 10.1016/j.jrurstud.2023.01.006

Nafulah, M. (2020). Digitalisation of agriculture: How to align and engage partners to move the industry into precision farming. Case study, maize farming in Belgium.

Nagaraja, G., Shoba, H., Ms, S., & Pn, K. (2024). The impact of robotics and drones on agricultural efficiency and productivity. *International Journal of Research in Agronomy, 7*(9), 1001–1009.

Nagarsheth, S., Agbossou, K., Henao, N., & Bendouma, M. (2025). The Advancements in Agricultural Greenhouse Technologies: An Energy Management Perspective. *Sustainability, 17*(8), 3407.

Naim, M., Rizzo, D., Sauvée, L., & Medici, M. (2025). Advancing agroecology and sustainability with agricultural robots at field level: A scoping review. *Computers and Electronics in Agriculture*, *237*, 110650. 10.1016/j.compag.2025.110650

Nakalembe, C., Becker-Reshef, I., Bonifacio, R., Hu, G., Humber, M. L., Justice, C. J., Keniston, J., Mwangi, K., Rembold, F., Shukla, S., Urbano, F., Whitcraft, A. K., Li, Y., Zappacosta, M., Jarvis, I., & Sanchez, A. (2021). A review of satellite-based global agricultural monitoring systems available for Africa. *Global Food Security*, *29*, 100543. DOI: 10.1016/j.gfs.2021.100543

Nalakurthi, N. V. S. R., Abimbola, I., Ahmed, T., Anton, I., Riaz, K., Ibrahim, Q., Banerjee, A., Tiwari, A., & Gharbia, S. (2024). Challenges and Opportunities in Calibrating Low-Cost Environmental Sensors. *Sensors (Basel)*, *24*(11), 3650.

Naorem, L. D., Muthaiyan, M., & Venkatesan, A. (2019). Identification of dysregulated miRNAs in triple negative breast cancer: A meta-analysis approach. *Journal of Cellular Physiology*, *234*(7), 11768–11779.

Naqvi, S. M. Z. A., Tahir, M. N., Raghavan, V., Awais, M., Hu, J., Said, Y., Othman, N. A., Ashurov, M., & Khan, M. I. (2025). AI-enhanced IoT sensors for real-time crop monitoring: An era towards self-monitored agriculture. *Telecommunication Systems*, *88*(3), 100. DOI: 10.1007/s11235-025-01326-7

Narayana, T. L., Venkatesh, C., & Kiran, A. J, C. B., Kumar, A., Khan, S. B., Almusharraf, A., & Quasim, M. T. (2024). Advances in real time smart monitoring of environmental parameters using IoT and sensors. *Heliyon, 10*(7), e28195. 10.1016/j.heliyon.2024.e28195

Narayanan, S. (2022). Data, power, and agrarian transformation in India. *The Journal of Peasant Studies*, *49*(6), 1271–1293. DOI: 10.1080/03066150.2021.1996358

Nath, D. (2023). Smart Farming: Automation and Robotics in Agriculture. In. Nations), F. F. a. A. O. o. t. U. (2022). Understanding the context. pests and pesticide management. Food and Agriculture Organization of the United Nations. https://www.fao.org/pest-and-pesticide-management/about/understanding-the-context/en/?utm_source=chatgpt.com

Nath, B., Chen, G., O'Sullivan, C. M., & Zare, D. (2024). Research and Technologies to Reduce Grain Postharvest Losses: A Review. *Foods*, *13*(12). Advance online publication. DOI: 10.3390/FOODS13121875 PMID: 38928816

Natsir, M. H., Mahmudy, W. F., Tono, M., & Nuningtyas, Y. F. (2025). Advancements in artificial intelligence and machine learning for poultry farming: Applications, challenges, and future prospects. *Smart Agricultural Technology*, *12*, 101307. DOI: 10.1016/J.ATECH.2025.101307

Nautiyal, M., Joshi, S., Hussain, I., Rawat, H., Joshi, A., Saini, A., Kapoor, R., Verma, H., Nautiyal, A., Chikara, A., Ahmad, W., & Kumar, S. (2025). Revolutionizing agriculture: A comprehensive review of artificial intelligence applications in enhancing the properties of agricultural produce. *Food Chemistry: X*, *29*, 102748. DOI: 10.1016/J.FOCHX.2025.102748

Nava-Pintor, J. A., Alcalá-Rodríguez, U. E., Guerrero-Osuna, H. A., Mata-Romero, M. E., Lopez-Neri, E., García-Vázquez, F., & Luque-Vega, L. F. (2025). Development and Evaluation of Solar Radiation Sensor Using Cost-Effective Light Sensors and Machine Learning Techniques. *Technologies*, *13*(5), 182. https://www.mdpi.com/2227-7080/13/5/182

Nega, A. (2025). Climate Change Impacts on Agriculture: A Review of Plant Diseases and Insect Pests in Ethiopia and East Africa, With Adaptation and Mitigation Strategies. *Advances in Agriculture*, *2025*(1), 5606701. 10.1155/aia/5606701

Nelson, A. (2025). The impact of digital technologies in the agro-allied sector: A systematic review of smart farming barriers in small and medium-scale enterprises.

Nguyen, C., Sagan, V., Maimaitiyiming, M., Maimaitijiang, M., Bhadra, S., & Kwasniewski, M. T. (2021). Early Detection of Plant Viral Disease Using Hyperspectral Imaging and Deep Learning. *Sensors (Basel)*, *21*(3), 742. DOI: 10.3390/s21030742

Nikolinakos, N. T. (2023). Ethical Principles for Trustworthy AI. In Nikolinakos, N. T. (Ed.), *EU Policy and Legal Framework for Artificial Intelligence, Robotics and Related Technologies - The AI Act* (pp. 101–166). Springer International Publishing., DOI: 10.1007/978-3-031-27953-9_3

Nimmo, J. R., Wiekenkamp, I., Araki, R., Groh, J., Singh, N. K., Crompton, O., Wyatt, B. M., Ajami, H., Giménez, D., Hirmas, D. R., Sullivan, P. L., & Sprenger, M. (2025). Identifying preferential flow from soil moisture time series: Review of methodologies. *Vadose Zone Journal*, *24*(2), e70017. DOI: 10.1002/VZJ2.70017

Nitti, A., de Tullio, M. D., Federico, I., & Carbone, G. (2025). A collective intelligence model for swarm robotics applications. *Nature Communications*, *16*(1), 6572. DOI: 10.1038/s41467-025-61985-7 PMID: 40670391

Nižetić, S., & Šolić, P., López-de-Ipiña González-de-Artaza, D., & Patrono, L. (2020). Internet of Things (IoT): Opportunities, issues and challenges towards a smart and sustainable future. *Journal of Cleaner Production*, *274*, 122877. DOI: 10.1016/j.jclepro.2020.122877

Nkwocha, C. L., Adewumi, A., Folorunsho, S. O., Eze, C., Jjagwe, P., Kemeshi, J., & Wang, N. (2025). A Comprehensive Review of Sensing, Control, and Networking in Agricultural Robots: From Perception to Coordination. *Robotics (Basel, Switzerland)*, *14*(11), 159.

Nkwocha, C. L., & Chandel, A. K. (2025). Towards an End-to-End Digital Framework for Precision Crop Disease Diagnosis and Management Based on Emerging Sensing and Computing Technologies: State over Past Decade and Prospects. *Computers*, *14*(10), 443. DOI: 10.3390/computers14100443

Nsoh, B., Katimbo, A., Guo, H., Heeren, D. M., Nakabuye, H. N., Qiao, X., Ge, Y., Rudnick, D. R., Wanyama, J., Bwambale, E., & Kiraga, S. (2024). Internet of Things-Based Automated Solutions Utilizing Machine Learning for Smart and Real-Time Irrigation Management: A Review. *Sensors 2024, Vol. 24, 24*(23). DOI: 10.3390/S24237480

Nsoh, B., Katimbo, A., Guo, H., Heeren, D. M., Nakabuye, H. N., Qiao, X., & Kiraga, S. (2024). Internet of Things-Based Automated Solutions Utilizing Machine Learning for Smart and Real-Time Irrigation Management: A Review. *Sensors (Basel)*, *24*(23). Advance online publication. DOI: 10.3390/s24237480

Nurmalitasari, Nurchim, & Lestari, R. D. (2025). Artificial intelligence-driven solar smart irrigation for sustainable agriculture: Trends, challenges, and SDG implications – A systematic review. *Smart Agricultural Technology, 12*, 101665. 10.1016/j.atech.2025.101665

Nurmalitasari, N., Nurchim, N., & Lestari, R. D. (2025). Artificial intelligence-driven solar smart irrigation for sustainable agriculture: Trends, challenges, and SDG implications–A systematic review. *Smart Agricultural Technology*, •••, 101665.

Nyakuri, J., Nkundineza, C., Gatera, O., Nkurikiyeyezu, K., & Mwitende, G. (2025). *AI-IoT driven system for agricultural pest outbreak risk prediction*. https://doi.org/ DOI: 10.21203/rs.3.rs-6881038/v1

Nyakuri, J., Nkundineza, P. C., Omar Gatera, O., Nkurikiyeyezu, K. K., & Mwitende, G. (2025). AI and IoT-powered edge device optimized for crop pest and disease detection. *Scientific Reports*, *15*(1). Advance online publication. DOI: 10.1038/s41598-025-06452-5

Nyathi, L. D., & Mutale, J. (2025). Leveraging smart technologies for sustainable economic development: Evidence from urban and rural communities in Bulawayo Province, Zimbabwe. *African Journal of Commercial Studies*, *6*(6), 21–34.

Nzimande, T. N. M., Chimonyo, V. G. P., Wimalasiri, E. M., & Mabhaudhi, T. (2025). Using AquaCrop, DSSAT and the SIMPLE to Estimate Water Use of Underutilised Cereal in South Africa. In Mabhaudhi, T., Chimonyo, V. G. P., Senzanje, A., & Chivenge, P. P. (Eds.), *Enhancing Water and Food Security Through Improved Agricultural Water Productivity: New Knowledge, Innovations and Applications* (pp. 201–221). Springer Nature Singapore., DOI: 10.1007/978-981-96-1848-4_9

Obaideen, K., & Hassoun, A. (2025). Industry 4.0 for sustainable agriculture and food industry. In *Food Industry 4.0 and Food Sustainability* (pp. 25-48). Springer Nature Switzerland.

Obaideen, K., Yousef, B. A. A., AlMallahi, M. N., Tan, Y. C., Mahmoud, M., Jaber, H., & Ramadan, M. (2022). An overview of smart irrigation systems using IoT. *Energy Nexus*, *7*, 100124. DOI: 10.1016/j.nexus.2022.100124

Office, U. S. G. A. 2024. Precision Agriculture: Benefits and Challenges for Technology Adoption and Use. U.S. Government Accountability Office, Washington, DC. Report No. GAO-24-105962.

Ofosu-Ampong, K. (2024). Artificial intelligence research: A review on dominant themes, methods, frameworks and future research directions. *Telematics and Informatics Reports, 14*, 100127. 10.1016/j.teler.2024.100127

Ogunleke, A., & Idowu, F. O. (2025). Climate-smart agriculture and youth empowerment. Available at *SSRN* 5172794.

OĞUZTÜRK, G. E. (2025). AI-driven irrigation systems for sustainable water management: A systematic review and meta-analytical insights. *Smart Agricultural Technology*, *11*, 100982. DOI: 10.1016/J.ATECH.2025.100982

Ojo, M. O., Giordano, S., Procissi, G., & Seitanidis, I. N. (2018). A review of low-end, middle-end, and high-end iot devices. *IEEE Access : Practical Innovations, Open Solutions*, *6*, 70528–70554.

Ojo, M. O., & Zahid, A. (2022). Deep Learning in Controlled Environment Agriculture: A Review of Recent Advancements, Challenges and Prospects. *Sensors (Basel)*, *22*(20), 7965. https://www.mdpi.com/1424-8220/22/20/7965 PMID: 36298316

Okoli, A. O., & Birkenberg, A. (2024). Monitoring soil carbon in smallholder carbon projects: Insights from Kenya. *Climatic Change*, *177*(9), 143.

Okonta, D. E., & Vukovic, V. (2024). Smart cities software applications for sustainability and resilience. *Heliyon, 10*(12), e32654. DOI: 10.1016/J.HELIYON.2024.E32654 PMID: 39183850

Okpala, C. C., & Nwamekwe, C. O. (2025). Artificial intelligence-augmented edge computing: Architectures, challenges, and future directions. *International Journal of Engineering Inventions, 14*(9), 18–27.

Olabanji, M. F., & Chitakira, M. (2025). The adoption and scaling of climate-smart agriculture innovation by smallholder farmers in South Africa: A review of institutional mechanisms, policy frameworks and market dynamics. *WORLD (Oakland, Calif.), 6*(2), 51.

Olliver, A., Schlingmann, N., & van der Vlugt, P. (2022). ISOBUS Technologies – The Standard for Smart Agriculture. In Zhang, Q. (Ed.), *Encyclopedia of Smart Agriculture Technologies* (pp. 1–18). Springer International Publishing., DOI: 10.1007/978-3-030-89123-7_117-2

Omia, E., Bae, H., Park, E., Kim, M. S., Baek, I., Kabenge, I., & Cho, B. K. (2023). Remote Sensing in Field Crop Monitoring: A Comprehensive Review of Sensor Systems, Data Analyses and Recent Advances. *Remote Sensing 2023, Vol. 15, 15*(2). DOI: 10.3390/RS15020354

Omotayo, A. O., Adediran, S. A., Omotoso, A. B., Olagunju, K. O., & Omotayo, O. P. (2025). Artificial intelligence in agriculture: Ethics, impact possibilities, and pathways for policy. *Computers and Electronics in Agriculture, 239*, 110927. DOI: 10.1016/J.COMPAG.2025.110927

Onunga, J. (2025). Responsible innovation in AI and data governance in Kenyan agriculture: A review article. *Regulation, 9*(6), 8–12.

OpokuDarko, R., Liu, J., Yuan, S., K. Sam-Amoah, L., & Yan, H. (2020). Irrigated agriculture for food self-sufficiency in the sub-saharan African region. *International Journal of Agricultural and Biological Engineering, 13*(3), 1–12.

Organisation for Economic Co-operation and Development (OECD). (2021). *AI policy observatory: Responsible AI practices.* OECD Publishing.

Organisation for Economic Co-operation and Development (OECD). (2022). *Artificial intelligence and agriculture: Opportunities and risks.* OECD Publishing., DOI: 10.1787/2baf5a32-

Osman, M., Xia, Y., Mahdi, M., & Ahmed, A. (2025). Hybrid VTOL UAV technologies: Efficiency, customization, and sector-specific applications. *Alexandria Engineering Journal, 120*, 13–49. 10.1016/j.aej.2024.12.087

Ou, G., Chen, Y., Han, Y., Sun, Y., Zheng, S., & Ma, R. (2025). Design and Experiment of an Internet of Things-Based Wireless System for Farmland Soil Information Monitoring. *Agriculture, 15*(5), 467.

Oyeyemi, B. B. (2022). Artificial Intelligence in Agricultural Supply Chains: Lessons from the US for Nigeria. *International Journal of Multidisciplinary Research and Growth Evaluation, 3*(1), 1064–1074. DOI: 10.54660/.IJMRGE.2022.3.1.1064-1074

Pace, R., Schiano Di Cola, V., Monti, M. M., Affinito, A., Cuomo, S., Loreto, F., & Ruocco, M. (2025). Artificial intelligence in soil microbiome analysis: A potential application in predicting and enhancing soil health—a review. *Discover Applied Sciences, 7*(2), 85. DOI: 10.1007/s42452-024-06381-4

Padhiary, M., Kumar, A., & Sethi, L. N. (2025). Emerging technologies for smart and sustainable precision agriculture. *Discover Robotics 2025 1:1, 1*(1), 6-. DOI: 10.1007/S44430-025-00006-0

Padhiary, M., Hoque, A., Prasad, G., Kumar, K., & Sahu, B. (2025). Precision agriculture and AI-driven resource optimization for sustainable land and resource management. In *Smart Water Technology for Sustainable Management in Modern Cities* (pp. 197–232). IGI Global Scientific Publishing.

Padhiary, M., Kumar, A., & Sethi, L. N. (2025). Emerging technologies for smart and sustainable precision agriculture. *Discover Robotics, 1*(1), 6. DOI: 10.1007/s44430-025-00006-0

Padhiary, M., Saha, D., Kumar, R., Sethi, L. N., & Kumar, A. (2024). Enhancing precision agriculture: A comprehensive review of machine learning and AI vision applications in all-terrain vehicle for farm automation. *Smart Agricultural Technology, 8*, 100483. DOI: 10.1016/J.ATECH.2024.100483

Pahlavan, K., & Krishnamurthy, P. (2021). Evolution and Impact of Wi-Fi Technology and Applications: A Historical Perspective. *International Journal of Wireless Information Networks, 28*(1), 3–19. DOI: 10.1007/s10776-020-00501-8

Pal, O., Shovon, M., Ph, D. M., & Shin, J. (2023). *A Comprehensive Review of AI-enabled Unmanned Aerial Vehicle: Trends, Vision, and Challenges.* /arXiv.2310.16360DOI: 10.48550

Palva, R., Kaila, E., García-Pascual, B., Bloch, V., Palva, R., Kaila, E., García-Pascual, B., & Bloch, V. (2024). Assessment of the Performance of a Field Weeding Location-Based Robot Using YOLOv8. *Agronomy 2024, Vol. 14, 14*(10). https://doi.org/DOI: 10.3390/AGRONOMY14102215

Pande, C. B., & Moharir, K. N. (2023). Application of Hyperspectral Remote Sensing Role in Precision Farming and Sustainable Agriculture Under Climate Change: A Review. In Pande, C. B., Moharir, K. N., Singh, S. K., Pham, Q. B., & Elbeltagi, A. (Eds.), *Climate Change Impacts on Natural Resources, Ecosystems and Agricultural Systems* (pp. 503–520). Springer International Publishing., DOI: 10.1007/978-3-031-19059-9_21

Pané, S., Wendel-Garcia, P., Belce, Y., Chen, X. Z., & Puigmartí-Luis, J. (2021). Powering and Fabrication of Small-Scale Robotics Systems. *Current Robotics Reports*, *2*(4), 427–440. DOI: 10.1007/s43154-021-00066-1 PMID: 35036926

Papadopoulos, G., Arduini, S., Uyar, H., Psiroukis, V., Kasimati, A., & Fountas, S. (2024). Economic and environmental benefits of digital agricultural technologies in crop production: A review. *Smart Agricultural Technology*, *8*, 100441.

Pareek, C. M., Tewari, V. K., & Nare, B. (2025). A mechatronic seed metering control system for improving sowing uniformity of planters. *Journal of Engineering Research*, *13*(2), 808–819. 10.1016/j.jer.2023.10.041

Parra-López, C., Ben Abdallah, S., Garcia-Garcia, G., Hassoun, A., Trollman, H., Jagtap, S., Gupta, S., Aït-Kaddour, A., Makmuang, S., & Carmona-Torres, C. (2025). Digital technologies for water use and management in agriculture: Recent applications and future outlook. *Agricultural Water Management*, *309*, 109347. DOI: 10.1016/J.AGWAT.2025.109347

Patel, S. A., & Kamrani, A. K. (1996). Intelligent decision support system for diagnosis and maintenance of automated systems. *Computers & Industrial Engineering*, *30*(2), 297–319. DOI: 10.1016/0360-8352(95)00173-5

Pathirana, S., Lambot, S., Krishnapillai, M., Cheema, M., Smeaton, C., & Galagedara, L. (2023). Ground-Penetrating Radar and Electromagnetic Induction: Challenges and Opportunities in Agriculture. *Remote Sensing*, *15*(11). Advance online publication. DOI: 10.3390/RS15112932

Pathmudi, V. R., Khatri, N., Kumar, S., Abdul-Qawy, A. S. H., & Vyas, A. K. (2023). A systematic review of IoT technologies and their constituents for smart and sustainable agriculture applications. *Scientific African*, *19*, e01577. DOI: 10.1016/j.sciaf.2023.e01577

Patil, S., & Deshmukh, R. (2021). Irrigation decision-making and digital advisory tools in India. *Water Policy*, *23*(5), 1103–1119. DOI: 10.2166/wp.2021.084

Patkar, V., & Mehendale, N. (2025). Autonomous ground vehicles: Technological advancements, implementation challenges, and future directions. *International Journal of Intelligent Robotics and Applications*, *9*(3), 1044–1081. DOI: 10.1007/s41315-024-00418-y

Paul-Chima Ugwu, O., Ogenyi, F. C., Ugo Alum, E., Hyginus Udoka Eze, V., Basajja, M., Nnenna Ugwu, J., Ugwu, C. N., Idu Ejemot-Nwadiaro, R., Ben Okon, M., Ikechukwu Egba, S., Daniel Ejim, U., & Daniel, U., paul-Chima Ugwu, okechukwu, ogenyi, F. C., Ugo alum, esther, hyginus Udoka eze, V., nnenna Ugwu, J., Ugwu, C., idu ejemot-nwadiaro, regina, … Daniel ejim, U. (2025). Implementing artificial intelligence and machine learning algorithms for optimized crop management: A systematic review on data-driven approaches to enhancing resource use and agricultural sustainability. *Cogent Food & Agriculture*, *11*(1), 2569982. DOI: 10.1080/23311932.2025.2569982

Paul, N., Sunil, G. C., Horvath, D., & Sun, X. (2025). Deep learning for plant stress detection: A comprehensive review of technologies, challenges, and future directions. *Computers and Electronics in Agriculture*, *229*, 109734. DOI: 10.1016/J.COMPAG.2024.109734

Peddicord, L., Xavier, A., Cryer, S., Barr, J., Heijden, G. van der, Peddicord, L., Xavier, A., Cryer, S., Barr, J., & Heijden, G. van der. (2025). Scalable Prediction of Northern Corn Leaf Blight and Gray Leaf Spot Diseases to Predict Fungicide Spray Timing in Corn. *Agronomy 2025, Vol. 15*, *15*(2). DOI: 10.3390/AGRONOMY15020328

Peet, M. M., & Welles, G. (2005). Greenhouse tomato production. In *Tomatoes* (pp. 257–304). https://doi.org/DOI: 10.1079/9780851993966.0257

Peladarinos, N., Piromalis, D., Cheimaras, V., Tserepas, E., Munteanu, R. A., & Papageorgas, P. (2023). Enhancing Smart Agriculture by Implementing Digital Twins: A Comprehensive Review. *Sensors (Basel)*, *23*(16), 7128. DOI: 10.3390/s23167128 PMID: 37631663

Penuelas, J., Coello, F., & Sardans, J. (2023). A better use of fertilizers is needed for global food security and environmental sustainability. *Agriculture & Food Security*, *12*(1). Advance online publication. DOI: 10.1186/S40066-023-00409-5

Pereira, L. S., Allen, R. G., Smith, M., & Raes, D. (2015). Crop evapotranspiration estimation with FAO56: Past and future. *Agricultural Water Management*, *147*, 4–20. DOI: 10.1016/j.agwat.2014.07.031

Pérez, M. Á. G., Guerrero-González, A., Rodríguez, F. J. C., Leon, I. M. M., & Abrisqueta, F. A. L. (2024). Precision Agriculture 4.0: Implementation of IoT, AI, and Sensor Networks for Tomato Crop Prediction. *Buletin Ilmiah Sarjana Teknik Elektro*, *6*(2), 172–181. DOI: 10.12928/BISTE.V6I2.10954

Petcu, M. A., Sobolevschi-David, M. I., Curea, S. C., & Moise, D. F. (2024). Integrating Artificial Intelligence in the Sustainable Development of Agriculture: Applications and Challenges in the Resource-Based Theory Approach. *Electronics 2024, Vol. 13, 13*(23). /DOI: 10.3390/ELECTRONICS13234580

Petrovic, B., & Csambalik, L. (2025). Enhancing precision agriculture for climate change mitigation in Visegrad countries: Factors shaping adaptation. *Land (Basel)*, *14*(2), 399.

Pienwisetkaew, T., Wongsaichia, S., & Ketkaew, C. (2025). Technology adoption in smart agricultural waste management among farmers: The role of simplicity and gamification. *Cleaner and Responsible Consumption*, *19*, 100344. DOI: 10.1016/J.CLRC.2025.100344

Pineda, M., Barón, M., & Pérez-Bueno, M. L. (2021). Thermal imaging for plant stress detection and phenotyping. *Remote Sensing*, *13*(1), 1–21. DOI: 10.3390/RS13010068

Pintus, M., Colucci, F., & Maggio, F. (2025). Emerging Developments in Real-Time Edge AIoT for Agricultural Image Classification. *IoT*, *6*(1), 13. DOI: 10.3390/iot6010013

Pitoro, V. S. J., de Amorim, C. A., Franco, J. R., & Sánchez-Román, R. M. (2025). Application of low-cost soil moisture sensors for irrigation management in Brassica oleracea var. acephala cultivation. *Smart Agricultural Technology*, *12*, 101596. DOI: 10.1016/J.ATECH.2025.101596

Popescu, D., Ichim, L., & Stoican, F. (2023). Orchard monitoring based on unmanned aerial vehicles and image processing by artificial neural networks: A systematic review. *Front Plant Sci*, *14*, 1237695. DOI: 10.3389/fpls.2023.1237695 PMID: 38089793

Popp, L., Zhumasheva, S., Ibrishev, N., Islamov, E., & Nurgalieva, R. (2025). Engineering solutions for sustainable development in the agro-industrial sector. In *International Conference Innovation in Engineering* (pp. 246-265). Springer Nature Switzerland.

Portela, F., Sousa, J. J., Araújo-Paredes, C., Peres, E., Morais, R., & Pádua, L. (2024). A Systematic Review on the Advancements in Remote Sensing and Proximity Tools for Grapevine Disease Detection. *Sensors (Basel)*, *24*(24), 8172. DOI: 10.3390/s24248172

Pourreza, A., Lee, W. S., Etxeberria, E., & Zhang, Y. (2016). Identification of citrus Huanglongbing disease at the pre-symptomatic stage using polarized imaging technique. *IFAC-PapersOnLine*, *49*(16), 110–115. DOI: 10.1016/j.ifacol.2016.10.021

Pozza, L. E., & Field, D. J. (2020). The science of Soil Security and Food Security. *Soil Security*, *1*, 100002. DOI: 10.1016/J.SOISEC.2020.100002

Prajapati, H., Yadav, K., Hanamasagar, Y., Kumar, M., Khan, T., Belagalla, N., Thomas, V., Jabeen, A. G. G., & Malathi, G. (2024). Impact of Climate Change on Global Agriculture: Challenges and Adaptation. *International Journal of Environment and Climate Change*, *14*, 372–379. DOI: 10.9734/ijecc/2024/v14i44123

Prasad, A., & Sharma, A. (2025). From lesion detection to outcome prediction: Artificial intelligence and deep learning applications in multiple sclerosis. *Neurological Sciences*, *47*(1), 46. DOI: 10.1007/s10072-025-08661-2

Prasanjana, V., Herath, I. S., Senadheera, U. E., Darshani, N. P., Suwan, B. V. S., Madusanka, C., Takimoto, Y., Amila, B., Hewawsam, C., Jayasanka, D. J., & Nakayama, T. (2025). Review of current robotic approaches for weed management in paddy cultivation. *Journal of Agriculture and Food Research*, *24*, 102516. DOI: 10.1016/J.JAFR.2025.102516

Premachandra, J., & Kumara, P. (2021). Mobile Applications for Precision Agriculture Practices: A Review. 14th International Research Conference, Security, Stability and National Development in the New Normal, Ratmalana, Sri Lanka.

Purcell, W., & Neubauer, T. (2023). Digital Twins in Agriculture: A State-of-the-art review. *Smart Agricultural Technology*, *3*, 100094. DOI: 10.1016/j.atech.2022.100094

Puspaningrum, A., Sumarudin, A., & Putra, W. P. (2022). Irrigation prediction using machine learning in precision agriculture. 2022 5th International Conference of Computer and Informatics Engineering (IC2IE).

Pustavrh, J., Hočevar, M., Podržaj, P., Trajkovski, A., & Majdic, F. (2023). *Comparative study of a hydraulic, pneumatic and electric linear actuator system*. DOI: 10.21203/rs.3.rs-3184855/v1

Qamar, A. (2024). Ethical Considerations in AI-Powered Decision-Making Systems. *Research Corridor Journal of Engineering Science, 1*(1), 90–102. https://researchcorridor.org/index.php/RCJES/article/view/236

Qayoom, K., & Manzoor, S. (2022). Digital Agriculture: An Emerging Approach Towards Sustainability in Agriculture and Rural Development. Journal of Community Mobilization and Sustainable Development, 1(Seminar Special Issue), 150-158.

Quy, V. K., Hau, N. V., Anh, D. V., Quy, N. M., Ban, N. T., Lanza, S., & Muzirafuti, A. (2022). IoT-enabled smart agriculture: Architecture, applications, and challenges. *Applied Sciences (Basel, Switzerland), 12*(7), 3396.

Quy, V. K., Hau, N. V., Anh, D. V., Quy, N. M., Ban, N. T., Lanza, S., Randazzo, G., & Muzirafuti, A. (2022). IoT-Enabled Smart Agriculture: Architecture, Applications, and Challenges. *Applied Sciences (Basel, Switzerland), 12*(7), 3396. https://www .mdpi.com/2076-3417/12/7/3396

Rabie, A. B., Elhag, M., Subyani, A., Rabie, A. B., Elhag, M., & Subyani, A. (2025). Remote Sensing, GIS, and Machine Learning in Water Resources Management for Arid Agricultural Regions: A Review. *Water 2025, Vol. 17, 17*(21). DOI: 10.3390/ W17213125

Ragazou, K., Garefalakis, A., Zafeiriou, E., & Passas, I. (2022). Agriculture 5.0: A new strategic management mode for a cut cost and an energy efficient agriculture sector. *Energies, 15*(9), 3113. DOI: 10.3390/en15093113

Rahu, M., Karim, S., Shams, R., Hoshu, A. A., & Chandio, A. (2022). Wireless Sensor Networks-based Smart Agriculture: Sensing Technologies, Application and Future Directions. *Sukkur IBA Journal of Emerging Technologies, 5*, 18–32. DOI: 10.30537/sjet.v5i2.1104

Raj, M., & Prahadeeswaran, M. (2025). Revolutionizing agriculture: a review of smart farming technologies for a sustainable future. *Discover Applied Sciences, 7*(9), 937-. /DOI: 10.1007/S42452-025-07561-6/TABLES/9

Raj, P., Sharma, S., Kapri, A., & Gulaiya, S. (2025). *Precision Agriculture: A Strategic Approach to Resource Efficiency and Sustainable Farming* (Vol. 31). https:// doi.org/DOI: 10.9734/jsrr/2025/v31i83381

Rajabzadeh, M., & Fatorachian, H. (2023). Modelling factors influencing IoT adoption: With a focus on agricultural logistics operations. *Smart Cities, 6*(6), 3266–3296.

Rajak, P., Ganguly, A., Adhikary, S., & Bhattacharya, S. (2023). Internet of Things and smart sensors in agriculture: Scopes and challenges. *Journal of Agriculture and Food Research, 14*, 100776. DOI: 10.1016/J.JAFR.2023.100776

Raji, E., Ijomah, T. I., & Eyieyien, O. G. (2024). Data-Driven decision making in agriculture and business: The role of advanced analytics. *Computer Science & IT Research Journal, 5*(7), 1565–1575. DOI: 10.51594/CSITRJ.V5I7.1275

Raj, M., & Prahadeeswaran, M. (2025). Revolutionizing agriculture: A review of smart farming technologies for a sustainable future. *Discover Applied Sciences, 7*(9), 937.

Rajoub, B. (2020). Supervised and unsupervised learning. In Zgallai, W. (Ed.), *Biomedical Signal Processing and Artificial Intelligence in Healthcare* (pp. 51–89). Academic Press., DOI: 10.1016/B978-0-12-818946-7.00003-2

Rajpurohit, T., Singh, D., & Kumar, R. (2023). Capacity Building in Agriculture. In (pp. 139–152).

Rajuroy, A. (2025). Autonomous Sprayers and Precision Pesticide Application.

Rajuroy, A., Alkhakani, A., & Navandar, P. (2025). *Advanced Smart Irrigation Systems: Leveraging AI.* IoT, and Nanotechnology for Sustainable Agriculture.

Ramana, M. V., Kumari, C. P., Karthik, R., Alibaba, M., Reddy, G. K., Chiranjeevi, K., Kumar, M. S., Chandra, M. S., Ravisankar, N., Bhatt, R., Gaber, A., & Hossain, A. (2025). Integrated Farming Systems Improve the Income of Small Farm Holdings—An Overview of Earlier Findings in the Indian Context. *Food and Energy Security, 14*(2). Advance online publication. DOI: 10.1002/FES3.70064

Ram, B. G., Oduor, P., Igathinathane, C., Howatt, K., & Sun, X. (2024). A systematic review of hyperspectral imaging in precision agriculture: Analysis of its current state and future prospects. *Computers and Electronics in Agriculture, 222*, 109037.

Ram, B. G., Oduor, P., Igathinathane, C., Howatt, K., & Sun, X. (2024). A systematic review of hyperspectral imaging in precision agriculture: Analysis of its current state and prospects. *Computers and Electronics in Agriculture, 222*, 109037. DOI: 10.1016/J.COMPAG.2024.109037

Rane, J., Kaya, Ö., Mallick, S. K., & Rane, N. L. (2024). *Smart farming using artificial intelligence, machine learning, deep learning, and ChatGPT: Applications, opportunities, challenges, and future directions.* Generative Artificial Intelligence in Agriculture, Education, and Business., DOI: 10.70593/978-81-981271-7-4_6

Rani, S. (2025). Precision Agriculture for Sustainable Development: Concepts and Applications. In Rani, S., Dutta, S., Rocha, Á., & Cengiz, K. (Eds.), *AI and Data Analytics in Precision Agriculture for Sustainable Development* (pp. 1–9). Springer Nature Switzerland., DOI: 10.1007/978-3-031-93087-4_1

Rashid, A. B., & Kausik, M. D. A. K. (2024). AI revolutionizing industries worldwide: A comprehensive overview of its diverse applications. *Hybrid Advances, 7*, 100277. 10.1016/j.hybadv.2024.100277

Rashid, A. B., Kausik, A. K., Khandoker, A., & Siddque, S. N. (2025). Integration of Artificial Intelligence and IoT with UAVs for Precision Agriculture. *Hybrid Advances, 10*, 100458. 10.1016/j.hybadv.2025.100458

Rashid, A. Bin, & Kausik, M. A. K. (2024). AI revolutionizing industries worldwide: A comprehensive overview of its diverse applications. *Hybrid Advances*, *7*, 100277. DOI: 10.1016/J.HYBADV.2024.100277

Rashid, A. Bin, Kausik, A. K., Khandoker, A., & Siddque, S. N. (2025). Integration of Artificial Intelligence and IoT with UAVs for Precision Agriculture. *Hybrid Advances*, *10*, 100458. DOI: 10.1016/J.HYBADV.2025.100458

Rastogi, E., Saxena, N., Roy, A., & Shin, D. R. (2020). Narrowband Internet of Things: A Comprehensive Study. *Computer Networks*, *173*, 107209. DOI: 10.1016/j.comnet.2020.107209

Rathinapriya, P., Maharajan, T., Jothi, R., Prabakaran, M., Lee, I.-B., Yi, P.-H., & Jeong, S. T. (2025). Unlocking biochar impacts on abiotic stress dynamics: A systematic review of soil quality and crop improvement. *Frontiers in Plant Science*, *15*. Advance online publication. DOI: 10.3389/fpls.2024.1479925

Rathor, A. S., Choudhury, S., Sharma, A., Nautiyal, P., & Shah, G. (2024). Empowering vertical farming through IoT and AI-Driven technologies: A comprehensive review. *Heliyon*, *10*(15), e34998. DOI: 10.1016/J.HELIYON.2024.E34998

Razu, Md. A. H., Md. Faruk Shah, & Muhammad Bayezid Husain. (2024a). A study on the impact of agricultural mechanization and practices of traditional knowledge in agriculture in jhenaidah district, Bangladesh. *Humanities Journal*, *2*(1), 2025002. DOI: 10.31893/HUMANITJ.2025002

Rehman, W. u., Koondhar, M. A., Afridi, S. K., Albasha, L., Smaili, I. H., Touti, E., Aoudia, M., Zahrouni, W., Mahariq, I., & Ahmed, M. M. R. (2025). The role of 5G network in revolutionizing agriculture for sustainable development: A comprehensive review. *Energy Nexus*, *17*, 100368. DOI: 10.1016/j.nexus.2025.100368

Reinking, J., Arnold, V., & Sutton, S. G. (2020). Synthesizing enterprise data through digital dashboards to strategically align performance: Why do operational managers use dashboards? *International Journal of Accounting Information Systems*, *37*, 100452. 10.1016/j.accinf.2020.100452

Rejeb, A., Rejeb, K., Zailani, S., Keogh, J. G., & Appolloni, A. (2022). Examining the interplay between artificial intelligence and the agri-food industry. *Artificial Intelligence in Agriculture*, *6*, 111–128.

Reppa, V., Polycarpou, M., & Panayiotou, C. (2016). Sensor Fault Diagnosis. *Foundations and Trends in Systems and Control, 3*, 1–248. DOI: 10.1561/2600000007

Reza, M. N., Lee, K. H., Karim, M. R., Haque, M. A., Bicamumakuba, E., Dey, P. K., Jang, Y. Y., & Chung, S. O. (2025). Trends of Soil and Solution Nutrient Sensing for Open Field and Hydroponic Cultivation in Facilitated Smart Agriculture. *Sensors 2025, Vol. 25, 25*(2). DOI: 10.3390/S25020453

Roccatello, E., Pagano, A., Levorato, N., & Rumor, M. (2025). State of the Art in Internet of Things Standards and Protocols for Precision Agriculture with an Approach to Semantic Interoperability. *Network 2025, Vol. 5, 5*(2). DOI: 10.3390/NETWORK5020014

Roccatello, E., Pagano, A., Levorato, N., & Rumor, M. (2025). State of the Art in Internet of Things Standards and Protocols for Precision Agriculture with an Approach to Semantic Interoperability. *Network (Bristol, England), 5*(2), 14.

Rogulski, M., & Badyda, A. (2020). Investigation of Low-Cost and Optical Particulate Matter Sensors for Ambient Monitoring. *Atmosphere, 11*, 1040. DOI: 10.3390/atmos11101040

Rojek, I., Prokopowicz, P., Piechowiak, M., Kotlarz, P., Náprstková, N., & Mikołajewski, D. (2025). The impact of data analytics based on Internet of Things, edge computing, and artificial intelligence on energy efficiency in smart environments. *Applied Sciences (Basel, Switzerland), 16*(1), 225.

Ronzhin, A., Figurek, A., Surovtsev, V., & Dibirova, K. (2025). Digital transformation and precision farming as catalysts of rural development. *Land (Basel), 14*(7), 1464.

Rostami, K., & Salehi, L. (2024). Rural cooperatives social responsibility in promoting Sustainability-oriented Activities in the agricultural sector: Nexus of community, enterprise, and government. *Sustainable Futures : An Applied Journal of Technology, Environment and Society, 7*, 100150. DOI: 10.1016/j.sftr.2023.100150

Roussaki, I., Doolin, K., Skarmeta, A., Routis, G., Lopez-Morales, J. A., Claffey, E., Mora, M., & Martinez, J. A. (2023). Building an interoperable space for smart agriculture. *Digital Communications and Networks, 9*(1), 183–193. DOI: 10.1016/j.dcan.2022.02.004

Roux, J., Escriba, C., Fourniols, J.-Y., & Soto-Romero, G. (2019). A new bi-frequency soil smart sensing moisture and salinity for connected sustainable agriculture. *Journal of Sensor Technology, 9*(3), 35–43.

Rovira-Sugranes, A., Razi, A., Afghah, F., & Chakareski, J. (2022). A review of AI-enabled routing protocols for UAV networks: Trends, challenges, and future outlook. *Ad Hoc Networks*, *130*, 102790. 10.1016/j.adhoc.2022.102790

Roy, P., Padhiary, M., Hoque, A., Sahu, B., Roy, D., & Kumar, K. (2025). Machine Learning for Precision Agriculture and Crop Yield Optimization: Techniques and Applications. In (pp. 189–233). DOI: 10.4018/979-8-3693-9208-9.ch007

Ruder, S.-L., & Wittman, H. (2025). Agricultural data governance from the ground up: Exploring data justice with agri-food movements. *Big Data & Society*, *12*(1), 20539517251330182. DOI: 10.1177/20539517251330182

S, E., Sandiya, H., Dharshini, D., Nanthini, D., Karthickraja, A., A, R., S, B., & Pasoubady, S. (2025). Revolutionizing Organic Farming With Artificial Intelligence: A Sustainable Future. In (pp. 113-121).

Sachithra, V., & Subhashini, L. D. C. S. (2023). How artificial intelligence is used to achieve the agriculture sustainability: Systematic review. *Artificial Intelligence in Agriculture*, *8*, 46–59. DOI: 10.1016/J.AIIA.2023.04.002

Saetra, H. S. (2021). AI in context and the sustainable development goals: Factoring in the Unsustainability of the Sociotechnical system. *Sustainability*, *13*(4), 1738.

Saggi, M. K., & Jain, S. (2022). A Survey Towards Decision Support System on Smart Irrigation Scheduling Using Machine Learning approaches. *Archives of Computational Methods in Engineering*, *29*(6), 4455–4478. DOI: 10.1007/s11831-022-09746-3 PMID: 35573028

Saha, S., Ghosh, B., Barman, B., & Satpathy, B. (2025). Predictions for the Future of Digital Agriculture and Potential Developments. In (pp. 423-451).

Saha, S., Kucher, O. D., Utkina, A. O., & Rebouh, N. Y. (2025). Precision agriculture for improving crop yield predictions: A literature review. *Frontiers in Agronomy*, *7*, 1566201.

Sahu, A., Acharya, B., & Sahoo, P. S. (2025). Agricultural farming decision support system using artificial intelligence: A comparative analysis. In *Optimizing Smart and Sustainable Agriculture for Sustainability*. CRC Press.

Sahu, B. (2024). Artificial intelligence and automation in smart agriculture: A comprehensive review of precision farming, all-terrain vehicles, IoT innovations, and environmental impact mitigation. *Int. J. Sci. Res. IJSR*, *13*(11), 656–665.

Said, Z., Vigneshwaran, P., Shaik, S., Rauf, A., & Ahmad, Z. (2025). Climate and carbon policy pathways for sustainable food systems. *Environmental and Sustainability Indicators, 27*, 100730. DOI: 10.1016/J.INDIC.2025.100730

Saiz-Rubio, V., & Rovira-Más, F. (2020). From Smart Farming towards Agriculture 5.0: A Review on Crop Data Management. *Agronomy 2020, Vol. 10, 10*(2). DOI: 10.3390/AGRONOMY10020207

Sajid, J., Hayawi, K., Malik, A. W., Anwar, Z., & Trabelsi, Z. (2023). A fog computing framework for intrusion detection of energy-based attacks on UAV-assisted smart farming. *Applied Sciences (Basel, Switzerland), 13*(6), 3857. DOI: 10.3390/app13063857

Saki, S., & Soori, M. (2026). Artificial intelligence, machine learning and deep learning in advanced transportation systems, a review. *Multimodal Transportation, 5*(1), 100242. DOI: 10.1016/J.MULTRA.2025.100242

Saleh, D., Milazzo, J., Adreit, H., Fournier, E., & Tharreau, D. (2014). South-East Asia is the center of origin, diversity and dispersion of the rice blast fungus, Magnaporthe oryzae. *The New Phytologist, 201*(4), 1440–1456. DOI: 10.1111/nph.12627

Salka, T. D., Hanafi, M. B., Rahman, S. M. S. A. A., Zulperi, D. B. M., & Omar, Z. (2025). Plant leaf disease detection and classification using convolution neural networks model: a review. *Artificial Intelligence Review 2025 58:10, 58*(10), 322-. /DOI: 10.1007/S10462-025-11234-6

Salomi, M., & Samsudeen, S. (2024). Agricultural Artificial Intelligence: Obstacles and Opportunities. In Chouhan, S. S., Saxena, A., Singh, U. P., & Jain, S. (Eds.), *Artificial Intelligence Techniques in Smart Agriculture* (pp. 55–66). Springer Nature Singapore., DOI: 10.1007/978-981-97-5878-4_4

Samreen, T., Ahmad, M., Baig, M. T., Kanwal, S., Nazir, M. Z., & Sidra-Tul-Muntaha. (2022). Remote Sensing in Precision Agriculture for Irrigation Management. *Environmental Sciences Proceedings, 23*(1), 31. Retrieved from https://www.mdpi.com/2673-4931/23/1/31

Samuel, M. K., Manickam, T., Omar, M. S., Zamir, M., Rashid, A., Sahil, N. M., ... & Farouk, N. (2025). From pedons to policy: The strategic role of soil mapping in Peninsular Malaysia's agricultural transformation.

Sanchez-Cubillo, J., Del Ser, J., & Martin, J. L. (2024). Toward Fully Automated Inspection of Critical Assets Supported by Autonomous Mobile Robots, Vision Sensors, and Artificial Intelligence. *Sensors (Basel), 24*(12). Advance online publication. DOI: 10.3390/s24123721 PMID: 38931502

Sanjay, K. (2024). Sensor Based Technology Development: Advancements In Precision Agriculture. *International Journal of Advances in Engineering & Scientific Research, 11*(1), 50–64. DOI: 10.53882/IJAESR.2023.1001004

Sankaran, S., Maja, J., Buchanon, S., & Ehsani, R. (2013). Huanglongbing (Citrus Greening) detection using visible, near infrared and thermal imaging techniques. *Sensors (Basel), 13*(2), 2117–2130. DOI: 10.3390/s130202117

Sankaran, S., Mishra, A., Ehsani, R., & Davis, C. (2010). A review of advanced techniques for detecting plant diseases. *Computers and Electronics in Agriculture, 72*(1), 1–13. DOI: 10.1016/j.compag.2010.02.007

Santaera, G., Zeni, V., Manduca, G., Canale, A., Mele, M., Benelli, G., Stefanini, C., & Romano, D. (2025). Development of an autonomous smart trap for precision monitoring of hematophagous flies on cattle. *Smart Agricultural Technology, 10*, 100842. DOI: 10.1016/J.ATECH.2025.100842

Sanyaolu, M., & Sadowski, A. (2024). The Role of Precision Agriculture Technologies in Enhancing Sustainable Agriculture. *Sustainability 2024, Vol. 16, 16*(15). DOI: 10.3390/SU16156668

Sanyaolu, M., & Sadowski, A. (2024). The Role of Precision Agriculture Technologies in Enhancing Sustainable Agriculture. *Sustainability, 16*(15), 6668.

Sathya, D., Thangamani, R., & Balaji, B. S. (2024). The Revolution of Edge Computing in Smart Farming. In Balasubramanian, S., Natarajan, G., & Chelliah, P. R. (Eds.), *Intelligent Robots and Drones for Precision Agriculture* (pp. 351–389). Springer Nature Switzerland., DOI: 10.1007/978-3-031-51195-0_17

Satoh, Y., & Kakiuchi, H. (2021). Calibration method to address influences of temperature and electrical conductivity for a low-cost soil water content sensor in the agricultural field. *Agricultural Water Management, 255*. Advance online publication. DOI: 10.1016/J.AGWAT.2021.107015

Savary, S., Ficke, A., Aubertot, J.-N., & Hollier, C. (2012). Crop losses due to diseases and their implications for global food production losses and food security. *Food Security, 4*(4), 519–537. DOI: 10.1007/s12571-012-0200-5

Sayekti, N. C., Fhaizal, M. A., Wijaya, W., Hendajany, N., Aprianti, I., &Kusmadi. (2024). Smart farming design with NodeMCU ESP32 for NFT hydroponic. *2024 18th International Conference on Telecommunication Systems, Services, and Applications (TSSA)*, 1-4.

Schweidtmann, A. M., Zhang, D., & von Stosch, M. (2024). A review and perspective on hybrid modeling methodologies. *Digital Chemical Engineering*, *10*, 100136. DOI: 10.1016/j.dche.2023.100136

Schweng, S., Bernardini, L., Keiblinger, K., & Kaul, H.-P.Jr, I. F., Lukač, N., Del Ser, J., & Holzinger, A. (2026a). What can artificial intelligence do for soil health in agriculture? *Computer Science Review*, *59*, 100832. DOI: 10.1016/J.COSREV.2025.100832

Selod, H., & Shilpi, F. (2021). Rural-urban migration in developing countries: Lessons from the literature. *Regional Science and Urban Economics*, *91*, 103713. 10.1016/j.regsciurbeco.2021.103713

Sen, A., Poddar, R., Sarkar, A., Saha, S., & Asif, S. M. (2025). Latest advancements in microbial inoculants for enhanced agricultural productivity. In Kumar, A., Singh, J., Queijeiro López, A. M., & Kharwar, R. N. (Eds.), *Microbial Inoculants* (pp. 181–201). Academic Press., DOI: 10.1016/B978-0-443-23560-3.00012-7

Sengupta, S., Ghosh, A., Mallick, T. K., Chanda, C. K., Saha, H., Bose, I., Jana, J., & Sengupta, S. (2021). Model based generation prediction of SPV power plant due to weather stressed soiling. *Energies*, *14*(17), 5305.

Senoo, E. E. K., Anggraini, L., Kumi, J. A., Karolina, L. B., Akansah, E., Sulyman, H. A., Mendonça, I., & Aritsugi, M. (2024). IoT Solutions with Artificial Intelligence Technologies for Precision Agriculture: Definitions, Applications, Challenges, and Opportunities. *Electronics 2024, Vol. 13, 13*(10). https://doi.org/DOI: 10.3390/ELECTRONICS13101894

Seppelt, R., Klotz, S., Peiter, E., & Volk, M. (2022). Agriculture and food security under a changing climate: An underestimated challenge. *iScience*, *25*(12), 105551. 10.1016/j.isci.2022.105551 PMID: 36458255

Setiawan, A., & Budiman, A. A. (2025). Optimization of Gray Level Co-occurrence Matrix (GLCM) Texture Feature Parameters in Determining Rice Seed Quality. *EMITTER International Journal of Engineering Technology*, *13*(1), 110–123. DOI: 10.24003/emitter.v13i1.928

Shafik, W. (2024). Barriers to implementing computational intelligence-based agriculture system. In *Computational Intelligence in Internet of Agricultural Things* (pp. 193–219). Springer Nature Switzerland.

Shah, M. H., Aktar, S. N., Pramanik, K., Pande, C. B., & Ali, G. T. (2024). The impact of climate change on plant diseases and food security. *Biocontrol Agents for Improved Agriculture*, 353–384. https://doi.org/DOI: 10.1016/B978-0-443-15199-6.00002-6

Shahab, H., Naeem, M., Iqbal, M., Aqeel, M., & Ullah, S. S. (2025). IoT-driven smart agricultural technology for real-time soil and crop optimization. *Smart Agricultural Technology, 10*, 100847. DOI: 10.1016/J.ATECH.2025.100847

Shah, R., Doss, A. S. A., & Lakshmaiya, N. (2025). Advancements in AI-enhanced collaborative robotics: Towards safer, smarter, and human-centric industrial automation. *Results in Engineering, 27*, 105704. DOI: 10.1016/J.RINENG.2025.105704

Shahria, M. T., Sunny, M. S. H., Zarif, M. I. I., Ghommam, J., Ahamed, S. I., & Rahman, M. H. (2022). A Comprehensive Review of Vision-Based Robotic Applications: Current State, Components, Approaches, Barriers, and Potential Solutions. *Robotics (Basel, Switzerland), 11*(6), 139.

Shahriar, S., Corradini, M. G., Sharif, S., Moussa, M., & Dara, R. (2025). The role of generative artificial intelligence in digital agri-food. *Journal of Agriculture and Food Research, 20*, 101787.

Shamshiri, R. R., Sturm, B., Weltzien, C., Fulton, J., Khosla, R., Schirrmann, M., Raut, S., Basavegowda, D. H., Yamin, M., & Hameed, I. A. (2024). Digitalization of agriculture for sustainable crop production: a use-case review [Review]. *Frontiers in Environmental Science, Volume 12 - 2024*. https://doi.org/DOI: 10.3389/fenvs.2024.1375193

Shamsoddini, A., Jelodar, S. M., & Rasouli, S. H. (2025). *Modeling opportunities and challenges for entrepreneurship development in the marketplaces of Sari County*. Village and Space Sustainable Development.

Sharafat, S., Das Kabya, N., Emu, R. I., Ahmed, M. U., Onik, J. C., Islam, M. A., & Khan, R. (2025). An IoT-enabled AI system for real-time crop prediction using soil and weather data in precision agriculture. *Smart Agricultural Technology, 12*, 101263. DOI: 10.1016/J.ATECH.2025.101263

Sharafi, S., Mottaghi, F., & Omidvari, F. (2025). Tracking agricultural transformation: Fertilizer efficiency and crop production trends worldwide (1961–2022). *Cleaner and Circular Bioeconomy, 12*, 100191. DOI: 10.1016/J.CLCB.2025.100191

Sharma, A., Hazarika, M., Heisnam, P., Pandey, H., Devadas, V. S., & Wangsu, M. (2024). Controlled Environment Ecosystem: A plant growth system to combat climate change through soilless culture. *Crop Design, 3*(1), 100044. 10.1016/j.cropd.2023.100044

Sharma, A., Rani, S., & Sharma, A. (2025). Sustainable intelligence: Ethical issues in the evolution of intelligent systems. *Mathematical Computing and Sustainability: Predictive Modeling and Ramifications of Intelligent Systems, 57*.

Sharma, H., Sidhu, H., Bhowmik, A., Sharma, H., Sidhu, H., & Bhowmik, A. (2025). Remote Sensing Using Unmanned Aerial Vehicles for Water Stress Detection: A Review Focusing on Specialty Crops. *Drones 2025, Vol. 9, 9*(4). DOI: 10.3390/DRONES9040241

Sharma, L., & Garg, P. K. (2025). *Digital twins and simulation technology: Concepts and applications.*

Sharma, P., Kumar, P., Singh, P. K., & Rana, A. K. (2024). "Precision agriculture farming" enhancing farming efficiency through technology integration. *2024 1st International Conference on Advances in Computing, Communication and Networking (ICAC2N)*, 365-370.

Sharma, S. K. (2025). Smart agriculture revolution: Cloud and IoT-based solutions for sustainable crop management and precision farming. In *AI-based advanced optimization techniques for edge computing* (pp. 253-288).

Sharma, A., Jain, A., Gupta, P., & Chowdary, V. (2021). Machine learning applications for precision agriculture: A comprehensive review. *IEEE Access : Practical Innovations, Open Solutions, 9*, 4843–4873.

Sharma, D. K., Chatterjee, M., Kaur, G., & Vavilala, S. (2022). 3—Deep learning applications for disease diagnosis. In Gupta, D., Kose, U., Khanna, A., & Balas, V. E. (Eds.), *Deep Learning for Medical Applications with Unique Data* (pp. 31–51). Academic Press., DOI: 10.1016/B978-0-12-824145-5.00005-8

Sharma, K., & Shivandu, S. K. (2024). Integrating artificial intelligence and Internet of Things (IoT) for enhanced crop monitoring and management in precision agriculture. *Sensors International, 5*, 100292. DOI: 10.1016/J.SINTL.2024.100292

Sharma, M., Tomar, A., & Hazra, A. (2024). Edge computing for industry 5.0: Fundamentals, applications, and research challenges. *IEEE Internet of Things Journal, 11*(11), 19070–19093.

Sharma, N. K., Anand, A., Budhlakoti, N., Mishra, D. C., & Jha, G. K. (2024). Artificial Intelligence and Machine Learning for Rice Improvement. In Singh, A., Singh, S. K., & Shrestha, J. (Eds.), *Climate-Smart Rice Breeding* (pp. 273–300). Springer Nature., DOI: 10.1007/978-981-97-7098-4_11

Sharma, R. D., & Gawade, S. (2025). Advancements in precision agriculture: A literature review of machine learning applications for crop monitoring and yield prediction. *Environmental Claims Journal*, •••, 1–45.

Sharma, R., Kumar, N., & Sharma, V. (2025). Computer Vision for Disease Detection—An Overview of How Computer Vision Techniques Can Be Used to Detect Diseases in Medical Images, Such as X-Rays and MRIs. In *AI in Disease Detection* (pp. 77–98). John Wiley & Sons, Ltd., DOI: 10.1002/9781394278695.ch4

Shawon, S. M., Barua Ema, F., Mahi, A. K., & Mohsin Sarker Raihan, M. (2023). Crop Yield Prediction: Robust Machine Learning Approaches for Precision Agriculture. *2023 26th International Conference on Computer and Information Technology, ICCIT 2023*. /DOI: 10.1109/ICCIT60459.2023.10441634

Shawon, S. M., Ema, F. B., Mahi, A. K., Niha, F. L., & Zubair, H. T. (2025). Crop yield prediction using machine learning: An extensive and systematic literature review. *Smart Agricultural Technology, 10*. Advance online publication. DOI: 10.1016/J.ATECH.2024.100718

Shiba, M., & Miwa, H. (2022). Path Control Algorithm for Weeding AI Robot. In *International Conference on Emerging Internetworking, Data & Web Technologies* (pp. 375-385). Springer International Publishing.

Shinde, G. U., Agrawal, R. C., Mani, I., Agrawal, A., Khodke, U. M., Muley, S. V., Tekale, D. D., Bhalerao, S. R., & Kakade, O. D. (2024, 2024//). Digital Farming Solution by Automation Using Agri-Bot, Agri-Drone, and Agri-AGV for Organic Farming Practices. Agriculture Digitalization and Organic Production, Singapore.

Shin, J., Chang, Y. K., Heung, B., Nguyen-Quang, T., Price, G. W., & Al-Mallahi, A. (2021). A deep learning approach for RGB image-based powdery mildew disease detection on strawberry leaves. *Computers and Electronics in Agriculture*, *183*, 106042.

Shi, P., Li, J., & Zhang, Y. (2023). A fast LiDAR place recognition and localization method by fusing local and global search. *ISPRS Journal of Photogrammetry and Remote Sensing*, *202*, 637–651. 10.1016/j.isprsjprs.2023.07.008

Shneiderman, B. (2020). Human-centered artificial intelligence: Reliable, safe & Trustworthy. *International Journal of Human-Computer Interaction*, *36*(6), 495–504.

Shoaib, M., Sadeghi-Niaraki, A., Ali, F., Hussain, I., & Khalid, S. (2025). Leveraging deep learning for plant disease and pest detection: A comprehensive review and future directions. *Frontiers in Plant Science*, *16*, 1538163. DOI: 10.3389/FPLS.2025.1538163/BIBTEX PMID: 40061031

Shu, M., Harfouche, A. L., Trtílek, M., Panzarová, K., Alasia, O. F., Lagergren, J. H., Labbé, A., Engle, N. L., Clark, M. M., Chen, J. G., Tuskan, G. A., & Tschaplinski, T. J. (2025). Leveraging hyperspectral phenotyping for accurate, non-destructive prediction of metabolite profiles in poplar under drought stress. *Environmental and Experimental Botany, 237*, 106218. DOI: 10.1016/J.ENVEXPBOT.2025.106218

Siegfried, J., Rajan, N., Adams, C. B., Neely, H., Hague, S., Hardin, R., & Thomasson, A. (2024). High-accuracy infrared thermography of cotton canopy temperature by unmanned aerial systems (UAS): Evaluating in-season prediction of yield. *Smart Agricultural Technology, 7*, 100393. DOI: 10.1016/j.atech.2023.100393

Simbeye, D. S., Mkiramweni, M. E., Karaman, B., & Taskin, S. (2023). Plant water stress monitoring and control system. *Smart Agricultural Technology, 3*, 100066. 10.1016/j.atech.2022.100066

Singh, A., Singh, K. M., Kumari, T., Kumari, S., & Biswas, P. (2025). Artificial Intelligence in Agriculture: Case Studies and Success Stories. *Transforming Agriculture through Artificial Intelligence for Sustainable Food Systems*, 191–199. / DOI: 10.1007/978-981-96-4795-8_11

Singh, A., & Haghverdi, A. (2023). Development and evaluation of temperature-based deep learning models to estimate reference evapotranspiration. *Artificial Intelligence in Agriculture, 9*, 61–75. DOI: 10.1016/j.aiia.2023.08.003

Singh, A., Rajput, V. D., Varshney, A., Ghazaryan, K., & Minkina, T. (2023). Small Tech, Big Impact: Agri-nanotechnology Journey to Optimize Crop Protection and Production for Sustainable Agriculture. *Plant Stress, 10*, 100253.

Singh, D., Sobti, R., Jain, A., Malik, P., & Le, D.-N. (2022). LoRa based intelligent soil and weather condition monitoring with internet of things for precision agriculture in smart cities. *IET Communications, 16*. Advance online publication. DOI: 10.1049/cmu2.12352

Singh, R., Krishnan, P., Singh, V. K., & Das, B. (2023). Estimation of yellow rust severity in wheat using visible and thermal imaging coupled with machine learning models. *Geocarto International, 38*(1), 2160831. DOI: 10.1080/10106049.2022.2160831

Singh, S., Khatana, K., Singh, Y., & Mishra, A. K. (2025). Enhancing Resilience and Sustainability in Farming Practices. In Mishra, A. K., Sharma, S., Mishra, A., & Roy, A. (Eds.), *Transition to Regenerative Agriculture: Principles and Indicators of Soil Health Management* (pp. 187–203). Springer Nature Singapore., DOI: 10.1007/978-981-96-1421-9_9

Sinha, A. K., Vaid, A., Gupta, V., Sharma, N., Gupta, S., Sharma, V. K., & Shalini, K. (2025). Productivity and economics of maize cultivation after wheat crop by different sowing methods under rain-fed conditions in Samba district of J&K. *Journal of Community Mobilization and Sustainable Development*, *20*(2), 642–646.

Sirish Kumar, P., & Srilatha Indira Dutt, V. B. S. (2020). The global positioning system: Popular accuracy measures. *Materials Today: Proceedings*, *33*, 4797–4801. 10.1016/j.matpr.2020.08.380

Sivagnanapalani, P., Ansari, N. I., & Panda, P. (2021). Nd2Ti2O7 (NTO) with high Curie temperature (TC) for high temperature sensor applications. *Sensors International*, *2*, 100093.

Skendžić, S., Zovko, M., Živković, I. P., Lešić, V., & Lemić, D. (2021). The Impact of Climate Change on Agricultural Insect Pests. *Insects*, *12*(5). Advance online publication. DOI: 10.3390/insects12050440 PMID: 34066138

Sobolev, A. V., Pakhomov, V. M., Brilon, A. V., Konnova, O. A., & Minchenckova, I. V. (2021). Cooperative Platform in the Modern Economy. In Bogoviz, A. V., Suglobov, A. E., Maloletko, A. N., Kaurova, O. V., & Lobova, S. V. (Eds.), *Frontier Information Technology and Systems Research in Cooperative Economics* (pp. 25–31). Springer International Publishing., DOI: 10.1007/978-3-030-57831-2_3

Soori, M., Jough, F. K. G., Dastres, R., & Arezoo, B. (2026). AI-based decision support systems in Industry 4.0, a review. *Journal of Economy and Technology*, *4*, 206–225. 10.1016/j.ject.2024.08.005

Soto-Silva, W. E., González-Araya, M. C., Oliva-Fernández, M. A., & Plà-Aragonés, L. M. (2017). Optimizing fresh food logistics for processing: Application for a large Chilean Apple supply chain. *Computers and Electronics in Agriculture*, *136*, 42–57.

Sousa, M., Alves, A., Antunes, R., Aguiar, M. L., Gaspar, P. D., & Pereira, N. (2026). Interoperable IoT/WSN sensing station with edge AI-enabled multi-sensor integration for precision agriculture. *Agriculture*.

Soussi, A., Zero, E., Sacile, R., Trinchero, D., & Fossa, M. (2024). Smart Sensors and Smart Data for Precision Agriculture: A Review. *Sensors (Basel)*, *24*(8), 2647. PMID: 38676264

Sreejith, S., Leo Joseph, L. M. I., Kollem, S., Vijumon, V. T., & Ajayan, J. (2023). Biodegradable sensors: A comprehensive review. *Measurement*, *219*, 113261. DOI: 10.1016/j.measurement.2023.113261

Srivastav, A. K., & Das, P. (2025). Challenges and Barriers to Adoption. In *Biotechnology and IoT in Agriculture and Food Production: Green Innovation* (pp. 161-168). Springer.

Srivastav, A. K., & Das, P. (2025). Agricultural Robotics and Autonomous Systems. In Srivastav, D. A. K., & Das, D. P. (Eds.), *Biotechnology and IoT in Agriculture and Food Production: Green Innovation* (pp. 201–225). Apress., DOI: 10.1007/979-8-8688-1469-3_22

Stasiak-Betlejewska, R., & Potkány, M. (2015). Construction Costs Analysis and its Importance to the Economy. *Procedia Economics and Finance*, *34*, 35–42. 10.1016/S2212-5671(15)01598-1

Stolarz, M. (2025). Integration of Plant Electrophysiology and Time-Lapse Video Analysis via Artificial Intelligence for the Advancement of Precision Agriculture. *Sustainability*, *17*(12), 5614.

Strange, R. N., & Scott, P. R. (2005). Plant Disease: A Threat to Global Food Security. *Annual Review of Phytopathology*, *43*(Volume 43, 2005), 83–116. https://doi.org/DOI: 10.1146/annurev.phyto.43.113004.133839

Subeesh, A., & Chauhan, N. (2026). Agricultural digital twin for smart farming: A review. *Green Technologies and Sustainability*, *4*(2), 100299. 10.1016/j.grets.2025.100299

Subeesh, A., & Chauhan, N. (2025). Agricultural digital twin for smart farming: A review. *Green Technologies and Sustainability*, *100299*. Advance online publication. DOI: 10.1016/J.GRETS.2025.100299

Subeesh, A., & Mehta, C. R. (2021). Automation and digitization of agriculture using artificial intelligence and internet of things. *Artificial Intelligence in Agriculture*, *5*, 278–291. DOI: 10.1016/J.AIIA.2021.11.004

Sullivan, C. S., Gemtou, M., Anastasiou, E., & Fountas, S. (2024). Building trust: A systematic review of the drivers and barriers of agricultural data sharing. *Smart Agricultural Technology*, *8*, 100477.

Sun, B., Yu, J., Khattak, S. I., Tariq, S., & Zahid, M. (2025). Digital innovation, business model transformations, and agricultural SMEs: A PRISMA-based review of challenges and prospects. *Systems*, *13*(8), 673.

Sungheetha, A. (2022). State of art survey on plant leaf disease detection. *Journal of Innovative Image Processing*, *4*(2), 93–102.

Surendran, U., Nagakumar, K. C. V., & Samuel, M. P. (2024). Remote sensing in precision agriculture. In *Digital agriculture: A solution for sustainable food and nutritional security* (pp. 201–223). Springer International Publishing.

Surur, F. M., Mamo, A. A., Gebresilassie, B. G., Mekonen, K. A., Golda, A., Behera, R. K., & Kumar, K. (2025). (in press). Unlocking the power of machine learning in big data: A scoping survey. *Data Sci. Manag.*

Tadala, N., & Fanny, C. (2024). IoT-Enabled crop storage monitoring system. *I-Manager's. Journal of Computational Science, 11*(4), 43. DOI: 10.26634/JCOM.11.4.20654

Taha, M. F., Mao, H., Zhang, Z., Elmasry, G., Awad, M. A., Abdalla, A., Mousa, S., Elwakeel, A. E., & Elsherbiny, O. (2025). Emerging Technologies for Precision Crop Management Towards Agriculture 5.0: A Comprehensive Overview. *Agriculture, 15*(6), 582.

Taherdoost, H. (2023). Deep Learning and Neural Networks: Decision-Making Implications. *Symmetry, 15*(9), 1723. https://www.mdpi.com/2073-8994/15/9/1723

Talaeizadeh, A., Sharifi, I., Alasty, A., & Ghatrehsamani, S. (2025). Agricultural spraying drones: A comprehensive review. *Smart Agricultural Technology, 12*, 101519. 10.1016/j.atech.2025.101519

Talaviya, T., Shah, D., Patel, N., Yagnik, H., & Shah, M. (2020). Implementation of artificial intelligence in agriculture for optimisation of irrigation and application of pesticides and herbicides. *Artificial Intelligence in Agriculture, 4*, 58–73. DOI: 10.1016/J.AIIA.2020.04.002

Taleb, M. A., Korsoveczki, G., & Husi, G. (2025). Automotive navigation for mobile robots: Comprehensive review. *Results in Engineering, 27*, 105837. 10.1016/j.rineng.2025.105837

Tanaka, T. S., Heuvelink, G. B., Mieno, T., & Bullock, D. S. (2024). Can machine learning models provide accurate fertilizer recommendations? *Precision Agriculture, 25*(4), 1839–1856. DOI: 10.1007/s11119-024-10136-x

Tang, Y., Chen, M., Wang, C., Luo, L., Li, J., Lian, G., & Zou, X. (2020). Recognition and Localization Methods for Vision-Based Fruit Picking Robots: A Review. *Front Plant Sci, 11*, 510. DOI: 10.3389/fpls.2020.00510 PMID: 32508853

Tang, Y., Zhao, W., Zhu, G., Tan, Z., Huang, L., Zhang, P., Gao, L., & Rui, Y. (2024). Nano-Pesticides and Fertilizers: Solutions for Global Food Security. *Nanomaterials (Basel, Switzerland), 14*(1), 90.

Tarafdar, R. (2025). *Human-AI Collaboration in Workflow Optimization: A Framework for Hybrid Decision Systems in Automation-Heavy Industries* (Vol. 11). https://doi.org/DOI: 10.32628/CSEIT251112384

Taye, M. M. (2023). Understanding of Machine Learning with Deep Learning: Architectures, Workflow, Applications and Future Directions. *Computers 2023, Vol. 12, 12*(5). DOI: 10.3390/COMPUTERS12050091

Tayler, M. (2012). New Technologies in Arable Farming. https://ruralleaders.co.nz/wp-content/uploads/2025/09/2012-Michael-Tayler.pdf

Theodorakopoulos, L., Theodoropoulou, A., & Stamatiou, Y. (2024). A State-of-the-Art Review in Big Data Management Engineering: Real-Life Case Studies, Challenges, and Future Research Directions. *Eng, 5*(3), 1266-1297. https://www.mdpi.com/2673-4117/5/3/68

Theodorou, C., Velisavljevic, V., Dyo, V., & Nonyelu, F. (2022). Visual SLAM algorithms and their application for AR, mapping, localization and wayfinding. *Array (New York, N.Y.), 15*, 100222. 10.1016/j.array.2022.100222

Thi Thoa, M., Nguyen, T.-D., Dang, H.-K., Nguyen, D.-T., & Nguyen, X.-T. (2024). Intelligent agricultural robotic detection system for greenhouse tomato leaf diseases using soft computing techniques and deep learning. *Scientific Reports*, *14*, 23887. DOI: 10.1038/s41598-024-75285-5

Thilakarathne, N. N., Abu Bakar, M. S., Abas, P. E., & Yassin, H. (2025). Internet of things enabled smart agriculture: Current status, latest advancements, challenges and countermeasures. *Heliyon*, *11*(3), e42136. DOI: 10.1016/J.HELIYON.2025.E42136 PMID: 39959477

Thilakarathne, N., Yassin, H., Abu Bakar, M., & Abas, P. E. (2021). *Internet of Things in Smart Agriculture: Challenges.* Opportunities and Future Directions.

Titirmare, S., Margal, P., Gupta, S., & Kumar, D. (2024). AI-Powered Predictive Analytics for Crop Yield Optimization. In (pp. 89-110). https://doi.org/DOI: 10.1201/9781003570219-5

Tituaña, L., Gholami, A., He, Z., Xu, Y., Karkee, M., & Ehsani, R. (2024). A small autonomous field robot for strawberry harvesting. *Smart Agricultural Technology*, *8*, 100454. 10.1016/j.atech.2024.100454

Todmal, A. (2023). Precision Agriculture: Technology-Driven Approaches for Efficient Farming. In (pp. 46–53). DOI: 10.25215/9358791152.07

Tomashuk, I., Koliadenko, S., & Burdìak, M. (2025). The impact of digital innovations on the development of agricultural business. *Baltic Journal of Economic Studies, 11*(1), 361–375.

Tornese, I., Matera, A., Rashvand, M., & Genovese, F. (2024). Use of Probes and Sensors in Agriculture—Current Trends and Future Prospects on Intelligent Monitoring of Soil Moisture and Nutrients. *AgriEngineering, 6*(4), 4154-4181. https://www.mdpi.com/2624-7402/6/4/234

Touch, V., Tan, D. K. Y., Cook, B. R., Liu, D. L., Cross, R., Tran, T. A., Utomo, A., Yous, S., Grunbuhel, C., & Cowie, A. (2024). Smallholder farmers' challenges and opportunities: Implications for agricultural production, environment and food security. *Journal of Environmental Management, 370*, 122536. 10.1016/j.jenvman.2024.122536 PMID: 39299125

Trapani, N., & Longo, L. (2023). Fault Detection and Diagnosis Methods for Sensors Systems: A Scientific Literature Review. *IFAC-PapersOnLine, 56*(2), 1253–1263. DOI: 10.1016/j.ifacol.2023.10.1749

Tsouros, D., Bibi, S., & Sarigiannidis, P. (2019). A Review on UAV-Based Applications for Precision Agriculture. *Information (Basel), 10*, 349. DOI: 10.3390/info10110349

Tudi, M., Daniel Ruan, H., Wang, L., Lyu, J., Sadler, R., Connell, D., Chu, C., & Phung, D. T. (2021). Agriculture Development, Pesticide Application and Its Impact on the Environment. *International Journal of Environmental Research and Public Health, 18*(3). Advance online publication. DOI: 10.3390/ijerph18031112 PMID: 33513796

Tufail, R., Tassinari, P., & Torreggiani, D. (2025). Deep Learning Applications for Crop Mapping Using Multi-Temporal Sentinel-2 Data and Red-Edge Vegetation Indices: Integrating Convolutional and Recurrent Neural Networks. *Remote Sensing 2025, Vol. 17, 17*(18). /DOI: 10.3390/RS17183207

Tullo, E., Finzi, A., & Guarino, M. (2019). Review: Environmental impact of livestock farming and Precision Livestock Farming as a mitigation strategy. *The Science of the Total Environment, 650*, 2751–2760. 10.1016/j.scitotenv.2018.10.018 PMID: 30373053

Turner, C. J., Ma, R., Chen, J., & Oyekan, J. (2021). Human in the loop: Industry 4.0 technologies and scenarios for worker mediation of automated manufacturing. *IEEE Access : Practical Innovations, Open Solutions, 9*, 103950–103966.

Tushar, & Pradhan, N. (2024). Parametric analysis of DGS based sub-6 GHz antenna for WiFi communication. *2024 International Conference on Control, Computing, Communication and Materials (ICCCCM)*, 234-241. https://doi.org/DOI: 10.1109/icccm61016.2024.11039972

Tushar, D., Pradhan, N., & Singh, S. (2025a). A design of 2 × 2 element MIMO antenna with novel gain for 28/38 GHz 5G applications. *Lecture Notes in Electrical Engineering*, 257-270. https://doi.org/DOI: 10.1007/978-981-95-0203-5_22

Tushar, J. P., Shrivastava, P. C., & Pradhan, N. (2025b). *Carbon capture, utilization, and storage (CCUS) technologies: Net-zero strategies*. In *The future of green energy: Storage, materials, alternative fuels, and net-zero strategies* (pp. 251–281). IGI Global Scientific Publishing. https://doi.org/DOI: 10.4018/979-8-3373-3226-0

Tushar, P. N., Jaiswal, P., Shrivastava, P. C., Kumar, M., & Sundar, S. (2026b). *Precision farming and smart irrigation systems for energy management*. In B. Soufiene (Ed.), *Enhancing autonomous and adaptive systems with AI and IoT* (pp. 251–288). IGI Global Scientific Publishing. https://doi.org/DOI: 10.4018/979-8-3373-3146-1.ch009

Tushar, P. N., Shrivastava, P. C., Jaiswal, P., Mishra, S., & Yadav, A. K. (2026d). AI Regulation in K-12 and Higher Education Institutions. In K. Carlson (Ed.), *Ethical, Legal, and Pedagogical Perspectives on AI in Education* (pp. 1-32). IGI Global Scientific Publishing. https://doi.org/DOI: 10.4018/979-8-3373-3000-6.ch001

Tushar, P. N., Shrivastava, P. C., Jaiswal, P., Singh, S., & Mishra, V. (2026a). *Legal challenges in cloud computing*. In I. Oncioiu (Ed.), *Law and regulation of AI, blockchain, and digital rights* (pp. 63–102). IGI Global Scientific Publishing. https://doi.org/DOI: 10.4018/979-8-3373-3186-7.ch003

Tushar, P. N., Shrivastava, P., Jaiswal, P., Mishra, S., & Yadav, A. (2026c). *AI regulation in K–12 and higher education institutions*. In K. Carlson (Ed.), Ethical, legal, and pedagogical perspectives on AI in education. IGI Global Scientific Publishing.

Tyagi, A. K. (2025). Blockchain and digital twin applications in smart agriculture. In Tiwari, S. (Ed.), *CRC Press*.

Tzachor, A., Devare, M., King, B., & Avin, S., & Ó hÉigeartaigh, S. (2022). Responsible artificial intelligence in agriculture requires systemic understanding of risks and externalities. *Nature Machine Intelligence*, *4*(2), 104–109.

Uddin, M., Chowdhury, A., & Kabir, M. A. (2024). Legal and ethical aspects of deploying artificial intelligence in climate-smart agriculture. *AI & Society*, *39*(1), 221–234.

Ugenti, A., Galati, R., Mantriota, G., & Reina, G. (2023). Analysis of an all-terrain tracked robot with innovative suspension system. *Mechanism and Machine Theory*, *182*, 105237. 10.1016/j.mechmachtheory.2023.105237

Unasylva - Vol. 11, No. 1 - Shifting cultivation. (n.d.). Retrieved November 27, 2025, from https://www.fao.org/4/x5382e/x5382e03.htm

Upadhyay, A., G C, S., Zhang, Y., Koparan, C., & Sun, X. (2024). Development and evaluation of a machine vision and deep learning-based smart sprayer system for site-specific weed management in row crops: An edge computing approach. *Journal of Agriculture and Food Research*, *18*, 101331. DOI: 10.1016/J.JAFR.2024.101331

Upadhyay, A., Chandel, N. S., Singh, K. P., Chakraborty, S. K., Nandede, B. M., Kumar, M., Subeesh, A., Upendar, K., Salem, A., & Elbeltagi, A. (2025). Deep learning and computer vision in plant disease detection: A comprehensive review of techniques, models, and trends in precision agriculture. *Artificial Intelligence Review*, *58*(3), 92. DOI: 10.1007/s10462-024-11100-x

Upadhyay, A., Zhang, Y., Koparan, C., Rai, N., Howatt, K., Bajwa, S., & Sun, X. (2024). Advances in ground robotic technologies for site-specific weed management in precision agriculture: A review. *Computers and Electronics in Agriculture*, *225*, 109363.

Vaishnavi, P. (2025). Strengthening climate-smart agriculture through digital extension: A systematic review of technologies, adoption, and policy implications. *International Journal of Advanced Biochemistry Research*, •••, 785–791.

Vargas, C., Gamache, S., Henao, N., Agbossou, K., & Nagarsheth, S. (2024). A comprehensive cost mapping of digital technologies in greenhouses. *Smart Agricultural Technology*, *9*, 100634. DOI: 10.1016/J.ATECH.2024.100634

Varsamis, G., Merou, T., Alexandropoulou, I., Menti, C., Karapatzak, E., Varsamis, G., Merou, T., Alexandropoulou, I., Menti, C., & Karapatzak, E. (2025). Seed Germination Ecology and Dormancy Release in Some Native and Underutilized Plant Species with Agronomic Potential. *Agriculture 2025, Vol. 15*, *15*(20), 2139. DOI: 10.3390/AGRICULTURE15202139

Varzaru, A. A. (2025). Digital revolution in agriculture: Using predictive models to enhance agricultural performance through digital technology. *Agriculture*, *15*(3), 258.

Vatistas, C., Avgoustaki, D. D., & Bartzanas, T. (2022). A Systematic Literature Review on Controlled-Environment Agriculture: How Vertical Farms and Greenhouses Can Influence the Sustainability and Footprint of Urban Microclimate with Local Food Production. *Atmosphere*, *13*(8), 1258. https://www.mdpi.com/2073-4433/13/8/1258

Veeramachaneni, V. (2025). Edge computing: Architecture, applications, and future challenges in a decentralized era. *Recent Trends in Computer Graphics and Multimedia Technology*, *7*(1), 8–23.

Verdouw, C., Sundmaeker, H., Tekinerdogan, B., Conzon, D., & Montanaro, T. (2019). Architecture framework of IoT-based food and farm systems: A multiple case study. *Computers and Electronics in Agriculture*, *165*, 104939. DOI: 10.1016/j.compag.2019.104939

Verdouw, C., Tekinerdogan, B., Beulens, A., & Wolfert, S. (2021). Digital twins in smart farming. *Agricultural Systems*, *189*, 103046. 10.1016/j.agsy.2020.103046

Vermelho, A. B., Moreira, J. V., Akamine, I. T., Cardoso, V. S., & Mansoldo, F. R. P. (2024). Agricultural Pest Management: The Role of Microorganisms in Biopesticides and Soil Bioremediation. *Plants*, *13*(19), 2762. DOI: 10.3390/PLANTS13192762 PMID: 39409632

Vermesan, O., & Friess, P. (2013). *Internet of things: converging technologies for smart environments and integrated ecosystems*. River publishers.

Viana, C. M., Freire, D., Abrantes, P., Rocha, J., & Pereira, P. (2022). Agricultural land systems importance for supporting food security and sustainable development goals: A systematic review. *The Science of the Total Environment*, *806*, 150718. DOI: 10.1016/J.SCITOTENV.2021.150718

Vieira, G. H. S., & Ferrarezi, R. S. (2021). Use of Thermal Imaging to Assess Water Status in Citrus Plants in Greenhouses. *Horticulturae 2021, Vol. 7*, *7*(8). DOI: 10.3390/HORTICULTURAE7080249

Vijayakumar, S., Murugaiyan, V., Ilakkiya, S., Kumar, V., Sundaram, R. M., & Kumar, R. M. (2025). Opportunities, challenges, and interventions for agriculture 4.0 adoption. *Discover Food*, *5*(1), 1–35.

Vijayakumar, V., Ampatzidis, Y., Schueller, J. K., & Burks, T. (2023). Smart spraying technologies for precision weed management: A review. *Smart Agricultural Technology*, *6*, 100337. 10.1016/j.atech.2023.100337

Villa-Henriksen, A., Edwards, G. T., Pesonen, L. A., Green, O., & Sørensen, C. A. G. (2020). Internet of Things in arable farming: Implementation, applications, challenges and potential. *Biosystems Engineering*, *191*, 60–84.

Villalba, M., & Alarcón, D. (2022). Trust, technology, and participatory governance in agricultural innovation. *Technology in Society*, *68*, 101865. DOI: 10.1016/j.techsoc.2021.101865

Vinholis, M. D. M. B., Souza Filho, H. M., Carrer, M. J., Eusebio, G. D. S., & David, L. C. D. S. (2025). Twin transition in beef livestock farming: Adoption of digital technologies in Brazilian feedlots. *Journal of Agribusiness in Developing and Emerging Economies*, •••, 1–19.

von der Goltz, J., Dar, A., Fishman, R., Mueller, N. D., Barnwal, P., & McCord, G. C. (2020). Health Impacts of the Green Revolution: Evidence from 600,000 births across the Developing World. *Journal of Health Economics*, *74*. Advance online publication. DOI: 10.1016/J.JHEALECO.2020.102373 PMID: 33002797

Vrochidou, E., Tziridis, K., Nikolaou, A., Kalampokas, T., Papakostas, G. A., Pachidis, T. P., Mamalis, S., Koundouras, S., & Kaburlasos, V. G. (2021). An Autonomous Grape-Harvester Robot: Integrated System Architecture. *Electronics (Basel)*, *10*(9), 1056. DOI: 10.3390/electronics10091056

Wang, C., Zhao, T., Cheng, W., Ni, Z., & Xiang, N. (2024). Microfluidic strategies in soft robotics: Actuators, control systems, and pumps. *Device*, *2*(9), 100551. DOI: 10.1016/j.device.2024.100551

Wang, J., Wang, Y., Li, G., & Qi, Z. (2024). Integration of Remote Sensing and Machine Learning for Precision Agriculture: A Comprehensive Perspective on Applications. *Agronomy (Basel)*, *14*(9), 1975. https://www.mdpi.com/2073-4395/14/9/1975

Wang, L., Gao, J., & Qureshi, W. A. (2025). Evolution and Application of Precision Fertilizer: A Review. *Agronomy (Basel)*, *15*(8), 1939.

Wang, L., Lian, G., Harris, Z., Horler, M., Wang, Y., & Chen, T. (2023). The controlled environment agriculture: a sustainable agrifood production paradigm empowered by systems engineering. In Kokossis, A. C., Georgiadis, M. C., & Pistikopoulos, E. (Eds.), *Computer Aided Chemical Engineering* (Vol. 52, pp. 2167–2172). Elsevier., 10.1016/B978-0-443-15274-0.50345-0

Wang, S., Xu, D., Liang, H., Bai, Y., Li, X., Zhou, J., Su, C., & Wei, W. (2025). Advances in Deep Learning Applications for Plant Disease and Pest Detection: A Review. *Remote Sensing*, *17*(4), 698. DOI: 10.3390/rs17040698

Wang, X., Zhang, J.-H., Xun, L., Wang, J., Wu, Z., Henchiri, M., Zhang, S., Zhang, S., Yun, B., Yang, S., Li, S., & Yu, X. (2022). Evaluating the Effectiveness of Machine Learning and Deep Learning Models Combined Time-Series Satellite Data for Multiple Crop Types Classification over a Large-Scale Region. *Remote Sensing*, *14*, 2341. DOI: 10.3390/rs14102341

Wang, Y., Wang, Y., Mushtaq, R. T., & Wei, Q. (2024). Advancements in Soft Robotics: A Comprehensive Review on Actuation Methods, Materials, and Applications. *Polymers, 16*(8). Advance online publication. DOI: 10.3390/polym16081087 PMID: 38675005

Wang, Z., Dai, Y., Yang, L., & Yu, Z. (2025). Barriers to climate-smart agriculture adoption in Northeast China's Black Soil Region: Insights from a multidimensional framework. *Agriculture, 15*(21), 2236.

Wang, Z., & Li, D. (2025). Big data and artificial intelligence-driven risk assessment and response for water resources: Model applications and future perspectives. *Advances in Resources Research, 5*(1), 414–434.

Wang, Z., Walsh, K. B., & Verma, B. (2017). On-tree mango fruit size estimation using RGB-D images. *Sensors (Basel), 17*, 2738. DOI: 10.3390/s17122738

Waqas, M., Naseem, A., Humphries, U. W., Hlaing, P. T., Dechpichai, P., & Wangwongchai, A. (2025). Applications of machine learning and deep learning in agriculture: A comprehensive review. *Green Technologies and Sustainability, 3*(3), 100199. 10.1016/j.grets.2025.100199

Waqas, M., Naseem, A., Humphries, U. W., Hlaing, P. T., Dechpichai, P., & Wangwongchai, A. (2025a). Applications of machine learning and deep learning in agriculture: A comprehensive review. *Green Technologies and Sustainability, 3*(3), 100199. DOI: 10.1016/J.GRETS.2025.100199

Watson, E. 2025. CH4 Global teams up with Mitsubishi on livestock methane reduction. AgFunderNews. Available online: https://agfundernews.com/ch4-global-teams-up-with-mitsubishi-on-livestock-methane-reduction (accessed on 17 February 2026).

Wecht, C., Nachtmann, M., & Koppenhagen, F. (2021). BASF: Precision Farming with Lark Bread Initiative. In Gassmann, O., & Ferrandina, F. (Eds.), *Connected Business: Create Value in a Networked Economy* (pp. 315–325). Springer., DOI: 10.1007/978-3-030-76897-3

Wen, T., Li, J.-H., Wang, Q., Gao, Y.-Y., Hao, G.-F., & Song, B.-A. (2023). Thermal imaging: The digital eye facilitates high-throughput phenotyping traits of plant growth and stress responses. *The Science of the Total Environment, 899*, 165626. 10.1016/j.scitotenv.2023.165626 PMID: 37481085

Werulkar, J., & Pande, M. (2025). The adoption and efficacy of precision agriculture tools and technology in Amravati division: Exploring the insights of farmers and agtech industry. *Lex Localis, 23*(S6), 3949–3985.

Wijerathna-Yapa, A., & Pathirana, R. (2022). Sustainable agro-food systems for addressing climate change and food security. *Agriculture, 12*(10), 1554.

Wiley, V., & Lucas, T. (2018). Computer Vision and Image Processing: A Paper Review. *International Journal of Artificial Intelligence Research, 2*(1), 29–36. DOI: 10.29099/ijair.v2i1.42

Wiranata, A. D., Syahri, R., Jusman, Y., Saifuddin, S., & Borman, R. I. (2025). Hybrid Deep Learning Model with GLCM and CIELAB Feature Fusion for Tea Leaf Disease Classification. *BIO Web of Conferences, 199*, 01008. https://doi.org/ DOI: 10.1051/bioconf/202519901008

Wollschlaeger, M., Sauter, T., & Jasperneite, J. (2017). The future of industrial communication: Automation networks in the era of the Internet of things and industry 4.0. *IEEE Industrial Electronics Magazine, 11*(1), 17–27.

Wright, S., & Chan, L. (2022). Participatory AI and data justice. *AI & Society, 37*(4), 1521–1535. DOI: 10.1007/s00146-021-01278-8

Wu, J., Ping, L., Ge, X., Wang, Y., & Fu, J. (2010). Cloud storage as the infrastructure of cloud computing. 2010 International conference on intelligent computing and cognitive informatics, (ICICCI), Kuala Lumpur, Malaysia.

Wu, H., Ji, P., Ma, H., & Xing, L. (2023). A Comprehensive Review of Digital Twin from the Perspective of Total Process: Data, Models, Networks and Applications. *Sensors (Basel), 23*(19). Advance online publication. DOI: 10.3390/s23198306 PMID: 37837136

Wu, K., Ji, Z., Wang, H., Shao, X., Li, H., Zhang, W., Kong, W., Xia, J., & Bao, X. (2025). A Comprehensive Review of AI Methods in Agri-Food Engineering: Applications, Challenges, and Future Directions. *Electronics (Basel), 14*(20), 3994. https://www.mdpi.com/2079-9292/14/20/3994

Wulandari, D., Fajarianto, O., Prayitno, P. H., Kusuma, F. I., Gunawan, A., Syafruddin, A. B., & Basuki, A. (2025). Precision agriculture ecosystem innovation through IoT technology: A study of learning factory development for farmer empowerment in Ngajum Village. *Jurnal Syntax Imperatif: Jurnal Ilmu Sosial dan Pendidikan, 6*(5), 1299-1323.

Wu, P., & Zhong, Y. (2025). Artificial intelligence in sustainable agriculture: Towards a socio-technical roadmap. *Smart Agricultural Technology, 12*, 101578. DOI: 10.1016/J.ATECH.2025.101578

Wu, X., Xiao, L., Sun, Y., Zhang, J., Ma, T., & He, L. (2022). A survey of human-in-the-loop for machine learning. *Future Generation Computer Systems, 135*, 364–381. 10.1016/j.future.2022.05.014

Xiang, M., Qu, M., Wang, G., Ma, Z., Chen, X., Zhou, Z., Qi, J., Gao, X., Li, H., & Jia, H. (2024). Crop detection technologies, mechanical weeding executive parts and working performance of intelligent mechanical weeding: a review [Review]. *Frontiers in Plant Science, Volume 15 - 2024*. DOI: 10.3389/fpls.2024.1361002

Xu, J., Cui, Y., Zhang, S., & Zhang, M. (2024). The evolution of precision agriculture and food safety: A bibliometric study. *Frontiers in Sustainable Food Systems, 8*, 1475602.

Xu, X., Lu, Y., Vogel-Heuser, B., & Wang, L. (2021). Industry 4.0 and industry 5.0-Inception, conception and perception. *Journal of Manufacturing Systems, 61*, 530–535.

Xu, Y., Liu, X., Cao, X., Huang, C., Liu, E., Qian, S., Liu, X., Wu, Y., Dong, F., Qiu, C. W., Qiu, J., Hua, K., Su, W., Wu, J., Xu, H., Han, Y., Fu, C., Yin, Z., Liu, M., & Zhang, J. (2021). Artificial intelligence: A powerful paradigm for scientific research. *The Innovation, 2*(4), 100179. DOI: 10.1016/J.XINN.2021.100179

Yakkala, V. S., Nusimala, K. V., Gayathri, B., Kanamarlapudi, S., Aravinth, S. S., Salau, A. O., & Srithar, S. (2025). Deep learning-based crop health enhancement through early disease prediction. *Cogent Food & Agriculture, 11*(1), 2423244. DOI: 10.1080/23311932.2024.2423244

Yang, D., Wu, J., & He, Y. (2024). Optimizing the Agricultural Internet of Things (IoT) with Edge Computing and Low-Altitude Platform Stations. *Sensors (Basel), 24*(21), 7094.

Yang, Q., Du, X., Wang, Z., Meng, Z., Ma, Z., & Zhang, Q. (2023). A review of core agricultural robot technologies for crop productions. *Computers and Electronics in Agriculture, 206*, 107701. 10.1016/j.compag.2023.107701

Yan, H.-F., Zhao, Y.-Q., Chan, J. C.-W., Kong, S. G., Ei-Bendary, N., & Reda, M. (2025). Hyperspectral and multispectral image fusion: When model-driven meet data-driven strategies. *Information Fusion, 116*, 102803. 10.1016/j.inffus.2024.102803

Yan, J., Wu, H., Diao, Z., Miao, Y., Zhang, B., & Zhao, C. (2025). Recent Developments and Applications of Crop Disease Detection, Prediction, and Early Warning: A review. *Engineering*. Advance online publication. DOI: 10.1016/j.eng.2025.10.032

Yan, J., Wu, H., Diao, Z., Miao, Y., Zhang, B., & Zhao, C. (2025). Recent Developments and Applications of Crop Disease Detection, Prediction, and Early Warning: A review. *Engineering*. Advance online publication. DOI: 10.1016/J.ENG.2025.10.032

Yassein, M. B., Hmeidi, I., Meqdadi, O., Alghazo, F., Odat, B., AlZoubi, O., & Smairat, A. (2020). *Challenges and techniques of constrained application protocol (CoAP) for efficient energy consumption.* Paper presented at the 2020 11th International Conference on Information and Communication Systems (ICICS).

Ye, Z., Yin, S., Cao, Y., & Wang, Y. (2024). AI-driven optimization of agricultural water management for enhanced sustainability. *Scientific Reports*, *14*(1), 25721. DOI: 10.1038/s41598-024-76915-8

Yin, H., Cao, Y., Marelli, B., Zeng, X., Mason, A., & Cao, C. (2021). Smart Agriculture Systems: Soil Sensors and Plant Wearables for Smart and Precision Agriculture (Adv. Mater. 20/2021). *Advanced Materials*, *33*, 2170156. DOI: 10.1002/adma.202170156

Younas, M., Nadeem Akhtar, M., Batool, S., Owais, M., Sahar, S., & Anum, W. (2025). The Integration of Artificial Intelligence in Agriculture: Emerging Trends, Benefits and Challenges. *Journal of Asian Development Studies*, *14*(1), 1316–1333. DOI: 10.62345/JADS.2025.14.1.105

Yu, P., Teng, F., Zhu, W., Shen, C., Chen, Z., & Song, J. (2025). Cloud–edge–device collaborative computing in smart agriculture: architectures, applications, and future perspectives [Review]. Frontiers in Plant Science, Volume 16 - 2025. DOI: 10.3389/fpls.2025.1668545

Yuan, Y., & Sun, Y. (2024). Practices, Challenges, and Future of Digital Transformation in Smallholder Agriculture: Insights from a Literature Review. *Agriculture; Basel*, *14*(12).

Yu, P., Teng, F., Zhu, W., Shen, C., Chen, Z., & Song, J. (2025). Cloud–edge–device collaborative computing in smart agriculture: Architectures, applications, and future perspectives. *Frontiers in Plant Science*, *16*, 1668545. DOI: 10.3389/FPLS.2025.1668545/XML

Yusuf, A., Li, M., Zhang, S. Y., Odedishemi-Ajibade, F., Luo, R. F., Wu, Y. X., Zhang, T. T., Yunusa Ugya, A., Zhang, Y., & Duan, S. (2025). Harnessing plant-microbe interactions: Strategies for enhancing resilience and nutrient acquisition for sustainable agriculture. *Front Plant Sci*, *16*, 1503730. DOI: 10.3389/fpls.2025.1503730

Zaigham Abbas Naqvi, S. M., Hussain, S., Awais, M., Tahir, M. N., Saleem, S. R., Al-Yarimi, F. A. M., Ashurov, M., Saidani, O., Khan, M. I., Wu, J., Wei, Z., & Hu, J. (2025). Climate-resilient water management: Leveraging IoT and AI for sustainable agriculture. *Egyptian Informatics Journal, 30*, 100691. DOI: 10.1016/J.EIJ.2025.100691

Zainab, Z., & Astiko, W. (2025). Assessment and analysis of sustainable agroecosystem management through the implementation of an integrated farming system (IFS) in Lempenge Hamlet, Rempek Village. *Jurnal Biologi Tropis, 25*(4b), 559–567.

Zangana, H. M., Luckyardi, S., Mustafa, F. M., & Li, S. (2025). Ethical and secure AI applications in agricultural research: Challenges and opportunities. In *Ensuring secure and ethical STM research in the AI era* (pp. 43-76).

Zarbakhsh, S., Fakhrzad, F., Rajkovic, D., Niedbała, G., & Piekutowska, M. (2025). Approaches and challenges in machine learning for monitoring agricultural products and predicting plant physiological responses to biotic and abiotic stresses. *Current Plant Biology, 43*, 100535. 10.1016/j.cpb.2025.100535

Zeng, Y., Hao, D., Huete, A., Dechant, B., Berry, J., Chen, J. M., Joiner, J., Frankenberg, C., Bond-Lamberty, B., Ryu, Y., Xiao, J., Asrar, G. R., & Chen, M. (2022). Optical vegetation indices for monitoring terrestrial ecosystems globally. *Nature Reviews. Earth & Environment, 3*(7), 477–493. DOI: 10.1038/s43017-022-00298-5

Zhang, F., Zhang, W., Luo, X., Zhang, Z., Lu, Y., Wang, B., Zhang, F., Zhang, W., Luo, X., Zhang, Z., Lu, Y., & Wang, B. (2022). Developing an IoT-Enabled Cloud Management Platform for Agricultural Machinery Equipped with Automatic Navigation Systems. *Agriculture 2022, Vol. 12, 12*(2). /DOI: 10.3390/AGRICULTURE12020310

Zhang, D., & Wei, B. (2017). A review on model reference adaptive control of robotic manipulators. *Annual Reviews in Control, 43*, 188–198. 10.1016/j.arcontrol.2017.02.002

Zhang, L., Du, Q., Lu, L., & Zhang, S. (2023). Overview of the Integration of Communications, Sensing, Computing, and Storage as Enabling Technologies for the Metaverse over 6G Networks. *Electronics (Basel), 12*(17), 3651.

Zhang, P., Lu, B., Ge, J., Wang, X., Yang, Y., Shang, J., La, Z., Zang, H., & Zeng, Z. (2025). Using UAV-based multispectral and RGB imagery to monitor above-ground biomass of oat-based diversified cropping. *European Journal of Agronomy, 162*, 127422. 10.1016/j.eja.2024.127422

Zhang, R., Zhu, H., Chang, Q., & Mao, Q. (2025). A Comprehensive Review of Digital Twins Technology in Agriculture. *Agriculture, 15*(9), 903. https://www.mdpi.com/2077-0472/15/9/903

Zhang, X., Feng, G., & Sun, X. (2024). Advanced technologies of soil moisture monitoring in precision agriculture: A Review. *Journal of Agriculture and Food Research, 18*, 101473. DOI: 10.1016/j.jafr.2024.101473

Zhang, Y., Yasutake, D., Hidaka, K., Okayasu, T., Kitano, M., & Hirota, T. (2022). Crop-localised CO2 enrichment improves the microclimate, photosynthetic distribution and energy utilisation efficiency in a greenhouse. *Journal of Cleaner Production, 371*, 133465. 10.1016/j.jclepro.2022.133465

Zhang, Z., & Furman, A. (2021). Soil redox dynamics under dynamic hydrologic regimes - A review. *The Science of the Total Environment, 763*. Advance online publication. DOI: 10.1016/J.SCITOTENV.2020.143026

Zhang, Z., Luo, J., & Chen, B. (2020). Spatially explicit quantification of total soil erosion by RTK GPS in wind and water eroded croplands. *The Science of the Total Environment, 702*, 134716. 10.1016/j.scitotenv.2019.134716 PMID: 31726344

Zhang, Z., & Zhu, L. (2023). A Review on Unmanned Aerial Vehicle Remote Sensing: Platforms, Sensors, Data Processing Methods, and Applications. *Drones (Basel)*, 7(6), 398. https://www.mdpi.com/2504-446X/7/6/398

Zhao, J., Zhao, W., Deng, B., Wang, Z., Zhang, F., Zheng, W., Cao, W., Nan, J., Lian, Y., & Burke, A. F. (2024). Autonomous driving system: A comprehensive survey. *Expert Systems with Applications, 242*, 122836. 10.1016/j.eswa.2023.122836

Zhao, L., Walkowiak, S., & Fernando, W. G. D. (2023). Artificial intelligence: A promising tool in exploring the phytomicrobiome in managing disease and promoting plant health. *Plants*. Advance online publication. DOI: 10.3390/plants12091852

Zhao, Q., Yu, L., Du, Z., Peng, D., Hao, P., Zhang, Y., & Gong, P. (2022). An Overview of the Applications of Earth Observation Satellite Data: Impacts and Future Trends. *Remote Sensing, 14*(8), 1863. https://www.mdpi.com/2072-4292/14/8/1863

Zhao, X., Zhu, A. L., Liu, X., Li, H., Tao, H., Guo, X., & Liu, J. (2025). Current status, challenges, and opportunities for sustainable crop production in Xinjiang. *iScience*.

Zheng, H., Shen, L., Tang, A., Luo, Y., Hu, H., Du, B., Wen, Y., & Tao, D. (2025). Learning from models beyond fine-tuning. *Nature Machine Intelligence*, 7(1), 6–17. DOI: 10.1038/s42256-024-00961-0

Zheng, S., Xu, S., & Rai, R. (2025). Trailer-referenced autonomous navigation of agricultural tractor–trailer systems. *Computers and Electronics in Agriculture*, *237*, 110514. 10.1016/j.compag.2025.110514

Zhou, X.-G., Zhang, D., Lin, F., Zhou, X.-G., Zhang, D., & Lin, F. (2021). UAV Remote Sensing: An Innovative Tool for Detection and Management of Rice Diseases. In *Diagnostics of Plant Diseases*. IntechOpen. https://doi.org/DOI: 10.5772/intechopen.95535

Zhou, X., English, B. C., Larson, J. A., Lambert, D. M., Roberts, R. K., Boyer, C. N., & Martin, S. W. (2017). Precision Farming Adoption Trends in the Southern U.S. *Journal of Cotton Science*, *21*(2), 143–155.

Zhou, Y. (2025). Artificial Intelligence-Based Crop Disease Identification Technology: Applications, Challenges, and Future Prospects. *Journal of Electronic Research and Application*, *9*(6), 287–293. DOI: 10.26689/JERA.V9I6.13182

Zhuang, Y., Sun, X., Li, Y., Huai, J., Hua, L., Yang, X., Cao, X., Zhang, P., Cao, Y., Qi, L., Yang, J., El-Bendary, N., El-Sheimy, N., Thompson, J., & Chen, R. (2023). Multi-sensor integrated navigation/positioning systems using data fusion: From analytics-based to learning-based approaches. *Information Fusion*, *95*, 62–90. 10.1016/j.inffus.2023.01.025

Zhu, Y. (2024). Application of a QPSO-optimized CNN-LSTM model in water quality prediction. *Discover Water*, *4*(1), 100.

Zhu, Y., Irmak, S., Jhala, A., Vuran, M., & Diotto, A. (2019). Time-domain and Frequency-domain Reflectometry Type Soil Moisture Sensor Performance and Soil Temperature Effects in Fine- and Coarse-textured Soils. *Applied Engineering in Agriculture*, *35*, 117–134. DOI: 10.13031/aea.12908

Zia, A., & Haleem, M. (2025). Bridging research gaps in industry 5.0: Synergizing federated learning, collaborative robotics, and autonomous systems for enhanced operational efficiency and sustainability. *IEEE Access : Practical Innovations, Open Solutions*.

Zimdahl, R. L. (2022). The conduct of agricultural science. *Agriculture's Ethical Horizon*, 17–33. DOI: 10.1016/B978-0-12-823667-3.00006-9

Zinjurke, S., Bhaltilak, O., Vidhate, R., & Nikumbh, J. (2025). Leveraging Computer Vision for Crop Health Monitoring, Automated Yield Estimation, and Enhanced Disease Detection in Precision Agriculture. *International Journal of Research and Analytical Reviews*, *12*. Advance online publication. DOI: 10.56975/ijrar.v12i2.309776

Zreikat, A. I., AlArnaout, Z., Abadleh, A., Elbasi, E., & Mostafa, N. (2025). The Integration of the Internet of Things (IoT) Applications into 5G Networks: A Review and Analysis. *Computers*, *14*(7), 250. https://www.mdpi.com/2073-431X/14/7/250

Zualkernan, I., Abuhani, D. A., Hussain, M. H., Khan, J., ElMohandes, M., Zualkernan, I., Abuhani, D. A., Hussain, M. H., Khan, J., & ElMohandes, M. (2023). Machine Learning for Precision Agriculture Using Imagery from Unmanned Aerial Vehicles (UAVs): A Survey. *Drones 2023, Vol. 7*, *7*(6). https://doi.org/DOI: 10.3390/DRONES7060382

About the Contributors

Rashid Abbas Khan is a passionate botanist and researcher specializing in plant stress physiology, seed priming, and sustainable agriculture. He holds degrees from the University of Sargodha and the University of Education, Lahore, and has served as a Lecturer in Botany, contributing to both teaching and research. His work focuses on developing innovative, eco-friendly approaches to enhance crop resilience, improve soil health, and promote sustainable agricultural practices for long-term food security.

Vishnu D. Rajput working as "Principal Scientist" and Head of `International Laboratory on Nanobiotechnology' at Southern Federal University, Rostov-on-Don, Russia. Dr. Rajput comprehensively detailed the state of research in environmental science about "how nanoparticles/heavy metals interact with plants, soil, microbial community and the larger environment. In recent work, he is working on nano-bioengineering for soil rhizospheric microbes and enhanced remediation. Dr. Rajput is included in list of 2% world's potential scientists prepared by Stanford University. He is serving as "Highly Qualified Specialist", Russian Federation, and obtained 2nd rank in Environmental and Earth Science. Dr. Rajput received numerous appreciation certificates/letters. He has published 569 articles, H-Index 67 with 16,206 citations as per Scopus database. Dr. Rajput acknowledge the support by Strategic Academic Leadership Program of the Southern Federal University ("Priority 2030").

Asghari Bano is a Professor of Biosciences at the University of Wah, Pakistan, and currently resides in Mississauga, Ontario, Canada. She earned her Ph.D. in Plant Physiology from the University of Glasgow, UK, following an M.Sc. in Plant Physiology, Microbiology, and Plant Pathology from the University of Dhaka, Bangladesh. Over her distinguished career spanning more than 37 years, she has held key academic and leadership positions, including Dean, Chairperson, and Professor at Quaid-i-Azam University, Islamabad. She has completed prestigious postdoctoral fellowships in Germany (Alexander von Humboldt Fellow), the USA (Fulbright

Fellow), and Canada, with research focused on plant physiology, root nodulation, plant–microbe interactions, bioactive metabolites, allelopathy, and nanotechnology. Dr. Bano has authored an extensive body of peer-reviewed publications, contributed book chapters, and supervised numerous postgraduate theses, significantly advancing knowledge in plant sciences. Her pioneering contributions to stress physiology, plant–microbe communication, and eco-friendly agricultural solutions underscore her role as an internationally recognized scientist and mentor.

***

Md. Shoeab Akhter is a passionate researcher, research associate, and practicum student with a B.Sc. in the Agriculture Department of the College of Agricultural Sciences at IUBAT—International University of Business Agriculture and Technology, Dhaka 1230, Bangladesh. He is doing an internship at the Bangladesh Atomic Energy Commission. His research title is 'Green Synthesis of Nanoparticles for Agricultural Use.' Loved by nature and with a passion for the environment, his pursuits include areas of interest like agricultural sciences, food security and food science, environmental science, waste management, green energy, AI, IoT-based farming, and agritourism, among others. Along with his objective of creating an impact through writing and research, he combines new, innovative, suitable technologies and nature conservation with real-world issues. He has contributed book chapters (Emerald, CABI, Bentham Science, Springer, Cambridge Scholars) and has participated in various major projects on research. With the hope of bringing about a change in the world, he aspires to contribute to global knowledge and create a lasting impact on sustainability in society.

Md. Shoeab Akhter is a passionate researcher and a graduate of the B.Sc. program in the Department of Agriculture at the College of Agricultural Sciences, and Research Associate at IUBAT—International University of Business Agriculture and Technology, Dhaka 1230, Bangladesh. He completed his internship at the Bangladesh Atomic Energy Commission. His research focuses on "Green Synthesis of Nanoparticles for Agricultural Use." Driven by a deep love for nature and a strong commitment to environmental sustainability, his areas of interest include agricultural sciences, food security, food science, environmental science, waste management, green energy, AI- and IoT-based farming, and agritourism, among others. With the objective of creating meaningful impact through writing and research, he integrates innovative, appropriate technologies with nature conservation to address real-world challenges. He has contributed book chapters published by Emerald, CABI, Bentham Science, Springer, and Cambridge Scholars and has actively participated in several major research projects. With the hope of bringing about positive change, he aspires to contribute to global knowledge and create a lasting impact on sustainable development and society.

Md. Sakibul Hasan Dipto, a Bangladesh-based tech enthusiast and innovator passionate about drone avionics, flight controllers, loT systems, and embedded electronics. From developing cost-effective agricultural automation solutions to building high-performance UAV platforms to designing products. He enjoys turning complex engineering ideas into efficient, scalable, and commercially viable products.

Faiz Ul Hassan is a distinguished researcher and academic currently affiliated withUniversiti Kebangsaan Malaysia (UKM),where he is pursuing a research-based master's degree. His work focuses on various aspects of advancing sustainability, including energy efficiency, waste management, water conservation, transportation, and community engagement, all in alignment with the United Nations' 2030 Agenda for Sustainable Development. Mr. Hassan research interests entail the establishment of integrated frameworks that promote environmental sustainability in urban and university environments. He intends to resolve critical challenges in the reduction of environmental impact and the promotion of community awareness and engagement in sustainable practices by analyzing and implementing resource-efficient practices. His research involves exploring innovative approaches to reduce carbon emissions and optimize energy use, enhancing both environmental and economic resilience. Mr. Hassan is dedicated to long-term success and researches advanced waste reduction and water resource management solutions, focusing on conservation measures that promote ecosystem health.

Israt Jahan is a graduate student in Computer Science and Engineering at East West University, Dhaka, Bangladesh, where she specializes in Data Science. She is currently pursuing her M.Sc. degree. Israt has been actively involved in multiple research areas, including machine learning, deep learning, AI-driven healthcare applications, and nanotechnology. Her recent work focuses on AI applications in cancer diagnosis and treatment, utilizing advanced techniques in nanotechnology to enhance precision and effectiveness. She has worked as a research assistant at various institutions, including the Core Research & Testing Laboratory (USA) and the CHOWDHURY AI and Cybersecurity Research Lab (Australia). Israt's publications include journal articles in IEEE Access and Springer Nature, with a focus on sleep disorder diagnosis, AI in health telemedicine, and clinical trials for rare diseases. She is also a peer reviewer for C5K Research Publication and a program committee member for IICAIET. Israt has presented at multiple international conferences and is a recipient of the Best Paper Award at the International Conference on Data Mining and Information Security. She is dedicated to exploring explainable AI techniques and bridging the gap between research and practical applications in healthcare, agriculture, and cancer research.

Pooja Jaiswal received the B.Sc. degree in Science from J.K. Institute of Applied Physics & Technology, University of Allahabad (U.P.) in 2007. She received the B.Tech & M.Tech degree in Electronics and communication from J.K. Institute of Applied Physics & Technology, University of Allahabad (U.P.) in 2010 & 2012. Her research interest includes Quantum computing and Image Processing with Deep learning and their applications. Presently she is pursuing PhD from J.K. Institute of Applied Physics & Technology, University of Allahabad, India.

Dilsad Jahan Jui, a final-year student at Chattogram Veterinary and Animal Sciences University under the Faculty of Food Science and Technology. A strong academic foundation has been built by her, and a deep passion for research in the fields of food science, technology, and nutrition is being pursued. Her research interests are focused on the areas such as functional foods and nutraceuticals, nutrigenomics, food microbiology, food waste management, food fortification and biofortification, infant and maternal nutrition, sustainable diets, gut microbiome studies, clinical and therapeutic nutrition, food safety and quality assurance, novel preservation methods, smart packaging, alternative proteins, sustainable food processing, and nanotechnology-based food preservation. Recently, a commitment to scientific writing has been demonstrated by her through contributing several scientific book chapters, many of which have already been accepted for publication in reputed journals, while several others are under progress. Additionally, professional service to the scientific community is being provided by her through serving as a reviewer for Open Veterinary Journal (Q3, Scopus-indexed), where she reviewed a manuscript in 2025.A determination to her passion continues to be expressed by her through focused research works, academic engagements, and professional contributions, aiming to advance knowledge and innovation in the field of food science, technology, and nutrition.

Burhan Khalid is an emerging agronomy professional with a strong passion for crop science and sustainable agricultural practices. He is dedicated to improving crop production, soil health, and modern farming techniques. Burhan is committed to applying scientific knowledge to support farmers, enhance food security, and promote environmentally friendly agriculture.

Amjid Khan is a Post-Doctoral Research Fellow at the UNESCO–UNISA Africa Chair in Nanosciences and Nanotechnologies (U2ACN2), College of Graduate Studies, University of South Africa, Pretoria, South Africa. His research integrates green nanotechnology, plant–microbe interactions, and sustainable agriculture. He focuses on the biosynthesis of plant-mediated nanoparticles for applications as nano-fertilizers and biomedical agents. His work supports sustainable crop productivity, stress mitigation, and health-related nanomaterial development aligned with key UN SDGs.

Afnan Shinwari is a plant biotechnology researcher focused on plant–microbiome interactions and the role of silicon in improving plant tolerance to environmental stress. His work integrates molecular biology, gene expression, and physiological analysis to understand how plants adapt to challenging conditions. He is particularly interested in applying biotechnology to enhance crop resilience and productivity. His long-term goal is to contribute to sustainable agriculture and food security.

Andrew Tapiwa Kugedera is a PhD in Agronomy holder with experience in soil fertility management, rainwater harvesting techniques, agroforestry, and soil and water conservation. He has published over 100 articles including book chapters. He has been reviewing articles for several reputable journals. His research interests include agroforestry, Agronomy, climate smart Agriculture, climate change, and soil fertility.

Malik Maaza is Chair of the UNESCO–UNISA Africa Chair in Nanosciences and Nanotechnologies at the University of South Africa (UNISA), South Africa. He is an internationally recognized physicist known for contributions to nanosciences, nanophotonics, and biogenic advanced materials. He is a Fellow of several prestigious bodies, including the African Academy of Sciences, Royal Society of Chemistry (London), American Physical Society, and the American Association for the Advancement of Science. His research bridges fundamental physics and sustainable technological innovation.

Hanan Ibrahim Mudawi is an Associate Professor at the National Center for Research, Environment and Natural Resources and Desertification Research Institute. With more than 20 years of experience in biopesticides and biofertilizers research, her work focuses on developing environmentally friendly pest-management solutions and advancing integrated agricultural systems, driven by a strong commitment to promoting environmental sustainability. She earned her Ph.D. in Environmental Sciences, following an M.Sc. in Plant Protection and a B.Sc. in Agricultural Sciences. She currently serves as the Director of Scientific Affairs at the National Centre for Research (NCR). Dr. Mudawi has published in reputable scientific journals and serves as a reviewer for several high-impact journals. She is highly skilled in teaching, mentorship, and academic research. She is actively involved in multiple national and international projects as both a team leader and member, spanning areas such as plant biocontrol, organic farming, bio-compost production, weed management, and the formulation of natural biocides. Her research is recognized for bridging scientific innovation with practical field applications, particularly in low-input farming systems. She has participated in numerous national and international training programs and conferences. Dr. Mudawi also contributes to scientific and policy-making committees and is an active member of multiple scientific societies. Beyond her research contributions, she provides advisory and consultancy services to agribusiness companies in plant pathology and nematology, as well as training for women and farmers on eco-friendly pest-management practices.

Randa Osman associate professor with more than 29 years of experience in environmental microbiology. I am currently a researcher at the National Center for Research (NCR)—Sudan. The core of my work is on biofertilizers, biocontrol of pests and biodegradation of pollutants with the insight of sustainable agricultural production and environmental resource preservation, with special emphasis on dry lands. I graduated from the faculty of agriculture (Biochemistry) where I also had my M.Sc. in soil microbiology. I hold a PhD in microbiology from the University of Khartoum, Faculty of Science. I have my research work published in several reputable journals and I delivered presentations at various scientific platforms. I thrive in team and collaborative work environments of diverse backgrounds, aided by profound skills in proposal writing and fundraising.

Sakibul Islam Ratul is a passionate researcher and undergraduate student in the Department of Agricultural Science at Daffodil International University, Bangladesh. He is highly academic and committed towards research-driven innovation. His scholarly and research work is on sustainable agrifood solutions, especially in post-harvest technology, plant pathology, soil health, and climate-resilient agrifood systems. He is also pursuing more advanced research on solar-biomass hybrid drying systems, soil microplastic pollution and its effects on water retention, carbon sequestration measures, and plant extract to combat fungal pathogens, UV light treatment of farm soils, and microbial interactions of soil ecologies. In addition to coursework, he has acquired hands-on experience in fieldwork, laboratory work, and academic writing, which has allowed him to relate theoretical learning to practical agricultural problems. His interests lie in supporting a science-based approach to solving the problems of smallholder and off-grid farm communities in South Asia. He would like to advance to a new stage of education and to participate in the research work of the whole world, and in the long run, he hopes to develop sustainable technologies in farming that will enable the world to be free of food security problems and environmental issues.

Prabhat Chandra Shrivastava is an Assistant Professor in the Department of Electronics and Communication at the University of Allahabad, Prayagraj. He completed his Ph.D. and M.Tech from MNNIT Allahabad and his B.E. from RGPV Bhopal. He has more than 10 years of experience in teaching and research. His main areas of work include Architecture of Digital Filter Design, Digital Signal Processing, Artificial Intelligence (AI), Machine Learning (ML), and Smart Agriculture. He has worked on useful projects like predicting plant health using AI, analyzing ECG signals for heart problems, and recognizing Indian classical music with deep learning. Dr. Shrivastava has published many research papers in international journals and conferences. He has also qualified the GATE exam five times and received fellowships during his M.Tech and Ph.D. studies. He is passionate about using technology to solve real-life problems in agriculture.

Donald Smith During his 40 years at McGill he has conducted research in the production and physiology of crop plants, with an emphasis on plant-microbe interactions. The areas of research investigation have been: nitrogen metabolism, nitrogen fixation, root zone temperature stress and nodulation, development of methods for injection of metabolites into plants, barley production, use of plant growth regulators, intercropping, the dynamics of inter-plant competition, plant-microbe signaling, plants and climate change, biofuel crops, crop stress responses and biochar as a soil amendment. The work has involved 20 crop species, the most recent being Cannabis. He has trained 84 graduate students, 2/3 at the Ph.D. level. This research activity has resulted over 410 journal publications, fourteen patents, a spin-off company (Bios Agriculture Inc.) and products applied to over 100 million ha of crop land per year. He has been principal investigator on research grants totaling over $80 million. He currently leads the AAFC funded ($2.3 million per year for 5 years) Biomass Canada Research Cluster and the Eastern Canadian Oilseed Development Initiative network (AAFC funded at $1 million per year). He was head of the NSERC Strategic Green Crop Network ($1 million per year for 5 years) and the NCE funded BioFuelNet ($10 million per year for 5 years). He has had international collaborations with the US, India, China, Russia and Brazil. Dr. Smith has served as chair of his department, is a founding member of a Canada-US think tank on climate change and agriculture in North Eastern North America, is a Distinguished James McGill professorship and was a New Sun Professor before that. He has served on the NSERC Plant Biology and Food Discovery grant selection committee. He is a past recipient of the Young Agronomist Award. He is a past president and current fellow of the Canadian Society of Agronomy, and a member of the Clean50. He has been cited over 26,000 times and has an H-index (Research Gate) of 81.

Farjana Sultana joined the IUBAT in 2010, as a Faculty member. Since coming to IUBAT, she has been teaching Plant Pathology, Plant Physiology and Educational Planning. Dr. Sultana is specialized on biological control of plant diseases, an exceptionally environment friendly technique, and is trying to establish this system in Bangladesh. Dr. Sultana achieved her bachelor degree from Bangladesh Agricultural University, Mymensingh (BAU) in 2003 and the same year she went to Japan for post graduation. She did her MS and PhD at Gifu University, Japan in 2006 and 2009, respectively. During the course of her MS and PhD research, she had identified several genes which are responsible for inducing systemic resistance in plants by Plant Growth Promoting Fungus (PGPF). Immediately after her graduation, she came back to Bangladesh only for her strong patriotism. She is continuing the same research here. She has already published 12 scientific papers in different prestigious peer reviewed journal with the total impact factor of 18. She is personally working with the concept of KBAD (Knowledge Based Area Development) and trying to help the needy people. She believes that every people should extend their hand to the nearest people so that poverty alleviation from this country may come to fruition.

Index

E

F

I

M

P

R

S

T

W

www.ingramcontent.com/pod-product-compliance
Lightning Source LLC
LaVergne TN
LVHW081319300926
849161LV00008B/1815